S0-CAH-302

Studies in Natural Products Chemistry

Volume 2

Structure Elucidation (Part A)

Studies in Natural Products Chemistry
edited by Atta-ur-Rahman

Studies in Natural Products Chemistry

Volume 2

Structure Elucidation (Part A)

Edited by

Atta-ur-Rahman

H.E.J. Research Institute of Chemistry, University of Karachi, Karachi 32, Pakistan

ELSEVIER

Amsterdam — Oxford — New York — Tokyo 1988

ELSEVIER SCIENCE PUBLISHERS B.V.
Sara Burgerhartstraat 25
P.O. Box 211, 1000 AE Amsterdam, The Netherlands

Distributors for the United States and Canada:

ELSEVIER SCIENCE PUBLISHING COMPANY INC.
655, Avenue of the Americas
New York, NY 10010, U.S.A.

ISBN 0-444-43038-5 (Vol. 2)
ISBN 0-444-42971-9 (Series)

Printed in The Netherlands

FOREWORD

The tremendous advances made during the last two decades in spectroscopic techniques have accelerated the pace of research in the field of isolation and structure elucidation of complex natural products. These advances have been particularly spectacular in the fields of NMR and mass spectroscopy, which have considerably extended the capabilities of these techniques. For instance, the advent of pulse NMR techniques and mini-computers led to the development of two-dimensional NMR spectroscopy, which has now come to be routinely employed in laboratories around the world for unravelling complex structural problems. Less dramatic, though no less important, have been the advances in the field of mass spectroscopy, where the development of new ionization techniques such as negative and positive fast atom bombardment, field desorption and chemical ionization, etc., have followed the acquisition of mass spectra of non-volatile compounds which do not respond well to the standard electron impact ionization method.

These developments have provided a fresh impetus in the area of natural product chemistry, which is reflected by the rapidly growing unber of new natural products being discovered from terrestrial and marine plant and animal kingdoms. In view of the growing importance of natural products to man, it was felt that there was a strong need for a series of volumes which would highlight the latest developments in natural product chemistry, with particular reference to new techniques being developed for isolation and structure elucidation and their applications to the solution of structural problems, and which would focus attention on other related areas such as biosynthesis, cell structure, biotechnology, etc. The field of stereoselective synthesis of natural products is deliberately avoided in this volume as a separate sub-series on this topic is being published under my editorship by Elsevier as a part of the same series.

The present volume covers developments on a broad front of natural product chemistry. The first several chapters are concerned with developments in mass spectroscopy, NMR spectroscopy and circular dichroism, highlighting their applications to structure elucidation of natural products. The next few chapters present work carried out on the isolation and structure elucidation of new natural products from medicinal plants and liverworts. The last three chapters are devoted to polysaccharides from protozoa, biotechnology, and biosynthetic studies in plant tissue culture. All the authors are well known for their prolific and important contributions in this field, and the articles should prove to be of interest to a large number of organic chemists, phytochemists, medicinal chemists and biochemists.

I wish to express my gratitude to Miss Khurshid Zaman for her assistance in preparing the index of the manuscript and Mr Mahmood Alam for secretarial assistance.

August 1988 Atta-ur-Rahman, Editor

CONTENTS

CONTRIBUTORS

Yoshinori Asakawa	Faculty of Pharmacognosy, Tokushima Bunri University, Yamashiro-Cho, Tokushima 776, Japan
Kaleab Asres	Department of Chemistry, School of Pharmacy, University of Addis Ababa, Ethiopia
Johnson Foyere Ayafor	Department de Chimie Organique, Faculté des Sciences, Université de Yaounde, B.P.812, Yaounde, Cameroon
H. Budzikiewicz	Institute of Organic Chemistry, University of Cologne, Greinstrasse 4, D-5000 Cologne 41, F.R.G.
M. Iqbal Choudhary	H.E.J. Research Institute of Chemistry, University of Karachi, Karachi-32, Pakistan
Alice M. Clark	Department of Pharmacognosy, University of Mississippi, University, MS 38677, U.S.A.
Kimberly L. Colson	Department of Chemistry, The Pennsylvania State University, 152 Davey Laboratory, University Park, PA 16802, U.S.A.
Joseph D. Connolly	Department of Chemistry, University of Glasgow, Glasgow G12 8QQ, Scotland, U.K.
Alan J. Freyer	Department of Chemistry, The Pennsylvania State University, 152 Davey Laboratory, University Park, PA 16802, U.S.A.
William A. Gibbons	Department of Pharmaceutical Chemistry, School of Pharmacy, University of London, 29-39 Brunswick Square, London WC1N 1AX, England
Philip A.J. Gorin	Departamento de Bioquimica, Universidade Federal do Paraná, CP 19046, 81.504 Curitiba, Pr., Brazil

Helene Guinaudeau	Department of Chemistry, The Pennsylvania State University, 152 Davey Laboratory, University Park, PA 16802, U.S.A.
A.A. Leslie Gunatilaka	Department of Chemistry, University of Peradeniya, Peradeniya, Sri Lanka
Charles D. Hufford	Department of Pharmacognosy, University of Mississippi, University, MS 38677, U.S.A.
Lloyd M. Jackman	Department of Chemistry, The Pennsylvania State University, 152 Davey Laboratory, University Park, PA 16802, U.S.A.
Tikam Jain	Research and Development Division, Smith Kline and French Laboratories, Swedeland, PA 19479, U.S.A.
James P. Kutney	Department of Chemistry, The University of British Columbia, 2036 Main Mall, Vancouver, B.C., Canada V6T 1Y6
Stefano Mangani	Istituto de Chimica Generale, Università di Siena, Siena, I-53100 Siena, Italy
Paolo Mascagni	Department of Pharmaceutical Chemistry, School of Pharmacy, University of London, 29-39 Brunswick Square, London WC1N 1AX, England
Bonaventure T. Ngadjui	Department de Chimie Organique, Faculté des Sciences, Université de Yaounde, B.P.812, Yaounde, Cameroon
Neri Niccolai	Istituto di Chimica Generale, Università di Siena, I-53100 Siena, Italy
John H. Pazur	Department of Molecular and Cell Biology, Paul M. Althouse Laboratory, The Pennsylvania State University, University Park, PA 16802, U.S.A.

J.David Phillipson
Department of Pharmacognosy, School of Pharmacy, University of London, 29-39 Brunswick Square, London WC1N 1AX, England

Atta-ur-Rahman
H.E.J. Research Institute of Chemistry, University of Karachi, Karachi 32, Pakistan

David S. Rycroft
Department of Chemistry, University of Glasgow, Glasgow G12 8QQ, Scotland, U.K.

Maurice Shamma
Department of Chemistry, The Pennsylvania State University, 152 Davey Laboratory, University Park, PA 16802, U.S.A.

David L. Smith
Department of Medicinal Chemistry and Pharmacognosy, School of Pharmacy and Pharmacal Sciences, Purdue University, West Lafayette, IN 47907, U.S.A.

Gunther Snatzke
Abteilung für Chemie - Struktur Chemie, Ruhr-Universität Bochum, Postfach 10 21 48, D-6340 Bochum 1, F.R.G.

Beibam Lucas Sondegam
Department de Chimie Organique, Faculté des Sciences, Université de Yaounde. B.P.812, Yaounde, Cameroon

Motoo Tori
Faculty of Pharmaceutical Sciences, Tokushima Bunri University, Yamashiro-Cho, Tokushima 770, Japan

Luiz R. Travassos
Departamento de Micologia, Escola Paulista de Medicina, 04023 Sao Paulo, Sp., Brazil

Zhongrui Zhou
Department of Medicinal Chemistry and Pharmacognosy, School of Pharmacy and Pharmacal Sciences, Purdue University, West Lafayette, IN 47907, U.S.A.

Structure Elucidation

CHEMICAL IONIZATION MASS SPECTROMETRY WITH NITRIC OXIDE (NO) AS REAGENT GAS*

H. BUDZIKIEWICZ

1. INTRODUCTION

"Preliminary results indicate that it may be possible to identify the functional group present in many organic compounds from chemical ionization mass spectra recorded with nitric oxide as the reagent gas". This statement was made in 1972 by D.F. Hunt (ref. 2), one of the pioneers of the practical application of Chemical Ionization (CI) mass spectrometry. Now, 15 years later one can say that the importance of NO as a reagent gas does not lie in the identification of functional groups but rather in the location of double and triple bonds in aliphatic compounds. In addition, NO demonstrates perhaps more than any other gas used in CI the potential strengths and weaknesses of this technique.

2. PROPERTIES OF THE REAGENT GAS

NO produces the plasma ions NO^+ and $(NO)_2^{+\cdot}$. It has an ionization energy (IE) of 9.25 eV, the recombination energy of NO^+ has, however, been estimated as 8.3 eV or possibly about 0.5 eV higher (ref. 3). Therefore, ionization by charge exchange (CE) is to be expected for compounds with an IE of about 9 eV or below (ref. 3). CE can be fostered by mixing NO with N_2 (or another inert gas) (refs. 4-6) or by using N_2O (refs. 7,8) (which also produces NO^+ but has a rather high IE of 12.9 eV). In this way $M^{+\cdot}$ and characteristic EI-type fragment ions can be obtained. Examples have been reported from the steroid and alkaloid field. NO^+ reacts as an electrophile; thus $[M + NO]^+$ will usually be observed with compounds possessing either a π-system (refs. 2,7,9,12) or a nonbonding electron pair (refs. 2,9,17,18) (olefins, alkynes, alcohols, ethers, esters). The $[M + NO]^+$ ions may, however, be rather unstable and, therefore, of low intensity or even missing in the mass spectrum. In addition, having a hydride affinity of 1028 kJ/mole (ref. 7) NO can abstract H^- (resulting in $[M - H]^+$) even from alkanes.

NO has oxidizing properties, thus species arising from the loss of H_2 have been reported frequently (e.g., $[M - 3H]^+$, i.e. $C_nH_{2n+2} \rightarrow C_nH_{2n} \rightarrow C_nH_{2n-1}^{\ddagger}$ from

---

* Part XV of the series "Studies in Chemical Ionization Mass Spectrometry." For part XIV see (ref. 1).

alkanes) (ref. 10). It has been suggested that primary ($-CH_2OH \rightarrow -CHO \rightarrow -C{\equiv}O^+$, i.e. $[M - 3H]^+$), secondary ($>CHOH \rightarrow >C{=}O \rightarrow [M - 2H + NO]^+$) and tertiary alcohols ($\geq C-OH \rightarrow C^+$, i.e. $[M - OH]^+$) can be distinguished from each other in this way (refs. 2,17). These dehydrogenation reactions, however, depend strongly on experimental parameters (NO pressure, temperature, catalytic effects of metal surfaces), and with modern instruments frequently they are not observed at all.

3. REACTIONS OF NO^+ WITH FUNCTIONAL GROUPS

After a promising start (ref. 2) the reactions of the various functional groups have not been investigated systematically. From the limited literature data available the following conclusions can be drawn (the formation of $M^{+\cdot}$ by charge exchange and ions formed by the fragmentation of $M^{+\cdot}$ as well as oxidation reactions of the unionized compounds will not be mentioned):

Carboxylic acids and their methyl esters (refs. 11-13) give $[M + NO]^+$, $[M + NO - OH(OR)]^+$ and $[M - H]^+$, esters with higher alcohols also $[M + NO - ROH]^+$, $[M + NO - (R - H)]^+$ and, interestingly, $[M + NO - H_2O]^+$ (cf. Fig. 4).

Aldehydes (refs. 2,9,14) give $[M + NO]^+$ and $[M - H]^+$ (in most cases the H from the CHO-group is lost exclusively), while *ketones* (refs. 2,9) give $[M + NO]^+$.

Alcohols give $[M - H]^+$ (primary and secondary) (refs. 9,11) and/or $[M - OH]^+$ (secondary and tertiary) (refs. 2,9,17). A distinction by the dehydrogenation reactions mentioned above is not reliable (ref. 11).

Dialkyl ethers (ref. 9) give $[M - H]^+$, the only alkyl aryl ether mentioned in the literature, viz. anisole (ref. 9) $[M + NO]^+$.

Long chain aliphatic *epoxides* (refs. 15,16,18) besides the quasi molecular ions $[M + NO]^+$ and $[M - H]^+$ give two abundant fragments of the structure $R-C{\equiv}O^+$, the precursors of which are short-lived $[R-CH{=}ONO]^+$ ions (which are structurally identical with the adducts of NO to aldehydes) as could be shown by constant neutral loss and collision activation studies. The presence of OH, $OCOCH_3$ or CHO groups in the molecule does not interfere. CI (NO) may thus be used for the location of an epoxide ring in an aliphatic compound.

$$R-\overset{O}{CH-CH}-R' \rightarrow \begin{cases} RC(\overset{+}{O}NO)(H) \rightarrow RCO^+ \\ R'C(\overset{+}{O}NO)(H) \rightarrow R'CO^+ \end{cases}$$

4. REACTIONS OF NO^+ WITH HYDROCARBONS

4.1 General Remarks

Aromatic hydrocarbons (refs. 3,9,19) give $[M + NO]^+$ and/or (by charge exchange) $M^{+\cdot}$, alkanes (ref. 10) $[M - H]^+$ (only highly branched ones in addition to $M^{+\cdot}$ abundant fragment ions by cleavage of the most highly strained bond). From olefinic and acetylenic compounds, however, very characteristic fragments can be obtained and those compounds have, therefore, been investigated in considerable detail.

The reactions of NO^+ which may be used for the location of double or triple bonds in aliphatic compounds typically consist in an electrophilic addition to the π-system with subsequent hydrogen migrations frequently accompanied by more than one bond formation and bond cleavage step which may even involve the intervention of a functional group present in the molecule (in such cases the distance between the latter and the site of unsaturation plays an important role). These complex sequences of reactions involving cyclic transition states are of necessity rather slow processes. There is tentative evidence (ref. 20) that the residence time in the ion source influences the relative abundance of such ions. A major problem is a competing fragment formation starting from $M^{+\cdot}$ formed by CE. The recombination energy of NO (probably between 8.3 and 9.0 eV, *v. supra*) keeps CE and thus the formation of unspecific hydrocarbon ions at a relatively low level provided the source temperature is kept low. As can be seen from Fig. 1 by going from 60°C to 180°C the abundance of the N-containing ions formed from octadecene-5 drops almost to zero while that of the hydrocarbon ions goes up. Other examples will be mentioned below. The effect of the NO pressure is less pronounced; only at low values the rate of interaction of NO^+ with the substrate decreases drastically. Collision activation seems to be a promising technique to suppress the undesirable hydrocarbon ions (ref. 23), but little has been done in this respect so far.

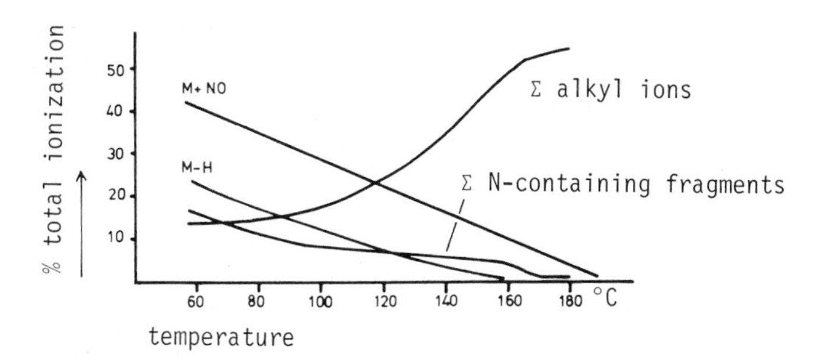

Fig. 1 Temperature dependance of the various types of ions in the CI(NO) spectrum of *n*-octadocene-5.

4.2 Compounds with double bonds

4.2.1 Alkenes with one double bond

For 1-alkenes a series of ions of the composition $(CH_2)_nNO$ of high abundance can be observed the intensity maximum of which shifts from n = 4 (*m/z* 86) to n = 6 (*m/z* 114) with increasing chain length (refs. 7,21) (see Fig. 2). For the genesis of these ions Markovnikov-oriented addition of NO^+ to the olefin followed by a series of hydride shifts, cyclisation and olefin elimination by a McLafferty rearrangement has been suggested (ref. 21).

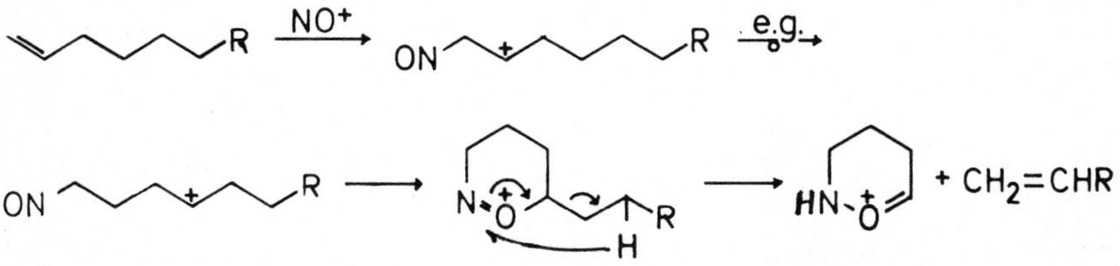

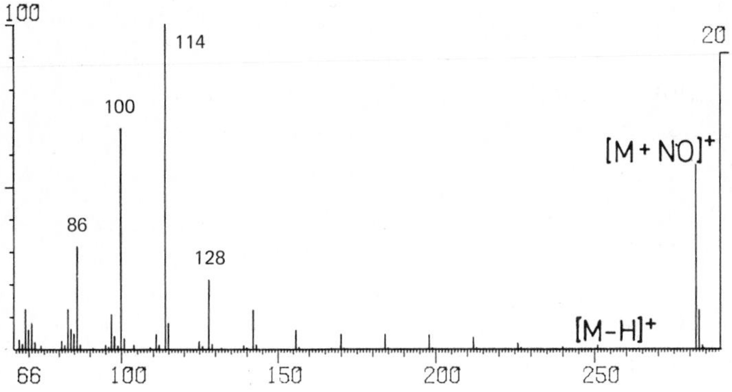

Fig. 2 CI(NO) spectrum of octadocene-1

This mechanism has not been substantiated by D-labelling (cf., however, below), but it is in agreement with the observation (ref. 21) that branching of the hydrocarbon chain results in a stabilization of the carbenium ion at the branching site which makes further hydride shifts less favourable. Thus, for 2-methyl alkenes the characteristic fragments are of low abundance (O-attack at C-2 in the cyclization step would result in a four-membered ring), for 3-methyl alkanes formation of a five-membered ring is enhanced, etc.

For alkenes with in-chain double bonds (ref. 7) (Fig. 3) by the same mechanism two series of $(CH_2)_nNO$ ions are obtained by addition of NO^+ to either end of the double bond. Here the high specificity of the H transfer during the olefin elimination has been demonstrated by labelling studies. Two further ions also belong formally to the $(CH_2)_nNO^+$ series; they are formed by highly specific McLafferty rearrangement reactions resulting in the cleavage of the vinylic bonds (a, *m/z* 128 and 212 in Fig. 3).

$$R-\underset{\underset{H}{|}}{CH}-CH_2-\underset{\underset{N=\bar{O}}{|}}{CH}-\overset{+}{C}H-R' \longrightarrow R-CH=CH_2 + \underset{\underset{H\overset{+}{O}=N}{|}}{CH}=CH-R' \quad \underline{\underline{a}}$$

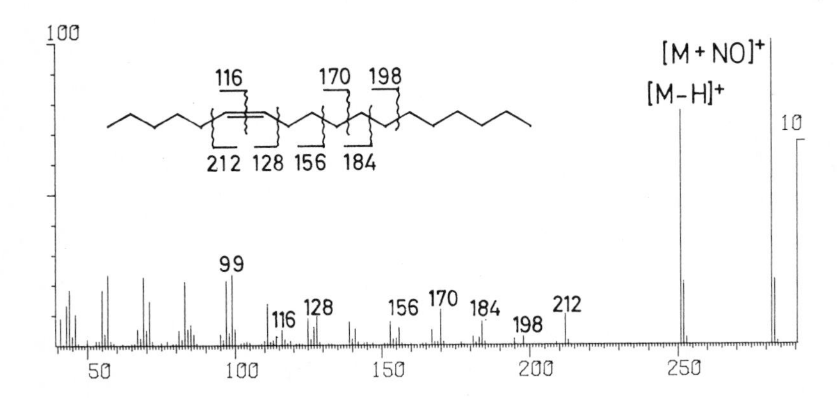

Fig. 3 CI(NO) spectrum of Z-octadocene-6

In addition, cleavage of the double bond by a rearrangement process with one specific and one unspecific H-transfer leads to a fragment of the composition $C_nH_{2n+2}NO^+$ the bigger part of the molecule being lost preferentially (refs. 7,24) (b, *m/z* 116 in Fig. 3).

$$C_5H_{11}-\underset{\underset{N=O}{|}}{CH}-\overset{+}{C}H-\underset{\underset{H}{|}}{CH}-(CH_2)_7-CH_3 \longrightarrow C_5H_{11}-\underset{\underset{\overset{+}{N}H-OH}{|}}{CH}-CH=CH-(CH_2)_7-CH_3$$

$$\longrightarrow C_5H_{11}-CH=\overset{+}{N}HOH + C_{11}H_{20} \quad \underline{\underline{b}}$$

The intensity of the fragments mentioned before is relatively low and, therefore, they are easily obscured by hydrocarbon ions at higher source temperatures.

One further ion characteristic for the position of the double bond has to be discussed here. It is formed by cleavage of the double bond and has the structure of an acyl species ($\underline{\underline{c}}$, *m/z* 99 in Fig. 3).

$$\begin{array}{c} ^{+}O=N \\ | \quad | \\ R-CH-CH-R' \end{array} \longrightarrow \begin{array}{c} ^{+}O-N \\ \| \quad \| \\ R-CH \quad CH-R' \end{array} \longrightarrow R-CO^{+} + \underset{\underline{\underline{c}}}{HN=CHR'}$$

Generally, the larger portion of the molecule is being lost more readily. The mechanism of formation of this ion has been corroborated by exact mass measurements and labelling studies. This fragment is, however, one of the most elusive ions as its abundance depends on experimental parameters even more drastically than that of the ions discussed sofar. Not only may increased source temperature reduce its intensity (under favourable circumstances it is the most abundant fragment of all (*m/z* 99 in Fig. 3), under unfavourable ones it is practically absent; in (refs. 7,20) CI(NO) spectra of Z-octadecene-6 are shown where *m/z* 99 is by no means conspicuous), even the type of instrument used (residence time in the ion source?) seems to play an important role (cf. also below) (ref. 20).

For olefins with an in-chain double bond having less than 10 C-atoms the N-containing fragments are hardly recognizable (those the genesis of which requires a minimum chain length are even missing); the same is true for olefins with tri- or tetra-substituted double bonds (refs. 7,21) (steric hindrance of the approach of NO^{+}). There is also a considerable difference in the intensity ratio $[M + NO]^{+}/[M - H]^{+}$ for Z- (1.4 to 2.5) and E-isomers (0.6 to 0.9), at least for C_{16} to C_{18} alkenes (ref. 7). Again steric effects (less hindered elimination of the allylic H in the case of the E-isomers) offer an explanation.

4.2.2 Polyenes

1,3- and 1,4-Dienes fragment as shown for terminal mono-olefins, the masses of the cyclic ions being 2 u lower (ref. 21). The picture changes when the second double bond moves farther to the center of the chain (ref. 11):1,Z-8-pentadecadiene and 1,Z-8-heptadecadiene yield M^{+} with 100% and $[M + NO]^{+}$ with ca. 15% rel. int. and in addition a series of ions *m/z* 67, 81, 96, 110, 124, 138 (most abundant species in the respective clusters), but the data available are too incomplete for definitive conclusions. For the characteristic fragmentation pattern of homoconjugated tri- and higher polyenes see below section 4.2.4.

4.2.3 Mono-olefins with functional groups

Investigated were the following groups: $-CH_2OH$, $-CH_2OAc$, -CHO, -COOH(R), Cl, Br. Two publications, one on alcohols, acetates and aldehydes (ref. 22) and the other one on acids and esters (ref. 13) lead to a somewhat puzzling situation: For the former the acyl ion c (absent in the spectra of the latter) had been described as characteristic, for the latter - in contrast - the MacLafferty-ion a (absent in the spectra of the former). The *prima facie* obvious role of the functional groups was not understood readily. More detailled recent studies (ref. 20) shed some light on the situation: Double bond cleavage of the $[M + NO]^+$ ion under formation of acyl ions (c) is an ubiquitous process unless the double bond is too close to the functional group. Their intensity, however, goes down drastically with increasing source temperature. In addition, $C_nH_{2n+1}CO^+$ ions being isobaric with $C_{n+2}H_{2n+5}{}^+$(= alkyl) ions they can be camouflaged by the latter especially when of low abundance. The influence of the type of mass spectrometers used became evident when the mass spectra of esters obtained with a Finnigan 3200 quadrupole and a Kratos MS-25 magnet instrument were compared: Acyl ions were not recognizable in the Finnigan and pronounced in the Kratos spectra. Whether one (stemming from the alkyl end) or both acyl ions can be observed, seems to depend on the nature of the functional group as well as on the position of the double bond. In any case, the acyl ion(s) are highly characteristic for the position of a double bond provided they can be identified unambiguously.

The results described in the pertinent papers will be discussed separately. For reasons given above they are to be taken with some reservation as far as reproducibility with another system (compound class as well as mass spectrometer) is concerned.

(i) n-Alkenoic acids and their esters when measured (ref. 13) with a Finnigan 3200 instrument (inlet: gas chromatograph) gave the following results: The behaviour of Δ^2-, of Δ^4- and of $\Delta^{\omega-1}$ compounds equals that of saturated acids (esters) (*v. supra*), *i.e.* one observes degradation of the ester group only. Δ^3-compounds yield with high abundance a fragment $C_3H_6NO_3$ (*m/z* 104) for acids which is shifted accordingly in mass (+ 14 u) for esters. It corresponds to ion b observed with alkenes. Acids and esters with a double bond between Δ^5 and $\Delta^{\omega-2}$ (both included, investigated up to Δ^{11}) give an ion which after elimination of H_2O (ROH) from $[M + NO]^+$ is formed by a McLafferty rearrangement (cf. ion a from alkenes) with elimination of the alkyl end of the chain (*m/z* 140 in Fig. 4) (the alternative ion formed by loss of the carboxyl end is not of any significance). This ion is of medium to low (especially with increasing chain length) abundance and it is accompanied by a species arising from a further loss of H_2O (*m/z* 122 in Fig. 4). Esters with a terminal double bond give with high abundance ions of the sequence

$$[M + NO]^+ \rightarrow [M + NO - ROH]^+ \rightarrow [M + NO - ROH - H_2O]^+ \rightarrow$$

$$[M + NO - ROH - H_2O - CO]^+.$$

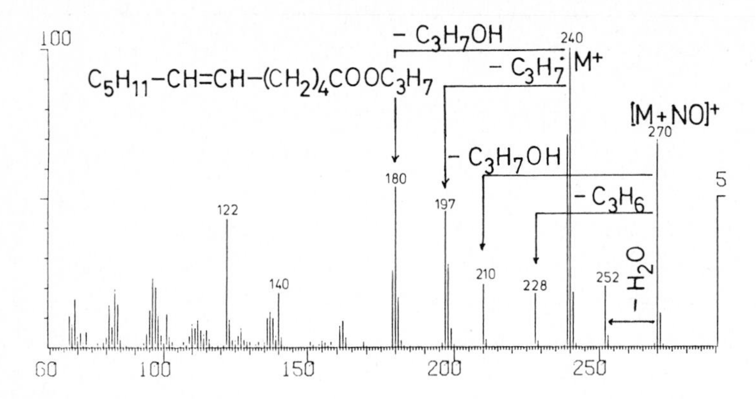

Fig. 4 CI(NO) spectrum of E-6-dodecenoic acid isopropyl ester

(ii) Alkenols, alkenyl acetates and alkenals (ref. 22). Representatives with the double bond position Δ^4 and beyond yield the acyl ion $\underline{\underline{c}}$ derived from the hydrocarbon part of the molecule; the complementary ion comprising the functional group is absent. Δ^2- and Δ^3- acetates give *m/z* 84 by the following sequence

$$[M + NO]^+ \xrightarrow{-H_2O} [R\text{-}CH{=}CH\text{-}CH{=}CH_2 + NO]^+ \rightarrow HONCH\text{-}CH{=}CH\text{-}CH_2^{\;+} \quad (m/z\ 84)$$

From Δ^4- acetates the homologous ion *m/z* 98 is obtained. Experimental details have not been given.

(iii) Alkenols, alkenyl acetates, methyl ethers, halides and alkenoic acid methyl esters (when measured with a Kratos MS-25 mass spectrometer (ref. 20) both with a direct and a GC inlet system) provided the source temperature is low give with high intensity the acyl ion $\underline{\underline{c}}$ derived from the alkyl part of the molecule. At elevated temperatures ion $\underline{\underline{c}}$ is burried, however, underneath the homologous series of alkyl ions. Whether the supplementary ion containing the functional group can be observed seems to depend on the nature of the functionality and the position of the double bond. In addition, ion $\underline{\underline{b}}$ occurs with an appreciable intensity if the double bond is located towards the alkyl end of the chain.

4.2.4 Compounds with three and more homoconjugated double bonds (refs. 25,26)

Compounds with three homoconjugated double bonds (sofar only all-*cis* representatives have been investigated) give four fragments of high to medium abundance:

d: C_6H_7R'

"d + 18": C_5H_7NOR'

e: C_6H_7R

"e + 18": C_5H_7NOR

The ions d and e (*m/z* 194 and 150 in Fig. 5) stem possibly (ref. 27) from $M^{+\cdot}$ rather than from $[M + NO]^+$; the formation of "d + 18" and "e + 18" (*m/z* 212 and 168 in Fig. 5) is initiated by addition of NO^+ to the central double bond.

"e + 18"

The formation of the four characteristic ions is independent from the nature of R^1 and R^2: Hydrocarbons, alcohols, halides, aldehydes, acids and esters (Fig. 5) as well as compounds with additional double bonds have been investigated and only two exceptions have been encountered: For R^1 = H (terminal double bond) d and "d + 18" are missing and for R^2 = CH_2COOCH_3 other processes prevail (ref. 25).

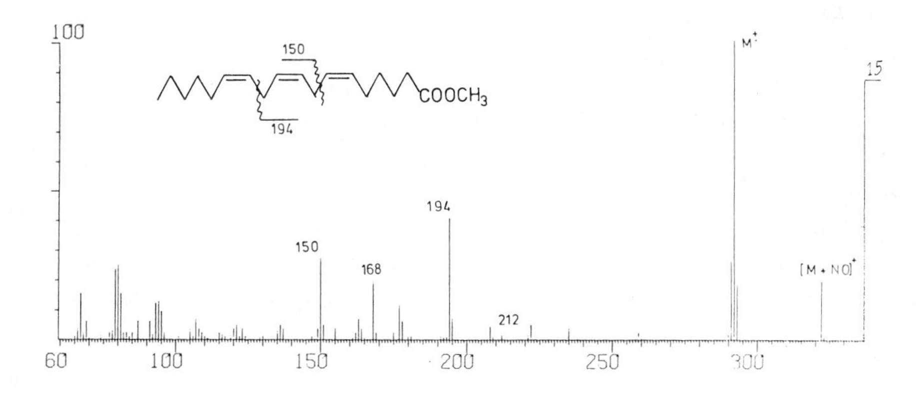

Fig. 5 CI(NO) spectrum of linolenic acid methyl ester

Compounds with more than three homoconjugated double bonds can be considered as superimposed triene systems. For each additional double bond, thus, two pairs of additional fragments will be observed (Fig. 6). With increasing unsaturation the spectra obviously become increasingly complicated.

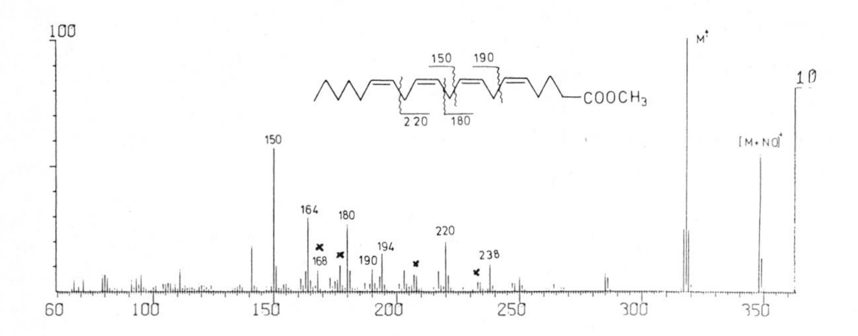

Fig. 6 CI(NO) spectrum of arachidonic acid methyl ester. The "+ 18" ions are marked with x.

The mass spectra of compounds with modifications in the vicinity of the homoconjugated double bond system as they are found, e.g., in the arachidonic acid metabolism, have not been investigated in detail. Obviously competing degradation reactions may complicate the picture as illustrated by 15-hydroxy-Z-5,Z-8,Z-11,E-13-eicosatetraenic acid methyl ester (ref. 11): Only one of the characteristic fragments (*m/z* 180 comprising the ester group - as shown by an appropriate shift to *m/z* 166 in the spectrum of the free acid) occurs with high abundance. The formation of other fragments is initiated by the NO-group.

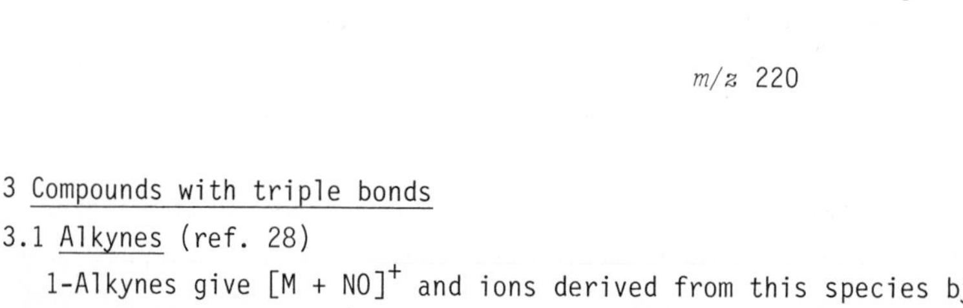

m/z 220

4.3 Compounds with triple bonds

4.3.1 Alkynes (ref. 28)

1-Alkynes give $[M + NO]^+$ and ions derived from this species by loss of ·OH or H_2O as well as $[M - H]^+$ of only low abundance (~10% rel. int.). Up to hex-1-yne only hydrocarbon fragments can be observed, starting from hept-1-yne series of homologous ions $[C_nH_{2n-2}NO]^+$ ($n \geqslant 4$), i.e., *m/z* 84, 98, 112 ..... are formed by elimination of alkene moieties (cf. below). Typical for all 1-alkynes is an ion *m/z* 72 (C_3H_6NO) of unknown genesis.

Alkynes with an in-chain triple bond form very abundant $[M + NO]^+$ ions and $M^{+\cdot}$ of low intensity (10% rel. int.). Hydrocarbon ions (mainly $C_nH_{2n-3}^+$ and $C_nH_{2n-1}^+$) crowd the lower mass range especially at elevated source temperatures. As in the case of alkenes three types of fragments can be observed as depicted schematically.

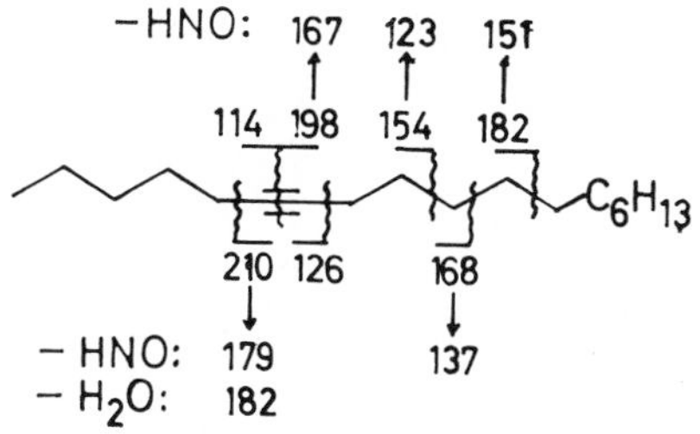

−HNO: 167 123 151
114 198 154 182
C_6H_{13}
210 126 168
− HNO: 179 137
− H_2O: 182

Labelling studies suggest the following mechanism for the series of $C_nH_{2n-2}NO$ ions (*m/z* 154, 168, 182 in Fig. 7) which decompose further by elimination of HNO. The smallest neutral to be expelled is C_3H_6, the minimum ring size is pyrrolidine.

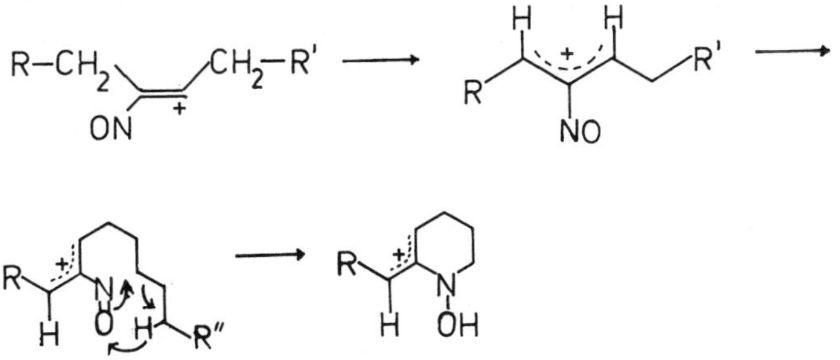

R−CH_2 CH_2−R' ON + → H H R + R' NO →

R + N O H H R'' → R + N H OH

Two ions belong to the same series which are formed by cleavage of the bond next to the triple bond. They remind of ions $\underline{\underline{a}}$ observed with olefins, but the H-transfer is much less specific than observed for $\underline{\underline{a}}$. These fragments (*m/z* 126 and 210 in Fig. 7) may lose HNO and/or H_2O.

Cleavage of the triple bond results in ions of the composition $C_nH_{2n}NO^+$ (*m/z* 114 and 198 in Fig. 7), again the larger portion of the molecule being lost preferentially.

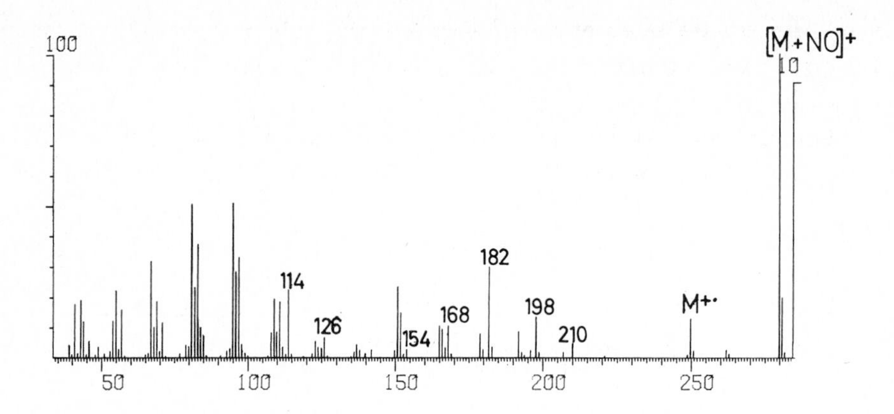

Fig. 7 CI(NO) spectrum of octadecyne-6

4.3.2 Alkynols

$[M + NO]^+$ ions are of medium to high intensity. The fragments described for alkynes are also observed in the mass spectra of alkyn-1-ols (refs. 23,29), but their relative abundance depends upon the distance between the triple bond and the hydroxyl group. This suggests that mechanisms of formation involving the latter or the double bond formed by H_2O elimination are responsible for their genesis. Thus, *m/z* 142 ($C_7H_{12}NO_2^+$, base peak) accompanied by a somewhat less abundant *m/z* 128 (both formed by olefin elimination) are characteristic for 2-yn-1-ols. The intensity of the homologous ion *m/z* 156 for 3-yn-1-ols is reduced and *m/z* 170 becomes negligible for 4-yn-1-ols.

For 5-yn-1-ols *m/z* 110 ($C_6H_8NO^+$) is responsible for the base peak. It is formed by the loss of the alkyl end of the chain and of H_2O. The homologous ion *m/z* 124 has still a relative intensity of 50% for 6-yn-1-ols (Fig. 8), but again with increasing distance of the hydroxyl group from the triple bond these ions become less and less characteristic (which suggests an intra-molecular substitution mechanism), especially since ions of the series $[C_nH_{2n-4}NO]^+$ of comparable abundance may also be formed by non-specific processes which have not been observed for alkynes.

$$C_9H_{19}-C\equiv C-(CH_2)_4-CH_2OH \xrightarrow[-\,H_2O]{+\,NO^+} \text{(cyclic ion, } C_7H_{15}\text{, H, N=O)} \xrightarrow{-\,C_9H_{18}} \text{(cyclic ion, =NOH)}$$

m/z 124

It should be noted that ions of high abundance (50% rel. int.) of masses *m/z* 110 and 124 are also formed from 10- and 11-yn-1-ols, resp. A concerted mechanism resulting in the expulsion of a cyclopentadienyl radical from the end of the chain has been suggested (ref. 29) for their genesis. Hence, these fragments can only be used for corroborative evidence for the location of triple bonds.

Most characteristic for the position of the triple bond starting from 5-yn-1-ols of the general formula $H(CH_2)_n$-C≡C-$(CH_2)_m OH$ are two odd-mass ions which have not been observed with alkynes and which are formed by the cleavage of the triple bond, viz. $C_{n+1}H_{2n+1}^{+}$ (*m/z* 139 in Fig. 8) form the alkyl and $C_{m+1}H_{2m-1}$ (*m/z* 81 in Fig. 8) form the hydroxyl end of the chain. Although homologs of these two ions may be present in the spectrum the position of the triple bond can be determined easily by multiplying the intensities of all pairs of ions adding up to $C_{n+m+2}H_{2n+2m}$. The product of those ions formed by cleavage of the triple bond will show a clear-cut maximum (Fig. 9).

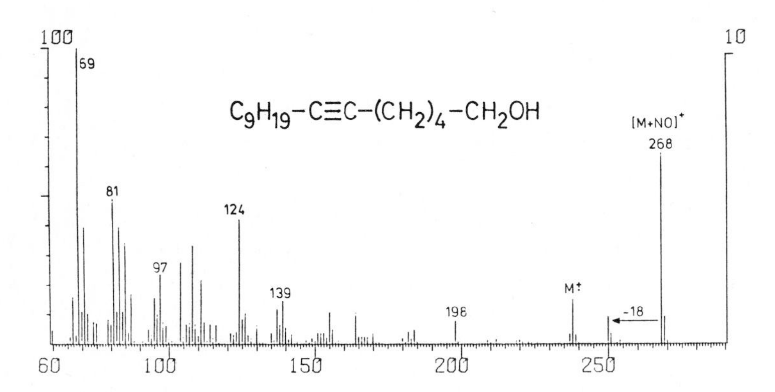

Fig. 8 CI(NO) spectrum of 6-hexadecyn-1-ol.

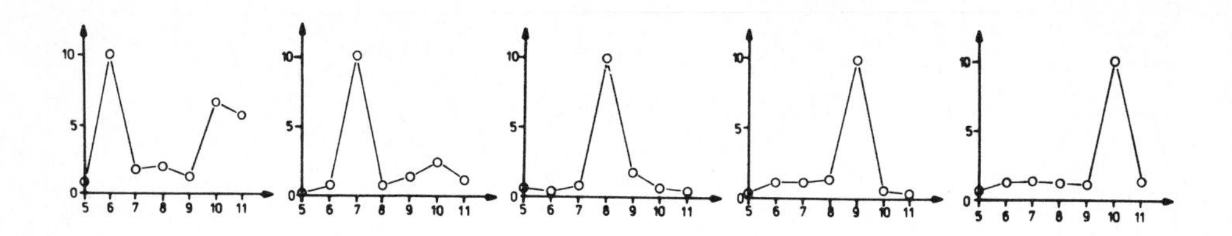

Fig. 9 Products of the intensities of the corresponding pairs of ions $[C_{n+1}H_{2n+1}]^+$ and $[C_{m+1}H_{2m-1}]^+$ plotted against the position of the triple bond (Δ^5 till Δ^{10}).

4.3.3 Alkynoic acids and their methyl esters (refs. 12,23).

Here also the distance between the triple bond and the carboxyl group plays an important role. It should be noted that also here representatives up to about C_{10} behave atypically as rearrangement processes do need a minimum chain length. The results with long chain compounds (C_{12} - C_{18} and Δ^2 - Δ^{10} having been investigated) can be summarized as follows:

(i) $[M + NO]^+$ ions are of medium to high intensity.

(ii) Δ^2-Compounds in addition to $[M + NO - {}^{\bullet}OH(CH_3)]^+$ give *m/z* 156 and 170 (base peak formed by alkene alimination; cf. alkynes above) which occur, however, also in the spectra of Δ^3-compounds (partial isomerization to 2,3-allene systems?).

(iii) Esters with the triple bond in positions Δ^4 to Δ^6 give abundant ions by loss of the alkyl residue next to the triple bond (*m/z* 87, 101 and 115, resp.), Δ^6-esters in addition an abundant ion *m/z* 111 of unknown origin.

(iv) Esters with triple bond positions between Δ^7 and Δ^{10} fragment by cleavage of the triple bond as described for alkynols (the correct pair of fragments can be determined as described there). The ion containing the ester group loses ${}^{\bullet}CH_3$ and HNO in either sequence, and finally CO.

(v) Esters with triple bond positions between Δ^6 and Δ^{10} yield fragments of the composition $C_nH_{2n-8}NO$ (*m/z* 120 in Fig. 10) the formation of which has been investigated in some detail (refs. 11,30) by labelling experiments. They are formed by loss of the ester alkoxyl group, of H_2O and of the carbon chain beyond the triple bond probably accompanied by hydrogen rearrangements. The H-atoms in α-position to the ester group are retained, those located two carbon atoms beyond the triple bond are lost (which excludes the elimination of an alkene by a McLafferty rearrangement analogous to the formation of ion $\underline{\underline{a}}$).

$$CH_3\text{-}(CH_2)_4\text{-}\overset{NO^+}{\overset{\vdots}{C}}\equiv C\text{-}(CH_2)_4\text{-}COOCH_3 \xrightarrow[-\ H_2O,\ -\ CH_3OH]{-\ C_5H_{10}} m/z\ 120$$

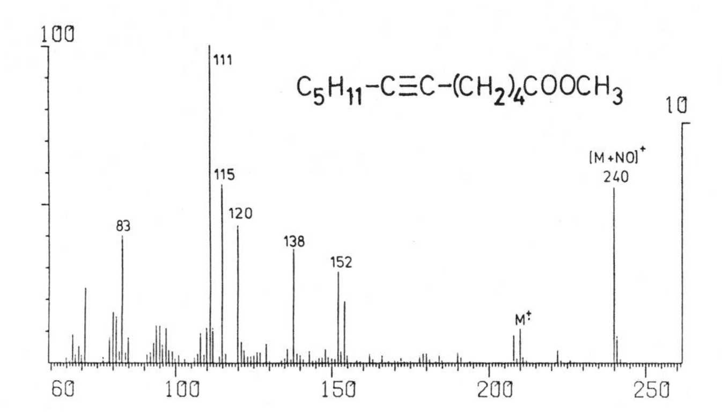

Fig. 10 CI(NO) spectrum of 6-dodecynoic acid methyl ester.

There are other fragmentation processes which seem to be characteristic for certain triple bond positions. Here the original literature should be consulted.

5. CONCLUSION

From the foregoing discussion it can be seen that NO^+ is a very useful reagent gas for the location of double and triple bonds in aliphatic compounds, but a certain amount of experience is necessary to avoid pitfalls, and further work will be necessary to elucidate the different fragmentation mechanisms in detail so that the various competing factors complicating the interpretation are better understood.

REFERENCES

1 E. Schröder and H. Budzikiewicz, The Influence of Source Parameters on DCI Spectra,, Adv. Mass Spectrom. (1986) 1599-1600.
2 D.F. Hunt and J.F. Ryan, Chemical Ionization Mass Spectrometry Studies. Nitric Oxide as a Reagent Gas, J.C.S. Chem. Comm. (1972) 620-621.
3 N. Einolf and B. Munson, High Pressure Charge Exchange Mass Spectrometry, Int. J. Mass Spectrom. Ion Phys. 9 (1972) 141-160.
4 B. Jelus, B. Munson and C. Fenselau, Charge Exchange Mass Spectra of Trimethylsiyl Ethers of Biological Important Compounds: An Analytical Technique, Anal. Chem. 46 (1974) 729-730.
5 B. Jelus, B. Munson and C. Fenselau, Reagent Gases for G.C.-M.S. Analyses, Biomed. Mass Spectrom. 1 (1974) 96-102.
6 I. Jardine and C. Fenselau, Charge Exchange Mass Spectra of Morphine and Tropane Alkaloids, Anal. Chem. 47 (1975) 730-733.
7 H. Budzikiewicz and E. Busker, CI-Spectra of Olefins, Tetrahedron 36 (1980) 255-266.
8 Ch.W. Polley, Jr. and B. Munson, Nitrous Oxide as Reagent Gas for Positive Ion Chemical Ionization Mass Spectrometry, Anal. Chem. 55 (1983) 754-757.

9 D.F. Hunt, Reagent Gases for Chemical Ionization Mass Spectrometry, Adv. Mass Spectrom. 6 (1974) 517-522.
10 D.F. Hunt and T.M. Harvey, Nitric Oxide Chemical Ionization Mass Spectra of Alkanes, Anal. Chem. 47 (1965) 1965-1969.
11 unpublished results.
12 A. Brauner and H. Budzikiewicz, CI(NO)-Spectra of *n*-Alkynoic Acids and their Methyl Esters, Spectros. Int. J. 2 (1983) 338-347.
13 A. Brauner, H. Budzikiewicz and W. Francke, Chemical Ionization (NO) Spectra of *n*-Alkenoic Acids and their Esters, Org. Mass Spectrom. 9 (1985) 578-581.
14 I. Jardine and C. Fenselau, The High Pressure Nitric Oxide Mass Spectra of Aldehydes, Org. Mass Spectrom. 10 (1975) 748-751.
15 J. Einhorn, C. Malosse, P. Wirsta and J.C. Tabet, CI/NO^+ Mass Spectrometry of Long Chain Aliphatic Epoxides, Adv. Mass Spectrom. (1986) 1367-1368.
16 J.C. Tabet and J. Einhorn, A Facile Structural Analysis of Long Chain Epoxides by Constant Neutral Mass Spectra under CI-NO^+ Conditions, Adv. Mass Spectrom. (1986) 1443-1444.
17 D.F. Hunt, T.M. Harvey, W.C. Brumley, J.F. Ryan III and J.W. Russel, Nitric Oxide Chemical Ionization Mass Spectrometry of Alcohols, Anal. Chem. 54 (1982) 492-499.
18 J.C. Tabet and J. Einhorn, The Use of Constant Neutral Spectra to Determine the Origin of the Acylium Ions in the CI/NO^+ Spectra of Aliphatic Epoxides, Org. Mass Spectrom. 20 (1985) 310-312.
19 D.F. Hunt, Ch.N. McEwen and T.M. Harvey, Positive and Negative Chemical Ionization Mass Spectrometry Using a Townsend Discharge Ion Source, Anal. Chem. 47 (1975) 1730-1734.
20 H. Budzikiewicz, B. Schneider, E. Busker, W. Boland and W. Francke, Are the Reactions of Aliphatic C,C-Double Bonds with NO^+ Governed by Remote Functional Groups?, Org. Mass Spectrom., 22 (1987) 458-461.
21 D.F. Hunt and T.M. Harvey, Nitric Oxide Chemical Ionization Mass Spectra of Olefins, Anal. Chem. 47 (1975) 2136-2141.
22 C. Malosse and J. Einhorn, Double Bond Location in Long Chain Mono-unsaturated Alcohols, Acetates and Aldehydes by CI-NO^+-MS, Adv. Mass Spectrom. (1986) 1369-1370.
23 H. Budzikiewicz, E. Busker and A. Brauner, Localization of Double and Triple Bonds by CI Mass Spectrometry, Adv. Mass Spectrom. 8 (1980) 713-722.
24 G.J. Bukovits and H. Budzikiewicz, The Cleavage of a CC-Double Bond after Chemical Ionization with NO^+ - A Complex Rearrangement Process, Org. Mass Spectrom. 19 (1984) 23-26.
25 A. Brauner, H. Budzikiewicz and W. Boland, Localization of Homoconjugated Triene and Tetraene Units in Aliphatic Compounds, Org. Mass Spectrom. 17 (1982) 161-164.
26 W. Boland, L. Jaenicke and A. Brauner, Vinyl-Olefins and Sesquiterpenes in the Root-Oil of *Senecio isatideus*, Z. Naturforsch. 37c (1982) 5-9.
27 H. Budzikiewicz, Structure Elucidation by Ion-Molecule Reactions in the Gas Phase: The Location of C,C-Double and Triple Bonds, Fresenius Z. Anal. Chem. 321 (1985) 150-158.
28 E. Busker and H. Budzikiewicz, *i*-C_4H_{10} and NO Spectra of Alkynes, Org. Mass Spectrom. 14 (1979) 222-226.
29 A. Brauner and H. Budzikiewicz, CI(NO) Spectra of *n*-Alkyn-1-ols, Org. Mass Spectrom. 18 (1983) 324-326.

IDENTIFICATION OF PROTEIN CROSS-LINKAGES BY FAST ATOM BOMBARDMENT MASS SPECTROMETRY

D.L. SMITH AND Z. ZHOU

1. INTRODUCTION

Important chemical and physical properties of proteins are often determined by covalent bonding between, or within, peptide chains. Thus, while recombinant DNA technology may be used to produce large quantities of protein with a specific amino acid sequence, the product may not have the desired biological function if the appropriate cross-linkages have not been formed. It follows that the great dividends promised by genetic engineering will not be fully realized until means are developed for inducing the required cross-linkages. Overcoming this challenge will require developing analytical methodology suitable for rapidly identifying and locating peptide cross-linkages. In some instances, the protein may be large (500 residues) and have many disulfide linkages.

Because of the importance of proteins, a variety of highly developed analytical methodologies are available for characterizing them and their constituent amino acids. These include highly sensitive amino acid analyzers and spinning-cup or gas phase sequenators, as well as a variety of purification methods such as isoelectric focusing, gel electrophoresis, and high performance liquid chromatography (HPLC). Unfortunately, the power of these traditional methods is substantially reduced in the case of chemically modified proteins. Although chromatography may be required to isolate a modified component of a protein, it is of little help in determining the elemental composition and structure of the modified component. This weakness of traditional analytical methods has greatly restricted investigation of protein cross-linkages since they often involve modified amino acids.

Because of the high selectivity and sensitivity of mass spectrometry, it has been the method of choice for identifying modified amino acids for the past two decades (ref. 1-3). Early mass spectrometric investigations used electron and chemical ionization to analyze volatile derivatives of amino acids and peptides having as many as ten residues. Although this approach has and continues to make many significant contributions to structure elucidation of modified amino acids, it suffers from reliance on an extensive, and often complex derivatization scheme. While derivatization has been successful under a variety of circumstances, there is always the concern that the structure to be determined may be inadvertently modified during derivatization.

With the advent of new ionization techniques (ref. 4), the role played by mass spectrometry in the investigation of proteins has been extended far beyond the highest goals of the previous decade (refs.5-7). For example, fast atom bombardment (FAB), more recently termed liquid secondary ion mass spectrometry (LSIMS) (ref. 8), is routinely used to determine the molecular weights of proteins with molecular weight up to 5,000. Heroic efforts with instruments specially designed for analysis of high-mass ions have demonstrated production of molecular ions above m/z 10,000.

Although early applications of LSIMS to peptides focused on demonstrating that the molecular weight of larger and larger peptides could be determined, more recent efforts have been directed toward exploring the role of this new technique, when used in conjunction with classical analytical methods, to investigate protein structure. For example, peptide molecular weight determination by LSIMS provides an excellent means of verifying that the amino acid composition of a protein matches that predicted from its cDNA (ref. 9). Modifications, such as phosphorylation, acetylation, and glycosylation, which may go undetected in conventional amino acid analyses, are readily detected by LSIMS (ref. 10). Sequence information based on predictable fragmentation patterns may also be gleaned from mass spectra of peptides which are blocked at the NH_2-terminus.

It has also been demonstrated that LSIMS is going to play an important role in identifying and locating protein cross-linkages (ref. 11-15). The purpose of this contribution is to acquaint the reader with salient features of LSIMS, as applied to the analysis of peptides, and to demonstrate how it can be used to investigate protein cross-linkages.

2. LSIMS OF PEPTIDES

The procedure for obtaining LSI mass spectra of small peptides (less than ten amino acids) is relatively easy. A few nanomoles of the peptide is dissolved in a few microliters of an acidic solution, such as acetic or formic acid, thoroughly mixed with a microliter of the dispersing matrix, and placed on the probe tip (ref. 16). Dispersing matrices such as glycerol, 1-thioglycerol, and a 3:1 mixture of dithiothreitol (DTT) and dithioerythritol (DTE) have proven most useful for positive ion mass spectra of peptides. It is important to note that, although many peptides give satisfactory spectra in any of these matrices, others yield good spectra with only a specific matrix.

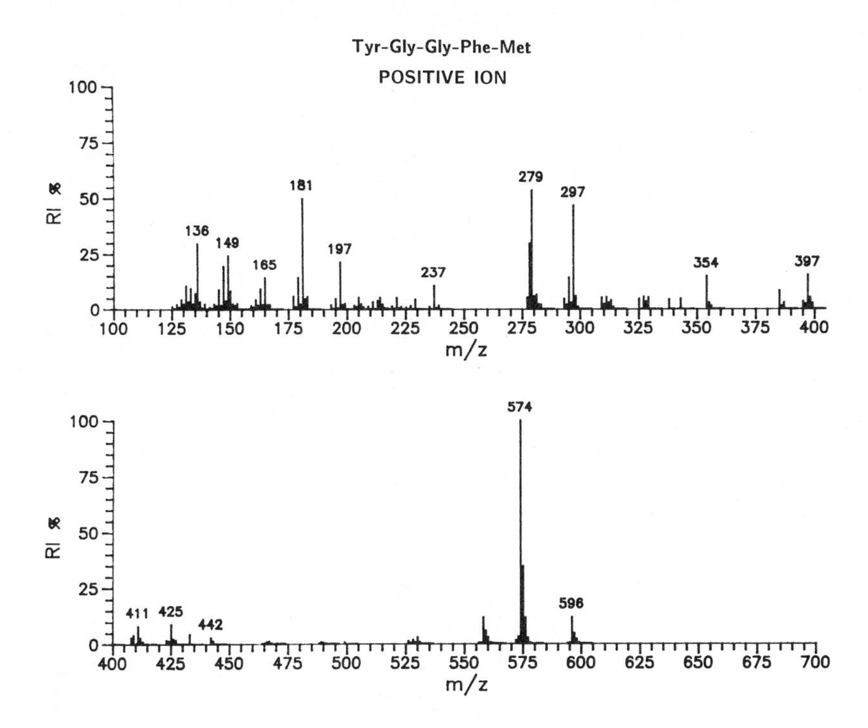

Figure 1. The positive ion LSI mass spectrum of methionyl-enkephalin obtained for 2 nanomole of peptide dissolved in 1-thioglycerol and 5% acetic acid. Spectrum taken on a Kratos MS-50 mass spectrometer with a xenon energy of 8 keV.

The basic features of positive and negative ion LSI mass spectra are illustrated in Figures 1 and 2, which were obtained with methionyl-enkephalin. In general, the protonated molecule ion, MH^+, is the largest high-mass peak in the positive ion spectrum. This feature is evident for methionyl-enkephalin which has a nominal molecular weight of 573 mu (MH^+ 574). In addition to the MH^+ ions, positive ion LSI mass spectra usually have a sodium cationated peak, MNa^+. Although usually somewhat less intense, negative ion LSI mass spectra of peptides have an $M\text{-}H^-$ ion, as indicated in Figure 2. As will be discussed below, determining the molecular weights of peptides is one of the most important uses of LSI mass spectra.

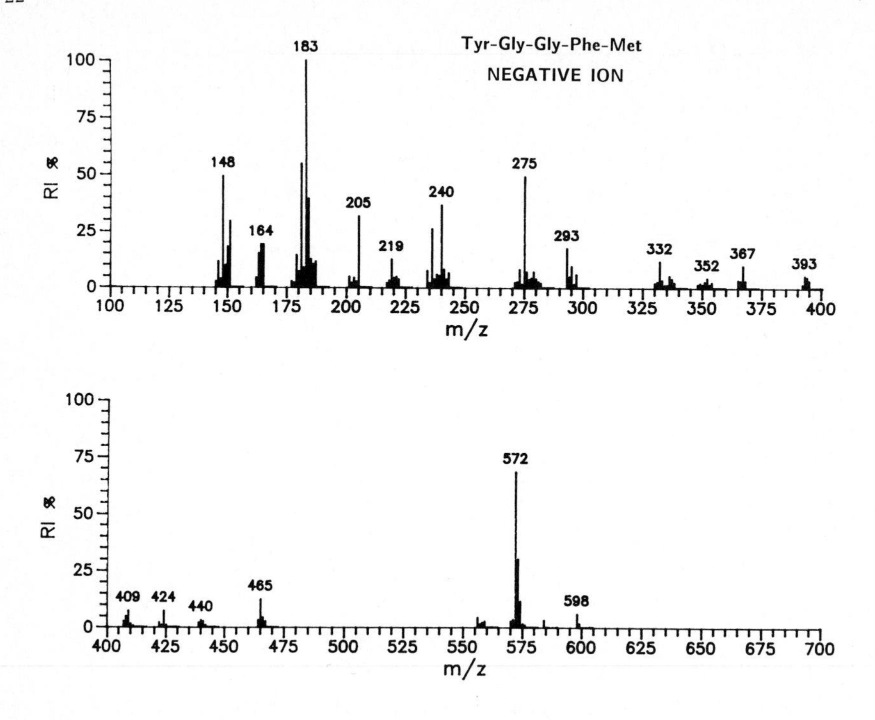

Figure 2. Negative ion LSI mass spectrum of methionyl-enkephalin obtained for 2 nanomoles of peptide dissolved in glycerol and 5% acetic acid.

In addition to indicating the molecular weight of a peptide, LSI mass spectra may be an important source of sequence information. This approach to amino acid sequencing is based on the fact that most of the amino acids have different molecular weights (leucine and isoleucine are the exceptions), and that the peptide chain undergoes cleavage at specific points, as indicated for methionyl-enkephalin in Figure 3. One series of fragment ions contains the NH_2-terminus plus some portion of the chain, another series contains the CO_2-terminus. Ions in the NH_2-terminal series of fragments resulting from cleavage of the amide bond often lose CO to give a peak 28 mu lower. It is important to note that some members of a series are not apparent in the spectrum. Although this method can often be used to sequence small peptides, as in the case of methionyl-enkephalin, it has had limited success for large peptides because of a scarcity of diagnostic fragment ions. However, in the case of peptides which are blocked on the NH_2-terminus, LSIMS may be the method of choice.

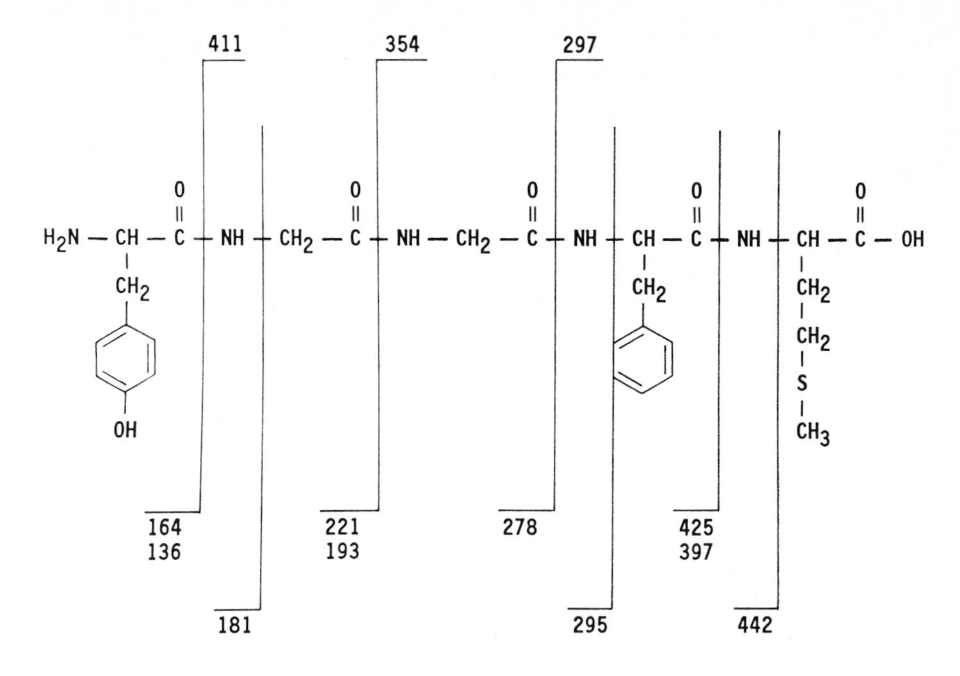

Figure 3. Illustration of fragment ions expected, and found in the positive ion LSIMS mass spectrum of methionyl-enkephalin.

Although LSIMS does give important structural information for small peptides, most applications focus on determining the molecular weights of high-mass peptides. The utility of LSIMS for determining the molecular weight of a high-mass peptide is illustrated in Figure 4 which shows the molecular ion region of the oxidized B chain of bovine insulin. This peptide has 30 amino acids and is representative of peptides which can be readily detected by LSIMS. In addition to intense protonated molecule ions, LSI mass spectra usually have background ions at every mass, often hiding weak fragment ions diagnostic of structure and sequence. One objective of research in the area is to find matrices which give intense molecular and fragment ions that are distinct from the background, and diagnostic for the structure of the peptide. The intensities of background ions are especially high at low mass, decrease monotonically with increasing mass, and are usually low for masses substantially above the molecular weight of the peptide. Detailed mechanisms responsible for production of the background ions are not completely understood, but presumably involve a vast array of non-specific addition and elimination reactions. The importance of rendering the sample salt-free is illustrated in Figure 4b which was obtained by adding 0.1 nanomole NaCl to the sample. Because of the deleterious effect of salt on the LSI mass spectra of peptides, it is usually advantageous to purify samples by reversed-phase HPLC using volatile buffers.

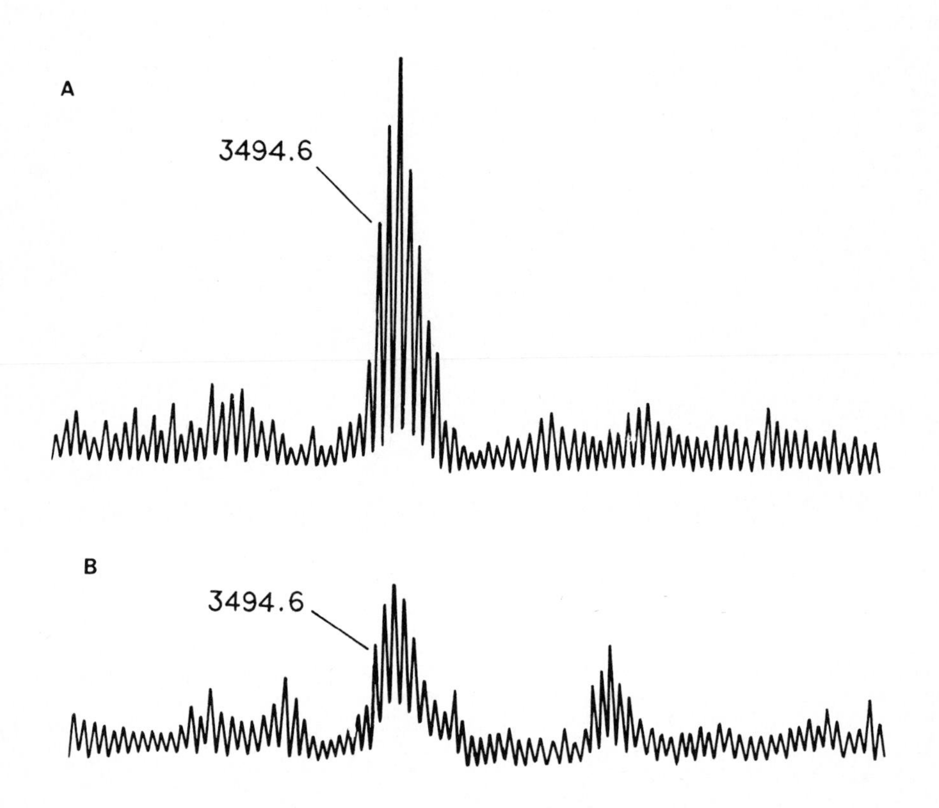

Figure 4. (a) The positive ion LSI mass spectrum of oxidized B chain of bovine insulin obtained using 1-thioglycerol as the matrix. Sample (0.3 nanomole) was dissolved in 1-thioglycerol and 5% acetic acid. (b) Same as (a) but after addition of 0.1 nanomole of NaCl.

The molecular ion region of a high mass substance is usually a series of peaks with an appearance quite different from that of low-mass substances. New terminology appropriate for describing the mass and intensity distribution of high-mass ions, as discussed by Yergey (ref. 17), will be illustrated here for the B chain of insulin. For low mass ions the lowest mass isotopic peak is usually the largest since, for carbon, hydrogen, nitrogen and oxygen, the lowest mass, naturally occurring isotope is also the most abundant. For example, ^{12}C is more abundant than ^{13}C. However, in the case of high mass ions the number of atoms is large and it is unlikely that the average molecule contains only the lowest mass isotope of each element. Hence the first, second or third isotope peaks may be the largest. This is illustrated in Figure 5 which gives the theoretical isotopic pattern for the MH^+ ion of the oxidized B chain of insulin. These results show that the second isotopic peak should be the most intense, as was found experimentally (Figure 4a).

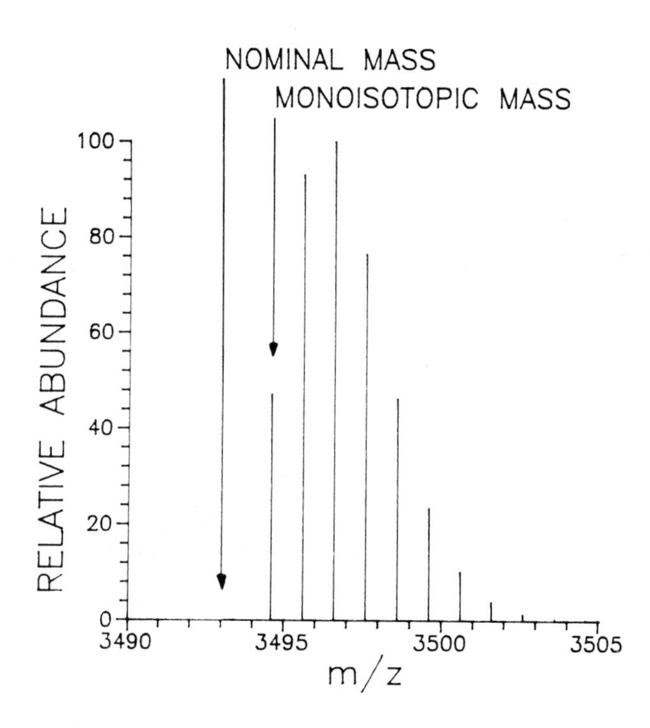

Figure 5. Theoretical isotopic distribution of the molecular ion (MH^+) of oxidized insulin B chain, $C_{157}H_{232}N_{40}O_{47}S_2$. The nominal mass is 3493, the exact monoisotopic mass is 3494.6.

New terminology has also been introduced to describe the molecular ions of high mass substances (ref. 17). The Nominal Mass is calculated from the empirical formula using integer atomic masses for the elements. For example, carbon is 12, hydrogen is 1, nitrogen is 14 and oxygen is 16. The Monoisotopic Exact Mass, calculated with exact atomic mass values of 12.0000, 1.007825, 14.0034, and 15.9959 for carbon, hydrogen, nitrogen, and oxygen, is the exact mass of the ion. Alternatively, the Average Mass can be calculated from the mass and abundance of each of the isotopes. This is synonymous with the formula weight normally used by chemists.

Quantitation of mixtures by LSIMS has been particularly frustrating because the signal usually has a nonlinear response with sample concentration, and is often depressed by presence of extraneous substances, such as excess salt, as illustrated in Figure 4. The mechanism by which one peptide suppresses desorption of another has been explained by their relative hydrophobicities (ref. 18). According to the proposed model, more hydrophilic peptides are driven away from the vacuum-matrix interface by hydrophobic peptides. Since ions are sputtered only from the surface of the matrix, hydrophilic peptides may not be observed until the concentration of the hydrophobic peptide on the surface has been reduced by sputtering. This phenomena gives rise to a temporal dependence of the relative intensities of hydrophobic and hydrophilic peptides when analyzed together. An example of this peculiar behavior is illustrated in Figure 6 which gives the intensities of two peptides as a function of the analysis time. At time t = 20 sec., only the peptide at m/z 1804 has a signal above the background. As the analysis time increases, its signal decreases and the peptide at m/z 1908 becomes dominant. It follows that a complete analysis of peptide mixtures must include recording spectra at different times.

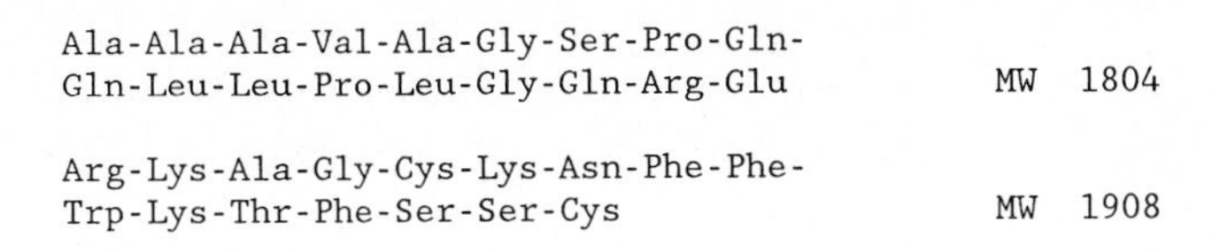
Ala-Ala-Ala-Val-Ala-Gly-Ser-Pro-Gln-
Gln-Leu-Leu-Pro-Leu-Gly-Gln-Arg-Glu MW 1804

Arg-Lys-Ala-Gly-Cys-Lys-Asn-Phe-Phe-
Trp-Lys-Thr-Phe-Ser-Ser-Cys MW 1908

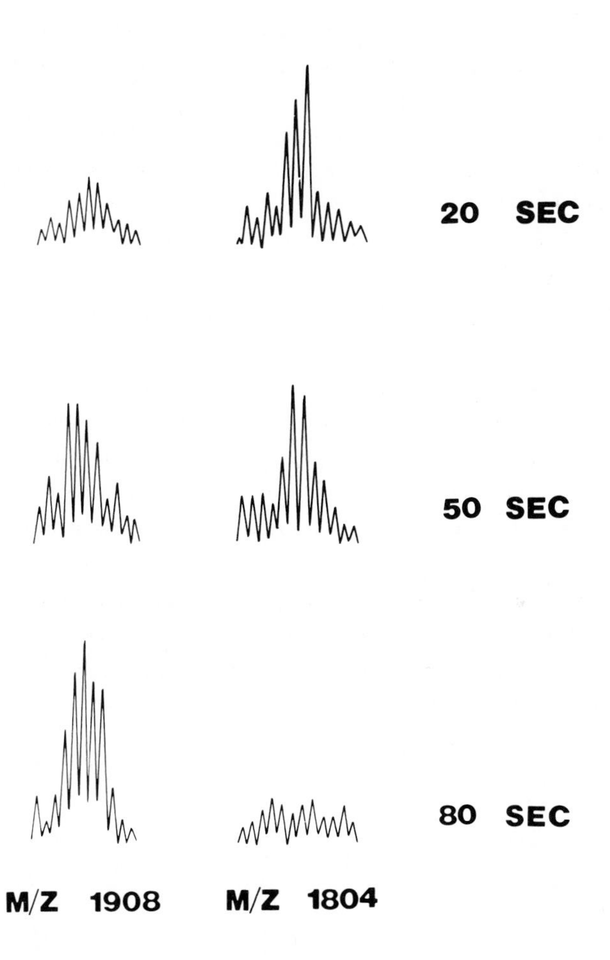

Figure 6. The variation of the relative intensities of the MH^+ ions of two peptides with analysis time. S. aureaus V-8 protease fragments from a pancreatic piscine somatostatin desalted by reversed phase liquid chromatography. (Sample provided by P.W. Andrews, Purdue University)

Another way in which LSIMS differs from conventional mass spectrometry is that the analyte may become chemically modified during analysis. Although detailed mechanisms for these modifications are not well understood, they are most likely due to highly reactive species such as free radicals, free electrons and ions which are produced during bombardment of the matrix and react with the analyte. Sethi et. al. (ref. 19) have shown that the halogen in halogenated nucleosides is partially replaced with hydrogen to give the unmodified nucleoside. One would normally avoid conditions which lead to modification of the analyte during analysis.

In some instances, however, such reactions can be used advantageously. For example, Yazdanparast et. al. (ref. 12) have shown that disulfide bonds in peptides can be detected and located by noting changes in the mass spectrum occuring during the analysis. The analog recordings of the molecular region of the positive ion FAB mass spectrum of [Arg^8]-vasopressin taken at the begining of the analysis, and 6 min later are given in Figure 7. It is apparent that, while the series of peaks in the first spectrum is similar to that expected for a natural abundance of isotopes, new products which contribute to ion intensity two mu above the original MH^+ are produced during the analysis. This unexpected behavior is attributed to reduction of the disulfide bond in [Arg^8]-vasopressin to give a new MH^+ ion two mu higher. Thus, there are two MH^+ peaks, one for the original, nonreduced form, and one for the reduced form of the peptide. It is important to note that this behavior was also observed in the sodium cationated peaks (MNa^+), and when glycerol, 1-thioglycerol, or a mixture of dithiothreitol and dithioerythritol were used as matrices.

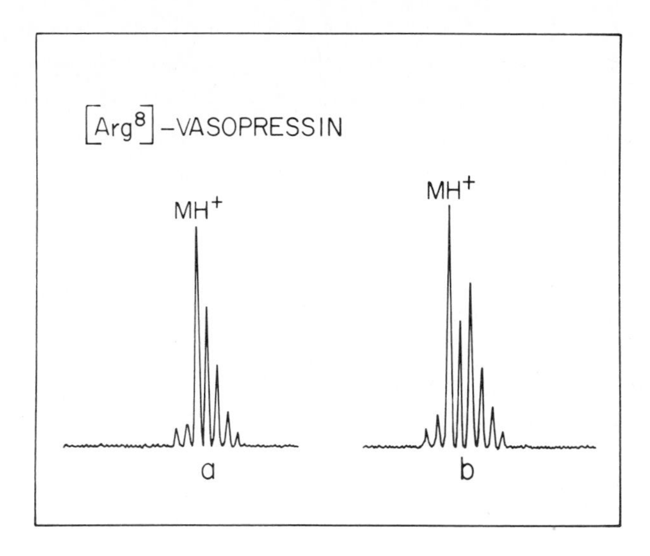

Figure 7. Mass spectrum of the molecular ion region of [Arg^8]-vasopressin in DTT/DTE: (a) zero time; (b) after 6 min of continuous bombardment by the xenon beam (Reproduced from Ref. 12 with permission).

The relative importance of reduction of the disulfide bond by the matrix and by factors related to the xenon beam is suggested by the spectra in Figure 8. These results give the ion intensitiy pattern for the molecular ions of oxytocin under different conditions. Spectrum (a), obtained immediately after dissolving the peptide in DTT/DTE, has an approximately normal isotopic pattern indicating that the disulfide bond is intact. Spectrum (b), obtained 8 hr later shows that the disulfide bond has undergone partial reduction. Spectrum (c) was obtained 5 min after spectrum (b). During this interval, the xenon beam was off and the sample was left in the vacuum chamber. The intensity pattern is essentially the same in spectra (b) and (c), indicating that significant reduction has not occured during the 5 min period in which the sample was in the vacuum chamber. Spectra (d) and (e), obtained 1 and 3 min later with the xenon beam on continously, show that considerably more reduction has occured. These results show that reduction occured during the 8-hour incubation period, and while the sample was bombarded by the xenon beam. Since a normal FAB analysis is completed in less than 5 min, it follows that the xenon beam, or reactive species formed by it, is the primary factor responsible for changing ion intensities during analysis. Although similar behavior was found for several model peptides, it should be noted that the rate of reduction is not the same for all peptides, and likely depends on the reduction potential of the disulfide bond.

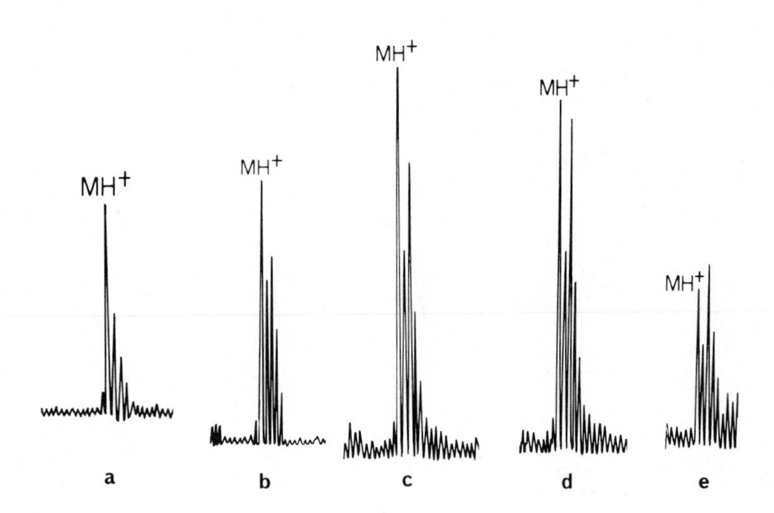

Figure 8. Mass spectra of the molecular ion region of oxytocin in DTT/DTE: (a) zero time without incubation in DTT/DTE; (b) after an 8 hr incubation; (c) after 5 min in the ion source with the xenon beam off; and (d) 1 min and (e) 3 min after continuous bombardment with the xenon atom beam (Reproduced from Ref. 12 with permission).

3. IDENTIFICATION AND LOCATION OF PROTEIN CROSS-LINKS

The utility of LSIMS for identifying peptide fragments which may be used to determine disulfide linkages has been demonstrated using hen egg-white lysozyme (ref. 13,15). This protein has a molecular weight of 14,600 and contains 129 amino acids, eight of which are cysteines that are normally connected via disulfide bonds. In the first step of this procedure, specific cleavage reactions are used to degrade the protein into peptides or disulfide bonded peptide pairs which ideally contain only one cystine and no cysteines. The molecular weights of these fragments are then determined by LSIMS and used with the known amino acid sequence to identify the peptides with respect to their location in the protein. Once the peptide fragments have been assigned to the protein, the disulfide linkages are determined directly.

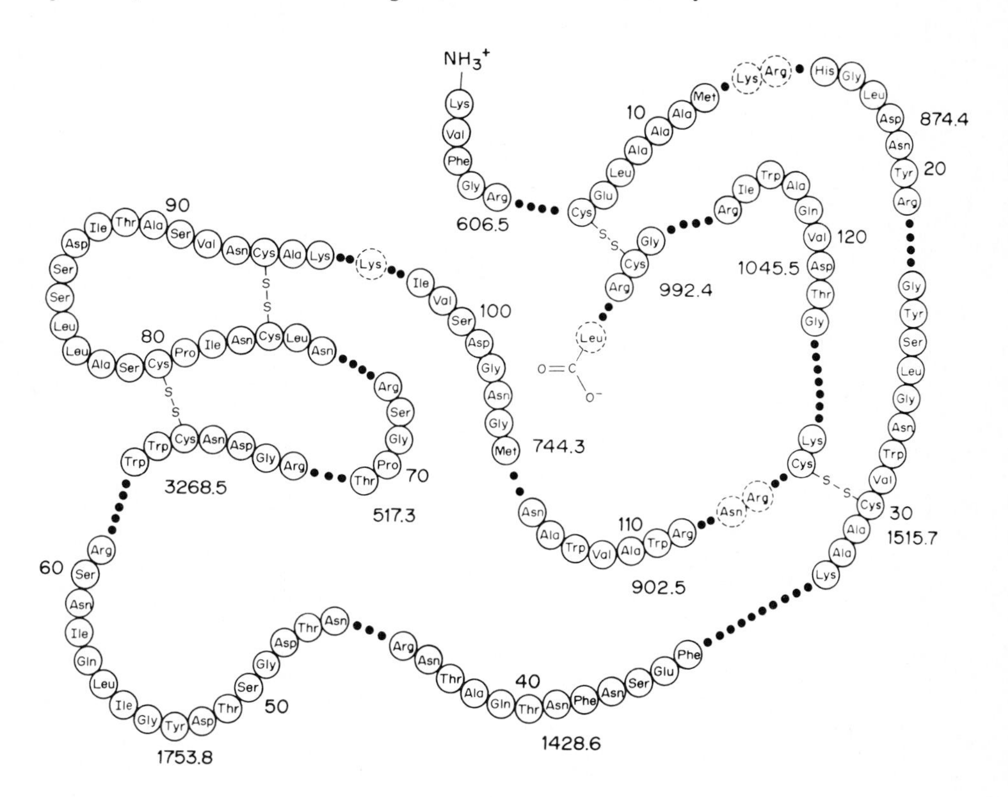

Figure 9. Peptide products produced by digestion of hen egg-white lysozyme with CNBr followed by trypsin. The numbers indicate monoisotopic exact masses of the protonated molecule ions, (MH^+), of each tryptic fragment. Methionyl residues are indicated as Met, not as homoserine (Reproduced from Ref. 13 with permission).

Hen egg-white lysozyme has two methionyl residues (12 and 105) which were cleaved with cyanogen bromide (CNBr). This step was useful because of the specificity of the reaction, and because it tends to open or unfold the protein rendering it more susceptible to enzymatic cleavage. After chemical cleavage, the protein was degraded further with trypsin which is highly specific, cutting the protein on the carboxyl side of all arginine and lysine residues. The structure and CNBr/tryptic fragments of hen egg-white lysozyme are illustrated in Figure 9. Since the mass spectrum of the digest contained many peaks, a protocol based on knowing the amino acid sequence, and CNBr/tryptic cleavage points was used to identify ions diagnostic of molecular structure. In most cases peaks due to protonated molecule ions, (MH^+) could be delineated from fragment ion peaks by the presence of a sodium cationated ion (MNa^+) 22 mu higher. Although the number of peaks in the mass spectrum depends on many factors, including the purity and composition of the protein, as well as the presence of other peptides and salts, all prominent peaks above m/z 300 could usually be identified.

The masses of non-cystine containing peptides, as well as their assignments to specific portions of the protein are given in Table I. Similar data are presented for cystine/cysteine-containing peptides in Table II. Results presented in Table II are given for an aliquot of the original digest (nonreduced state) and for a second aliquot which had been purposely reduced with a mixture of dithiothreitol and dithioerythritol. The assignments were verified by subjecting a third aliquot of the original digest to a one-step manual Edman degradation to remove a single amino acid from the NH_2-terminus of each peptide. The molecular weights of all but two of the peptides changed as anticipated.

Table I

Non-Cystine Containing Peptides Found in the Tryptic Digest of CNBr-Treated Hen Egg-White Lysozyme Before and After Chemical Reduction.

Residues	Mass of MH^+
1-5	606
2-5	478
6-12	660
15-21	874
22-33	1268
34-45	1428
46-61	1753
62-68	936
69-73	517
74-96	2336
98-105	745
106-112	902
113-114	289
117-125	1045
126-128	335

Two disulfide-containing peptides, 115/23-33 and 6/127-128, were not observed following Edman degradation. In both cases, a cystinyl residue constitutes one of the NH_2-terminal residues of each peptide before Edman degradation. It is therefore expected that the 6/127-128 peptide has a MH^+ at m/z 514 because of formation of the phenylthiohydantoin (PTH) derivative. A molecular ion at m/z 1601 is expected for the 115/23-33 peptide which is 270 mu higher than expected. This change in molecular weight is anticipated because of formation of both phenylthiocarbamoyl (ϵ-PITC) and phenylthiohydantion (PTH) conjugates of Lys-33 and Cys-115, respectively.

The peptides having molecular weights of 3266, 1514 and 991 mu (Table II) are of particular importance because they contain intact disulfide bonds and establish the existence of peptide linkages between Cys-6 and Cys- 127, and between Cys-30 and Cys-115. In addition to having the molecular weights expected from the amino acid sequence of the protein, these ions were not observed in the mass spectrum of the chemically reduced digest. These assignments were further substantiated by the fact that the intensity of these ions decreased relative to the protonated molecule ions corresponding to their constituent peptides as analysis time increased. For example, the intensity of the MH^+ peak for the 62-68/74-96 peptide decreased relative to the intensities of the MH^+ peaks for the 62-68 and 74-96 peptide, as would be expected if the disulfide bond were undergoing reduction during the analysis. It is interesting to note that the mass spectra taken prior to chemical reduction also have these ions. This partial reduction may occur prior to mass spectral analysis, or , as demonstrated by Yazdanparast (ref. 12), may occur during mass spectral analysis. The important point here is that conditions have been chosen to prevent total reduction of disulfide bonds so that the peptide pairs diagnostic for disulfide linkages remain intact.

Although all of the peptide fragments expected to be present in the digest were detected by LSIMS analysis of the crude digest, it was anticipated that the mass spectra of digests of larger proteins may be so complex that some of the peaks cannot be identified. To demonstrate the generality of the method, an aliquot of the digest was separated by HPLC into fractions which were then analyzed by LSIMS. Although the separation was incomplete, as indicated in the chromatogram in Figure 10,the number of peptide fragments in any fraction is less than in the crude digest.

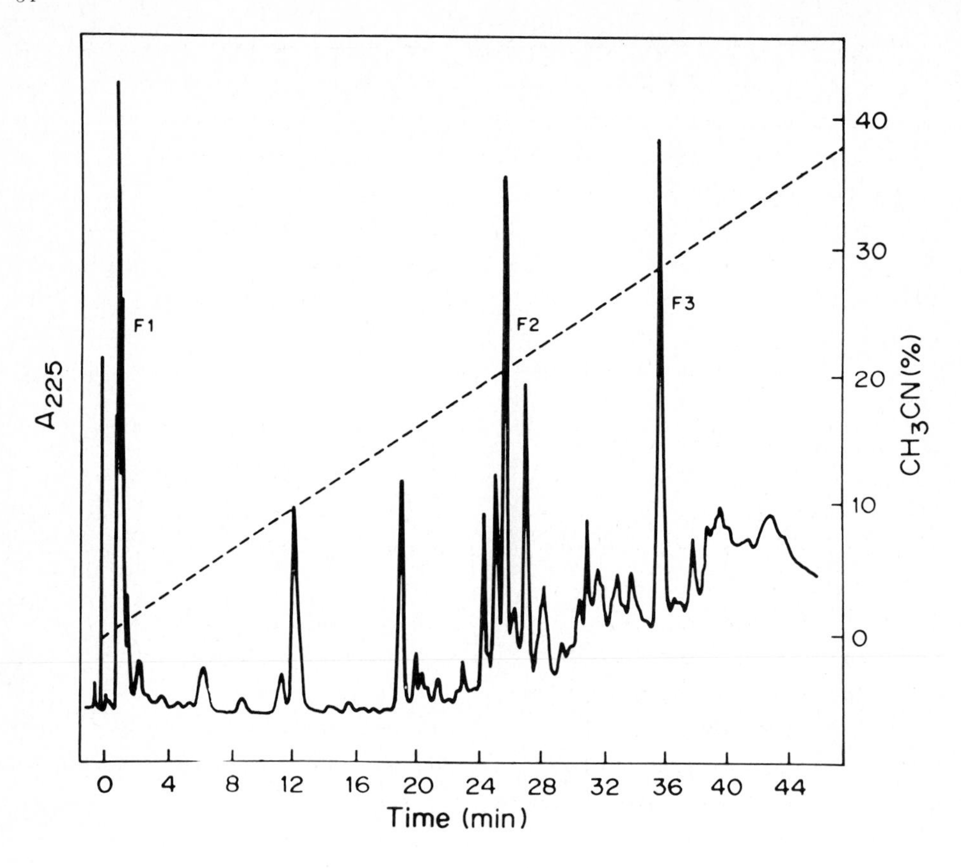

Figure 10. HPLC separation of the tryptic peptides of CNBr-treated hen egg-white lysozyme. Two hundred μg of the digest was subjected to reversed-phase HPLC (C-8 reversed phase 50 x 4.1 mm column, water/acetonitrile plus 0.1% trifluoroacetic acid). F1, F2, and F3 represent the disulfide-containing peptides: 6-12/126-128, 22-33/115-116, and 62-68/74-96, respectively (Reproduced from Ref. 13 with permission).

Mass spectral analysis of fractions 1-3 showed that they contained the three disulfide bonded peptide pairs originally found in the crude digest. In addition to the MH^+ ions of the peptide pairs, peaks corresponding to each of the constituent peptides were also present. For example, the mass spectrum of fraction 1 (nonreduced) had major peaks at m/z 992, 660 and 335 corresponding to MH^+ ions of peptide fragments 6-12/126-128, 6-12 and 126-128, respectively. After the sample was chemically reduced, the ion corresponding to the disulfide bonded peptide pair (m/z 992) was no longer present and only ions indicative of its two constituent peptides were observed. Since all three peptides would not co-elute in the same fraction, it is clear that some reduction occurred during the mass spectral analysis. Similar analyses of fractions 2 and 3 showed that they contain the peptides 22-33/115-116 and 62-68/74-96, respectively.

Although the results in Table II, as well as those of Takao et. al (15) clearly established disulfide cross-linkages between Cys-6 and Cys-127, and Cys-30 and Cys-115, there was no information about linkages between cysteins at positions 64, 76, 80, and 94. These linkages were, however, located by a further degradation of the peptide fragment 62-68/74-96 such that new peptides which had only a single disulfide bond.

Several enzymatic reactions (elastase,α-chymotrypsin and proline endopeptidase) failed to produce useful peptide fragments, presumably because the cleavage sites were sterically hindered by the disulfide knot. Peptide fragments suitable for locating the disulfide linkages were produced, however, by mild acid hydrolysis. Since acid hydrolysis is not highly specific, the mass spectra contained many peaks. Mass spectra of the crude hydrolysate were recorded before and after chemical reduction to delineate ions corresponding to disulfide bonded peptide pairs. Masses of protonated molecule ions of peptides diagnostic of disulfide linkages, as well as their assignments to the original protein, are illustrated in Figure 11. These assignments were confirmed by presence of protonated molecule ions of the constituent peptides in the mass spectrum of the reduced hydrolysate.

Table II

Cystine/Cysteine-Containing Peptides Found in the Tryptic Digest of CNBr-Treated Hen Egg-White Lysozyme Before and After Chemical Reduction.

Residues	Mass of MH^+	
	Nonreduced	Reduced
62-68/74-96	3267	2336
		935
22-33/115-116	1515	1268

6-12/126-128	992	660
		335

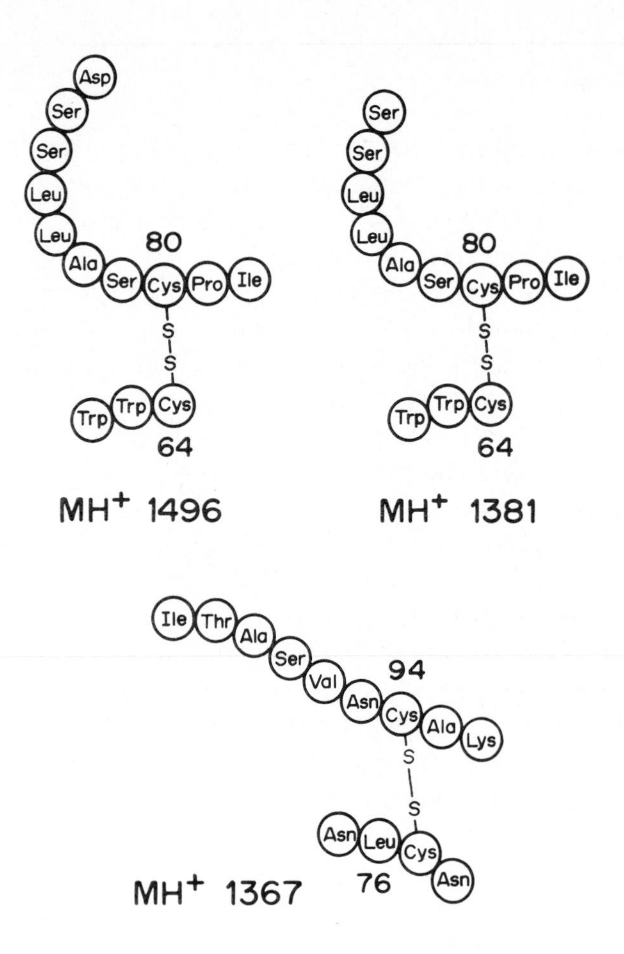

Figure 11. Three of the disulfide-containing peptides from partial acid of the 62-68/74-96 tryptic peptide of hen egg white-lysozyme (Reproduced from Ref. 13 with permission).

As is typical for the conditions used in this study, peaks for most of these peptides were also present in the mass spectrum of the nonreduced hydrolysate. A unique combination of amino acids is attached to each of the disulfide bonded peptide pairs and leads directly to an assignment of the remaining disulfide linkages (64-80 and 76-94) in hen egg-white lysozyme.

Since assignment of peaks found in the mass spectrum to specific peptide fragments was made on the premise that the molecular weight of the peptide fragment is unique for that fragment, it is important to investigate the validity of this assumption. For example, the MH^+ ion of m/z 1381 (62-64/78-86) would also fit the peptide 74-80/62-65. This assignment was rejected, however because neither of the constituents of the peptide pair (74-80 or 62-65) were found. Similarly, the ion at m/z 1496 (62-64/78-87) is also consistent with another peptide, 62-66/74-80 which incorrectly predicts a disulfide bond between Cys-64 and Cys-76. Again, the erroneous assignment was rejected because neither of the constituent peptides (62-66 or 74-80) was present. Since NaOH was added during chemical reduction step, MH^+ ions were often replaced with MNa^+ ions which must be considered when assigning ions in the mass spectrum to specific parts of the protein.

It is important to note that, although highly purified peptides of known sequence can be identified by animo acid analysis, LSIMS is uniquely able to identify such peptides when present in complex mixtures. This feature is particularly important when the sample contains extraneous peptides in addition to the peptide of interest. It is also important when less specific enzymes are used to cleave the protein. For example, pepsin tends to cleave on the carboxyl side of aromatic, acidic and hydrophobic residues. Not only is the specificity of pepsin low, but the rates at which different amide bonds are cleaved is variable. As a result, peptic digests will normally contain a large number of fragments making identification by amino acid analysis difficult.

The ability to identify peptides in a peptic digest may be illustrated with N-Carbamoyl-Gly-1A human insulin. The structure of this modified insulin, as well as major points at which it is cleaved by pepsin, is illustrated in Figure 12. Since pepsin cleaves different sites at different rates, it was important to use an optimum digestion time. The HPLC chromatogram of the 3-hour digest, given in Figure 13, has about ten major peaks. Reducing the digestion time decreased the number of peaks, increasing it increased the number of peaks. After an initial survey in which the digestion time was varied from 30 min to 24 hr, it was concluded that 3 hr was optimum.

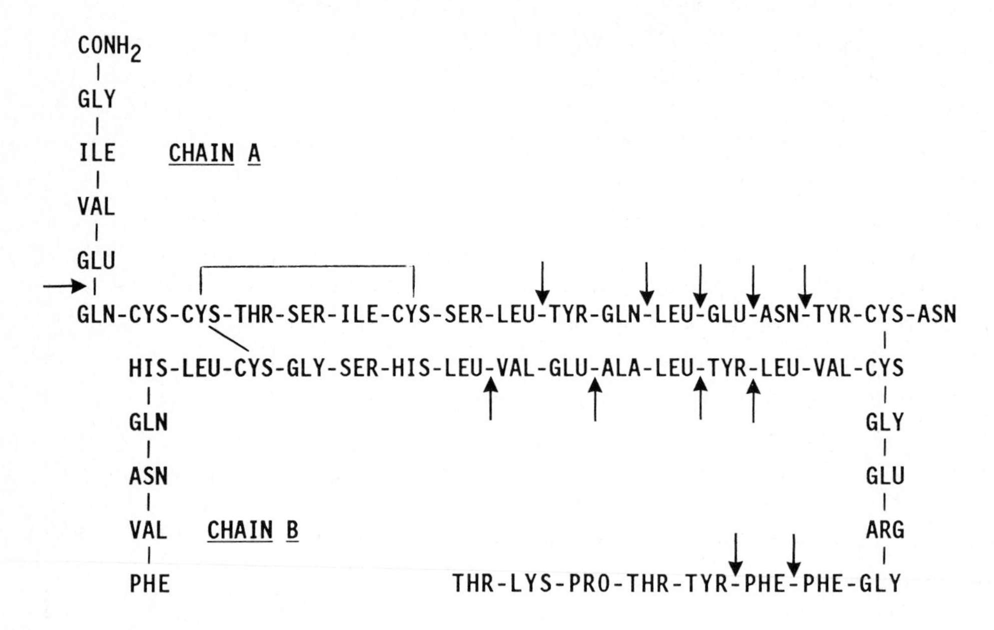

Figure 12. Primary structure of N-Carbamoyl-Gly-1A human insulin illustrating the locations of disulfide cross-linkages and important points at which the protein is cleaved by pepsin.

The numbered chromatographic fractions in Figure 13 were collected and analyzed by LSIMS. As in the case of hen egg-white lysozyme, only high mass ions which were accompanied by a sodium or potassium cationated ion were designated as molecular ions (MH^+). The mass spectral data given in Table III establish the identity of the enzymatic fragments comprising chromatographic fractions 1-8. Fractions 1 and 2 have peptides B26-30 and B25-30 which are derived from the NH_2-terminus of the the B chain. Fraction 3 contains the disulfide bonded peptide pair, A17-21/B17-25. The mass spectrum also has a peak corresponding to one of the constituent peptides, B17-25, which likely was produced by reduction of the disulfide bond by reactive species produced on the surface of the matrix by the incident xenon beam. It is interesting to note that, although the A chain constituent peptide, A17-21, is likely present at the same concentration as the B17-25 fragment, it was not observed in the mass spectrum. This is because the B17-25 peptide is substantially more hydrophobic and preferentially occupies the surface of the matrix from which ions are desorbed. The more hydrophilic A chain is driven away from the surface and is not observed.

Fractions 5-8 likewise have one or two disulfide bonded peptide pairs, as well as some of the constituent peptides. The structure of one of the components, A5-13/B1-11, found in fraction 5 is illustrated in Figure 14. From these results it is evident that structural modifications resulting in a change in mass in most parts of insulin could be identified and located by the combination of peptide mapping and LSIMS. A fragment corresponding to the NH_2-terminus of the A chain was not found, probably because it was not present in any of the collected fractions. This was anticipated since this peptide does not absorb at 254 nm.

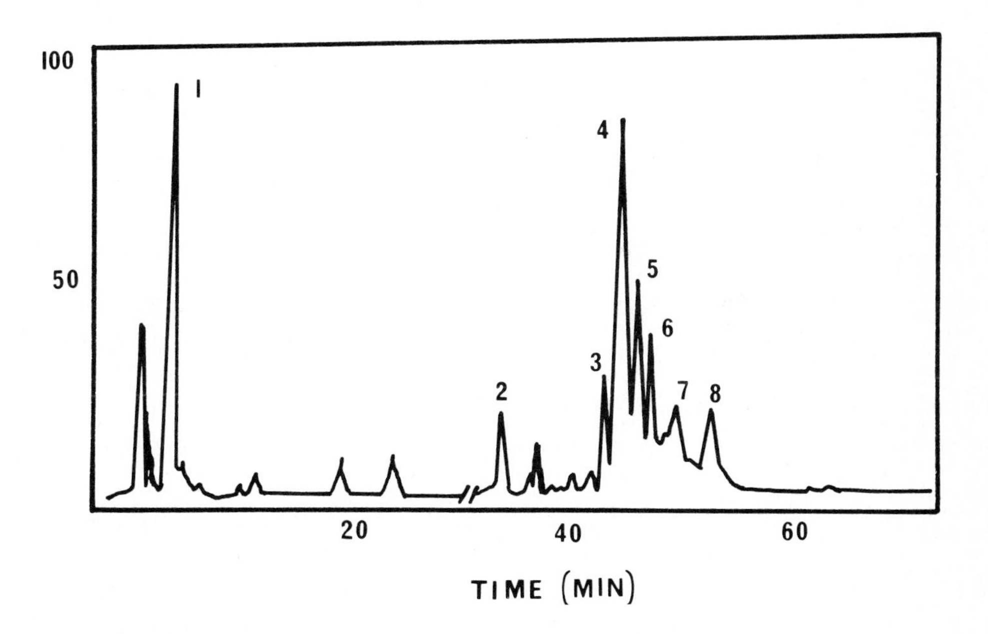

Figure 13 HPLC chromatogram of the 3 hr pepsin digest of N-Carbamoyl-Gly-1A human insulin. Conditions: C-18 reversed phase column, acetonitrile/water step gradient (10%, 25%), pH 2.0 buffered with 0.1 M NH_4COOH, UV detection at 254 nm (Ref. 20).

Table III

Summary of Peptides found by LSIMS in the
3 Hr Pepsin Digest of N-Carbamoyl-Gly-1A Human Insulin (ref. 20).
Chromatographic Fractions refer to Figure 13

Chromatographic Fraction	Peptide Mass	Assignment
1	608	B26-30
2	755	B25-30
3	1026	B17-25
	1665	A17-21/B17-25
4	1026	B17-25
	1422	A19-21/B17-25
	1536	A18-21/B17-25
5	957	A5-13
	1253	B1-11
	1481	B1-13
	2208	A5-13/B1-11
	2436	A5-13/B1-13
6	1026	B17-25
	1778	A16-21/B17-25
7	1189	B16-25
	1585	A19-21/B16-25
	1699	A18-21/B16-25
8	1026	B17-25
	2069	A14-21/B17-25

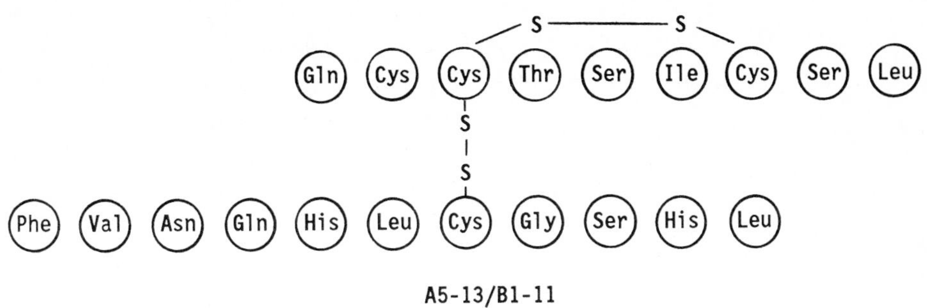

Figure 14. Structure of the A5-13/B1-11 fragment found in fraction 5 (Figure 13) of the peptic digest of N-Carbamoyl-Gly-1A human insulin. (Ref. 20).

These results demonstrate that a variety of peptide linkages can be determined by LSIMS of proteins fragmented by proteolytic and chemical methods. Partial separation of peptide fragments is a significant aid in improving the quality of the spectra and may be required when assigning cross-linkages in large proteins, or when less specific methods of cleavage are used. In addition to locating disulfide bonds, the method has very general applicability and may be used to locate and identify a wide variety of posttranslational modifications. Features warranting further improvement include a search for means of reducing the continuous background typical of LSIMS spectra, a means of moderating the relationship between peptide hydrophobicity and sensitivity, and better methods for fragmenting large proteins which are resistant to attack by highly specific proteases such as trypsin.

4. ACKNOWLEDGMENTS

The authors would like to thank Razieh Yazdanparast, Jack Dixon, Philip Andrews, Paul Toren, Stephen Byrn, Ronald Chance and James Hoffmann for their very important contributions to the original investigations described here.

REFERENCES

1 S.D. Putney, N.J. Royal, H.N de Vegvar, W.C. Herlihy, K. Biemann and P. Schimmel, Primary Structure of a Large Aminoacyl-tRNA Synthetase, Science, 213 (1981) 1497-1501.

2 P. Hoben, N. Royal, A. Cheung, F. Yamao, K. Biemann and D. Soll, Escherichia coli Glutaminyl-tRNA Synthetase, J. Biol. Chem., 257 (1982) 11644-11650.
3 S.A. Carr and K. Biemann, Identification of Posttranslationally Modified Amino Acids in Proteins by Mass Spectrometry, Methods in Enzymology 106 (1984) 29-58.
4 M. Barber, R.S. Bordoli, R.D. Sedgwick and A.N. Tyler, Fast Atom Bombardment of Solids (F.A.B.): A New Ion Source for Mass Spectrometry, J. Chem. Soc. Chem. Commun., (1981) 325-327.
5 H.R. Morris, M. Panico, M. Barber, R.S.Bordoli, R.D. Segwick and A. Tyler, Fast Atom Bombardment: A New Mass Spectrometric Method for Peptide Sequence Analysis, Biochem. Biophys. Res. Commun., 101 (1981) 623-631.
6 M. Barber, R.S. Bordoli, R.D. Sedgwick, A.N. Tyler, G.V. Garner, D.B. Gordon, L.W. Tetler, R.C. Hider, Fast Atom Bombardment Mass Spectrometry of Large Oligopeptides Melittin, Glucagon and the B chain of Bovine Insulin, Biomed. Mass Spectrom., 9 (1982) 265-268.
7 K. Eckart, H. Schwarz, M. Chorev and C. Gilon, Sequence Determination of N-terminal and C-terminal Blocked Peptides Containing N-alkylated Amino Acids and Structure Determination of these Amino Acid Constituents by using Fast-Atom-Bombardment/Tandem Mass Spectrometry, Eur. J. Biochem. 157 (1986) 209-216.
8 A.L. Burlingame, T.A. Baillie, and P.J. Derrick, Mass Spectrometry, Anal. Chem. 58 (1986) 165R-211R.
9 P.S. Andrews, D.H. Hawke, T.D. Lee, K. Legesse, B.D. Noe, J.E. Shively, Isolation and Structure of the Principal Products of Proglucagon Processing, Including an Amidated Glucagon-like Peptide, J. Biol. Chem. 261 (1986) 8128-8133.
10 B.W. Gibson and K. Biemann, Strategy for the Mass Spectrometric Verification and Correction of the Primary Structures of Proteins Deduced from their DNA Sequences, Proc. Natl. Acad. Sci. USA, 81 (1984) 1956-1960.
11 H.R. Morris and P. Pucci, A New Method for Rapid Assignment of S-S Bridges in Proteins, Biochem. Biophys. Res. Commun., 126 (1985) 1122-1128.
12 R. Yazdanparast, P. Andrews, D.L. Smith, and J.E. Dixon, A New Approach for Detection and Assignment of Disulfide Bonds in Peptides, Anal. Biochem., 153 (1986) 348-353.
13 R. Yazdanparast, P. Andrews, D.L. Smith and J.E. Dixon, Assignment of Disulfide Bonds in Proteins by Fast Atom Bombardment Mass Spectrometry, J. Biol. Chem., In Press.
14 P. Toren, D.L. Smith, R. Chance and J. Hoffmann, Determination of Peptide Linkages in Insulin and Related Materials by Fast Atom Bombardment Mass Spectrometry, Submitted for publication.
15 T. Takao, M. Yoshida, Y-M Hong, S. Aimoto and Y. Shimonishi, Fast Atom Bombardment (FAB) Mass Spectra of Protein Digests: Hen and Duck Egg-white Lysozymes, Biomed. Mass Spectrom., 11 (1984) 549-556.
16 S.A. Martin, C.E. Costello and K. Biemann, Optimization of Experimental Procedures for Fast Atom Bombardment Mass Spectrometry, Anal. Chem., 54 (1982) 2362-2368.
17 J. Yergey, D. Heller, G. Hansen, R. Cotter, and C. Fenselau, Isotopic Distributions in Mass Spectra of Large Molecules, Anal. Chem., 55 (1983) 353-356.
18 S. Naylor, A.F. Findeis, B.W. Gibson, and D.H. Williams, An Approach Toward the Complete FAB Analysis of Enzymic Digests of Peptides and Proteins, J. Am. Chem. Soc., 108 (1986) 6359-6363.
19 S.K. Sethi, C.C. Nelson, J.A. McCloskey, Dehalogenation Reactions in Fast Atom Bombardment Mass Spectrometry, Anal. Chem., 56 (1984) 1975-1977.
20 P.C. Toren, Ph. D. Thesis, 1985, Purdue University.

MASS SPECTROSCOPIC IONIZATION TECHNIQUES FOR ORGANIC COMPOUNDS OF LOW VOLATILITY

H. BUDZIKIEWICZ

1. INTRODUCTION

Both for electron impact (EI) and for chemical ionization (CI) mass spectrometry it is necessary to tranfer a compound to be analyzed into the gas phase prior to ionization. With other words, it is necessary to overcome the intermolecular attractive forces by thermal energy. This in turn may result both in a more or less severe pyrolytic decomposition (comprising the whole scale from defined elimination processes as loss of H_2O from alcohols or decarboxylation of carboxylic acids up to complete degradation), and due to excess vibrational energy in an enhanced fragmentation of the ions formed. Several techniques have been developed to reduce the strain imposed on molecules during the transfer into the gas phase and in the ionization process. It has to be pointed out, however, that there is no standard procedure which may be used for all types of compounds. Occasionally, even an experienced mass spectroscopist will have to try out several procedures to obtain optimal results.

2. GENERAL REMARKS

The various techniques will not necessarily give identical results. Especially one has to decide whether molecular weight or structural information is required: In cases where abundant (quasi-)molecular ions are formed fragment formation which reveals structural details is frequently of minor importance. Collision induced fragmentation may be envisaged but then special instrumentation (an MS/MS or - because of its low resolution less suitable - a linked scan system) must be available and the molecular ion current has to be stable for an extended period of time.

The mass range above 1000 which can easily be reached with the new ionization techniques requires a somewhat modified approach to the interpretation of the data obtained:

(i) Isotope pattern. Since carbon consists to 98.89% of ^{12}C and to 1.11% of ^{13}C starting from C_{90} the peak made up of ^{12}C is not any more the most abundant one in the isotopic ion clusters (see Fig. 1). Contributions of ^{2}H, ^{15}N, ^{17}O and ^{18}O etc. which are usually ignored at lower masses have to be taken into account also. In cases of unsufficiant resolving power of the mass spectrometer (less than unit resolution in the critical mass range) only the enveloping curve will be recorded (see broken line in Fig. 2) the *maximum of which cannot be taken as the mass of the molecular ion*. Computer evaluation (requiring at least a rough knowledge of the expected elemental composition) will be necessary. The situation may be aggravated by the formation of hydrogenation (+ 2H) and dehydrogenation (- 2H) products (cf. below) the isotopic clusters of which are superimposed (see, e.g., the calculated and observed isotopic pattern in Fig. 3).

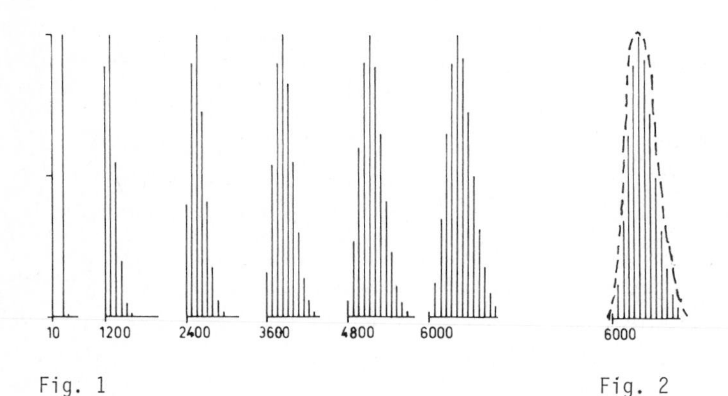

Fig. 1 Fig. 2

Fig. 1 Isotope patterns of C_1, C_{100}, C_{200}, C_{300}, C_{400} and C_{500}.

Fig. 2 Isotope cluster of C_{500} and enveloping curve.

(ii) Exact *vs.* nominal mass. Due to the relatively small mass defects (^1H ... 1.0078, ^{14}N ... 14.0031, ^{16}O ... 15.9949) nominal (based on the integers, i.e., e.g., 142 u for $C_{10}H_{22}$) and exact masses (142.1716 u for $C_{10}H_{22}$) differ at lower masses only by a few tenth of a mass unit, a difference which can be neglected unless elemental compositions are to be determined by exact mass measurements. At high masses they can amount to several u (Fig. 3: exact and nominal mass differ by 1.67 u). Since from the common elements H and N have a positive and O, P and S a negative mass defect it is necessary also here to know approximately the elemental composition. As an extreme example, a cyclopolysaccharide $(C_6H_{10}O_5)_{15}$ has an exact mass of 2430.7924, the hydrocarbon $C_{175}H_{330}$ 2432.5823 which makes a difference of almost 2 u at the same nominal mass of 2430! One should keep in mind that the *nominal* mass is obtained by "counting" a spectrum recorded with an analog (e.g., a UV-)recorder while a computer system will produce the *exact* mass, and that in the literature it is frequently not specified which type of mass one is dealing with (to enhance the confusion occasionally even the maximum of the enveloping curve - see (i) above - is given which coincides neither with the nominal nor with the exact mass!)

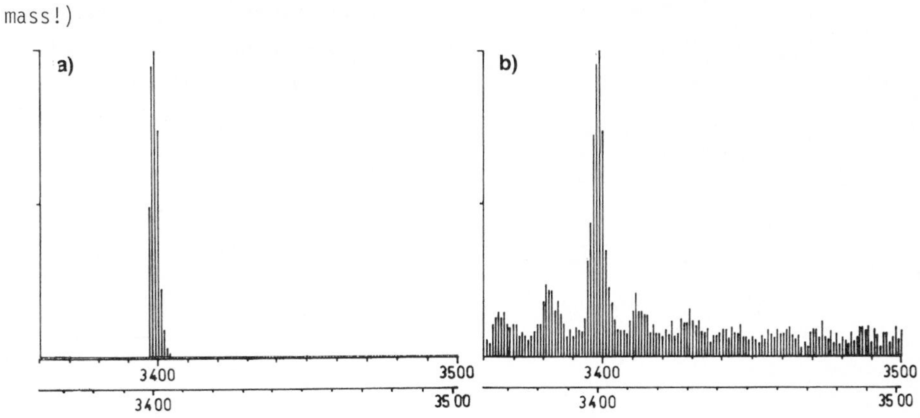

Fig. 3 Molecular ion region of the B chain of bovine insuline: a) calculated isotope pattern for $[M + H]^+$ ($C_{157}H_{233}N_{40}O_{41}S_2$), b) measured with FAB. Upper mass scale nominal, lower exact mass.

(iii) Reliability of the mass data. Mass spectrometry has the reputation that masses are being determined correctly at least to the integer (= nominal mass) and that one can, e.g., reliably distinguish between, e.g., ROH and RNH_2 (difference 1 u). At high masses both for technical reasons (calibration of the mass scale) and especially when dealing with unknown compounds - for reasons stated above under (i) and (ii) this accuracy cannot be expected for routine measurements. As a rule of thumb one should allow for an uncertainty of $\pm$ 1 for each 1000 mass units.

3. REDUCED STRAIN DURING EVAPORATION ("in-beam" EI and DCI)

Attempts in this direction go back to the early days of mass spectrometry of organic compounds. One way - though not altogether satisfactory because of the extra time and the amount of material needed - is to obtain volatile derivatives (acetates from alcohols, esters from carboxylic acid etc.). The other way is to reduce the amount of energy necessary for evaporation as well as to eliminate collisions especially with hot metal surfaces. Metal inlet systems were replaced by all-glass ones, direct inlet systems reduced the distance molecules had to travel and allowed evaporation at high vacuum. The last step in this direction was the "in-beam" ionization for EI and the "direct chemical ionization" for CI. In both cases the sample is deposited in the loop of a heatable wire ("emitter") which can be placed directly in the electron beam (EI) or in the reagent gas plasma (CI). In this way collisions are not to be expected and in addition, pyrolysis products will carry with them undecomposed molecules (in this case rapid heating is necessary, see Fig. 4). For DCI the plasma temperature and pressure have to be optimized emperically. Regarding the reagent gas see below (Section 5). From an experimental point of view both techniques are less demanding than those discussed in Section 4.

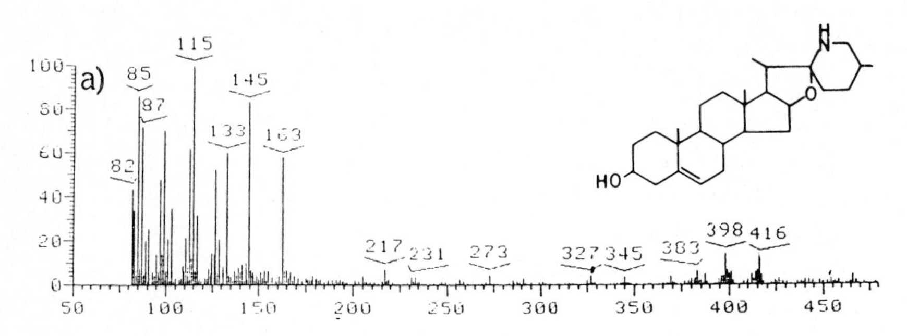

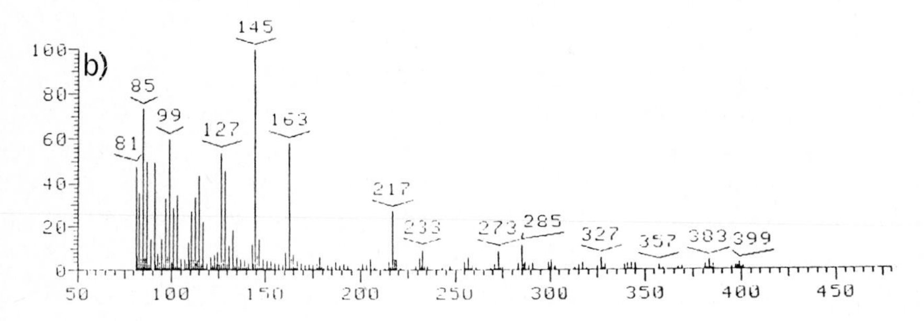

Fig. 4 DCI (i-C_4H_{10}) spectrum of tomatine a) fast heating: $[M + H]^+$ of the degradation product tomatidine (m/z 416) is visible; b) slow heating: no information can be obtained from the upper mass range.

4. SURFACE IONIZATION TECHNIQUES (FD, FAB, PD).

4.1 Field desorption (FD).

Here the substance to be analyzed is deposited on an emitter consisting of a fine tungsten wire carrying carbon needles like a bottle brush. In a special ion source between the emitter and the exit hole a potential difference of several kV is applied. In "true" FD the electron cloud of a molecule at the tip of one of the carbon needles is distorted in a way that an electron is transferred to the emitter and the cation thus formed is ejected and accellerated towards the exit hole of the ion source. In this case $M^{+\cdot}$ is formed and (due to the minimal excess energy transferred and the short time the ion remains in the ion source) hardly any fragments are observed. "True" FD is typical for non-polar compounds, especially hydrocarbons. Freshly activated emitters have to be used.

From polar compounds ions are formed in a different way: in a semiliquid system at the bottom of the carbon needles (therefore old emitters may be used)

H^+ (*m/z* 348 = $[M + H]^+$ in Fig. 5) or cations as Na^+, K^+ etc. are added to the molecules. The quasi-molecular ions thus formed are then absorbed. To provide a sufficient supply of cations a *small* amount of an inorganic salt may be added (an excess will result in the desorption of inorganic cations only; samples especially of biological origin contaminated with inorganic salts have, therefore, to be purified by passing through an ion exchange column), and to enhance ion mobility it is sometimes helpful to add substances such as tartaric acid. Side reactions during the formation of quasi-molecular ions are similar to those observed with FAB and will be discussed in Section 4.4.

FD spectra of the latter type frequently show abundant fragment ions the intensity of which can be influenced by the emitter temperature (Fig. 5); also the sample preparation, the state of the emitter etc. are of importance. The appearance of the spectra frequently changes from measurement to measurement or even within a series of consecutive recordings during one measurement. Occasionally, the ion currents of quasi-molecular or fragment ions are very shortlived and cannot be recorded *both* during the same measurement. Fragment formation occurs essentially by pyrolytic and protolytic processes in the condensed phase and, therefore, the rules known from EI cannot be applied to explain their formation. A typical example is adenosine 3'-monophosphate (1) (Fig. 5) from which ions formed by elimination of H_2O and H_3PO_4 as well as protonated adenine etc. are obtained (Scheme 1).

For FD a special ion source and a high voltage power supply are necessary.

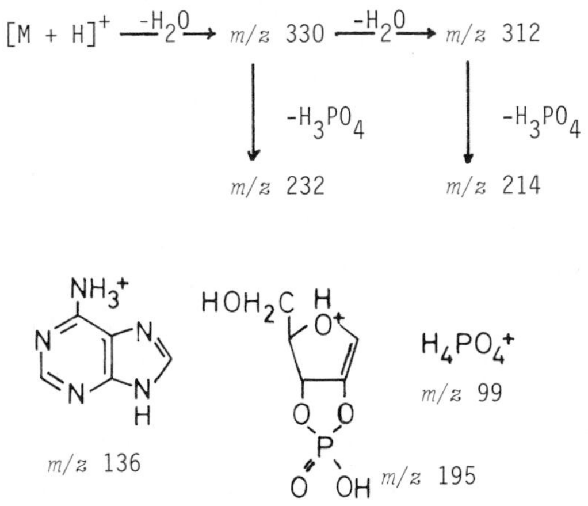

$$[M + H]^+ \xrightarrow{-H_2O} m/z\ 330 \xrightarrow{-H_2O} m/z\ 312$$

$$m/z\ 330 \xrightarrow{-H_3PO_4} m/z\ 232 \qquad m/z\ 312 \xrightarrow{-H_3PO_4} m/z\ 214$$

NH_3^+ (protonated adenine) *m/z* 136

HOH_2C ... O^+ ... O P O OH *m/z* 195

$H_4PO_4^+$ *m/z* 99

Scheme 1 Fragment ions observed in the FD spectrum of adenosine 3'-monophosphate (1).

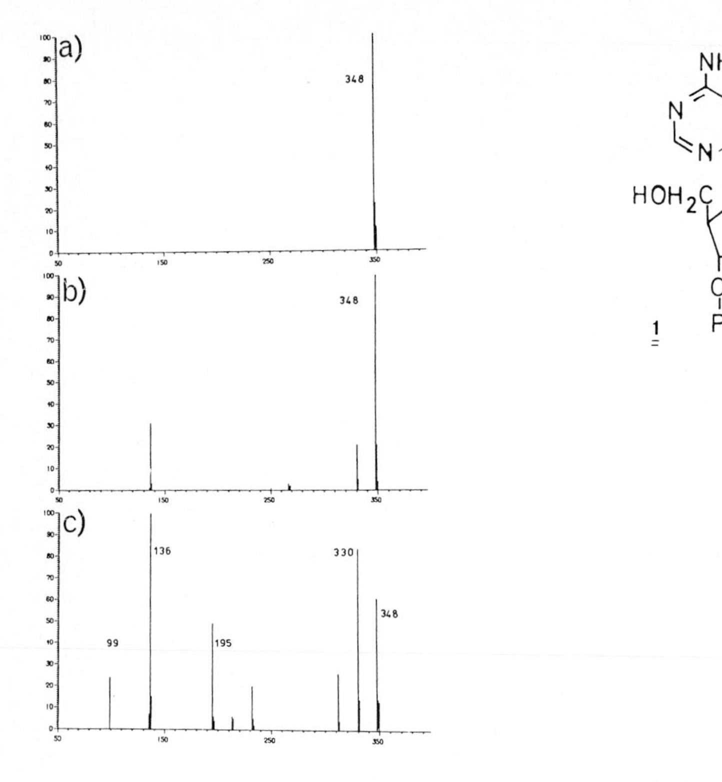

Fig. 5 FD spectrum of adenosine 3'-monophosphate obtained with different heating currents (a: 14, b: 18, c: 19 mA). For an interpretation see Scheme 1.

4.2 Fast atom bombardment (FAB).

For this ionization method the sample dissolved in a matrix (a high-boiling liquid as glycerol, monothiogylcerol, nitrobenzyl alcohol etc.) is deposited on a metal target. Since a *solution* (and not a dispersion) is *necessary* solvents may be added as well as inorganic salts for cationization of the molecules. One of the main purposes of the matrix is to transport substrate molecules to the surface. The matrix is being bombarded either with ions (e.g., Cs^+) accelerated by a potential difference of several kV or with fast-moving (~100 km/sec) atoms (Ar or Xe). In this way preformed ions as $[M + H]^+$, $[M + Na]^+$ etc. (cf. Sections 4.1 and 4.4) and/or neutral molecules are ejected from the surface; the latter may be ionized in the gas phase by CI where matrix ions play the role of the reagent gas plasma. The correct choice of the matrix material can be crucial as can be seen from Fig. 6 which shows the FAB spectra of glycerol trimyristate obtained with three different matrix systems: If

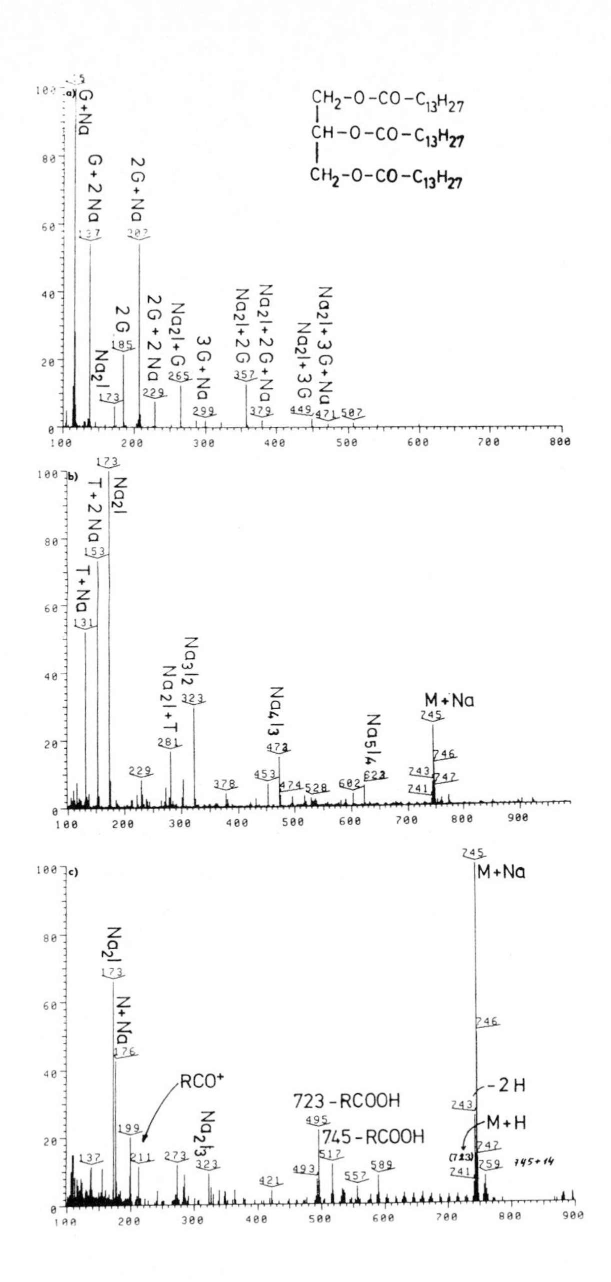

Fig. 6 FAB spectra of glycerol tri-myristate obtained with different matrices: a) glycerol (G)/CH_3OH/CH_2Cl_2/NaI, b) monothioglycerol (T)/CH_3OH/CH_2Cl_2/NaI, c) nitrobenzyl alcohol (N)/CH_3OH/CH_2Cl_2/NaI. Substrate ions are marked horizontally (R = $C_{13}H_{27}$), matrix ions vertically (G + Na means $[glycerol + Na]^+$, G + 2Na $[glycerol - H + 2Na]^+$ etc.).

glycerol is used only matrix ions are observed, monothioglycerol leads to a fairly abundant $[M + Na]^+$ ion, and with nitrobenzyl alcohol $[M + Na]^+$ is the most abundant ion and in addition fragment ions (loss of one molecule of myristic acid, $C_{13}H_{27}CO^+$) can be recognized. No reliable rules can be given regarding the choice of the best matrix system, but the importance of the substrate material being *dissolved* should be stressed.

Fragment formation does take place especially with compounds having weak bonds. Whether this is fragmentation of the quasi-molecular ions in the gas phase or protolysis in the matrix (cf. Section 4.1) is a moot point; probably both processes do occur. The utility of FAB spectra for the determination of the sequence of the sugars in glycosides, of amino acids in peptides and of nucleobases in nucleotides has been pointed out, but a fair amount of experience seems to be necessary to sort out the informative fragments. More promising seems to be to do chemistry on the FAB target: E.g., a peptide sample may be measured several times after a drop of acid or a proteolytic enzyme has been added. The appearance and disappearance of ions formed during the course of degradation can be monitored in this way.

For FAB a FAB gun with high voltage power supply is necessary.

4.3 Plasma desorption (PD)

The substance deposited on a target without a matrix is bombarded with high energy (MeV) fission products of ^{252}Cf. Singly and multiply charged $[M + H]^+$ or $[M + Na]^+$ ions as well as negative species are desorbed. In most cases only molecular weight information can be obtained. PD has the reputation of being the "softest" ionization method for high molecular weight compounds. It can, however, only be used if a special mass spectrometer specifically constructed for this method is available.

4.4 Formation of quasi-molecular ions and side reactions.

In Sections 4.1 and 4.2 it has been shown that positive quasi-molecular ions are formed by attachment of H^+ or of an alkali cation (Na^+, K^+ etc.).

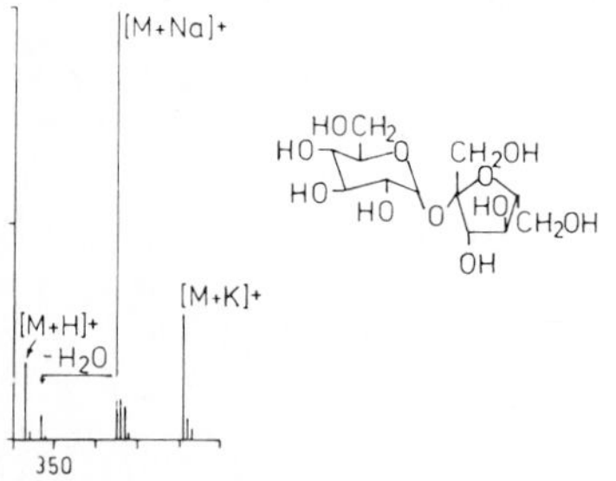

Fig. 7 Molecular ion region of a commercial sample of sucrose (FD).

Frequently, especially with natural products where alkali salts are carried on readily from the isolation procedures several quasi-molecular ions can be observed (see Fig. 7). The mass difference of 16 u between $[M + Na]^+$ and $[M + K]^+$ should not be confused with an admixture of an oxigenated compound (O ≙ 16 u!).

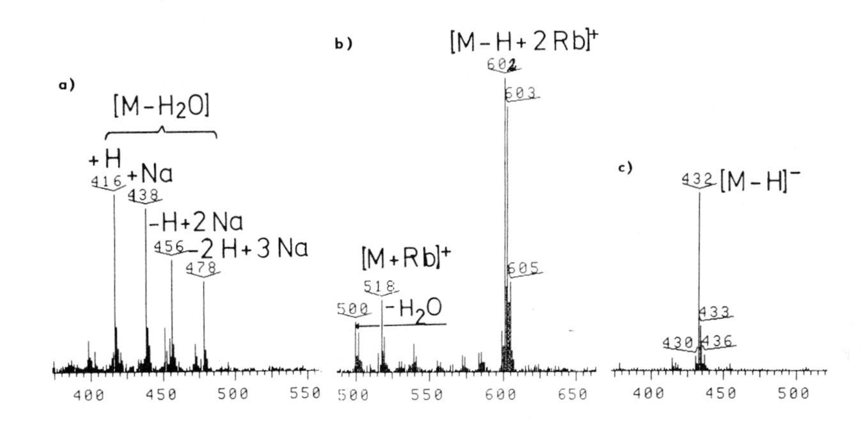

Fig. 8 Molecular ion region of a prostaglandine derivative (FAB)
a) matrix monothioglycerol/CH_3OH/NaI, positive ions, b) matrix glycerol/CH_3OH/RbI, positive ions, c) matrix glycerol/CH_3OH, negative ions.

Acidic hydrogens (acids, alcohols) can be replaced (partially) by alkali ions. In such cases one usually observes series as $[M + H]^+$ (not always present), $[M + Na]^+$, $[M - H + 2Na]^+$ etc., the differences in mass being 22 u for Na, 38 u for K etc. (see Figs. 6 and 8a). If there are doubts about the number of Na^+ in an ion one should repeat the measurement with another cation. RbI is very useful since the isotope pattern (^{85}Rb 72%, ^{87}Rb 28%) reveals immediately the number of Rb^+ present. Occasionally one observes also the addition of one (rarely two or more) molecules of the matrix (Fig. 9).

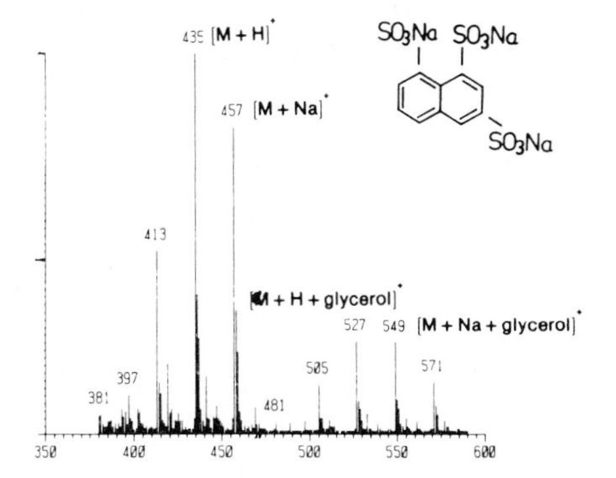

Fig. 9 Molecular ion region of Na naphtalene-1,3,8-trisulfonate (FAB, glycerol)

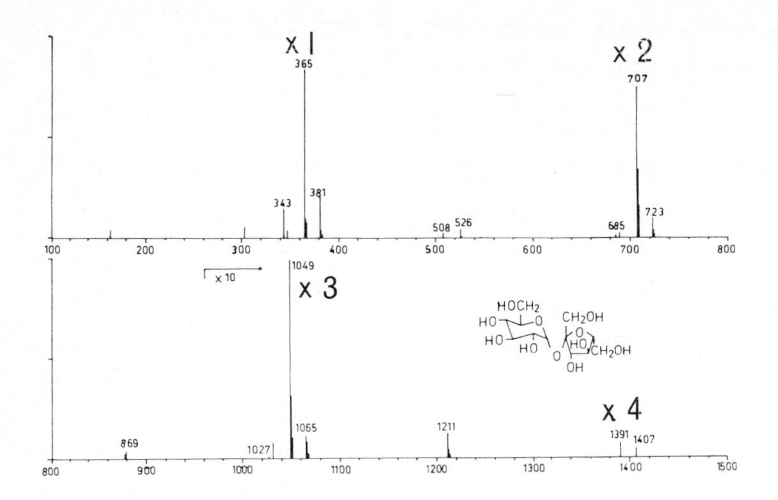

Fig. 10 FD spectrum of sucrose (for the various clusters see Fig. 7).

Not infrequently one observes an aggregation of two or more substrate molecules, i.e., the formation of , e.g., $[2M + Na]^+$, $[3M + Na]^+$ etc., as can be seen in Fig. 10 where ions up to the tetramer of sucrose can be discerned. It is not always easy to decide whether one is dealing with an aggregate or with a molecule with the double molecular weight. Also multiply charged ions as $[M + 2Na]^{++}$ or $[3M + 2Na]^{++}$ ($[2M + 2Na]^{++}$ coincides with $[M + Na]^+$) (Fig. 11) complicate occasionally the interpretation.

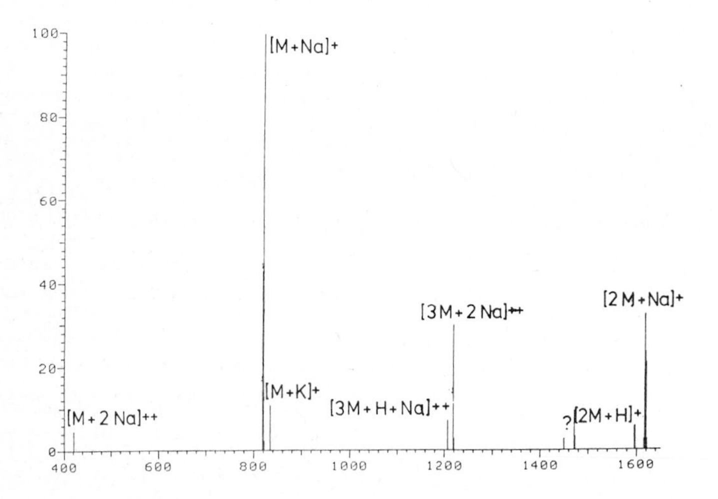

Fig. 11 Molecular ion region of a terpene glycoside (FD).

The molecular ion region of FAB spectra obtained in the negative mode is usually less complex. In many cases $[M - H]^-$ ions are formed (Fig. 8c). Also loss of H_2O is less pronounced than in the positive spectra - $R\text{-}OH_2^+$ is predestined to lose H_2O (Fig. 8a)!

More annoying are intermolecular reactions of the substrate (transacetylations and transalkylations in the condensed phase have been reported in the literature, see, e.g., Scheme 2) and reactions of the substrate with the matrix and its decomposition products. Dehydrogenation resulting in [M - 2H] (cf. Section 2 (i) and Fig. 6c) and hydrogenation products ([M + 2H]) are observed rather frequently. $[(M + 12u) + H]^+$ ions occasionally observed in the FAB spectra of peptides do stem from glycerol (probably CH_2O formed by decomposition of the latter reacts, e.g., with an NH_2-group to give a Schiff base). In cases of doubt the measurement should be repeated with a different matrix.

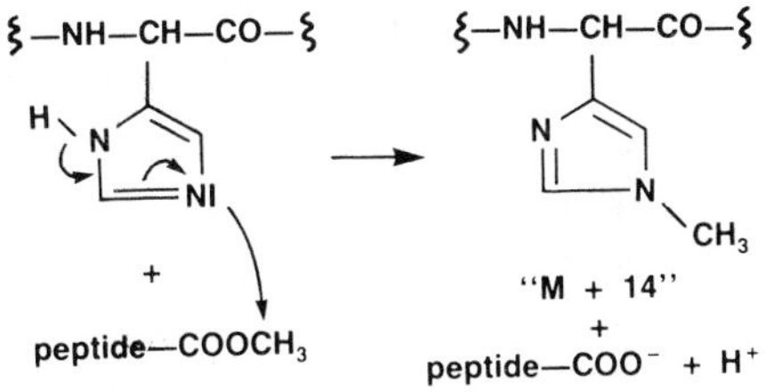

Scheme 2 Intermolecular transmethylation between two peptide molecules.

5. COMPARISON OF THE TECHNIQUES

5.1 Positive mode

The application of the "in-beam" and the DCI technique is limited to compounds which can be transferred to the gas phase at least partially undecomposed by flash evaporation. The ratio quasi-molecular ion *vs.* fragment ions can be influenced to some extent by a selection of the reagent gas. An example is given in Fig. 12: $t\text{-}C_4H_9^+$ (the main ion in the $i\text{-}C_4H_{10}$ plasma) readily protonates the glycosidic linkages which results in a degradation to the aglycon. NH_4^+ is a much weaker Brönsted acid; $[M + NH_4]^+$ and the consecutive losses of the three digitose units can be seen.

FD and FAB give comparable results though the experience teaches that occasionally one gets good results with one method but not with the other (one should also distinguish between "true" FD and cationization). Sample preparation (admixtures in FD, matrix in FAB) is still a matter of experience or even of trial and error. It influences the formation of quasi-molecular ions as well as that of fragments: The formation of $[M + H]^+$ in the FAB spectrum (Fig. 12c) induces loss of the digitose units, while the $[M + Na]^+$ ion formed in FD

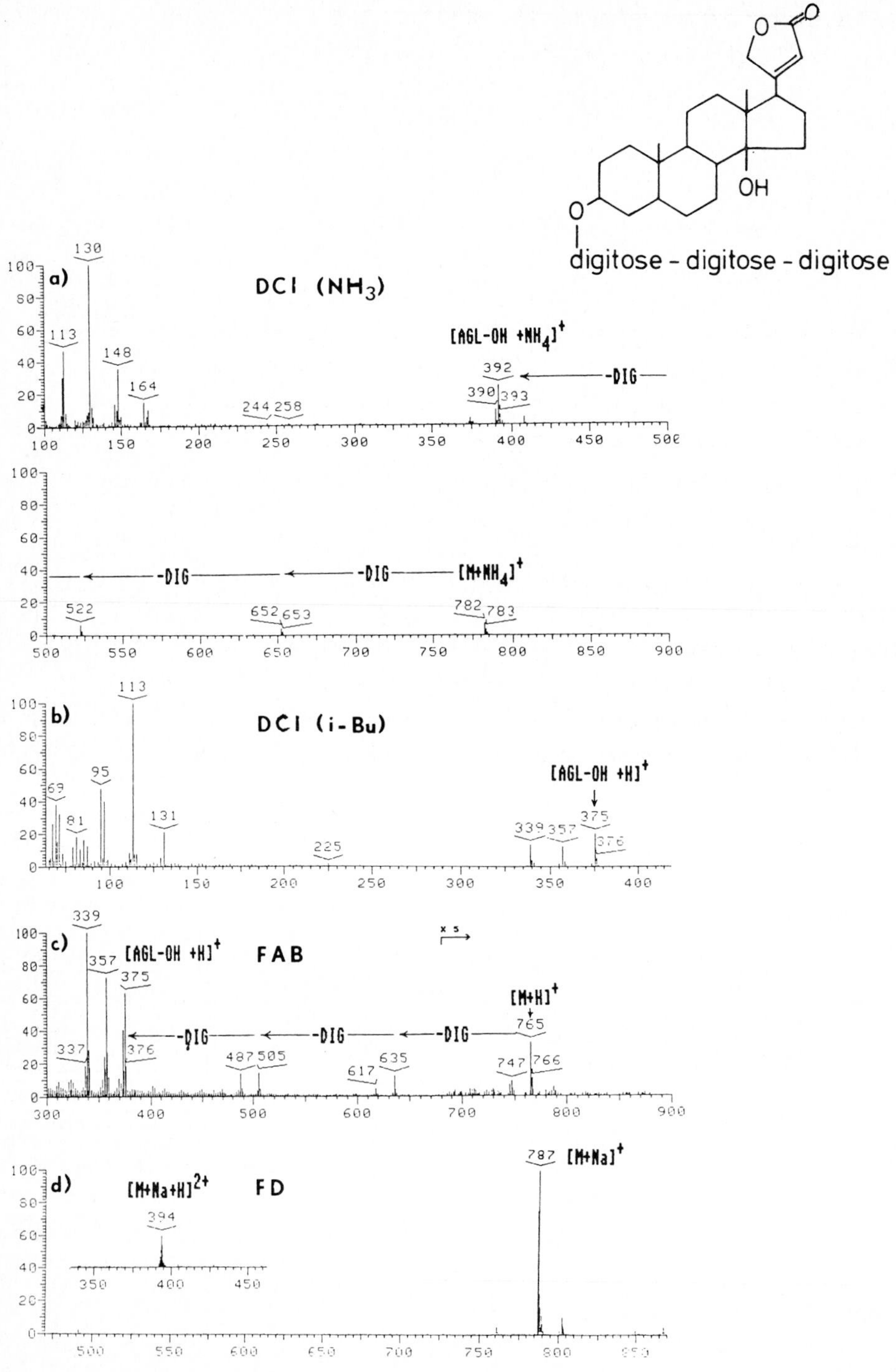

Fig. 12 Mass spectra of digitonin: a) DCI (NH_3), b) DCI (i-C_4H_{10}), c) FAB (glycerol), d) FD.

obviously does not undergo protolytic cleavages.

PD needs specific instrumentation. Its main advantage seems to lie in an easy and quick determination of molecular weights of large molecules as peptides.

5.2 Negative mode

For "in-beam" EI the same limitations pertain as for normal EI, hence this method is of no practical importance. Negative DCI has been used occasionally (Cl^- and OH^- as plasma ions). FD of preformed anions (salts) is possible but has been used only rarely (especially in the pre-FAB days). Negative FAB spectra can be obtained easily. The molecular ion region is frequently more straightforward than in the positive mode. Structural information which can be extracted from fragments may be complementary. Also in PD negative quasi-molecular ions can be used for confirmation purposes.

6. CONCLUSIONS

The surface ionization techniques have given a new dimension to mass spectrometry: Molecular weight and structural information can be obtained from compounds in a mass range of several thousand. But for these possibilities a price has to be paid: In this area mass spectrometry is not any more a routine method which gives clear-cut answers. Experience and chemical knowledge are essential and a fair amount of research has still to be done.

APPLICATION OF THE NUCLEAR OVERHAUSER EFFECT TO THE STUDY OF INTERNUCLEAR DISTANCES IN NATURAL PRODUCTS: THE SOLUTION AND CRYSTAL STRUCTURE OF CYTISINE

P. MASCAGNI , S. MANGANI , N. NICCOLAI , K. ASRES ,
J.D. PHILLIPSON AND W.A. GIBBONS

1 INTRODUCTION

1.1 The Nuclear Overhauser Effect

The nuclear Overhauser effect (NOE) is the change in the integrated nuclear magnetic resonance absorption intensity of a nuclear spin when the NMR absorption of another spin is saturated. The spins involved may be either heteronuclear or chemically shifted homonuclear spins.

To illustrate the phenomenon let us consider the system of two dipolarly coupled spin-1/2 nuclei, i and s, having the four energy levels illustrated in Fig. 1.1.

Let the transition probabilities W_2, W_0, W_i and W_s refer to dipole-dipole relaxation only, while all the other relaxation mechanisms are represented by the transition probabilities W^*_i and W^*_s. The NMR resonance of spin s will consist of two transitions. 1-2 and 3-4, which in the absence of J coupling between i and s will have the same frequency. Likewise the resonance of spin i will have two components, the 1-3 and 2-4 transitions. The absorption intensity of the spins i and s will be proportional to the population difference across the respective transitions.

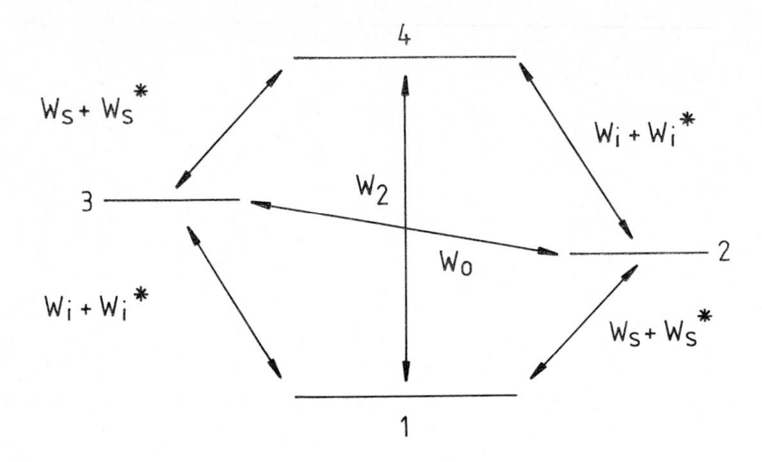

Fig. 1.1 Energy level diagram and transition probabilities for a system of two spin-1/2 nuclei i and s.

Suppose that the spin populations are perturbed by a selective rf field applied at the resonance frequency of s: the populations in the levels linked by W_s are rapidly equalised but <u>initially</u> there is no change in the magnetisation of spin i since levels 1 and 3 have lost equal amounts of population and levels 2 and 4 gained equal amounts, the population difference across transitions 1-3 and 2-4 being unchanged immediately after the start of irradiation.

A new population distribution is generated however by relaxation through W_2 and W_0 as irradiation of s is continued. W_2 increases the intensity of the i transition by attempting to establish an equilibrium (i.e. a Boltzmann distribution of spins) between levels 1 and 4, whilst W_0 decreases the intensity by equilibrating 2 and 3. The net change in the intensity of i which results is the sum of these two contributions: this is the nuclear Overhauser enhancement. It is clearly a kinetic effect since it takes time to develop after the start of the irradiation and time to decay again after the irradiation is turned off.

It can be readily shown (ref. 1) that the relative enhancement of the intensity of spin i upon irradiation of spin s (NOEi(s)) is given by:

$$NOE_i(s) = \frac{I_i^z - I_i^o}{I_i^o} = \frac{\sigma_{is}}{(\rho_{is} + \rho_i^*)} \frac{I_s^o}{I_i^o} \qquad (1.1)$$

where

$$\rho_{is} = 2\, W_i + W_2 + W_o \qquad (1.2)$$

$$\rho_i^* = 2\, W_i^* \qquad (1.3)$$

$$\sigma_{is} = W_2 - W_o \qquad (1.4)$$

I^o and I^z are the intensity of the magnetisation vector at time zero and the observable magnetisation respectively. The $_{is}$ term is often referred to as the cross-relaxation term.

If the relaxation is exclusively dipolar in character, ρ_i^* can be neglected in comparison with ρ_{is} and

$$NOE_i(s) = \frac{\sigma_{is}}{\rho_{is}} \frac{I_s^o}{I_i^o} \qquad (1.5)$$

The analysis of a multiple spin system has been treated in detail by Noggle and Schirmer (ref. 2). Following their analysis the NOE of spin i coupled to a group of spins j, when spin s (with $\gamma_{s,j} = \gamma_i$) is saturated becomes

$$NOE_i(s) = \frac{\sigma_{is}}{R_i} - \sum_{n=i,s} \left[\frac{\sigma_{in}}{R_i} - NOE_i(n) \right] \qquad (1.6)$$

R_i $(= \sum_{m=i,s} \rho_{ij} + \rho_m^*)$ contains the sum of all ρ_{ij} dipolar contributions to the spin-lattice relaxation $(\sum \rho_{ij})$, as well as the non dipole-dipole interactions $(\sum \rho_m^*)$, such as the contribution from ^{14}N quadrupolar relaxation. The second term on the right of eqn. 1.6 accounts for the indirect contribution to an observed NCE that results from partial saturation of spins other than i and s. The contribution of this term to the total NOE for some of the protons of the molecule investigated in this paper was found to be less than 7% and hence not taken into account in the following discussion. The simplified eqn. 1.7 was used to relate observed NOE's to the other relaxation parameters, σ_{is} and R_i.

$$NOE_i(s) = \frac{\sigma_{is}}{\rho R_i} \qquad (1.7)$$

1.2 Cross relaxation rates and internuclear distances

For two dipolarly coupled spins-1/2 i and s, the relaxation parameters, σ_{is} and ρ_{is}, defined in Section 1.1, are (ref. 2)

$$\sigma_{is} = \frac{\hbar^2\gamma_i^2\gamma_s^2}{10r_{is}^6}\left[\frac{6\tau_c}{1+(\omega_i+\omega_s)^2\tau_c^2} - \frac{\tau_c}{1+(\omega_i-\omega_s)^2\tau_c^2}\right] \quad (1.8)$$

$$\rho_{is} = \frac{\hbar^2\gamma_i^2\gamma_s^2}{10r_{is}^6}\left[\frac{3\tau_c}{1+\omega_i^2\tau_c^2} + \frac{\tau_c}{1+(\omega_i-\omega_s)^2\tau_c^2} + \frac{6\tau_c}{1+(\omega_i+\omega_s)^2\tau_c^2}\right] \quad (1.9)$$

where τ_c is the correlation time, ω_i and ω_s the Larmor frequencies of spin i and s respectively and $\hbar = h/2$, h being Plank constant.

In the case of small and rigid molecules in non-viscous solvents, the extreme narrowing conditions, $(\omega_i+\omega_s)^2\tau_c^2 \ll 1$, apply and eqns. 1.8 and 1.9 become

$$\sigma_{is} = \frac{\hbar^2\gamma_i^2\gamma_s^2}{2r_{is}^6} * \tau_c \quad (1.10)$$

$$\rho_{is} = \frac{\hbar^2\gamma_i^2\gamma_s^2}{r_{is}^6} * \tau_c \quad (1.11)$$

Thus the dipole-dipole interaction constants are related quite simply to molecular geometry. However, the correlation time must be measured if the interatomic distance r_{is} is to be determined from cross relaxation rates.

In the following section we shall describe NMR methods which are commonly employed to measure τ_c. Here we discuss some of the currently available procedures to calculate cross relaxation rates.

a) The knowledge of NOE effects and R_i rates can be used to obtain cross-relaxation rates from eqns. 1.6 and 1.7.

b) Simultaneous, selective excitation of two nuclei, i and s (refs. 3,4) produces a new relaxation parameter, R^{BS}_{is}, which is related to the mono-selective relaxation rate (R^{SE}_i) (ref. 3) and the cross relaxation rate σ_{is}, by:

$$R^{BS}_{is} = \sigma_{is} + R^{SE}_i \quad (1.12)$$

c) The development of the NOE effect at spin i, expressed as function of the length of a saturating pulse applied to spin s $NOE^{t}_{i}(s)$, is (ref. 2):

$$NOE^{t}_{i}(s) = \frac{\sigma_{is}}{\rho_{is}} \frac{I^{o}_{s}}{I^{o}_{i}} (1-e^{-\rho_{is}t}) \tag{1.13}$$

For small t, the time course is linear with a gradient of $-\sigma_{is}$. Thus measuring the initial gradient of an NOE vs time gives the cross-relaxation rate (ref. 5).

d) Whereas conventional mono-dimensional NOE experiments have been applied extensively to the quantitation of internuclear distances (refs. 4,6-8) it is only in recent years that 2D NOE techniques (ref. 9) have been used to measure cross-relaxation rates (refs. 10-12).

One of the recent developments by Ernst and co-workers (ref. 9) utilises the initial buildup rate of the cross-peak intensity, $I_{is}(\tau_m)$, in 2D NOESY experiments to derive cross-relaxation rates from eqn. 1.14:

$$\frac{I_{is}(\tau_m)}{I_{ii}(0)} = \sigma_{is} * \tau_m \tag{1.14}$$

$I_{ii}(0)$ is the diagonal-peak intensity extrapolated to $\tau_m = 0$, and τ_m is the mixing time of the NOESY experiment.

1.3 Internuclear distances from NOE ratios (ref. 6)

In cases where irradiation of two different protons, say s and m, both produce NOE's at proton i and if the reorentation of the two vectors s-i and m-i is modulated by the same correlation time, then it is not necessary to measure R_i in eqns. 1.6 and 1.7 since

$$\left(\frac{\sigma_{is}}{R_i}\right)/\left(\frac{\sigma_{im}}{R_i}\right) = \frac{\sigma_{is}}{\tau_{im}} = \left(\frac{1}{r_{is}}\right)^6 / \left(\frac{1}{r_{im}}\right)^6 \tag{1.15}$$

If one of the internuclear distances in eqn. 1.15 is known, the other can be derived from experimental NOE's.

1.4 Calculation of the rotational correlation time

As discussed in Section 1.2 in order to interpret the cross-relaxation rate in terms of geometry and motion of the i-s system the rotational correlation time should be measured. Two methods for calculating τ_c are described here.

a) From ^{13}C spin-lattice relaxation measurements. The term R_i defined in Section 1.1 accounts for all spin relaxation contributions, i.e. intra- and inter-dipole-dipole, scalar coupling, spin rotation, chemical shift anisotropy etc. For ^{13}C spins the contribution arising from dipole coupling with their attached protons is the dominant and in most cases the only effective relaxation mechanism. If this latter case applies then experimental ^{13}C spin lattice relaxation rates, corrected for the number of protons attached, equal the R_i terms. Thus they can be used together with the C-H bond distance of 1.08 Å, to calculate the correlation time from eqns. 1.9 and/or 1.11.

b) From cross-correlation rates. In general τ_c can be calculated from cross relaxation rates if the internuclear distance, r_{is} in eqns. 1.8 and 1.10, is known from either x-ray data or bond angles and bond distances (e.g. geminal protons, aromatic protons etc.).

1.5 The real case: cytisine

Most of the techniques described in the previous sections were applied to cytisine (Fig. 1.2), an alkaloid of the quinolizidine family. Cytisine is well known for its toxicity to man and livestock (ref. 13). Some of its toxicological responses include nausea, convulsion and death by respiratory failure. The structure of cytisine has been elucidated by chemical degradation, synthesis and spectroscopy (refs. 14-16).

The strategy we adopted to describe the conformation of cytisine is as follows:

i) Proton and carbon spectral assignments
ii) Calculation of the rotational correlation time from C-13 relaxation times and broad-band NOE(NOE-bb)
iii) Measurements of R_i relaxation rates.
iv) Measurements of proton NOE's
v) Calculation of cross-relaxation rates from NOE's and R_i rates
vi) Calculation of r_{is} using eqns. 1-8 and 1-10

Parallel to the NMR analysis of cytisine an x-ray crystallography analysis was carried out and the results from the two sets of measurements then compared.

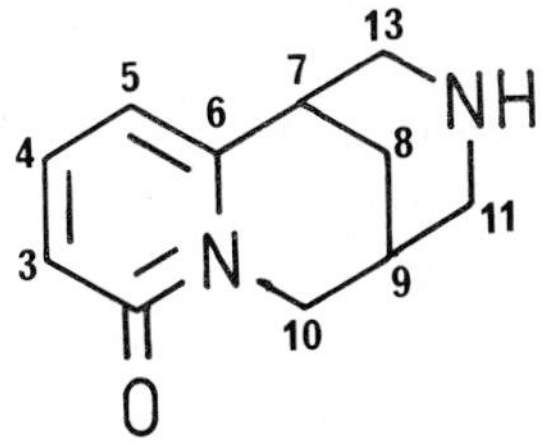

Fig. 1.2 Structure of cytisine

2 MATERIALS AND METHODS

Cytisine was purchased from Koch-Light Laboratories and purified by recrystallisation from benzene-hexane mixtures; the same solvent system was also used to obtain crystals for x-ray crystallography. The optical activity of cytisine was measured in ethanol: the $[\alpha]_D^{25}$ was -119°. 9 identical with that from the literature (ref. 15).

2.1 NMR analysis

5 mg of purified cytisine were dissolved in 1 ml of $CDCl_3$ and the solution bubbled with nitrogen.

1D- and 2D-NMR experiments were performed on a Varian XL-300 instrument. The C-13 relaxation measurements were also recorded on a Bruker WP-80-SY.

COSY and COSY-45 experiments were run using the t-90-tau-Pw-Acq pulse sequence with Pw=90 and 45 respectively (ref. 17). 512 increments of 1024 data points each were recorded in each case and the matrix thus obtained symmetrised with respect to the diagonal.

The NOESY (ref. 9) pulse sequence was modified according to a recent paper (ref. 18), in order to remove any residual scalar coupling constant contribution. The mixing time used was 0.2 s.

2D-J resolved and proton-carbon correlation spectra (refs. 19, 20) were run using 128 increments of 2K points in both cases.

Non-selective proton and carbon spin-lattice relaxation rates were measured with the inversion recovery method. Selective spin-lattice relaxation rates (ref. 3) were obtained by selectively inverting the proton magnetisation with the decoupler channel, and monitoring its recovery to equilibrium.

Quantitative, mono-dimensional NOE's were obtained saturating the proton of interest with a 5-10 s long, low-power irradiation. The interval between acquisitions was 20 s.

The gating of the decoupler channel prior to acquisition generated broad-band decoupled C-13 spectra with no Overhauser enhancement. Subtraction of these spectra from those obtained without decoupler gating yielded the NOE-bb.

2.2 X-ray crystallography

The x-ray diffraction measurements were done on a Phillips PW100 automatic diffratometer at the Department of General and Inorganic Chemistry of the University of Florence, Italy.

Unit cell parameters and other crystal data are reported in Table 2.1.

TABLE 2.1

Crystal data for cytisine

Molecular formula	$C_{11}H_{14}N_2O$
Space group	$P2_12_12_1$
Crystal size	0.5 * 0.2 * 0.1 mm
$\underline{a}$	9.978(5) Å
$\underline{b}$	26.615(8) Å
$\underline{c}$	7.217(4) Å
V	1916.6 $Å^3$
Z	8
F(000)	816
Radiation	MoK (λ = 0.7093 Å)
(MoK)	0.25 cm^{-1}
Molecular weight	190.2 amu
D calc.	1.322 g/cm^3
R^a	0.054
Rw^b	0.047

$^aR = \sum(|Fo| - |Fc|)/\sum |Fo|$

$^bRw = [\sum W(|Fo| - |Fc|)^2 / w(Fo)^2]^{\frac{1}{2}}$

Graphite monochromatised MoK radiation was used for measuring both cell dimensions and diffraction intensities. A least-square fit of 25 reflections was used to determine cell constants.

The quadrant of reciprocal space $\pm$ hkl was inspected in the range $5^{o} \leqslant 1 \Theta < 40^{o}$ and equivalent reflections were merged to give 1081 independent intensities.

The Θ -2 Θ scan technique was used with a scan speed of 0.08^{o} s^{-1} and a variable scan width of 1.1 + 0.3 tang$(\Theta)^{o}$. At each end of the scan the background was counted half the total scan time.

The 708 reflections with intensities greater than 2.5 σ (I) were used for the structure solution and refinement. The standard deviation of the intensity, σ (I), was computed as

$$[P + B_1 + B_2 + (0.031)^2]^{\frac{1}{2}} \qquad (2.1)$$

where P is the total integrated count, B_1 and B_2 are the background counts. I is the peak intensity after subtraction of the background and 0.031 is a correction factor for unrealistically small standard deviation in strong reflections.

The space group $P2_12_12_1$ was determined on the basis of systematic absences.

The structure of cytisine was solved by direct methods provided by the SHELX76 set of programs (ref. 21).

Origins, enanthiomorph and multisolution phases were chosen from reflections at the top of the convergence map. The E-map showed a recognizable fragment of both the molecules. A subsequent Fourier map revealed the rest of the non-hydrogen atoms.

Two full matrix least-squares cycles with isotropic temperature factors for all the atoms reduced the R factor to 0.085. Successive difference electron density syntheses revealed all hydrogen atoms. In the following cycles only the two N-12 protons were allowed to refine, the remainder being introduced in calculated positions and constrained to ride on their carbon atom.

All calculations were performed with the set of programs SHELX76 which uses the analytical approximation from the International Tables (ref. 22), for the atomic scattering factors. Plots were made using the program ORTEP (ref. 22).

3 RESULTS AND DISCUSSION

3.1 H-1 and C-13 spectral assignments

The assignment of the proton and carbon spectra of cytisine in deuterochloroform solutions was obtained by a combination of homo- and hetero-nuclear 2D experiments. In some instances (e.g. long range couplings of H-10 with H-8 and H-9) the coupling constants had to be derived from the signal line-width. Proton chemical shifts and coupling constants were refined by computer simulation and a complete list of the assignments is provided in Tables 3.1 and 3.2.

TABLE 3.1
Proton chemical shifts, configurations and relaxation parameters of cytisine in chloroform solution.

	ppm[a]	Configuration[b]	R_{NS}[c]	R_{SE}[d]	F_i[e]
H-3	6.18	-	0.14	0.10	1.33
H-4	7.02	-	0.19	0.14	1.30
H-5	5.73	-	0.25	0.19	1.31
H-7	2.69	e(α)	0.41	0.31	1.32
H-8	1.68	-	0.91	0.69[f]	1.35
H-8	1.68	-	0.91	0.69[f]	1.35
H-9	2.05	e(α)	0.48	0.36	1.33
H-10(d)	3.85	e(β)	0.73	0.53	1.38
H-10(u)	3.62	a(α)	0.73	0.52	1.40
H-11(d)	2.83	e(β)	0.91	0.69	1.32
H-11(u)	2.72	a(α)	-	-	-
N-H	1.17	-	-	-	-
H-13(d)	2.79	a(α)	-	-	-
H-13(u)	274	e(β)	-	-	-

[a]chemical shifts are in ppm and refer to TMS
[b]a=axial; e=equatorial; β=above the plane; α=below the plane
[c]$R_{NS} = 1/T_1^{NS}$; non-selective relaxation rates in sec.
[d]$R_{SE} = 1/T_1^{SE}$; selective relaxation rates in s^{-1}.

[e] $F_i = R_{NS}/R_{SE}$; see text for discussion.
[f] R_{SE} for H-8s were calculated from σ_{gem} (see text).

3.2 ^{13}C relaxation measurements

In order to use ^{13}C spin-lattice relaxation rates to calculate the rotational correlation time from eqns. 1.9 and/or 1.11 the prerequisite of the dipolar interaction as the sole mechanism of relaxation for protonated carbons must be demonstrated. The broad-band proton-carbon NOE provides a means for measuring the contribution to the C-13 spin-lattice relaxation rate arising from H-C dipolar interactions (ref. 24). Thus an NOE-bb of 1.988 indicates that the C-13 relaxation mechanism is overwhelmingly dominated by the latter interaction and also that the extreme narrowing conditions apply to the molecular rotation (ref. 24).

When the NOE-bb's were measured for cytisine we found (Table 3.2) that, with the exception of C-3 whose NOE-bb = 1.71, all protonated carbons had Overhauser contributions of 2.0 $\pm$ 0.1, proving that eqn. 1.11 can be used together with carbon R_i rates to obtain the correlation time.

The ^{13}C spin lattice relaxation rates were measured at two different spectrometer frequencies, 75 and 20 MHz respectively. This procedure allows an unequivocal assignment of the correlation time as shown in ref. 25.

TABLE 3.2

^{13}C chemical shifts and relaxation parameters for cytisine in chloroform solution.

	(ppm)[a]	R_1[b]	R_1[c]	NOE-bb[d]
C-2	163.6	-	-	0.26
C-3	116.6	0.44	0.34	1.75
C-4	138.7	0.37	0.31	2.00
C-5	104.9	0.34	0.29	2.00
C-6	151.1	-	-	0.4
C-7	35.6	0.36	0.30	2.00
C-8	26.3	0.37	0.27	2.00
C-9	27.7	0.33	0.31	2.00
C-10	49.7	0.36	0.27	2.00
C-11	54.0	0.34	0.29	1.93
C-13	53.0	0.32	0.27	1.88

[a] Chemical shifts refer to TMS.
[b] R_1 rates were measured at the spectrometry frequency of 75 MHz for carbon.
[c] Relaxation rates measured at the frequency of 20 MHz for carbon
[d] NOE-bb's were measured at the carbon frequency of 75 MHz.

When corrected for the number of protons directly bonded the rates of all protonated carbons were very similar at both frequencies (see Table 3.2). At 75 MHz all the rates ranged from 0.37 s^{-1} for C-4 to 0.32 s^{-1} for C-13 with the exception of C-3 which had a relaxation rate of 0.44 s^{-1}; the average value (calculated without taking into account the largest rate) was 0.32 s^{-1}. At 20 MHz the mean relaxation rate was 0.29 s^{-1} and also in this case the largest value was recorded for C-3 (0.34 s^{-1}) (Table 3.2).

In general the virtually identical relaxation rates at the two spectrometer frequencies indicated that (i) the carbon nuclei relax completely by dipolar interactions and (ii) the latter are modulated by molecular motions within the extreme narrowing conditions (ref. 24).

Relaxation mechanisms other than dipolar probably account for the anomalous behaviour of C-3. However this does not alter the conclusion that cytisine behaves in solution like a rigid rotor in isotropic motion.

The measured spin-lattice relaxation rates of all protonated carbons, with the exception of C-3, were hence used to extrapolate the correlation time from eqn. 1.11. The narrow range of values obtained, $\tau_c = 1.50 \pm 0.2 * 10^{-11}$ s confirmed that a single correlation time accounted for the fast molecular tumbling.

3.3 ^1H relaxation measurements

One of the prerequisite for the application of eqn. 1.6 to the calculation of σ_{is} and hence of r_{is}, is the existence of a dipolar pathway for the relaxation mechanism of spins i and s. Furthermore, a means of measuring or calculating R_i must be available. Freeman and co-workers (ref. 3) showed that spin-lattice relaxation rates obtained under selective excitation conditions (R^{SE}_i) can be used to measure (i) the dipolar contribution to the proton relaxation rate and (ii) the R_i term of eqn. 1.6. Thus for a spin i, in a multiple spin system at thermal equilibrium, the overall spin-lattice relaxation rate (R^{NS}_i) is expressed as:

$$R^{NS}_i = R_i + \sum_{i=j} \sigma_{ij} \qquad (3.1)$$

In condition of selective excitation and if measurements are restricted to short times after the pulse, the initial rate of recovery (R^{SE}_i) is no longer exponential and just equals R_i. Furthermore if the relaxation of spin i is solely by dipolar coupling to spin s and the extreme narrowing conditions apply then the ratio (F_i) of the observed rates,

$$F_i = \frac{R^{NS}_i}{R^{SE}_i} = 1.5 \qquad (3.2)$$

With the exception of the N-H and H-8's all other protons of cytisine had F_i ratios between 1.3 and 1.4 (Table 3.1). The lower value for the H-8's (1.18) was ascribed to spectral overlap of these two protons. Thus when the H-7's are irradiated, although the σ_{is} terms cancel, the cross relaxation contribution from the inter-methylene dipolar interaction (σ_{gem}) is still contained in the expression of the relaxation rate (refs. 1,26). We can then write

$$R^{SE}\ (CH_2) = R^*\ (CH_2) + \sigma_{gem} \qquad (3.3)$$

where the first term on the right of the equation refers to the effective relaxation rate. On the reasonable assumption that an average F_i=1.35 can be applied to protons H-8, R^* rate and $_{gem}$ were calculated using the experimental non-selective relaxation rate of 0.91 s^{-1}. The knowledge of the geminal cross-relaxation rate was then used to independently obtain the correlation time for the H8-H8 vector. This was found to be $1.15 * 10^{-11}$ s, a value close to that obtained from C-13 relaxation rates.

The dipolar interaction between ^{14}N and its directly bonded proton, as well as nitrogen inversion could account for the low F_i(N-H) ratio; the effects of the first interaction on the proton relaxation are well documented (ref. 27) while chemical exchange made it difficult to selectively invert the N-H proton magnetisation.

In conclusion, using 1.35 as an average F_i ratio, the dipolar interaction was found to account for 70% of the total relaxation mechanism of each proton of cytisine. A corollary of this is that eqn. 1.7 can be used to calculate cross-relaxation rates.

The deviation of the F_i ratios from 1.5 could be accounted for by relaxation mechanisms other than dipolar or by slow motion in solution. The former conclusion applies to cytisine since the ^{13}C data showed that cytisine undergoes fast rotational reorientation in solution.

3.4 Proton-proton Overhauser effects

At this stage we had obtained two of the four parameters needed for the application of eqns. 1.7–1.11 to the calculation of interatomic distances in cytisine. The next step in this procedure was then the measurement of Overhauser effects. These were obtained using long, low-power selective irradiations at selected proton frequencies. Two classes of NOE's were recorded: negative NOE's between H-3 and H-5 and positive ones at all the other protons. The experimental results are shown in Table 3.3 and were used to calculate interatomic distances as discussed in the following sections.

Some of the NOE's from Table 3.3 were also used for an independent calculation of the correlation time. For instance the Overhauser effects at spins 3 and 5 observed after irradiation of H-4, when used together with the appropriate relaxation rates and x-ray distances, yielded $\tau_{3-4} = 1.25 * 10^{-11}$ and $\tau_{4-5} = 1.55 * 10^{-11}$ s, values close to that calculated using ^{13}C relaxation parameters.

3.5 Interproton distances from NMR parameters

The negative NOE's observed at protons 3 and 5 were rationalised in terms of "linear three spin system (obtuse case)" (Ref. 2). Thus Noggle has shown that for such a system the effect recorded at an end spin (a) while the other end spin (c) is saturated has a negative sign and that, if the correlation time is constant over the whole system a-b-c, the interproton distances separating the three spins can be measured using the formula:

$$\frac{(r_{a-c})^6}{(r_{a-b})^6} = \frac{f_a(b) + [f_a(c) * f_c(b)]}{f_a(c) + [f_a(b) * f_b(c)]} \qquad (3.4)$$

Here $f_i(j)$ is the fractional Overhauser effect. From this equation one of the distances can be calculated if the other is known. In cytisine protons 3, 4 and 5 lie on the same plane, hence r_{3-4} can be calculated from standard interatomic distances. Using r_{3-4} =

TABLE 3.3

Proton-proton distances of cytisine from Overhauser effects and x-ray crystallography.

observed ^1H	irradiated ^1H	nOe %	σ^a	$r_{nOe}{}^b$	$r_{x\text{-}ray}{}^c$
H-3	H-4	17.8	1.6	2.54	2.46 (2.38)
H-5	H-4	11.2	2.3	2.39	2.49 (2.49)
H-3	H-5	-2.9	-	4.45	4.28 (4.24)
H-4	H-5	19.1	2.7	2.33	2.49 (2.49)
H-5	H-3	-3.0	-	-	4.28 (4.24)
H-4	H-3	13.2	1.8	2.48	2.46 (2.38)
H-9	H-11(d)	4.9	1.8	2.49	2.44 (2.42)
H-10(d)	H-11(d)	5.0	2.6	2.34	2.39 (2.38)
H-10(d)	H-9	2.9	1.5	2.56	2.61 (2.55)
H-11(d)	H-9	5.0	3.4	2.24	2.44 (2.42)
H-10(u)	H-9	7.3	3.7	2.21	2.26 (2.28)
H-11(u)	H-9	3.6	-	-	2.43 (2.38)
H-8	H-9	2.15	1.5	2.56	2.48 (2.50)
H-11(d)	H-10(d)	5.0	-	-	2.39 (2.38)
H-5	H-7	11.2	2.3	2.40	2.34 (2.46)
H-7	H-5	8.5	2.6	2.34	2.34 (2.46)
H-8	H-7	3.4	2.3	2.39	2.49 (2.53)

[a] sigma values were calculated from the Overhauser effects and the selective relaxation rates from Table 1. Values are in s x 10^{-2}.

[b] Distances in Å: they were calculated using equation (3) and (4) for the positive and negative nOes respectively and a correlation time of 1.5 x 10^{-11} s.

[c] Interproton distances (a) from x-ray data; figures in parenthesis refer to the second structure of cytisine.

2.4 Å we obtained r_{3-5} = 4.4 Å in good agreement with the crystallographic figure of 4.3 Å (see text below).

The selective relaxation rates from Table 3.1 and the NOE's from Table 3.3 were used in eqn. 1.7 to derive cross-relaxation rates. These and the correlation time of 1.5 * 10^{-11} s, when inserted in eqn. 1.10, gave the proton-proton distances of Table

3.3 where they are compared to the crystallographic figures.

3.6 The crystal structure of cytisine

The crystal structure of cytisine is depicted in two forms. Fig. 3.1 shows a perspective view of these two independent molecules in the asymmetric unit. Bond lengths and angles with their esd's are reported in Table 3.4. With the exception of N-12 proton, the fractional atomic coordinates for all the other hydrogens were obtained from calculated positions and hence the relative interatomic distances are affected by an estimated error of ± 0.15 Å.

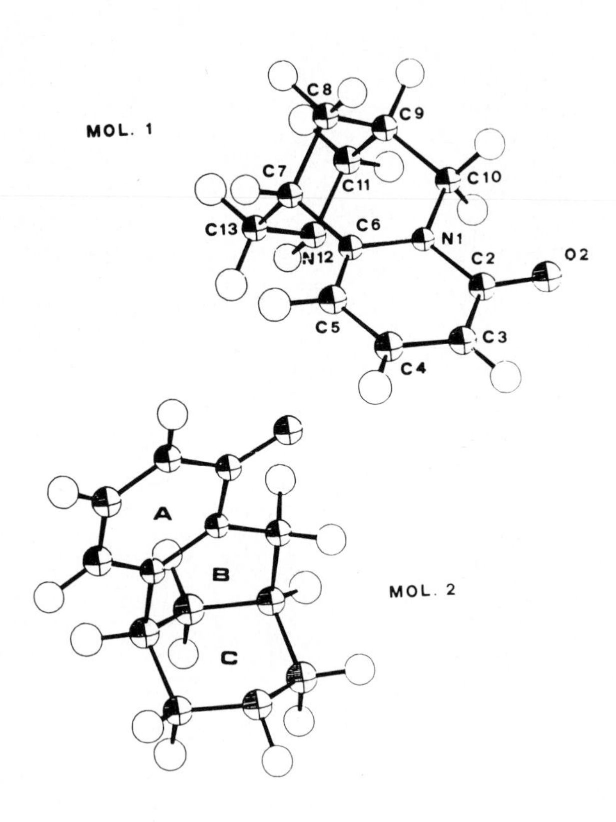

Fig. 3.1 ORTEP drawings of the two independent molecules of cytisine. Hydrogen atoms are omitted for clarity.

TABLE 3.4

Interatomic distances (Å) and angles (°) and their esds for the two independent molecules of cytisine in the unit cell.

Bond	Distance	Bond	Distance
N(1)-C(2)	1.428(1)	N(1)-C(6)	1.371(10)
N(1)-C(10)	1.495(10)	D(2)-C(2)	1.239(9)
C(2)-C(3)	1.408(11)	C(3)-C(4)	1.394(13)
C(4)-C(5)	1.402(11)	C(5)-C(6)	1.356(11)
C(6)-C(7)	1.516(11)	C(7)-C(8)	1.538(10)
C(7)-C(13)	1.536(11)	C(8)-C(9)	1.517(11)
C(9)-C(10)	1.542(11)	C(9)-C(11)	1.527(11)
C(11)-N(12)	1.466(10)	N(12)-H(1)	1.071(111)
N(12)-C(13)	1.488(10)	N(1')-C(2')	1.408(10)
N(1')-C(6')	1.369(10)	N(1')-C(10')	1.478(10)
D(2')-C(2')	1.24(10)	C(2')-C(3')	1.432(12)
C(3')-C(4')	1.332(11)	C(4')-C(5')	1.410(11)
C(5')-C(6')	1.379(12)	C(6')-C(7')	1.528(12)
C(7')-C(8')	1.532(12)	C(7')-C(13')	1.550(12)
C(8')-C(9')	1.501(12)	C(9')-C(10')	1.516(11)
C(9')-C(11')	1.532(12)	C(11')-N(12')	1.470(11)
N(12')-H(1')	0.924(114)	N(12')-C(13')	1.447(12)

Angle	Value	Angle	Value
C(2)-N(1)-C(6)	121.6(8)	C(2)-N(1)-C(10)	114.0(7)
C(6)-N(1)-C(10)	124.3(8)	D(2)-C(2)-N(1)	118.0(9)
D(2)-C(2)-C(3)	125.8(10)	N(1)-C(2)-C(3)	116.2(9)
C(2)-C(3)-C(4)	121.3(10)	C(3)-C(4)-C(5)	119.8(10)
C(4)-C(5)-C(6)	119.9(9)	N(1)-C(6)-C(5)	121.1(9)
N(1)-C(16)-C(7)	119.0(8)	C(5)-C(6)-C(7)	119.8(9)
C(6)-C(7)-C(8)	108.6(7)	C(6)-C(7)-C(13)	111.7(7)
C(8)-C(7)-C(13)	109.4(7)	C(7)-C(8)-C(9)	108.2(7)
C(8)-C(9)-C(10)	109.9(7)	C(8)-C(9)-C(11)	109.4(7)
C(10)-C(9)-C(11)	112.4(7)	N(1)-C(10)-C(9)	114.5(7)
C(9)-C(11)-N(12)	109.8(7)	C(11)-N(12)-C(13)	111.2(7)
C(11)-N(12)-H(1)	125.9(51)	C(13)-N(12)-H(1)	93.4(53)
C(7)-C(13)-N(12)	110.3(8)		
C(2')-N(1')-C(16')	122.6(8)	C(2')-N(1')-C(10')	115.0(7)
C(6')-N(1')-C(10')	122.4(7)	D(2')-C(2')-N(1')	118.5(9)
D(2')-C(2')-C(3')	126.2(9)	N(1')-C(2')-C(3')	115.3(8)
C(2')-C(3')-C(4')	122.1(9)	C(3')-C(4')-C(5')	121.1(10)
C(4')-C(5')-C(6')	118.8(10)	N(1')-C(6')-C(5')	120.1(8)
N(1')-C(6')-C(7')	119.8(9)	C(5')-C(6')-C(7')	120.1(9)

C(6')-C(7')-C(8')	109.6(9)	C(6')-C(7')-C(13')	109.8(7)
C(8')-C(7')-C(13')	108.7(8)	C(7')-C(8')-C(9')	107.5(8)
C(8')-C(9')-C(10')	109.3(8)	C(8')-C(9')-C(11')	110.9(8)
C(10')-C(9')-C(11')	113.4(8)	N(1')-C(10')-C(9')	115.6(8)
C(9')-C(11')-N(12')	109.6(8)	C(11')-N(12')-C(13')	111.8(8)
C(11')-N(12')-N(1')	107.3(77)	C(13')-N(12')-H(1')	98.8(81)
C(7')-C(13')-N(12')	110.8(8)		

Least-square planes with atomic deviations are listed in Table 3.5. Ring C exhibits a rigid chair conformation with atoms C-7, C-9, C-11 and C-13 lying in a plane. The latter makes an angle of 66.0^{o} (71.1^{o} in the second structure) with the A ring.

TABLE 3.5

Least square planes and atomic deviations from the planes of cytisine.

Least-squares Planes

Plane A : $-0.54435X + 0.80781Y + 0.22611Z = 12.14965$

Plane A': $0.77527X + 0.55303Y - 0.30515Z = 12.09000$

Plane C : $0.28269X + 0.82849Y - 0.48341Z = 9.05393$

Plane C': $-0.76225X + 0.60708Y + 0.22456Z = 10.78395$

Deviations of atoms from the planes (Å):

Plane	N(1) N(1')	C(2) C(2')	C(3) C(3')	C(4) C(4')	C(5) C(5')	C(6) C(6')
A	-.012	.008	.002	.007	.002	.008
A'	-.003	.006	.001	.009	.011	.006
C	-	-	-	-	-	-
C'	-	-	-	-	-	-

Plane	C(7) C(7')	C(9) C(9')	C(11) C(11')	C(13) C(13')	D(2) D(2')	C(10) C(10')
A	-0.33*	-.157*	-	-	.033*	-.094*
A'	-.066*	.091*	-	-	.025*	-.032*
C	-.010	.010	-.010	.010	-	-
C'	-.002	.002	-.001	.002	-	-

*Atoms not included in least-square plane calculation

The two independent molecules, in the asymmetric unit, are in perpendicular positions with respect to each other, the A and A' rings being 87.5^{O} apart. In the lattice, cytisine molecules are linkes together by the intramolecular hydrogen bonds shown in Fig. 3.2. Each molecule makes two hydrogen bonds of the type: O-2(x,y.z)--N'-12($\frac{1}{2}$+x,3/2-y,1-z) at a distance of 3.23 Å and N-12(x,y,z)--O-2'(1+x,y,z) at a distance of 3.20 Å.

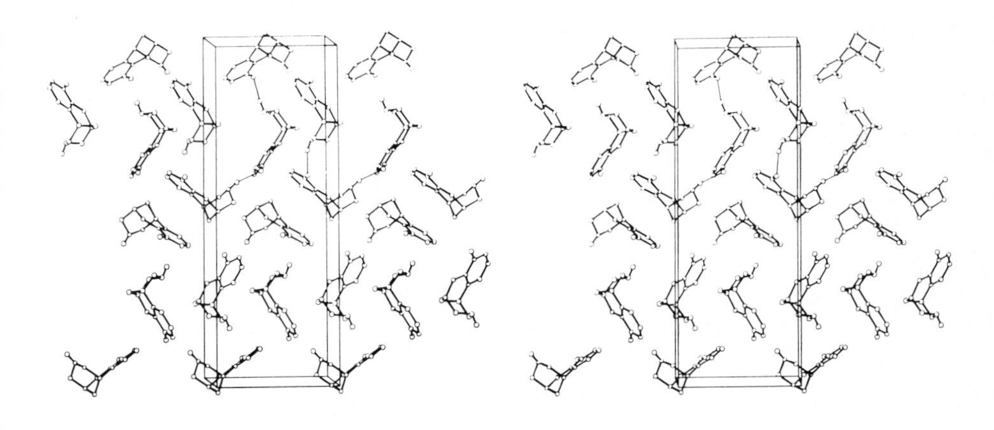

Fig. 3.2 Stereoview, approximately down C axis, of the molecular packing. Hydrogen bonds are shown only for the two molecules of the asymetric unit.

3.7 The solution conformation of cytisine

Molecular modelling based upon the interproton distances obtained from NMR measurements yielded a conformation for cytisine essentially identical to that obtained from x-ray data. This allowed unequivical assignments of dihedral angles from J coupling constants (ref. 28). A correlation of the latter with their respective theta angles is shown in Table 3.6 where the crystal angles are added for comparison purposes.

TABLE 3.6

Proton-proton coupling constants and dihedral angles of cytisine.

$^{1}H-^{1}H$	J(Hz)	θ_J [a]	x-ray [b]
3-5	1.41	-	-0.14 (-2.24)
3-4	9.49	-	0.96 (0.49)
4-5	7.06	-	-1.04 (-1.82)
7-8	2.54	53	63.5 (60.20)
7-8'	2.54	53	-59.65 (-60.23)
7-13(u)	2.42	54	58.94 (58.37)
7-13(d)	2.42	54	-60.87 (-60.47)
13(u)-13(d)	12.09	-	- -
8-8'	15.00[c]	-	- -
8-9	2.5	53	58.36 (57.23)
8'-9	2.5	53	-62.23 (-63.36)
9-10(u)	7.62	19	30.78 (38.13)
9-10(d)	0.8	73	-87.75 (-80.25)
10(d)-8	0.8	17	-27.39 (-46.55)
10(u)-11(u)	1.25	28	-24.87 (-16.15)
10(u)-10(d)	15.63	-	- -
9-11(u)	2.23	55	58.77 (62.04)
9-11(d)	2.35	54	-61.35 (-57.48)
11(u)-11(d)	12.60	-	- -
11(d)-8	0.91	25	-2.63 (2.98)
13(u)-8	1.29	28	0.00 (-0.61)

[a]dihedral angles (degrees) from J coupling constants (ref. 28)
[b]dihedral angles (degrees) from x-ray data
[c]J_{8-8} obtained from the spectrum of cytisine trifluoroacetate

4 CONCLUSIONS

In this paper we have undertaken a detailed analysis of the structure and conformation of cytisine in solution and solid state, as part of a wider program which involves isolation, structure and conformational analysis and structure-reactivity relationship studies of quinolizidine alkaloids.

Conformational and dynamic parameters of cytisine in chloroform solutions were obtained according to an approach successfully experimented for other classes of natural compounds (ref. 4,6-8). The results were then correlated with the crystal structure of cytisine.

Findings from the two sets of data indicate that in both phases the three rings of the alkaloid have a planar, half-chair and chair conformation respectively, with the proton at N-12 switching from an alpha to beta orientation in the solution structure.

In solution cytisine behaves like a rigid rotor whose motion and individual proton and carbon relaxation parameters are modulated by an unique correlation time. The magnitude of the latter was calculated using several independent methods and is consistent with an overall fast tumbling rate of $6.7 \pm 0.2 * 10^{-11}$ s^{-1}.

The net of hydrogen bonding seen in the lattice of cytisine crystals was not detected in the solution phase except, perhaps, at higher concentrations. Thus the relaxation figures for the carbon atoms obtained at concentrations in the 0.1 M range, although very similar to those obtained at lower concentrations, were not always reproducible (unpublished observations). This reflects aggregation in solution which would then have considerable effects on the motion and hence on the spin-lattice relaxation rates.

REFERENCES

1 I. Solomon, Relaxation process in a system of two spins, Physics Rev., 99 (1955) 559-565.
2 J.H. Noggle and R.E. Schirmer, The Nuclear Overhauser Effect, Academic Press, New-York-London, 1971.
3 R. Freeman, H.D. Hill, B.L. Tomlinson and L.D. Hall, Dipolar contribution to NMR spin-lattice relaxation of protons, J. Chem. Phys., 61 (1974) 4466-4473.
4 N. Niccolai, H.K. Schnoes and W.A. Gibbons, Study of the stereochemistry, relaxation mechanisms and internal motion of natural products utilizing proton relaxation parameters: solution and crystal structures of saxitoxin, J. Am. Chem. Soc., 102 (1980) 1513-1517.
5 C.M. Dobson, E.T. Olejniczak, F.M. Poulsen and R.G. Ratcliff, Time development of proton nuclear Overhauser effects in protein, J. Magn. Reson., 48 (1982) 97-110.
6 C.R. Jones, C.T. Sikakana, S. Hehir, M. Kuo and W.A. Gibbons, The quantitation of nuclear Overhauser effect methods for total conformational analysis of peptides in solution. Application to gramicidin S, Biophys. J., 24 (1978) 815-832.
7 P. Mascagni, W.A. Gibbons, D.H. Rich and N. Niccolai, The solution structure of Ala-4 -desdimethylchlamydocin: a H(1) NMR relaxation study, J. Chem. Soc. Perkin Trans. I (1985) 245-250.
8 P. Mascagni, A. Prugnola, W.A. Gibbons and N. Niccolai, Synthesis and solution structure of Val-3 -HC Toxin by H(1) and C(13) NMR relaxation parameters, J. Chem. Soc. Perkin

Trans. II, (1986) 1015-1019.
9 S. Macura and R.R. Ernst, Elucidation of cross-relaxation in liquids by two-dimensional NMR spectroscopy, Mol. Phys., 41 (1980) 95-117.
10 A. Kumar, G. Wagner, R.R. Ernst and K. Wuthrich, Buildup rates of the nuclear Overhauser effect measured by two-dimensional proton nuclear magnetic resonance spectroscopy: implications for studies of protein conformation, J. Am. Chem. Soc, 103 (1981) 3654-3658.
11 G.B. Young and J.L. James, Determination of molecular structures in solution via two-dimensional nuclear Overhauser effect experiments: proflavine as a rigid molecule test case, J. Am. Chem. Soc., 106 (1984) 7986-7988.
12 S. Macura, K. Wuthrich and R.R. Ernst, Separation and suppression of coherent transfer effects in two-dimensional NOE and chemical exchange spectroscopy, J. Magn. Reson., 46 (1982) 269-282.
13 G.A. Cordell, An Introduction to Alkaloids: a Biogenetic Approach, J. Wiley and Sons, New York, 1981.
14 F. Bohlmann, N. Ottawa, R. Keller, I. Nebel and J. Politti, Contribution to the synthesis of Cytisine: structure of tetrahydroquinolizone and bispidine, Ann., 587 (1954) 162-176.
15 S. Okuda, K. Tsuda and H. Kataoka, Absolute configuration of (-)-Cytisine and of related alkaloids, Chem. Ind., (1961) 1751.
16 F. Bohlmann and R. Zeisberg, Lupine alkaloids. XLI.carbon-13 NMR spectra of lupine alkaloids, Chem. Ber., 108 (1975) 1043-1051.
17 W.P. Aue, E. Bartholdi and R.R. Ernst, Two-dimensional spectroscopy. Application to nuclear magnetic resonance, J. Chem. Phys., 64 (1976) 2229-2246.
18 M. Rance, G. Bodenhausen, G. Wagner, K. Wuthrich and R.R. Ernst, A systematic approach to the suppression of J cross peaks in 2D exchange and 2D NOE spectroscopy, J. Magn. Reson. 62 (1985) 497-510.
19 W.P. Aue, J. Karhan and R.R. Ernst, Homonuclear broad band decoupling and two-dimensional J-resolved NMR spectroscopy, J. Chem. Phys. 64 (1976) 4226-4227.
20 R. Freeman and G.A. Morris, Experimental chemical shift correlation in NMR spectroscopy, J. Chem. Soc. Chem. Commun., (1978) 684-686.
21 G.M. Sheldrick, Program for Crystal Structure Determination, Cambridge University, 1976.
22 International Tables for X-ray Crystallography, Vol. 4, Kynoch Press, Birmingham, 1974.
23 C.K. Johnson, ORTEP report ORNL-3494, Oak National Laboratories, Oak Ridge Tn, 1965.
24 K.F. Kuhlmann and D.M. Grant, The nuclear Overhauser enhancement of the C-13 magnetic resonance spectra of formic acid, J. Am. Chem. Soc., 90 (1968) 7355-7357.
25 D. Doddrell, V. Glushko and A. Allerhand, Theory of NOE and Carbon-13-proton dipolar relaxation in proton decoupled Carbon-13 NMR spectra of macromolecules, J. Chem. Phys. 56 (1972) 3683-3689.
26 L. Werbelow, and D.M. Grant, Multiple relaxation studies in anisotropic media, Chem. Phys. Lett., 54 (1978) 571-574.
27 H.E. Bleich, K.R.K. Easwaran and J.A. Glasel, The contributions to amide proton spin-lattice relaxation in small peptides, J. Magn. Reson., 31 (1971) 517-522.

28 M. Karplus, Vicinal proton coupling in nuclear magnetic resonance, J. Am. Chem. Soc., 85 (1963) 2870-2871.

APPLICATION OF 2D NMR TECHNIQUES TO STRUCTURE DETERMINATION OF NATURAL PRODUCTS

Motoo Tori

1. INTRODUCTION

The recent development of superconducting magnets and computer techniques has enabled us to apply complicated NMR methods (ref.1). By combination of pulse sequences for both ^{1}H and ^{13}C atoms as well as other nuclei it has been possible to record not only homo- but also hetero-nuclear correlated spectra. These days, homo-nuclear shift correlation spectra are used routinely for structure determination of natural products. However, the number of examples of hetero-nuclear correlation spectra applied to this problem is increasing at an accelerated pace. In this report, practical applications of these methods to the field of natural products are described (ref.2).

2. STRUCTURE OF A DITERPENE HEMIACETAL, SACCULAPLAGIN

Liverworts are rich sources of various types of terpenoids and/or aromatic compounds (ref.3). Plagiochila acanthophylla subsp. japonica has once been studied by a Japanese group (ref.4). They reported relatively less polar chemical constituents such as maalioxide, bicyclohumulenone, cyclocolorenone. We have reinvestigated this species in order to look for more polar components. Extraction of fresh material with methanol, re-extraction with ethyl acetate from this crude methanol extract, and repeated column chromatography over silica gel led to the isolation of five new compounds. Four of these five components, fusicoplagins A - D, have been revealed to possess the fusicoccane skeleton and the results were published in 1985 (ref.5).

The final component, named sacculaplagin (**1**), mp 184-185°; $[\alpha]_D$=+25.9° ($CHCl_3$, c=1.16); IR (KBr) 3400, 1735, 1710, 1270,

1238, and 1030 cm^{-1}, showed the molecular formula of $C_{24}H_{38}O_7$ as judged by chemical ionization - high resolution mass spectroscopy (ref.6) suggesting a diterpene diacetate. The ^{1}H-NMR showed the presence of two acetoxyl groups, hemiacetal proton, exo methylenes, and other protons attached to the carbon bearing oxygen functions (Fig.1). However, the triacetate (2), $[\alpha]_D$=+28.6 ($CHCl_3$, c=2.17); IR (film) 3500, 1735, 1730, 1365, 1240, and 1225 cm^{-1}, was more easily purified by column chromatography of the acetylation products of the crude fraction. Therefore, the structure analysis has been performed on the triacetate (2). The ^{1}H-NMR spectrum showed three acetyl groups (δ 2.03, 2.07, and 2.09), exo methylene protons (δ 4.89 and 4.99), three t-methyl groups (δ 0.79, 1.18, and 1.18), and three methine protons attached to the carbon bearing oxygen functions. The complete decoupled ^{13}C NMR and the INEPT spectra indicated that there were six methyls (δ 18.9, 20.9, 21.5, 21.6, 25.4, and 26.4) including acetyl groups, seven methylenes (δ 18.7, 22.7, 36.8, 36.8, 38.4, 40.3, and 75.6), five methines (δ 49.5, 62.5, 70.2, 79.6, and 104.4), two olefinic carbons (δ 114.0 and 140.5), three quaternary

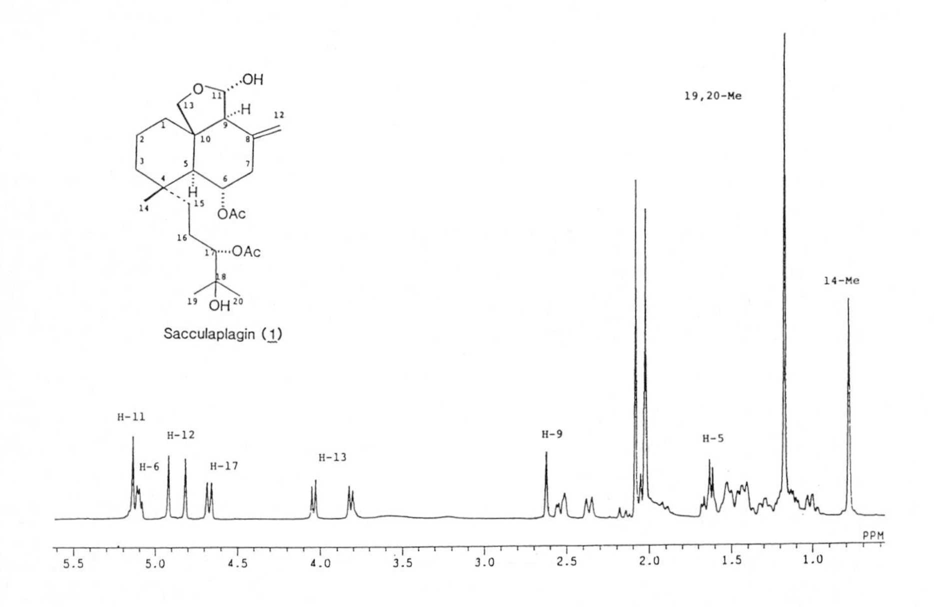

Fig. 1. The 400 MHz ^{1}H NMR of sacculaplagin (1).

carbons (δ 36.5, 44.8, and 72.1), and three carbonyl groups (δ 169.9, 170.3, and 171.0). As is evident from the COSY and NOESY spectra (Fig. 2 and 3), the methine proton resonating at δ 5.90 (s, H-11) was very close to the methine at δ 2.72 (s, H-9), which was also in the proximity of one of the exo methylene protons (δ 4.99, H-12). The fact that the proton at around δ 2.5, coupled with the methine at δ 5.12 (dd, J=7.6 and 6.1 Hz, H-6β), showed NOE with the other exo methylene proton (δ 4.89) suggests three

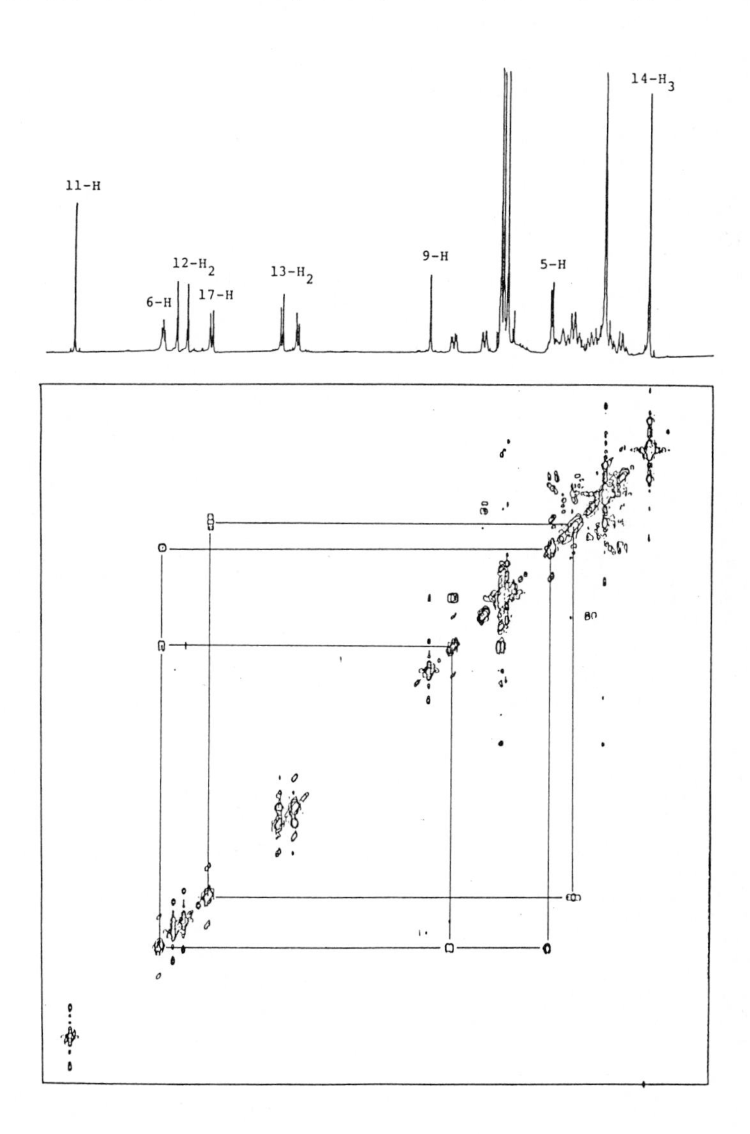

Fig. 2. The COSY spectrum of sacculaplagin triacetate (2).

partial structures A, B, and C, are connected to each other as shown (A-C) in Fig.4. These data can not be explained by the fusicoccane type skeleton isolated earlier(ref.5).

The $^{13}C-^{1}H$ correlation spectrum and the long-range $^{13}C-^{1}H$ correlation spectrum (J=10 Hz) are shown in Fig. 5 and 6, respectively. As the two t-methyl groups have long-range couplings with C-18 and C-17, these must be gem-dimethyls, attached

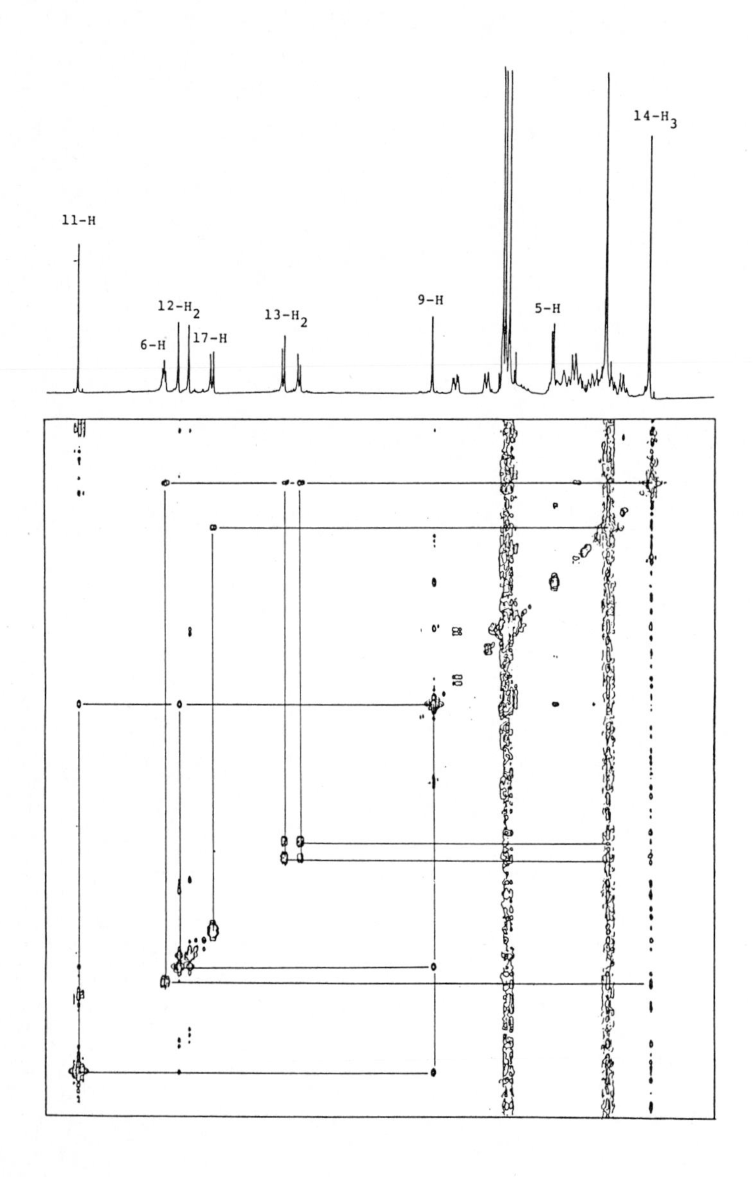

Fig. 3. The NOESY spectrum of sacculaplagin triacetate (2).

to C-18 (quaternary), and located near C-17 (methine). Thus, the partial structures, D and E, were connected. The other t-methyl group (C-14) has long-range couplings with C-5 (methine) and C-4 (quaternary). Similarly, the proton at H-5 showed long-range couplings with C-6 (methine), C-10 (quaternary), C-4 (quaternary), and C-14 (methyl). Further correlation between the proton at H-9 and C-8 (quaternary), C-12 (exo-methylene), and C-1 (methylene) indicated that the partial structures, A, B, and C, should be connected to A - C (Fig. 4) as discussed before. Although these spectra suggest the sacculatane skeleton, it is not fully proved. Therefore, we next planned to record the INADEQUATE spectrum.

```
 H    OAc          H H              OAc
 |     |         H  \/               |
■-C-O-C-         |  ||              -C-
 |     |        -C——C-               |
 H     H         |                   H

    A               B                C


  CH3             OAc
  |                |
 -C-OH           ■-C-
  |                |
  CH3              H

    D                E


 H    OAc  H   H H    H OAc  H
 |     |   |    \/    |  |   |
■-C-O-C————C————C————C-C————C-■
 |     |   |    ||    |  |   |
 H     H   ■          H  H   ■

            A - C
```

Fig. 4. Partial structures of sacculaplagin triacetate (2).

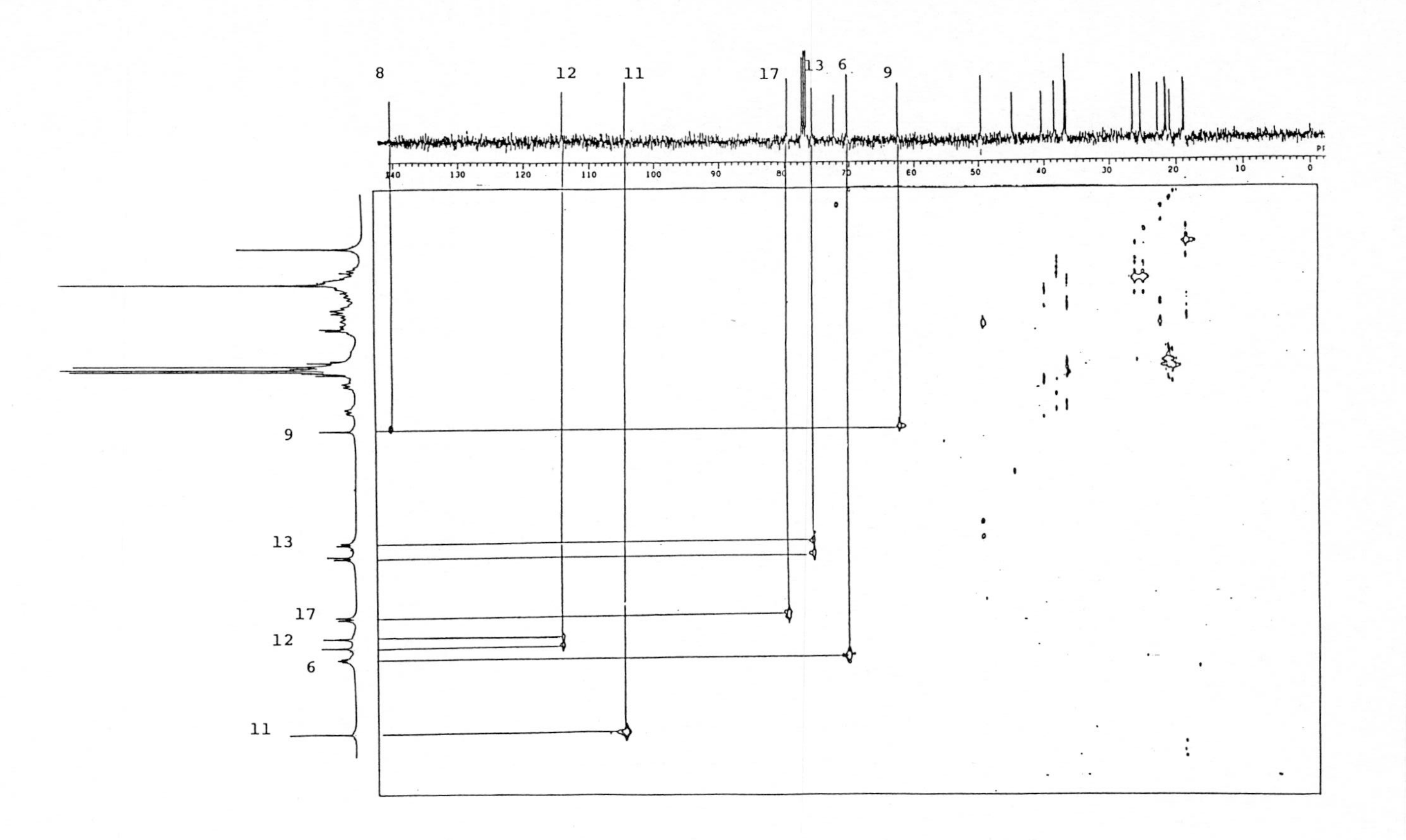

Fig. 5. The ^{13}C-^{1}H correlation spectrum of sacculaplagin triacetate (2).

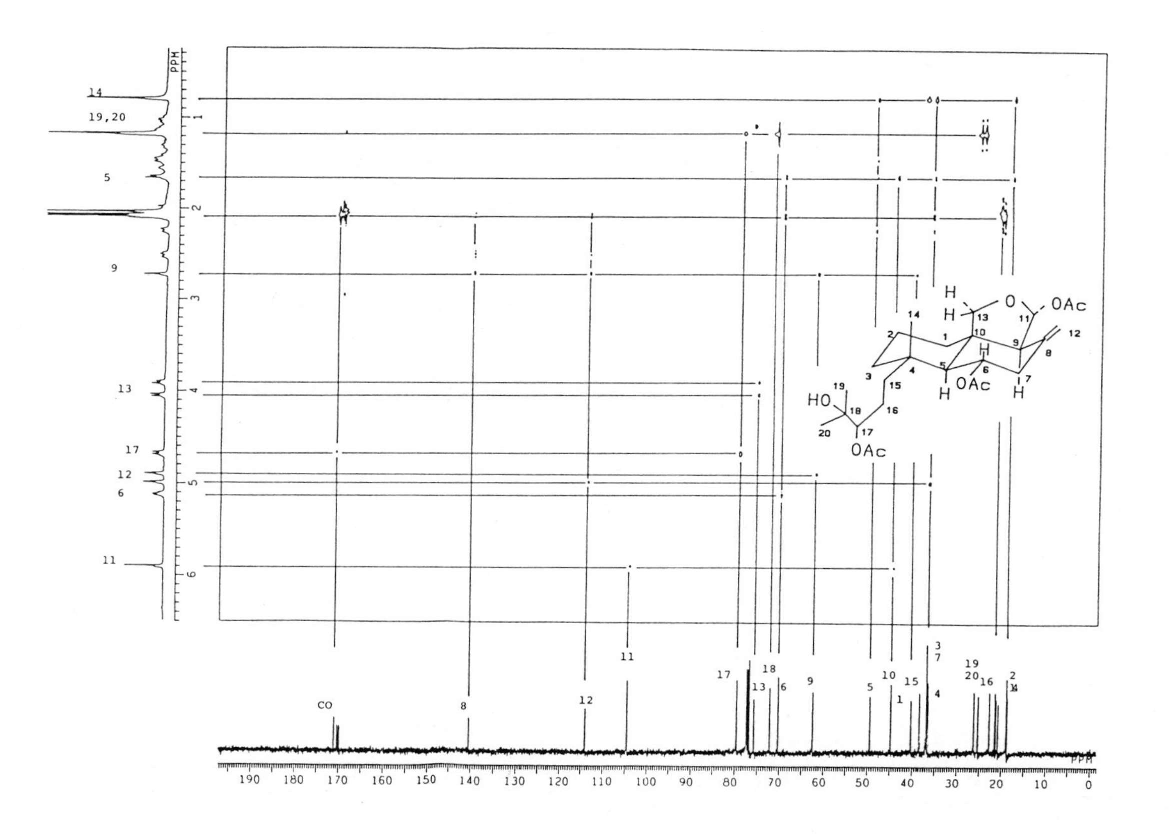

Fig. 6. The long-range $^{13}C-^{1}H$ correlation spectrum of sacculaplagin triacetate (2).

Collection of a large amount of the liverwort enabled us to isolate 110 mg of 2, with which 2D-INADEQUATE was performed successfully. The result is shown in Fig.7. The C-C connectivities were determined unambiguously except for the three carbons (δ 36.7, 36.9, and 36.9), which were congested in a narrow region. However, this ambiguity was clearly overcome with the aid of ^{13}C-^{1}H correlation and COSY spectra. Therefore, 2 was established to have the sacculatane skeleton as shown.

From the NOESY spectrum (Fig. 3), it was shown that the methyl group at δ 0.79 (Me-14) was located close to the acetal methylene protons (H-13) and to the methine at δ 5.12 (H-6β) and that the methine at δ 1.65 (H-5α) was very close to the methine at

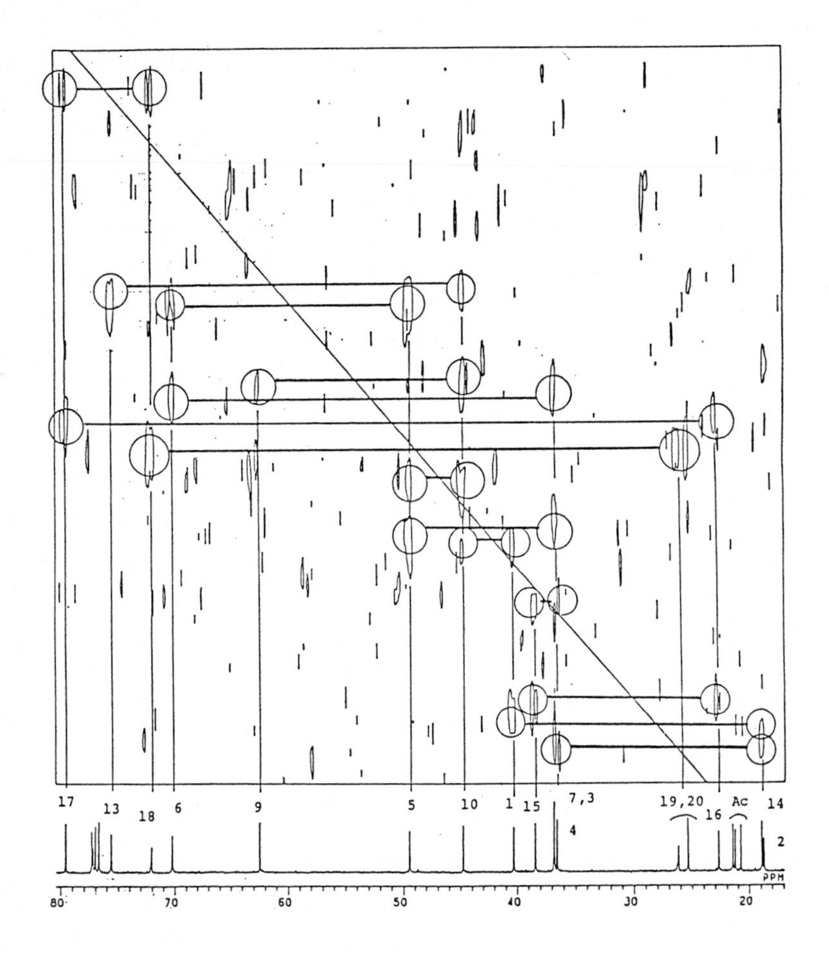

Fig.7. The 2D-INADEQUATE spectrum of sacculaplagin triacetate (2).

δ 2.72 (H-9α). These facts led to the conclusion that the stereostructure of 2 should be expressed as shown except for the configuration at C-17. The configuration at C-11 was determined to be α orientation, as the proton at C-11 was a singlet (90° to H-9α).

The absolute configuration was determined as shown because the lactone 3, mp 141-142°; $[\alpha]_D$=-41.0°($CHCl_3$, c=0.9); IR (KBr) 3525, 1780, 1735, 1375, and 1255 cm^{-1}, obtained by chromic acid oxidation of 1, showed a negative Cotton effect at λ_{max}231 nm ($\Delta\varepsilon$ -16.3). The final problem of the absolute configuration at C-17 was solved as follows. The triacetate 2 was converted to the diol 5 through the acetonide 4 (Fig. 8). The CD spectrum of 5 in CCl_4 in the presence of $Eu(fod)_3$ showed a positive Cotton effect at λ_{max}322 nm ($\Delta\varepsilon$ +7.13). Hence the absolute configuration at C-17 was assigned to be S according to Nakanishi's method (ref.7).

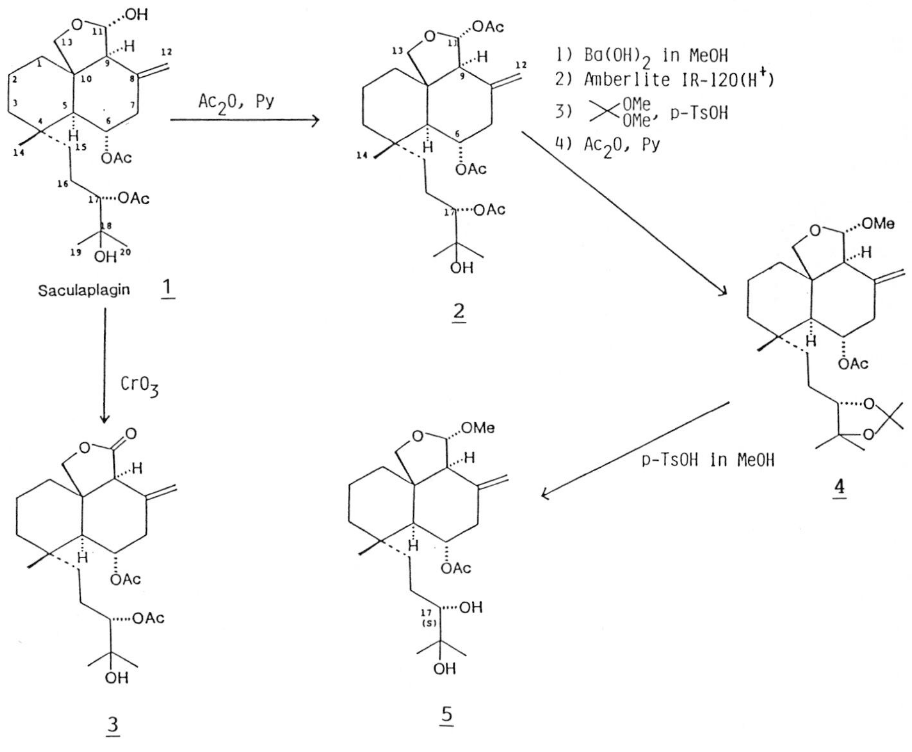

Saculaplagin 1

Ac_2O, Py

2

1) $Ba(OH)_2$ in MeOH
2) Amberlite IR-120(H^+)
3) $\times^{OMe}_{OMe}$, p-TsOH
4) Ac_2O, Py

CrO_3

3

5

p-TsOH in MeOH

4

Fig. 8. The derivatives of sacculaplagin (1).

Sacculatane diterpenoids have been isolated from the liverworts *Trichocoleopsis*, *Pellia*, and *Porella* species and they exhibit potent bitter taste and ichthyotoxicity. This is the first example of the sacculatane isolated from *Plagiochila* species and it is also interesting from the biogenetic point of view to note that both fusicoccane and sacculatane diterpenes were isolated from this liverwort.

3. STRUCTURE DETERMINATION OF OXIDATION PRODUCTS

3.1 Oxidation of unactivated carbon atoms by m-chloroperbenzoic acid

Recent applications of *m*-chloroperbenzoic acid (*m*CPBA) to oxidation of unactivated carbon atoms have appeared in the field of natural products (ref.8). The method used was very simple, safe, and applicable to a variety of compounds. When a compound having no double bond or carbonyl group was allowed to react with

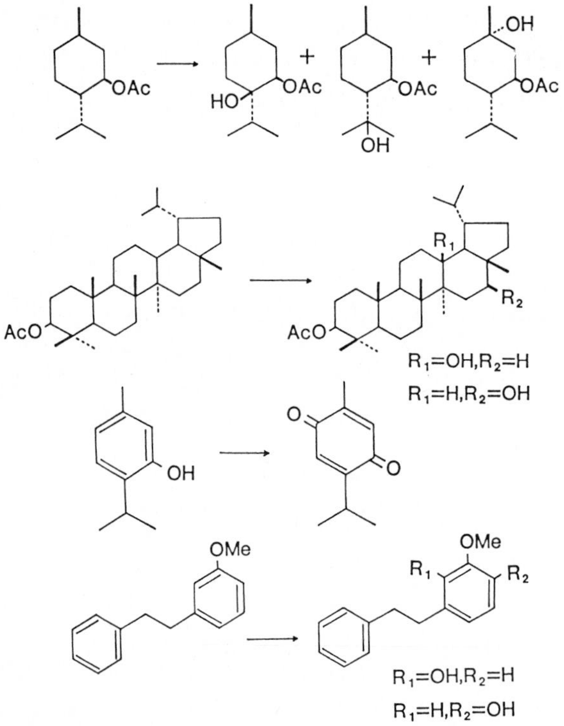

Fig. 9. Examples of *m*CPBA oxidation.

*m*CPBA in chloroform under reflux, introduction of a hydroxyl group into a tertiary position preferentially and/or a secondary position in some cases occurred. In the case of aromatic compounds, if it is a phenolic material, oxidation into a quinone was a major route. However, in the case of those protected by an ether, hydroxyl groups were introduced at *o*- or *p*-position of the protected hydroxyl groups. The secondary alcohols were sometimes further oxidized to give a ketone or a lactone in a cyclic case. Thus *l*-menthol gave three *t*-alcohols under these conditions as well as menthone and the corresponding lactone. However, *l*-menthyl acetate afforded only *t*-alcohols. Several examples were shown in Fig. 9.

3.2. Oxidation products of dammarane triterpenoids

Dammaran-20(*S*)-ol (6) and 20(*S*)-hydroxydammaran-3β-yl acetate (7) were allowed to react with *m*CPBA in chloroform under

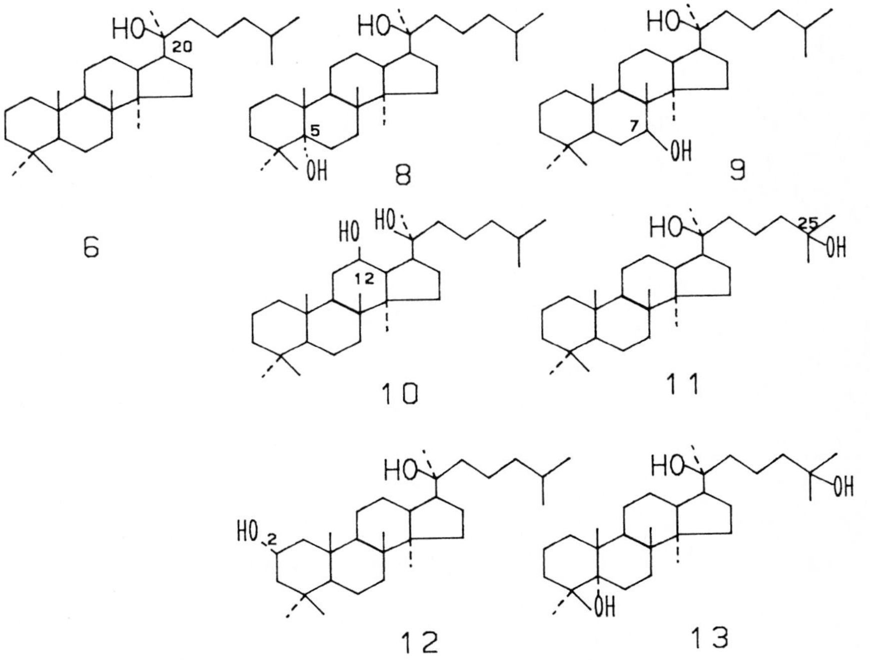

Fig. 10. Products from the oxidation reaction of dammaran-20(*S*)-ol (6) with *m*CPBA.

reflux for 6 h. After the usual work up and chromatography, alcoholic products were isolated. The products from **6** and **7** were summarized in Fig. 10 and 11. Although the structures were determined mainly by analysis of the ^{1}H NMR data, the ^{13}C NMR data were very important for these purposes. Therefore, the assignment of the ^{13}C NMR data of the parent compounds must be free of errors.

As indicated in the INEPT spectra of **6**, shown in Fig. 12, the carbons at δ 24.8, 25.6, and 27.6 were CH_2, CH_3, and CH_2, respectively. Tanaka et al. reported the assignment of dammarane triterpenes mainly by using off-resonance and shift reagent tech-

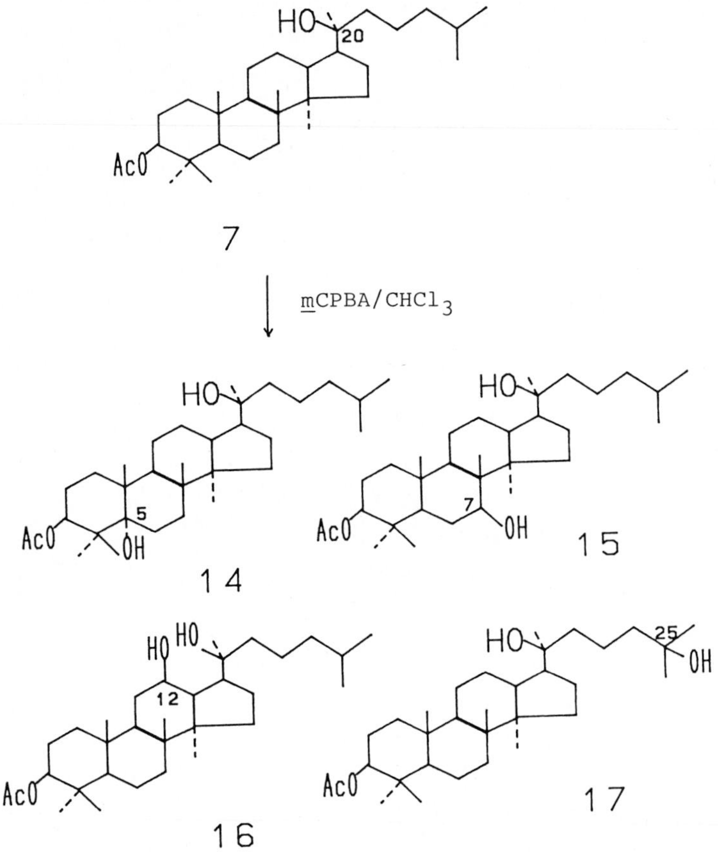

HO 20 AcO 7 mCPBA/$CHCl_3$ HO 5 AcO OH 14 HO 7 AcO OH 15 HO HO 12 AcO 16 HO 25 OH AcO 17

Fig. 11. Products from the oxidation reaction of 20(S)-hydroxydammaran-3β-yl acetate (7).

niques and the signals at δ 24.7 and 25.4 were assigned to C-21 and C-12, respectively (ref.9). It is now clear from the INEPT spectra that these assignments should be replaced. In the case of triterpenes most ^{13}C NMR signals appear in the highly congested region between 10 and 50 ppm and it is hence difficult to determine the multiplicities from the SFORD spectrum. The INEPT or DEPT pulse sequences are the best choice for such compounds.

These revised assignments look nice superficially. However, we thought that the complete analysis was necessary to be free of errors. The $^{13}C-^{1}H$ correlation and long-range $^{13}C-^{1}H$ correlation

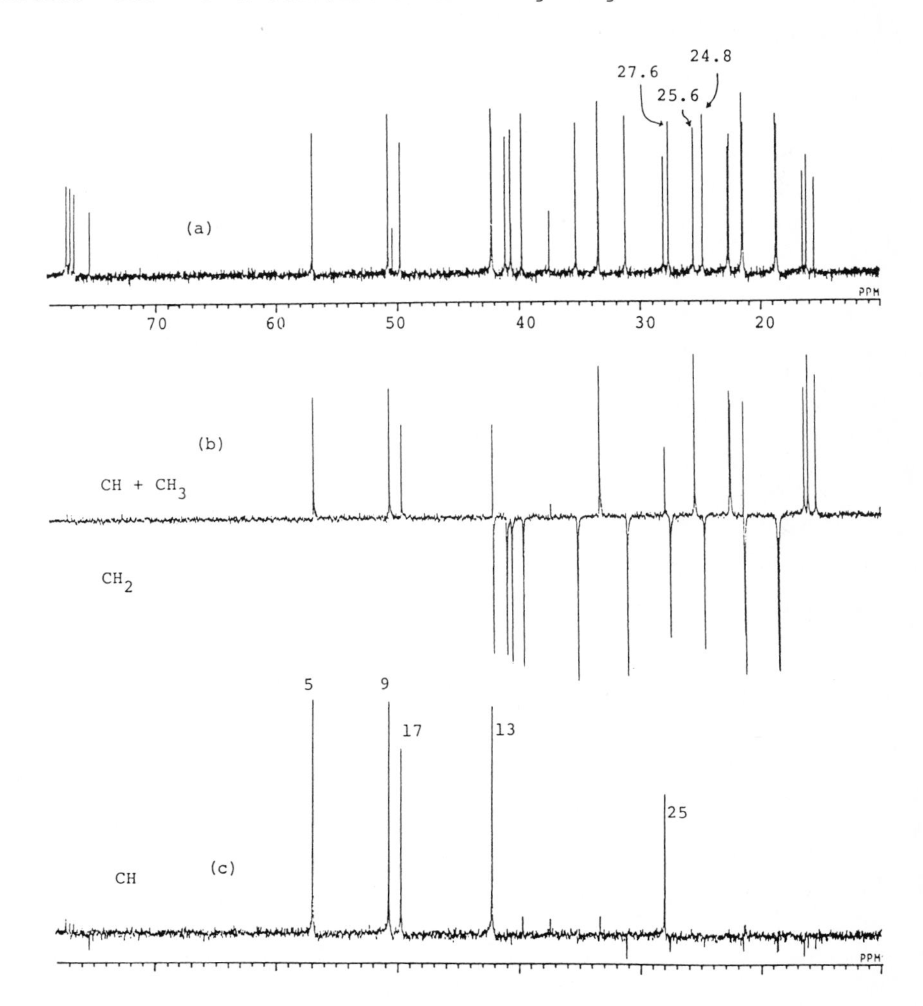

Fig. 12. (a) The complete decoupled ^{13}C NMR spectrum of dammaran-20(S)-ol (6). (b) The INEPT spectrum showing CH and CH_3 up and CH_2 down. (c) The INEPT showing CH up.

spectra of 20(S)-hydroxydammaran-3-one (18) are shown in the Fig. 13 and 14. The two methyl groups having long-range couplings with the carbonyl group at C-3 must be *gem*-dimethyls (C-28 and C-29). Then, the quaternary carbon having long-range couplings with these methyl groups should be assigned to C-4 and the methine to C-5. The methyl group (C-19) was assigned, because the protons of the C-19 methyl group had a long-range coupling with the C-5 signal. Hence, by similar reasoning assignments were made to the quaternary carbon at C-10, the methine at C-9, the quaternary carbon at C-8 etc. The carbons at C-1 and 2 were distinguished by the ^{13}C-^{1}H correlation spectrum, since the protons attached to those carbons had been already assigned by the COSY spectrum. However, the methylene carbons at C-6, C-11, C-12, C-16, and C-23 were not determined. Therefore, we have measured the 2D-

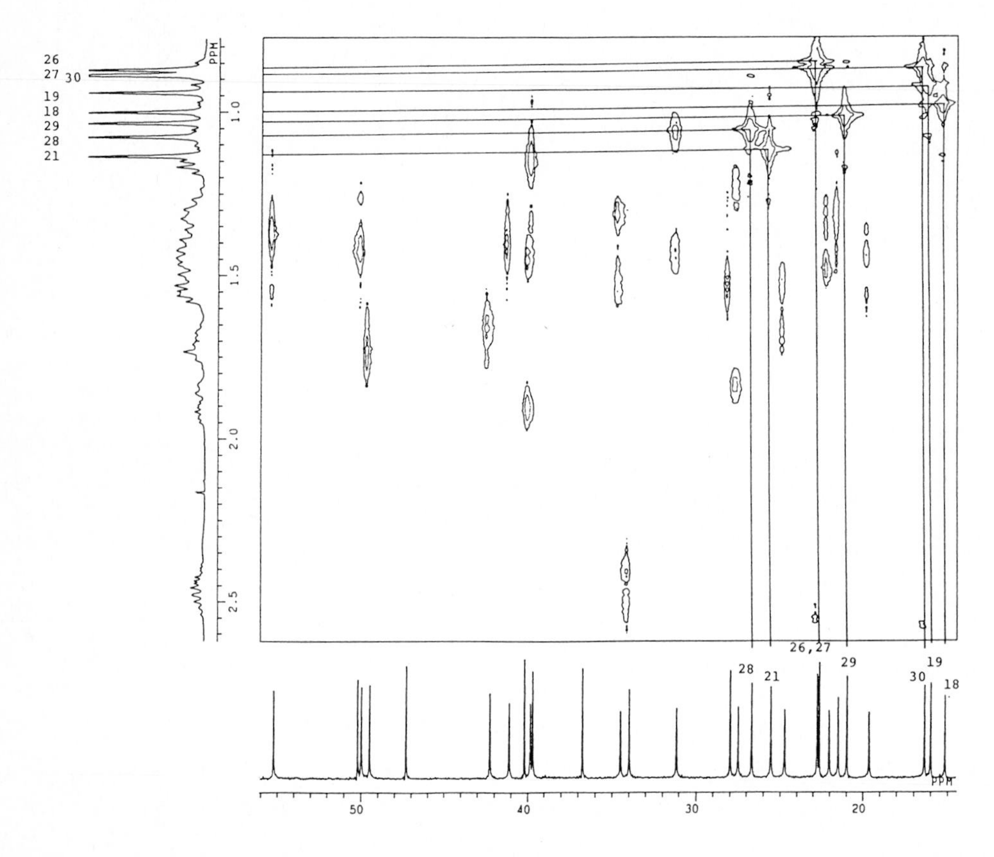

Fig. 13. The ^{13}C-^{1}H correlation spectrum of 20(S)-hydroxydammaran-3-one (18).

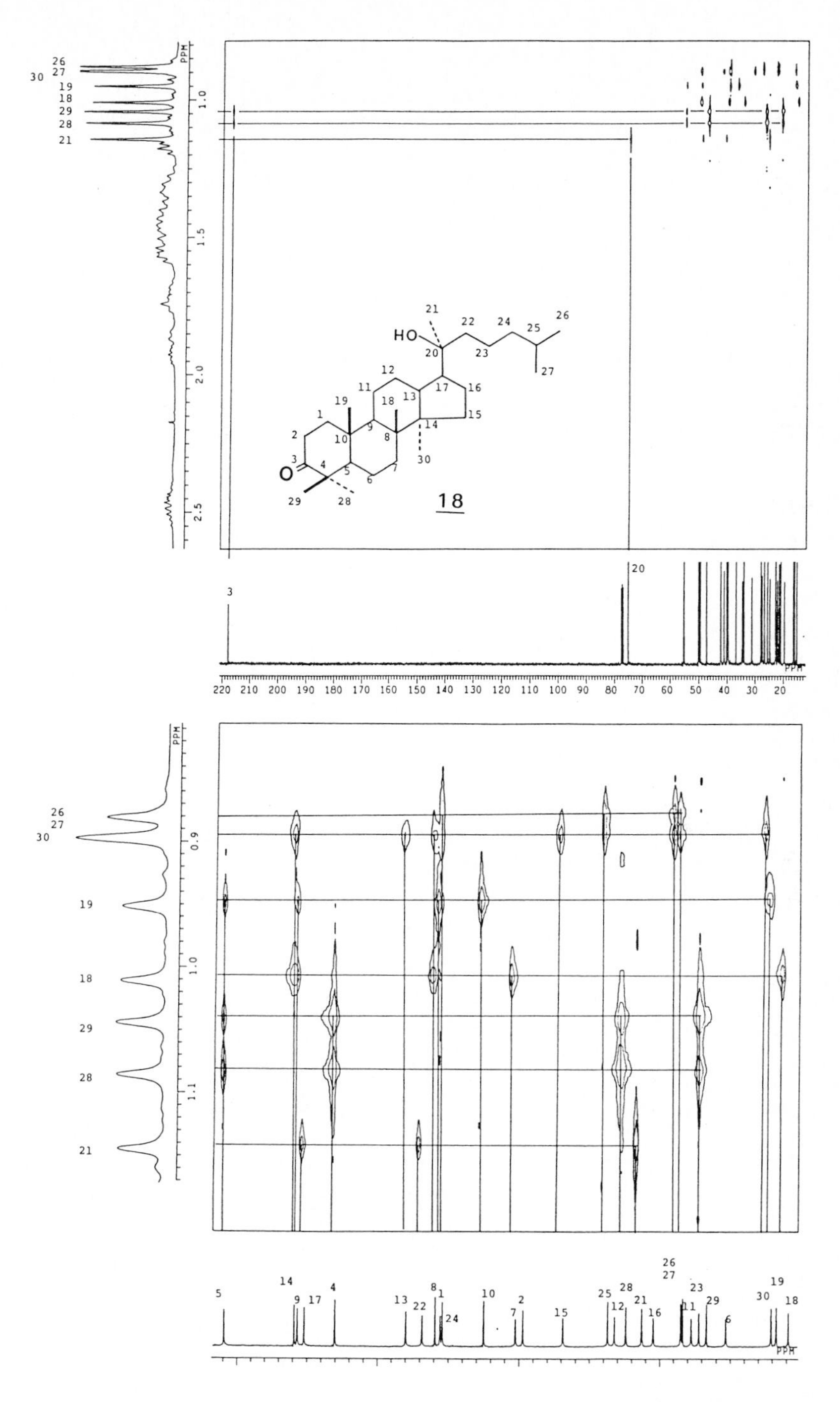

Fig. 14. The long-range $^{13}C-^{1}H$ correlation spectrum of 20(S)-hydroxydammaran-3-one (18) (J=10 Hz).

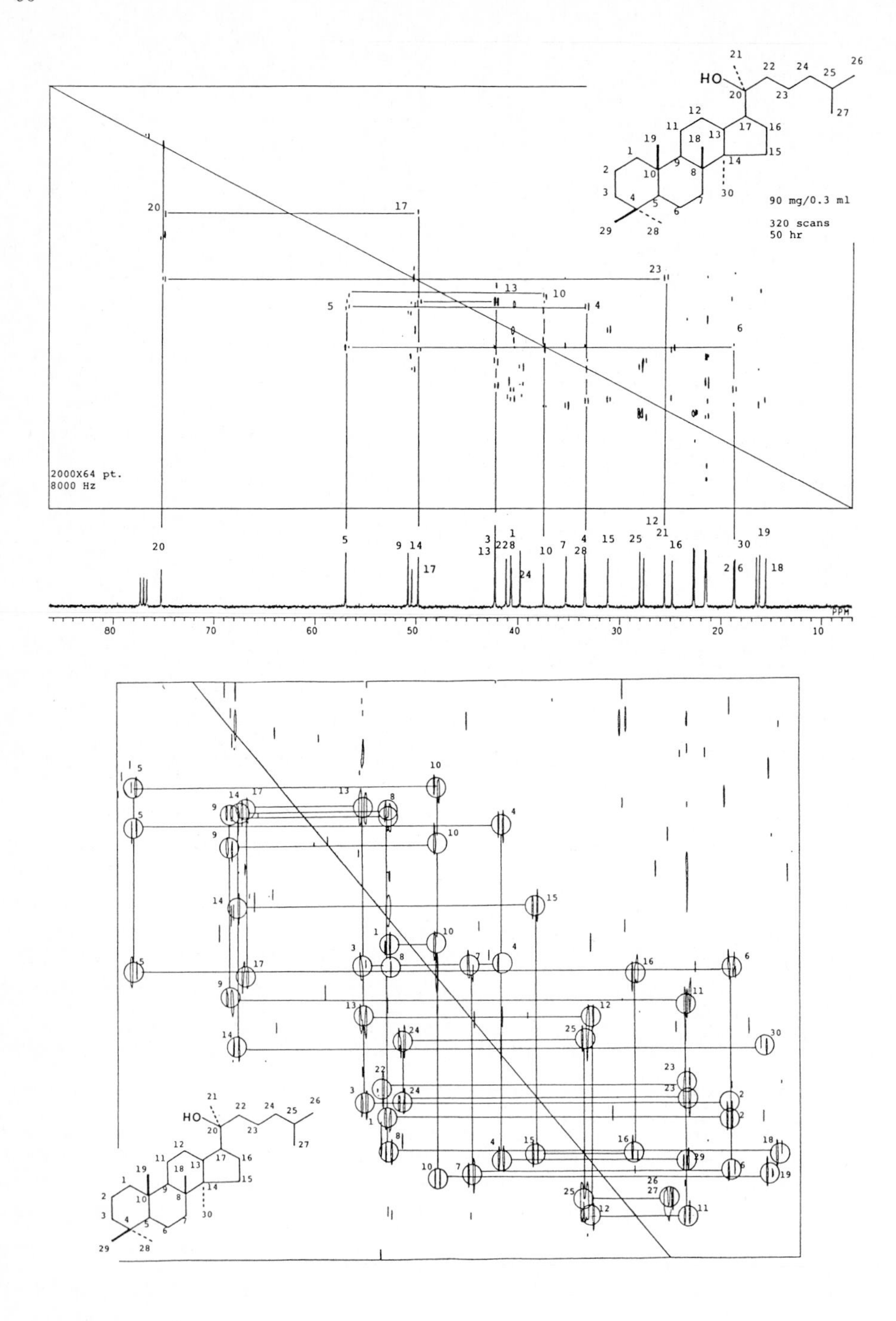

Fig. 15. The 2D INADEQUATE spectrum of dammaran-20(S)-ol (6).

TABLE 1.
^{13}C NMR data of dammarane derivatives.

	6	8	9	10	11	12	13	14	15	16	17	18
1	40.6	32.9	40.5	40.5	40.7	49.9	32.8	32.1	36.9	37.9	38.8	39.5
2	18.7	17.9	18.7	18.7*	18.7	65.3	17.9	23.6	21.4	23.7	23.7	33.8
3	42.2	36.2	41.9	42.1	42.2	51.2	36.1	77.2	80.6	80.8	81.0	217.6
4	33.4	38.5	33.2	33.3	33.4	39.7	38.4	39.3	38.5	38.6	38.0	47.1
5	57.0	77.3	54.3	57.0	57.0	56.4	77.4	78.9	53.3	55.9	56.0	55.1
6	18.7	23.0	28.2	18.6*	18.7	18.3	23.0	23.4	27.7	18.2	18.2	19.4
7	35.3	28.7	75.3	34.8	35.3	35.0	28.6	28.4	75.1	34.8	35.2	34.3
8	40.6	40.9	46.2	39.9	40.7	41.0	41.0	40.7	46.0	39.8	40.4	40.0
9	50.8	42.0	48.8	50.2	50.8	50.7	41.9	42.0	48.7	50.0	50.6	49.8
10	37.5	39.6	37.3	37.4	37.5	42.1	39.6	43.1	37.7	37.0	37.1	36.5
11	21.5	21.5	21.2	31.2	21.5	21.5	21.5	21.7	21.3	31.2	21.6	21.8
12	27.6	27.8	27.4	71.2	27.7	27.5	27.8	27.8	27.3	70.9	27.5	27.3
13	42.2	42.3	43.0	47.7	42.4	42.1	42.3	42.4	43.0	47.8	42.3	42.1
14	50.4	50.9	49.7	51.7	50.4	50.4	50.8	50.8	49.7	51.6	50.3	50.0
15	31.2	31.3	34.7	30.9	31.2	31.1	31.3	31.3	34.7	31.0	31.2	31.0
16	24.8	24.8	25.3	26.6	24.9	24.8	24.8	24.8	23.7	26.5	24.9	24.5
17	49.8	49.8	50.4	53.5	50.0	49.7	49.9	49.8	50.1	53.5	49.9	49.3
18	15.6	15.7	9.8	15.6	15.6	15.6	15.7	15.7	9.7	15.7	15.5	15.8
19	16.2	19.2	16.1	16.1	16.2	17.3	19.2	19.3	16.2*	16.2+	16.3+	15.0
20	75.3	75.4	75.4	74.4	75.4	75.4	75.5	75.4	75.4	74.5	75.4	74.9
21	25.6	25.6	25.7	27.1	25.6	25.6	25.5	25.7	25.8	27.2	25.6	25.4
22	41.1	41.1	40.9	35.1	41.1	40.5	40.9	41.2	40.9	35.0	41.1	40.9
23	21.5	21.5	21.5	21.3	18.5	21.6	18.5	21.5	21.5	21.3	18.5	21.3
24	39.8	39.8	39.7	39.8	44.6	39.1	44.5	39.8	39.7	39.8	44.5	39.6
25	28.0	28.1	28.0	28.2	71.1	28.0	71.1	28.1	28.0	28.2	71.0	27.8
26	22.7	22.6	22.6*	22.6+	29.3*	22.6*	29.3*	22.7*	22.6+	22.6*	29.3*	22.4*
27	22.7	22.6	22.7*	22.7+	29.5*	22.7*	29.4*	22.8*	22.7+	22.7*	29.5*	22.5*
28	33.4	27.9	33.3	33.4	33.4	33.5	27.9	22.6	28.0	28.0	28.0	26.5
29	21.5	23.9	21.5	21.5	21.6	22.4	23.9	18.5	16.5*	16.5+	16.5+	20.8
30	16.5	16.7	16.4	16.9	16.6	16.4	16.7	16.6	16.3*	16.8+	16.5+	16.2
CO								170.9	171.0	171.0	171.0	
CH_3								21.3	25.3	21.3	21.3	

*,+ Assignments may be reversed.

INADEQUATE spectrum of dammarane-20(S)-ol (6) (Fig. 15). This, of course, clearly established all the signal assignment as shown making another revision necessary. The signal at δ 24.8 must be due to C-16, not C-12, while the signal at δ 27.6 should be assigned to C-12. This means that the three signals assigned originally by Tanaka et al. must be revised as shown in Table 1 (ref. 10).

3.3 Oxidation products of dendropanoxide

Dendropanoxide is a pentacyclic triterpene oxide having a D:B-friedooleanane skeleton isolated from *Dendropanax trifidus*. Dendropanoxide (19) was treated with mCPBA in chloroform under reflux to afford four compounds, which were all secondary alcohols judging from IR, MS, and ^{1}H NMR data. The positions of the hydroxyl groups of the two of the four products were deduced to be

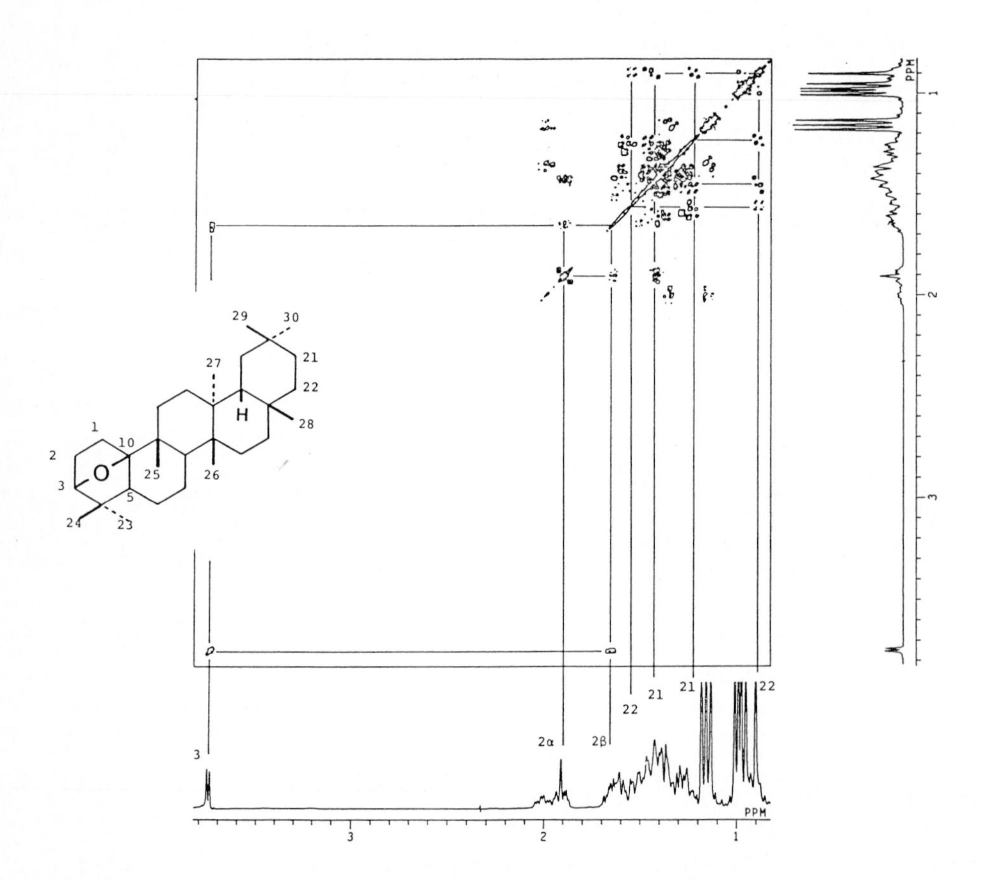

Fig. 16. The COSY spectrum of dendropanoxide (19).

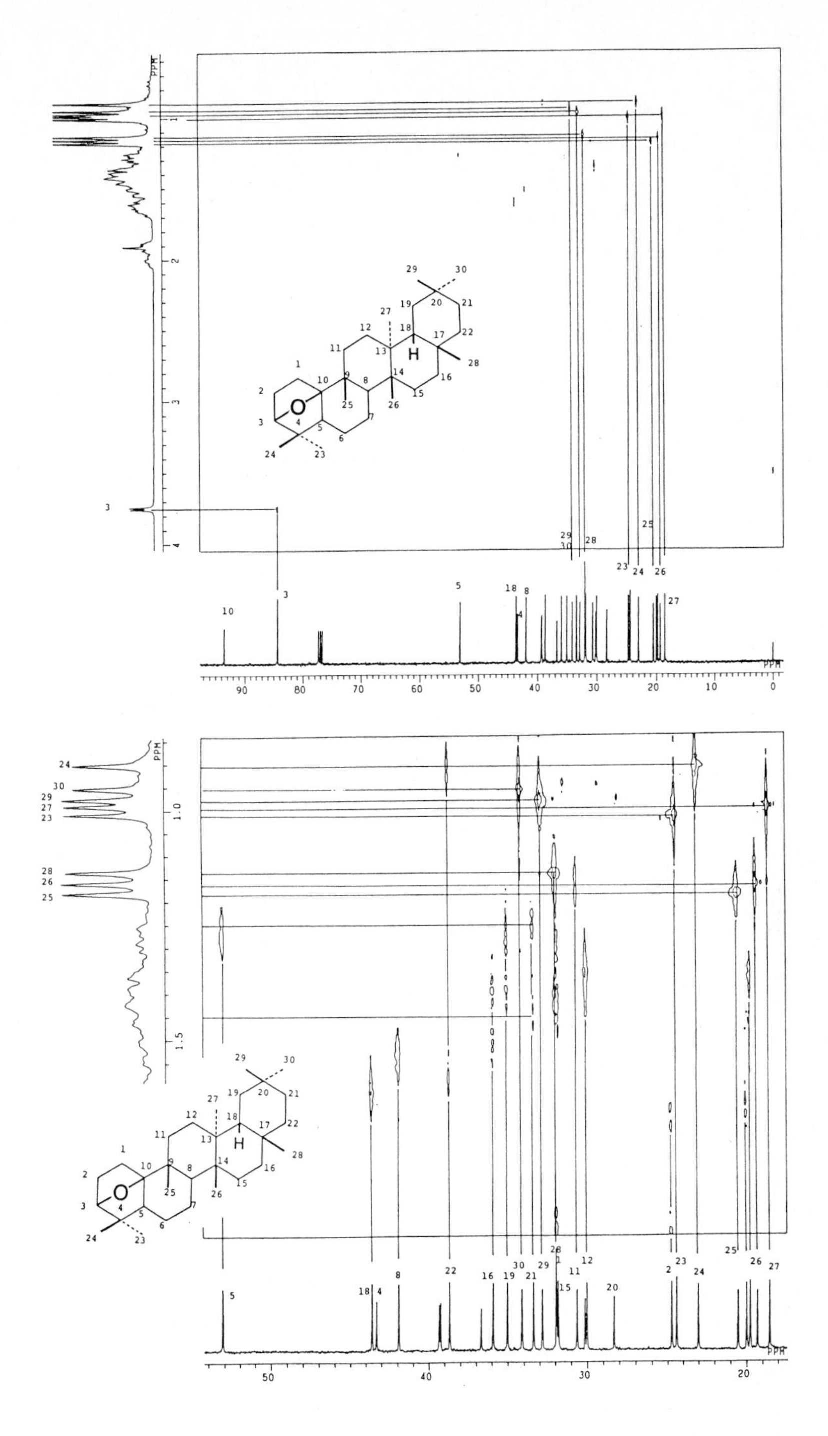

Fig. 17. The ^{13}C-^{1}H correlation spectrum of dendropanoxide (19).

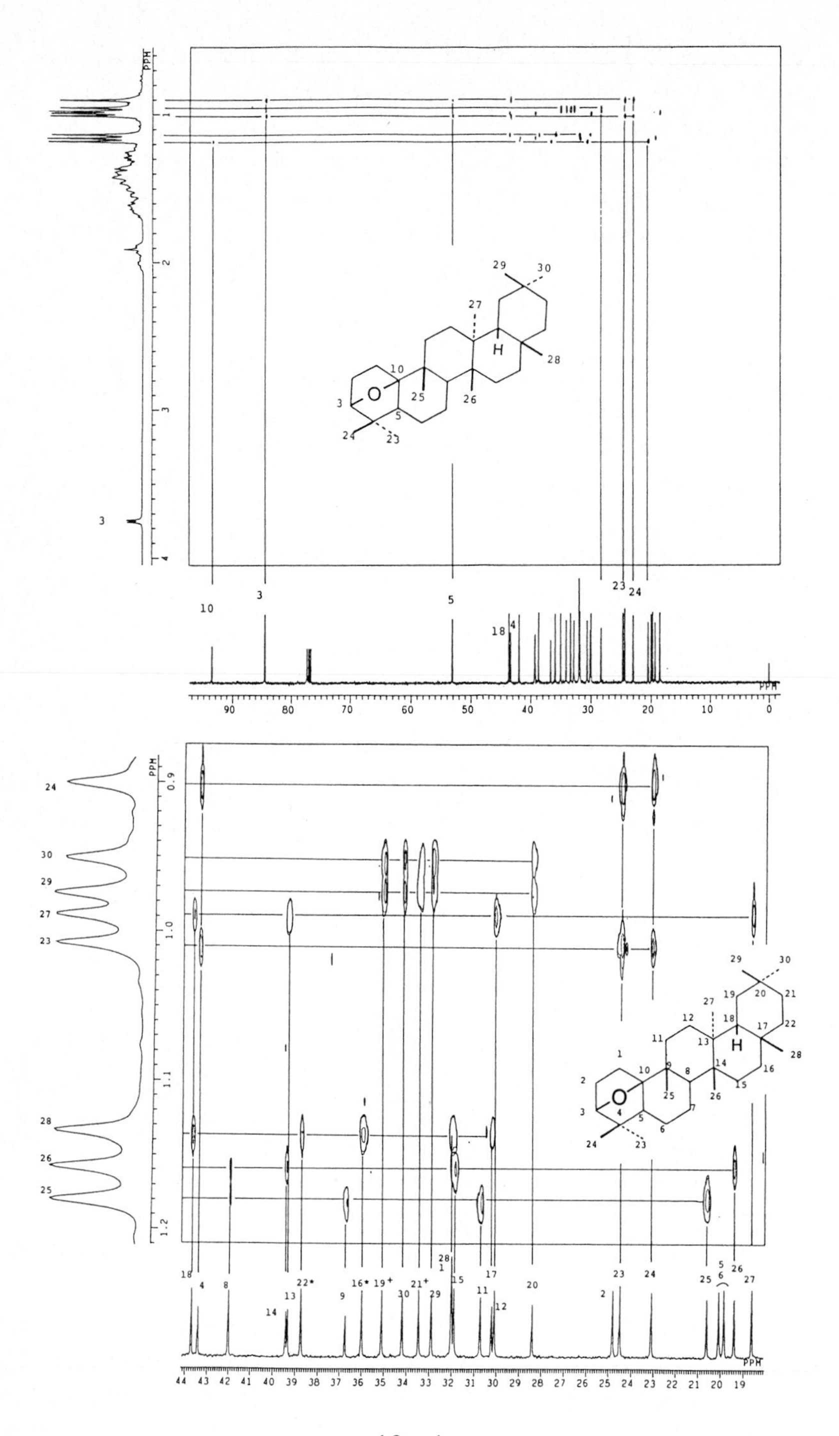

Fig. 18. The long-range $^{13}C-^{1}H$ correlation spectrum of dendropanoxide (19).

either C-6 or C-7 by analysis of the methine proton attached to the carbon bearing the hydroxyl group. By measuring the NOE, they were determined to be 6α-OH and 7α-OH derivatives. However, in the case of the other two products it was very difficult to elucidate the structures without the assignment of the ^{13}C NMR data of dendropanoxide (19). Therefore, the total assignment of the ^{13}C NMR of this compound was necessary for structure elucidation of the oxidation products.

The protons at C-1 and C-2 were revealed by the COSY spectrum (Fig. 16). The ^{13}C-^{1}H correlation (Fig. 17) and long-range ^{13}C-^{1}H correlation (Fig. 18) spectra indicated, first of all, the C-3, C-10, and the methyl groups having long-range couplings with H-3. When H-3 (δ 3.75) was irradiated, the methyl group at δ 0.90 showed an NOE in the difference spectrum. Therefore, the signal at δ 0.90 was assigned to C-24 (methyl group) and the signal at

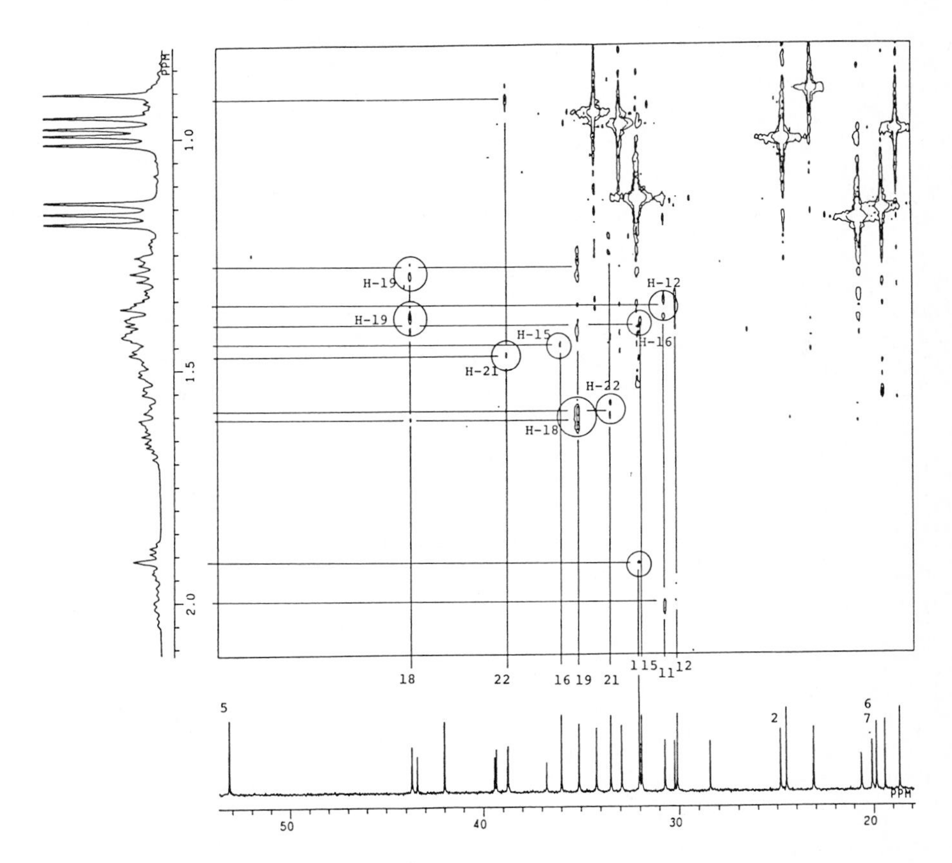

Fig. 19. The H-H-C relayed ^{13}C-^{1}H correlation spectrum of dendropanoxide (19).

δ 1.01 to C-23. The C-4 and C-5 carbons were easily assigned by cross peaks between these methyl groups. The methyl group showing a long-range coupling with C-10 must be C-25. This allowed C-8, C-9, and C-11 to be recognized. Since C-8 has a long-range coupling with the methyl group at C-26, C-13, C-14, and C-15 were automatically assigned. The carbons at C-12, C-17, C-18, and C-20 were also clear. However, the pair of carbons at C-19 and C-21, C-22 and C-16 were not distinguished from each other. The H-H-C relayed $^{13}C-^{1}H$ correlation method was utilized to solve this problem. As shown by the lines in Fig. 19, C-19 exhibited

TABLE 2.
^{13}C NMR Data of derivatives of dendropanoxide (19).

Carbons	Dendropanoxide 19	6α-OH 20	7α-OH 21	21α-OH 22	22β-OH 23
1	31.9	32.1	31.9	31.9	31.9
2	24.7	24.5	24.8	24.8	24.7
3	84.4	84.6	84.0	84.4	84.4
4	43.3	44.0	43.5	43.4	43.4
5	53.1	60.3	51.8	53.2	53.1
6	19.8*	67.5	31.5	19.8*	19.8*
7	20.0*	32.5	68.4	20.3*	20.0*
8	41.9	42.9	48.0	40.3	42.3
9	36.7	36.6	36.9	36.7	36.6
10	93.6	93.6	93.3	93.7	93.5
11	30.6	29.9	31.1*	29.7+	30.3+
12	30.0	29.9	30.0*	29.8+	30.9+
13	39.2	39.1*	39.3+	38.7#	38.9#
14	39.3	39.3*	40.4+	39.9#	39.4#
15	31.8	31.8	31.8	30.3	31.2
16	35.9	35.8	36.0	36.0	32.0
17	30.1	30.2	29.8	33.0	41.6
18	43.6	43.6	44.2	45.2	45.2
19	35.0	35.0	34.6	36.3	34.5
20	28.3	28.3	28.3	34.5	34.7
21	33.4	33.4	33.9	74.8	41.5
22	38.7	38.6	38.1	46.0	75.4
23	24.4	25.6	23.9	24.4	24.4
24	23.0	22.5	22.8	23.0	23.0
25	20.5	20.9	21.7	20.9	20.4
26	19.3	19.3	19.8	19.1	20.0
27	18.5	18.5	18.8	16.9	18.7
28	31.9	32.0	33.5	32.5	23.3
29	32.8	32.9	32.4	33.0	31.9
30	34.1	34.1	34.0	24.6	34.9

*,+,# signals may be interchanged in each vertical column.

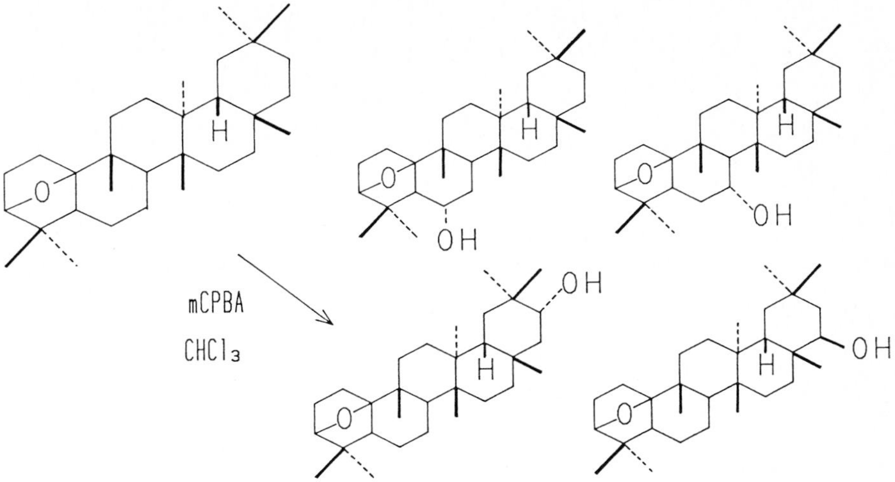

Fig. 20. Oxidation products of dendropanoxide (19).

relayed cross peaks with H-18, which had been located previously (see Fig. 17), and similarly C-18 with H-19. These facts enabled the assignments of C-19, C-21, and hence H-19 and H-21. With this information at hand, C-22 was found to show the relayed cross peaks with H-21 and C-16 with H-15. Thus all the carbon signals except for C-6 and C-7 were assigned unambiguously as shown in Table 2. However, for the time being, C-6 and C-7 were left undetermined.

By checking the ^{13}C NMR data of the oxidation products and also the NOE's, the structures of the four products were established as shown in Fig. 20 (ref. 10).

4. TOTAL ASSIGNMENT OF ^{1}H AND ^{13}C NMR OF MACROCYCLIC BIS(BIBENZYLS)

The marchantins are a new class of compounds isolated from the liverworts, *Marchantia polymorpha*, *M. paleacea* var. *diptera*, and *M. tosana*, and have been shown to possess cytotoxic, antifungal, and antimicrobial properties (ref. 3). The structure of marchantin A (24) was established as a macrocyclic bis(bibenzyl) by both chemical degradations and X-ray analysis of

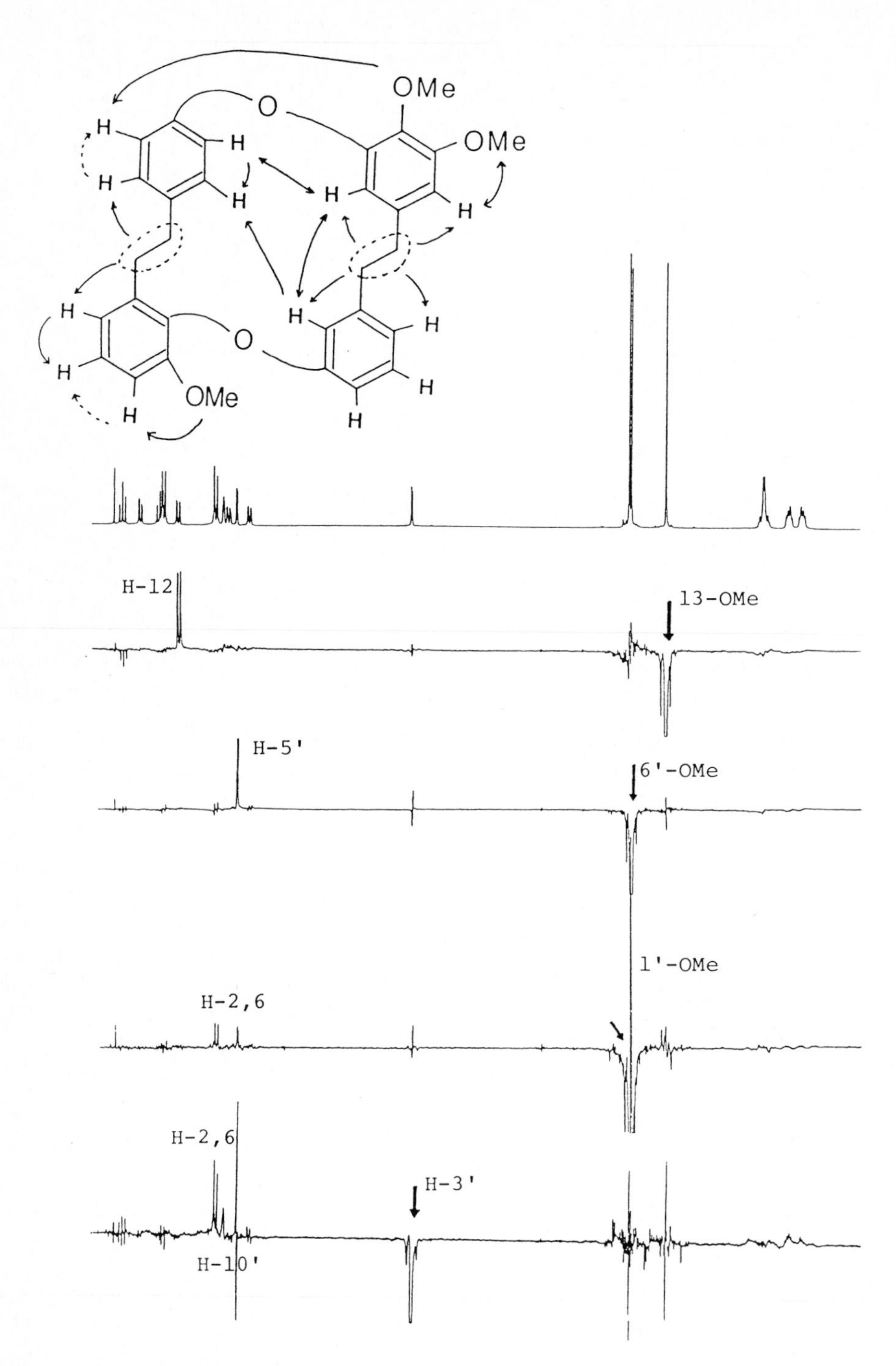

Fig. 21.1. The NOE difference spectra of marchantin A trimethyl ether (25).

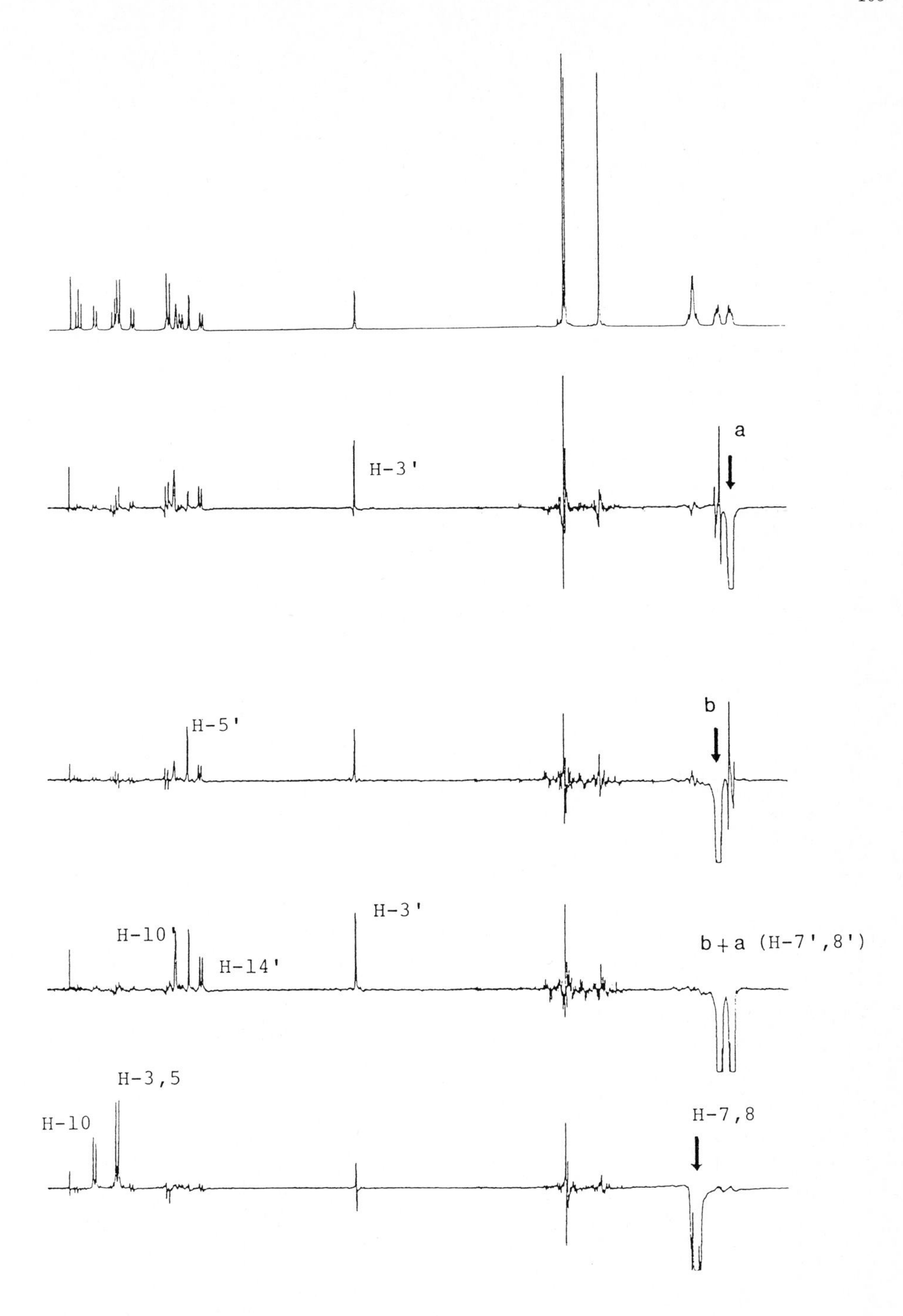

Fig. 21.2. The NOE difference spectra of marchantin A trimethyl ether (25).

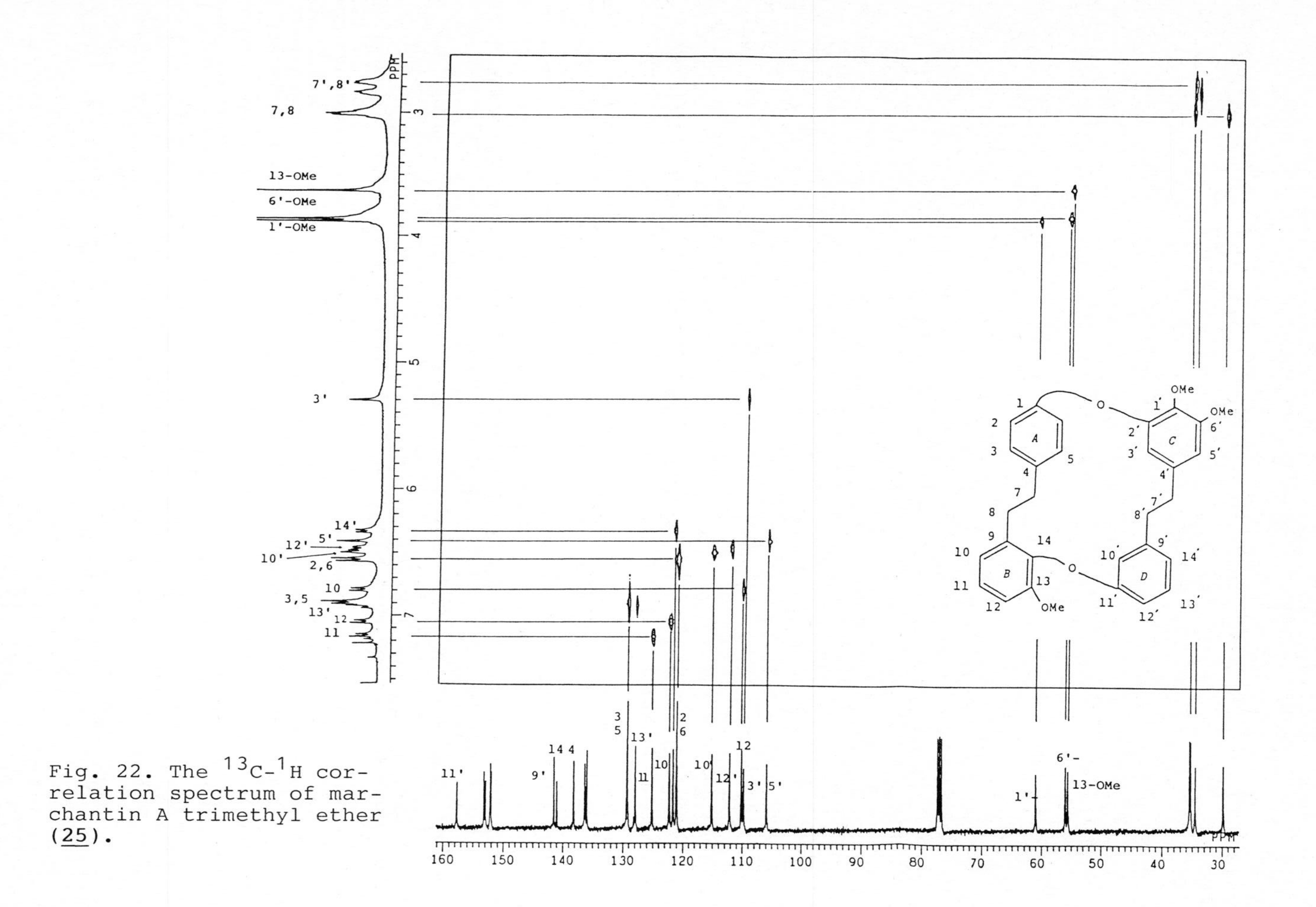

Fig. 22. The $^{13}C-^{1}H$ correlation spectrum of marchantin A trimethyl ether (25).

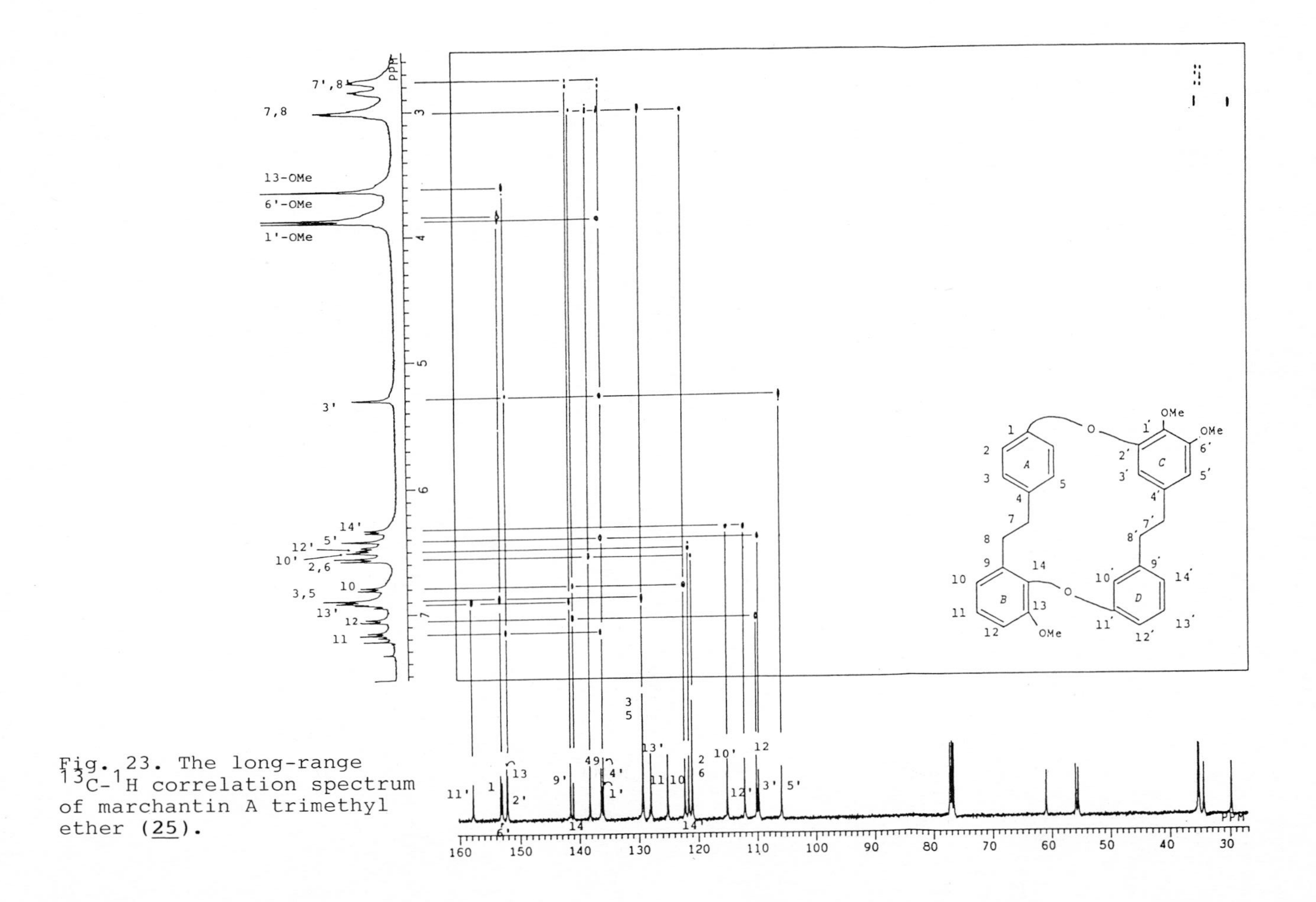

Fig. 23. The long-range ^{13}C-^{1}H correlation spectrum of marchantin A trimethyl ether (25).

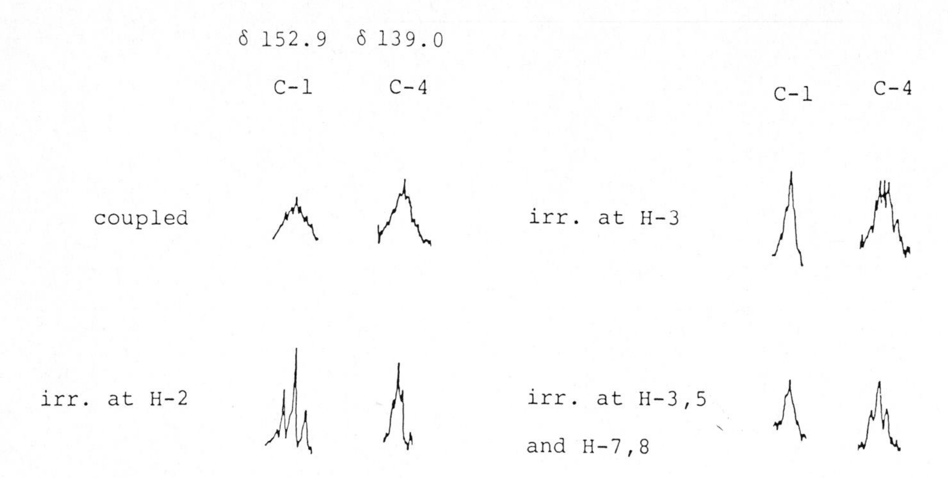

Fig. 24. The LSPD experiments for ring A carbons of marchantin A (<u>24</u>).

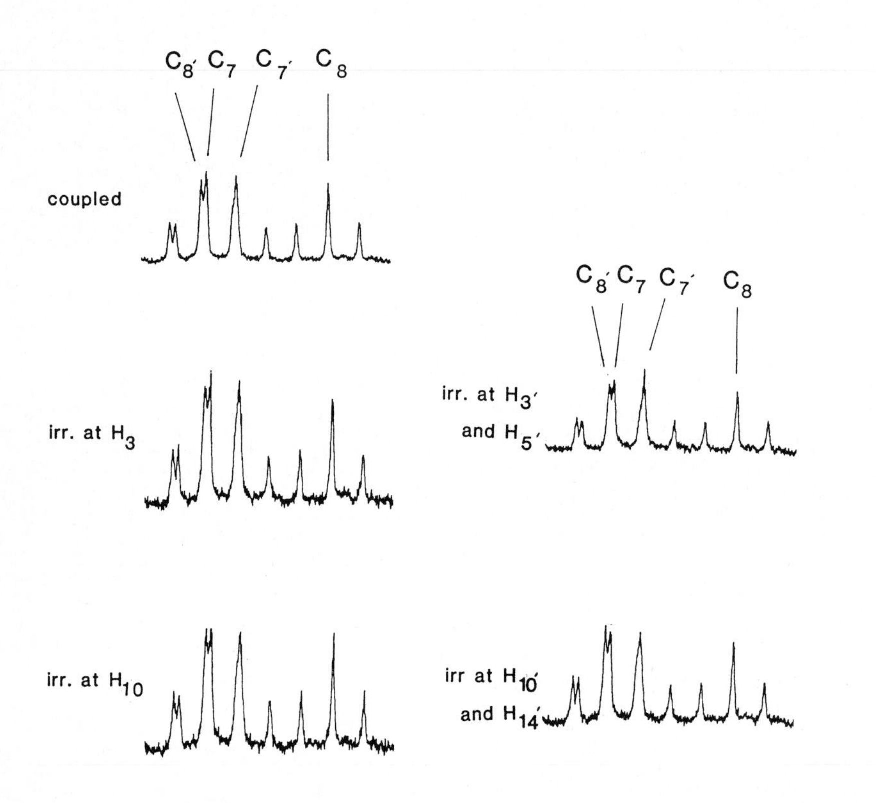

Fig. 25. The LSPD experiments for benzylic carbons of marchantin A (<u>24</u>).

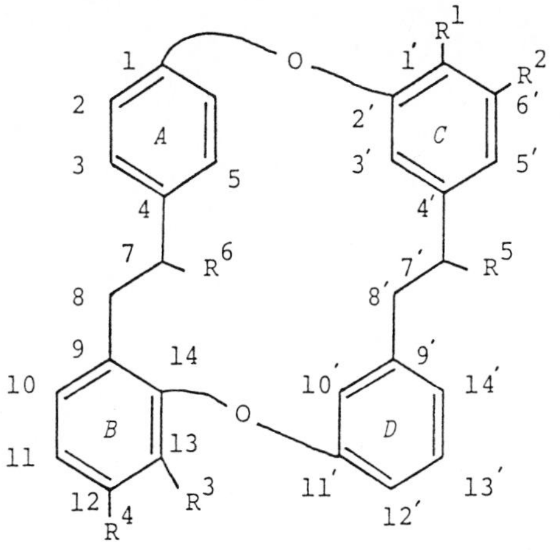

	compound	R^1	R^2	R^3	R^4	R^5	R^6
24	MA	OH	OH	OH	H	H	H
25	$MA(OMe)_3$	OMe	OMe	OMe	H	H	H
26	MB	OH	OH	OH	OH	H	H
27	$MB(OMe)_4$	OMe	OMe	OMe	OMe	H	H
28	MC	OH	H	OH	H	H	H
29	$MD(OMe)_3$	OMe	OMe	OMe	H	OH	H
30	$MD(OAc)_4$	OAc	OAc	OAc	H	OAc	H
31	$ME(OMe)_3$	OMe	OMe	OMe	H	OMe	H
32	$ME(OAc)_3$	OAc	OAc	OAc	H	OMe	H
33	MH	OH	H	OH	OH	H	H
34	MJ	OH	OH	OH	H	OEt	H
35	$MJ(OMe)_3$	OMe	OMe	OMe	H	OEt	H
36	$MJ(OAc)_3$	OAc	OAc	OAc	H	OEt	H
37	MK	OH	OH	OH	OH	OMe	H
38	$MK(OMe)_4$	OMe	OMe	OMe	OMe	OMe	H
39	ML	OH	OH	OH	H	H	OH
40	$ML(OMe)_3$	OMe	OMe	OMe	H	H	OH

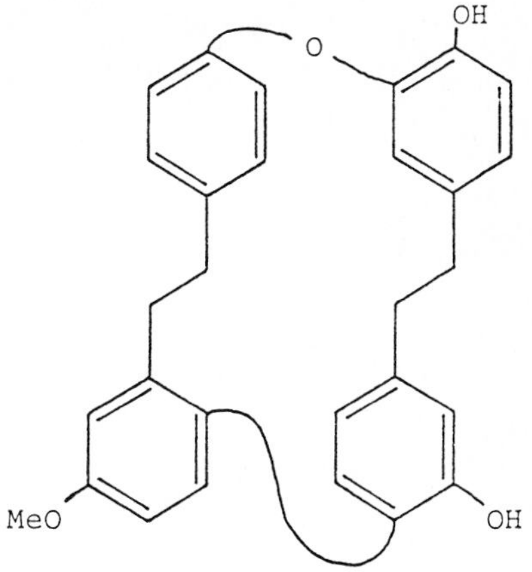

41 Riccardin A

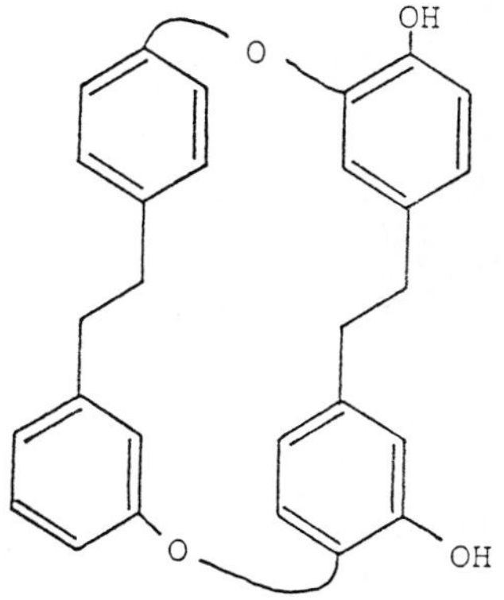

42 Riccardin B

Fig. 26. Structures of marchantins and riccardins.

TABLE 3.
^{1}H NMR Data of marchantins.

	MA (24)	MA(OMe)$_3$(25)	MB (26)	MC (28)	MD(OMe)$_3$(29)	MH (33)	MJ(OMe)$_3$(35)
2,6	6.58 d (8.5)	6.56 d (8.5)	6.58 d (8.5)	6.60 d (8.5)	6.51 d (8.3)	6.59 d (8.3)	6.52 brd (7.7)
3,5	6.93 d (8.5)	6.90 d (8.5)	6.93 d (8.5)	6.94 d (8.5)	6.88 d (8.3)	6.93 d (8.3)	6.89 brd (7.7)
7,8	2.96-3.01 m	2.99-3.04 m	2.91-2.99 m	2.97-3.03 m	2.97-3.10 m	2.91-2.99 m	2.93-3.17 m
10	7.02 dd (7.8,1.5)	7.06 dd (8.1,1.5)	6.92 d (8.5)	7.02 dd (7.8,1.6)	7.05 dd (7.8,1.5)	6.92 d (8.3)	7.07 dd (7.8,1)
11	7.15 t (7.8)	7.17 t (8.1)	6.85 d (8.5)	7.15 t (7.8)	7.19 t (7.8)	6.87 d (8.3)	7.21 t (7.8)
12	6.87 dd (7.8,1.5)	6.80 dd (8.1,1.5)	----	6.87 dd (7.8,1.6)	6.80 dd (7.8,1.5)	----	6.82 brd (7.8)
3'	5.13 d (2)	5.30 d (1.7)	5.13 d (1.8)	5.52 d (2)	5.06 d (2)	5.53 d (2)	5.02 d (1.5)
5'	6.47 d (2)	6.42 d (1.7)	6.47 d (1.8)	6.74 dd (8.1,2)	6.76 d (2)	6.73 dd (8.7,2)	6.70 d (1.5)
6'	----	----	----	6.88 d (8.1)	----	6.85 d (8.7)	----
7',8'	2.72-2.78 m	2.73-2.88 m	2.74-2.82 m	2.75-2.86 m	4.70 dd (10.4) 3.11 dd (12,4) 2.65 dd (12,10)	2.75-2.87 m	4.20 dd (11,3.9) 3.15 m 2.61 m
10'	6.85 t (2)	6.51 t (1.8)	6.56 dd (2.3,1.5)	6.62 dd (2.5,1.5)	6.66 dd (2,1)	6.61 dd (2.4,1.4)	6.72 t (1.5)
12'	6.55 ddd (7.8,2,1)	6.47 ddd (8.1,2.4,1)	6.56 ddd (7.8,2.3,1)	6.54 ddd (7.8,2.5,0.7)	6.36 ddd (7.8,2,1)	6.54 ddd (7.8,2.4,0.7)	6.34 ddd (8,2,1)
13'	6.98 t (7.8)	6.93 t (8.1)	6.99 t (7.8)	6.98 t (7.8)	6.83 t (7.8)	6.97 t (7.8)	6.81 t (8)
14'	6.41 brd (7.8)	6.34 brd (8.1)	6.41 ddd (7.8,1.5,1)	6.38 ddd (7.8,1.5,0.7)	6.06 ddd (7.8,1,1)	6.37 brd (7.8)	6.02 brd (8)
OMe		3.64 (13) 3.87 (6') 3.89 (1')			3.62 (1') 3.88 (13) 3.88 (6')		3.66 (13) 3.92 (1') 3.92 (6') 3.45 m (CH_2) 1.23 t (7.1)(CH_3)

TABLE 4.
^{13}C NMR Data of marchantins and riccardins.

Carbon	24	25	26	27	28	29	31	33	35	38	40	41	42
1	152.9	153.4	153.6	153.6	152.7	153.2	153.3	152.8	153.3	153.4	153.8	152.3	154.8
2	121.2	121.1	121.2	120.9	121.1	121.6	121.8	121.3	121.8	121.7	121.2	122.2	116.6
3	129.5	129.4	129.6	129.4	129.4	129.5	129.4	129.7	129.4	129.5	130.1	129.0	130.1
4	139.0	138.3	138.9	137.9	138.8	139.0	139.2	139.2	139.2	138.9	140.6	139.6	135.2
5	129.5	129.4	129.6	129.4	129.4	129.5	129.4	129.7	129.4	129.5	130.1	129.0	130.1
6	121.2	121.1	121.2	120.9	121.1	121.6	121.8	121.3	121.8	121.7	121.2	122.2	116.6
7	35.2	35.5	35.7	35.7	35.1	36.0	36.2	35.9*	36.1	36.4	63.9	38.3*	38.2*
8	30.2	29.9	29.7	29.4	30.1	30.3	30.3	30.0	30.3	29.2	43.8	35.3	37.5*
9	136.1	136.5	127.3	128.1	136.0	136.7	136.8	127.3	136.8	128.7	133.9	143.2	143.4
10	121.9	122.3	120.5	124.2	121.7	122.6	122.7	121.6	122.7	124.7	119.4	116.1	125.0
11	126.0	125.2	112.5	108.9	125.8	125.3	125.4	112.4	125.4	109.2	126.1	159.5	129.9
12	114.3	110.0	143.9	151.6	114.4	110.3	110.1	143.5	110.1	151.7	111.9	112.5	118.4
13	148.6	152.3	137.6	141.7	148.6	152.3	152.3	135.7	152.3	141.7	152.1	132.4	154.2
14	139.6	141.2	140.7	146.0	139.6	141.0	140.8	140.0	140.9	145.8	138.2	128.0	120.7
1'	130.6	136.5	131.9	136.1	143.1	137.3	137.1	143.5	137.1	137.3	136.4	143.1	145.5
2'	146.4	152.2	147.4	152.0	146.0	152.0	152.2	146.1	152.1	152.1	152.1	146.1	141.3
3'	107.9	109.8	108.4	109.9	115.5	108.1	109.3	115.6	109.2	109.5	109.6	115.9	121.0
4'	132.4	136.0	132.1	136.4	132.5	138.3	135.7	132.6	136.5	135.8	136.0	132.9	133.8
5'	109.3	106.0	109.5	106.0	122.2	102.2	101.6	122.4	101.6	101.7	106.0	122.0	125.2
6'	144.1	153.1	145.2	153.1	115.1	153.6	153.8	115.0	153.7	153.8	153.1	114.8	115.9
7'	34.0	34.6	34.2	34.4	33.7	74.9	84.8	34.2	82.9	84.7	34.1	37.9*	37.4*
8'	35.4	35.3	35.7	35.1	35.5	45.3	44.1	35.8*	44.4	44.0	35.0	37.2*	38.1*
9'	143.0	141.7	142.6	141.7	142.6	138.2	138.4	143.2	138.6	138.4	141.8	141.7	136.0
10'	115.4	115.2	115.7	115.4	115.4	116.2	116.3	115.6	116.4	116.4	114.9	121.5	121.0
11'	156.6	157.9	157.6	157.6	156.8	157.8	157.7	156.6	157.7	157.4	157.6	131.2	115.0
12'	112.0	112.2	112.6	112.9	111.9	112.3	112.0	112.0	112.0	112.9	112.6	124.3	143.0
13'	128.8	128.0	128.4	128.1	128.6	127.9	127.7	129.0	127.7	127.9	128.2	151.6	145.6
14'	123.1	121.6	122.4	121.8	122.9	123.2	123.5	123.5	123.5	123.7	121.8	115.9	115.8
12-OMe	---	---	---	56.1	---	---	---	---	---	---	---	---	55.4(11)
13-OMe	---	55.7	---	60.3	---	55.7	55.7	---	55.7	60.4	55.8	---	---
1'-OMe	---	61.0	---	61.1	---	61.0	61.1	---	61.1	61.1	61.1	---	---
6'-OMe	---	56.1	---	56.1	---	56.2	56.2	---	56.2	56.2	56.1	---	---
7'-OMe	---	---	---	---	---	---	56.8	---	64.2(CH_2) 15.5(CH_3)	56.2	---	---	---

*may be interchanged

its trimethyl ether (25). However, these methods can not be applied to other members of the marchantin series which are available only in small quantities and are generally non-crystalline. Therefore, the total assignment of the ^1H and ^{13}C NMR spectra of the marchantins would facilitate the structure determination of new members of this class.

The ^1H NMR spectrum of marchantin A trimethyl ether (25) was very easily assigned by both double resonance and NOE experiments (Fig. 21). The most characteristic signal appears at δ 5.13 (d, J=2 Hz) and is due to H-3', which is strongly shielded by ring A. The proton NMR data of representative marchantins are listed in Table 3. The proton-bearing aromatic carbons were identified from ^{13}C-^1H correlation spectrum (Fig. 22). The quaternary carbons were assigned by long-range ^{13}C-^1H correlation spectrum (Fig. 23), as ^3J were normally larger than ^2J in the case of aromatic cycles (ref. 11). However, some peaks congested in a narrow region (e.g. C-1' and C-9, C-2' and C-13, C-1 and C-6') were not fully distinguished. Therefore, the LSPD experiments were carefully carried out. For example, when H-3 and H-5 (which have the same chemical shifts) were irradiated with very weak power, the carbon at δ 153.4 (C-1) became a sharp triplet, whilst simultaneous irradiation of H-3 and H-7,8 reduced C-4 to a triplet (Fig. 24). Thus, by repeating these experiments, all the signals were assigned completely.

As the LSPD methods are applicable to benzylic methylenes, irradiation of H-3 and H-5 sharpened the carbon at C-7. Similarly C-8 was found by irradiation of H-10, C-7' by H-3' and H-5', and C-8' by H-10' and H-14' (Fig. 25). By applying these methods, all the signals were successfully assigned unambiguously and this methodology was used for determination of other marchantins (Fig. 26 and Table 4) (ref. 12).

By now, we have isolated about 20 macrocyclic bis(bibenzyls) including biphenyl type of compounds (Fig. 26) (ref. 13) and established the structures. The ^{13}C NMR signals, listed in Table 4, assigned by these methods show apparent similarities and differences with each other. It is notable that the chemical shifts are very close each other when the mode of cyclization of the marchantins is very similar. Thus riccardins show quite different chemical shifts. It is, of course, necessary to accumulate ^{13}C NMR data of other compounds having other modes of cyclization. We believe these assignments will be of considerable use in the structure elucidation of this class of compounds.

In this report, several practical applications of 2D NMR methods are described. By application of these methods, it is becoming increasingly easier to assign the ^{1}H and ^{13}C NMR data. They also enabled to determine very complicated structures of natural products. As shown in this report, the long-range ^{13}C-^{1}H correlation spectroscopy will be frequently used to solve these problems, although it can be used only for the coupling constants ranging between 5-10 Hz. It also depends on how long you can occupy the spectrometer, such as in the case of 2D-INADEQUATE. However, the sensitivity of the ^{13}C probes is becoming high and the calculation time is decreasing, for example, by introducing the array processor.

We thank Professor Y. Asakawa for helpful discussions and encouragement throughout the work. Thanks are also due to Dr. T. Hashimoto, Dr. R. Matsuda, Dr. M. Toyota, Miss K. Takikawa, Mr. M. Sono, Mr. T. Masuya, and Mr. Y. Kohama for their collaboration.

REFERENCES

1) R. Benn and H. Günter, Angew. Chem., Int. Ed. Engl., 22 (1983) 350-380.
2) All the NMR spectra were taken on a JEOL GX 400 spectrometer.
3) Y. Asakawa, Chemical Constituents of the Hepaticae, in: W. Herz, H. Griesebach, and G. W. Kirby (Ed.), Progress in the Chemistry of Organic Natural Products, Springer-Verlag, Wien New York, 1982, pp 1-285.
4) A. Matsuo, M. Nakayama, S. Sato, T. Nakamoto, S. Uto, S. Hayashi, Experientia, 30(4) (1974) 321-322; A. Matsuo, H. Nozaki, M. Nakayama, Y. Kushi, S. Hayashi, T. Komori, N. Kamijo, J. C. S. Chem. Commun., (1979) 174.
5) T. Hashimoto, M. Tori, Z. Taira, and Y. Asakawa, Tetrahedron Lett., 26 (1985) 6473-6476.
6) MS spectra were measured on a JEOL HX-100 spectrometer.
7) J. Dillon and K. Nakanishi, J. Am. Chem. Soc., 97 (1975) 5417-5422.
8) a) Y. Asakawa, R. Matsuda, and M. Tori, Experientia, 42(2) (1986) 201-203; b) M. Tori, R. Matsuda, and Y. Asakawa, Bull. Chem. Soc. Jpn., 58(9) (1985) 2523-2525; c) M. Tori, R. Matsuda, and Y. Asakawa, Tetrahedron Lett., 26(2) (1985) 227-230; M. Tori, R. Matsuda, and Y. Asakawa, Chemistry Lett., (2) (1985) 167-170; M. Tori, R. Matsuda, and Y. Asakawa, Tetrahedron, 42(5) (1986) 1275-1283; d) M. Tori, M. Sono, and Y. Asakawa, Bull. Chem. Soc. Jpn., 58(9) (1986) 2669-2672.
9) J. Asakawa, R. Kasai, K. Yamasaki, O. Tanaka, Tetrahedron, 33 (1977) 1935-1939.

10)in preparation.
11)J. B. Stothers, Carbon-13 NMR Spectroscopy, ORGANIC CHEMISTRY, Vol. 24, Academic Press, New York, 1972
12)M. Tori, M. Toyota, L. J. Harrison, K. Takikawa, and Y. Asakawa, Tetrahedron Lett.,26(39) (1985) 4735-4738.
13)Y. Asakawa, M. Toyota, R. Matsuda, K. Takikawa, and T. Takemoto, Phytochemistry, 22 (1983) 1413-1415; Y. Asakawa, Rev. Latinoamer. Quim., 14(3) (1984) 109-115; Y. Asakawa, Journ. Hattori Bot. Lab., 56 (1984) 215-219; Y. Asakawa, M. Toyota, Z. Taira, T. Takemoto, and M. Kido, J. Org. Chem., 48 (1983) 2164.

THE USE OF TWO-DIMENSIONAL LONG-RANGE δ_C/δ_H CORRELATION IN THE STRUCTURE ELUCIDATION OF SOME NOVEL AROMATIC NATURAL PRODUCTS

JOHNSON FOYERE AYAFOR , BONAVENTURE T. NGADJUI, B.L. SONDENGAM , JOSEPH D. CONNOLLY AND DAVID S. RYCROFT

INTRODUCTION.

While it has been long recognised {1,2} that long-range carbon-proton coupling constants are useful in assigning ^{13}C NMR spectra and establishing structural connectivity across heteroatoms and quaternary carbons there are few reports, other than in the peptide field {3,4} of their use to define an extended sequence of bond connectivities and hence for complete structural elucidation. Applications have generally been limited to defining a partial structure and have used time-consuming selective proton decouplings, SPI {4} or 2D selective J-resolved spectra {5} to obtain correlations and measure individual long-range couplings. In this communication we wish to demonstrate that the combination of 2D long-range δ_C/δ_H correlation and consideration of the 1D proton-coupled ^{13}C NMR spectrum can be sufficient to form a powerful method of structural eludication. In principle 2D non-selective heteronuclear long-range shift correlation methods {6} which also give the value of the coupling constants would render measurement of the proton-coupled ^{13}C NMR spectrum superfluous. In practice, however, we find that limited digital resolution is more a problem in 2D than 1D spectra and, in addition, it is common to find that a set of correlations is complete. The strategy outlined above is particularly suitable for aromatic compounds, but before we go on to apply it to the structural elucidation of some complex naturally occurring aromatic compounds on which we have been working recently, let's first of all look at the magnitude of long-range carbon-proton coupling constants in some aromatic compounds.

Long-range carbon-proton couplings in simple aromatic systems.

With aromatic compounds it is possible to observe a variety of long-range carbon-proton couplings with the carbon either as a substituent or as part of the aromatic ring. The size of the couplings depends on several factors including distance, angle (trans greater than cis), bond order and electronegativity of substituents. The book on C-C and C-H NMR couplings by Marshall {2} and the review by Gottlieb {1} provide good sources of model systems. A few examples will illustrate the variations. In benzoic acid the carboxyl group carbon has a long-range coupling to the ortho proton of 4.1 Hz (^3J) while the values of ^4J and ^5J are 1.1 Hz and 0.5 Hz respectively. The ^3J coupling of the ortho proton to the methyl carbon in toluene is of the order of 5 Hz.

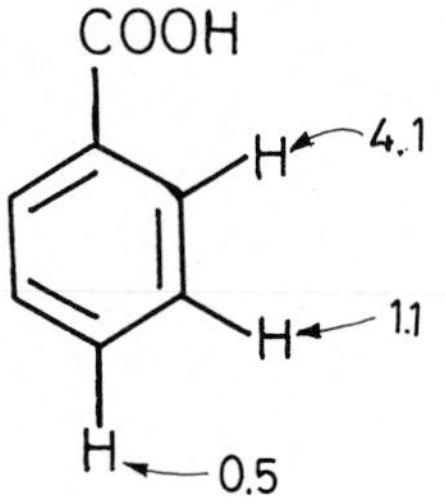

COOH

H 4.1

H 1.1

H 0.5

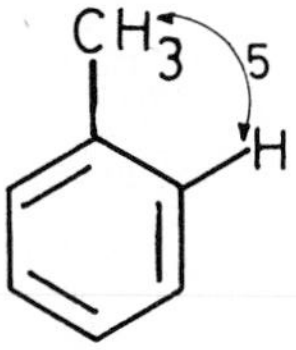

CH_3 5 H

For carbons within the aromatic ring the ^3J coupling is the largest. For benzene itself the values are as follows : ^2J 1.1 Hz, ^3J 7.6 Hz, ^4J 1.2 Hz. While these values can vary considerably as a result of substitution it is generally expected that $^3J_{CH}$ will be much greater than $^2J_{CH} \approx {}^4J_{CH}$.

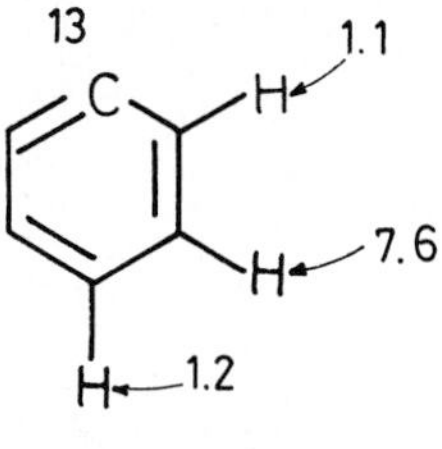

13 C

H 1.1

H 7.6

H 1.2

For carbon-proton couplings involving protons external to the aromatic ring, $^2J_{CH}$ may well be greater than $^3J_{CH-}$. Toluene provides a good example.

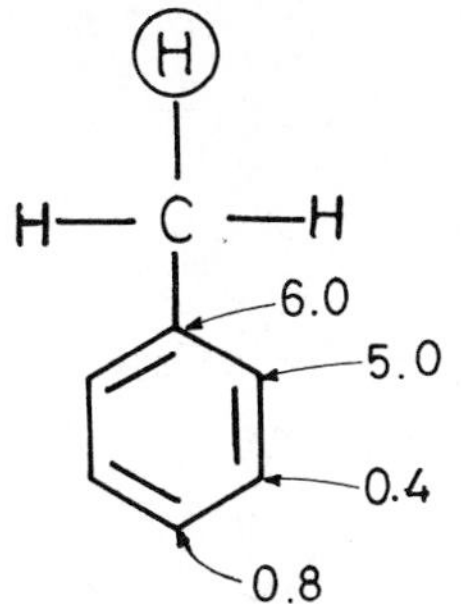

The variation resulting from bond order and electronegativity is well illustrated by benzaldehyde. In this case $^{2}J_{CH}$ is large and provides a ready method of identification of the ring carbon bearing the aldehyde function.

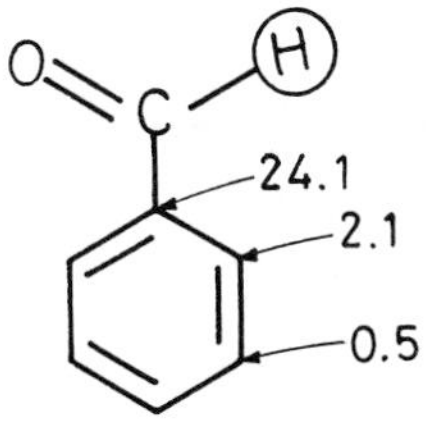

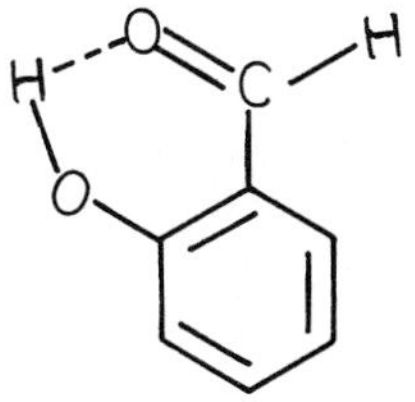

The external protons may be phenolic hydroxyl protons, methoxyl protons or N-methyl protons. In the case of methoxyl groups or N-methyl groups, the substituted carbon or carbons normally shows quartet splitting ($^{3}J_{CH} \approx 5$ Hz). If the phenolic hydroxyl proton is frozen in a particular conformation by hydrogen bonding the $^{3}J_{CH}$ values will differ depending on whether they are cis or trans. This is illustrated by the case of salicylaldehyde. These couplings are removed by addition of D_2O.

This background information of long-range couplings in simple aromatic systems coupled to the new strategy mentioned above enabled us to elucidate the structures of ekeberginine, hoslundin, isohoslundin, hoslundal and hoslundiol and to a lesser extent, vepridimerines C, D and E.

EKEBERGININE FROM EKEBERGIA SENEGALENSIS (MELIACEAE)

Several years ago, Professor Ekong and his collaborators examined the bark of Ekebergia senegalensis, (Meliaceae) growing in Nigeria and isolated some complex tetranortriterpenoid derivatives {7}and the coumarin, ekesenin {8}. Extraction of the bark and wood of *E. senegalensis* collected in the North-West province of Cameroon did not yield any of the compounds obtained by the Nigeria group. Instead, we isolated xanthoxyletin (1) and a new carbazole alkaloid which we named ekeberginine {9}.

Ekeberginine (2), $C_{19}H_{19}NO_2$, m.p. 230-231°, readily formed an N-methyl derivative (3), m.p. 155-157°, m/z 307.1589, which had resonances in its 200.13 MHz ^1H NMR spectrum for a dimethylallyl group, a methoxyl group, an N-methyl group, an aldehyde, an isolated aromatic proton, and an *ortho*-disubstituted benzene ring. These data coupled with the u.v. spectrum were consistent with the presence of a carbazole moiety with dimethylallyl, methoxyl and aldehyde substituents on one of the rings and suggested that ekeberginine is related to indizoline (4), m.p. 170-171°, from *Clausena indica* {10} and heptaphylline (5), m.p. 190-191°, from *C. heptaphylla* and *C. pentaphylla* {11}. Ekeberginine (2) differs from (4) and the *O*-methyl derivative of (5) in physical properties and in the absence of a strongly deshielded H-4 resonance in its ^1H NMR spectrum. The ^{13}C NMR spectrum of *N*-methyl ekeberginine was consistent with the carbazole ring system and the above susbtituents. 2D one-bond δ_C/δ_H correlation permitted the direct assignment of all the protonated carbons (see table 1). The structure (3) of *N*-methyl ekeberginine was then established unambiguously by comparing the pattern and size of the couplings with the observed qualitative correlations and considering the carbon resonances in an appropriate sequence. Although for ease of presentation the results are discussed in terms of the carbazole structure (3), it is important to realise that the bond connectivities obtained lead independently to this structure.

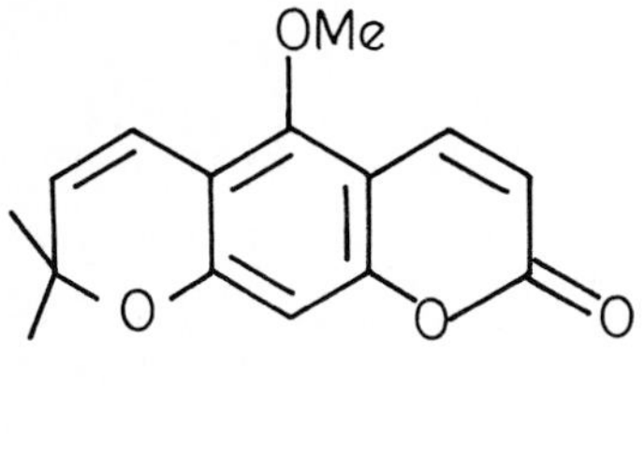

OMe

(1)

5′ 4′ 2′ 1′ 6 5a 4a 3 CHO 7 8a N 2 H 1 R OMe

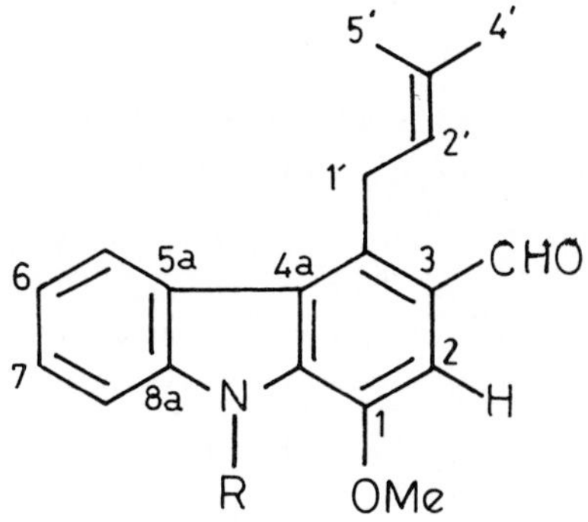

(2) R = H

(3) R = Me

Table 1 lists the ^{13}C chemical shifts, their assignments, the values of the direct and long-range carbon-proton couplings and the observed correlations. The unsubstituted nature of ring B, already defined by the ^{1}H NMR spectrum, was readily confirmed by the observation of correlations of C-5 with H-7, C-6 with H-8, C-7 with H-5, and C-8 with H-6. As each of these signals has only one large coupling and shows only one long-range correlation in the experiments performed (see table 1) it is reasonable to assume that the observed correlation arise through ^{3}J interactions {1,2}. The olefinic carbon C-3' of the dimethylallyl group was identified by its long-range correlations with the C-1' methylene protons and the methyl groups. Correlation with the N-methyl group identified C-1a and C-8a which were distinguished by the fact that C-8a has ^{3}J interactions with H-5 and H-7 while C-1a has a ^{3}J interaction with the aromatic proton on ring A. This indicates that the aromatic proton is attached either to C-4 or C-2. The assignment of the remaining ring junction carbons C-4a and C-5a followed readily from their correlations with protons in ring B. Thus C-5a has ^{3}J correlation with H-6 and H-8 while C-4a has ^{3}J correlations with H-4 and the C-1' methylene protons. It is clear from the last observation that the dimethylallyl group is attached to C-4 and therefore the isolated aromatic proton must be at C-2. The carbon bearing the aldehyde group, distinctive {2} because of its large ^{2}J interaction (22.6 Hz) with the aldehyde proton, must be C-3, as expected on biogenetic grounds, since it correlates with the C-1' methylene protons. The resonance at δ_C 136.3 is C-4 since it couples with the C-1' methylene protons. Finally the methoxyl group, which couples with the only remaining resonance, must be placed at C-1 which also shows correlations with H-2 and the aldehyde proton. Thus the structure of N-methyl ekeberginine is defined as (3) and hence ekeberginine has structure (2).

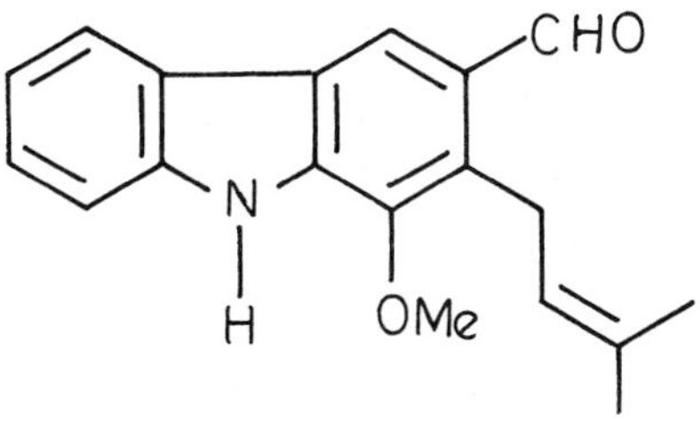

(4)

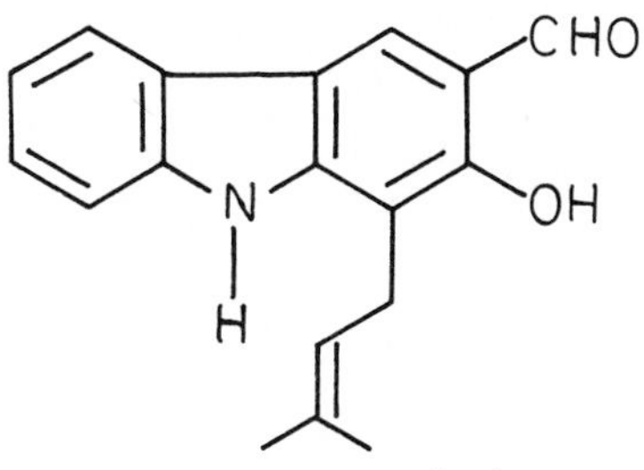

(5)

TABLE 1

50.325 MHz ^{13}C NMR data of N-Methyl Ekeberginine (3).

Carbon	δ_C [a]	$^1J_{CH}$	Long-range Couplings and Correlations [b]	
CHO	190.1	172.3	d (4.2)	H-2[c]
1	145.4	-	qdd (4.3, 2.9, 1.4)	OMe, H-2, CHO
8a	141.8	-	ddqq (9, 8, 3, 1)	H-5, H-7, NMe
4	136.3	-	bqd (6.5, 3.5)	2H-1', H-2
1a	134.3	-	dq (8.0, 2.5)	H-2, NMe
3'	132.7	-	septet t (6.2, 1.4)	3H-4', 3H-5', 2H-1'
3	125.7	-	dtd (22.6, 3.8, 1.2)	CHO, 2H-1'[c]
7	125.6	160.5	dt (7.9, 1.2)	H-5
5a	123.0	-	dddd (8.7, 4.9, 1.8, 1.0)	H-6, H-8
5	122.9	159.8	ddd (7.8, 2[d], 0.5[d])	H-7
4a	122.7	-	btdd (6, 2, 0.5)	2H-1', H-5[c]
2'	122.2	155.3	qqt (6, 5, 1)	3H-4', 3H-5', 2H-1'
6	120.2	160.5	dd (7.1, 1.2)	H-8
8	109.2	160.6	dt (8.2, 1.2)	H-6
2	105.5	159.0	d (3.1)	CHO
OMe	55.6	144.5	-	-
NMe	32.2	139.7	-	-
1'	26.5	127.0	d (4.0)	-
4'	25.6	125.7	dqt (7.0, 4.3, 1.3)	H-2'[c], 3H-5'[c]
5'	18.3	125.4	dqt (8.3, 4.2, 0.8)	H-2'[c], 3H-4'[c]

a Relative to $CDCl_3$ at δ 77.0

b The pulse sequence used {13a} $90°\{^1H\}-\frac{1}{2}\ t_1-180°\{^{13}C\}-\frac{1}{2}\ t_1\ -\tau_1-90°\{^1H\}90°\{^{13}C\}$ $-\tau_2-BB\{^1H\}FID\{^{13}C\}\ t_2$ with phase cycling to achieve quadrature detection in both dimensions {13b}. Two experiments were performed, with τ_1 = 40 ms, τ_2 = 20 ms and τ_1 = 80 ms, τ_2 = 40 ms.

c Only observed in the experiment with τ_1 = 80 ms, τ_2 = 40 ms.

d Approximate value of second order splittings.

The isolation of ekeberginine from E. senegalensis is of considerable taxonomic interest. The Meliaceae family, unlike the Rutaceae family, is a poor source of nitrogen-containing metabolites {12}. The isolation of the coumarin xanthoxyletin, known to be an insect growth inhibitor also corroborates the folkloric use of E. senegalensis for the preservation of grain against weevel attack.

THE VEPRIDIMERINES : STRUCTURE AND SYNTHESIS

A recent investigation of the stem bark constituents of *Vepris louisii*, *Oricia renieri* and *Araliopsis tabouensis* {14} resulted in the isolation of four isomeric dimeric quinolone alkaloids, vepridimerines-A-D, $C_{34}H_{38}N_2O_8$ {16}. These alkaloids occur together with veprisine {15}, another quinolone, and it is clear that they are dimers of veprisine. The vepridimerines are the most complex members of a small group of heptacyclic dimeric quinolone alkaloids found, to date, only in the Rutaceae and which includes pteledimerine and pteledimeridine from *Ptelea trifoliata* {17, 18} and paraensidimerins A, C, E, F and G. from *Euxylophora paraensis* {19}. The structures (7-10) have been assigned to the vepridimerines largely on the basis of spectroscopic data, extensive decoupling experiments in their 360 MHz ^{1}H NMR spectra playing a key role.

Vepridimerine A (7) has a *cis* ring junction (J 6.1 Hz). Models show that the cyclohexane ring exists in a chair conformation and this is confirmed by the observation of a ^{4}J coupling (2.2 Hz) between the α proton (Hf) and the bridge proton (Ha) which depicts an ideal W relationship between them. It is clear from the ring junction protons' coupling constant (J 12.5 Hz) of vepridimerine B(8) that it is *trans*. Again models show that with the arrangement shown in (8) the cyclohexane ring must adopt a boat conformation thus excluding the possibility of the ^{4}J coupling observed with (7). The alternative *trans* (*vide infra*) permits a chair conformation of the cyclohexane and the observation of a long range coupling would have been expected.

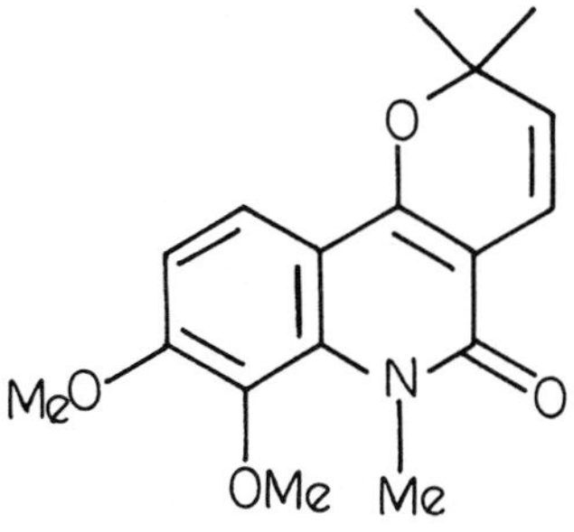

(6)

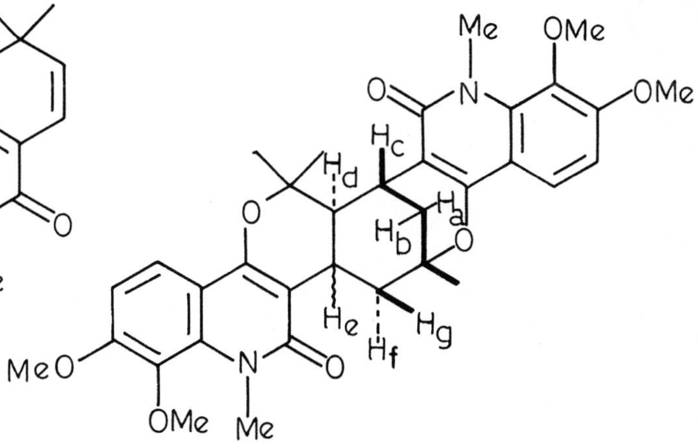

(7) α - He

(8) β - He

Vepridimerine C (9) and D (10), respectively, are the corresponding <u>cis</u> and <u>trans</u> isomers to A and B, but differ from the latter pair in the fact that the left hand unit in them is converted into a 4-quinolone. The 4-quinolone moiety was apparent from the NMR data. Thus in (9), for example, one of the H-5 resonances (δ_H 8.06) and one of the carbonyl resonances (δ_C 176.29) exhibit the large downfield shifts expected for a 4-quinolone {the corresponding 2-quinolone shifts in vepridimerine A (7) are δ_H 7.69 and δ_C 164.33}.

Further strong support for the heptacyclic structure and stereochemistry of vepridimerines A and B has come from the X-ray crystallographic confirmation of the structures of their lower analogues, paraensidimerins A and C (lacking the 4 OMe groups), by Jurd and co-workers {20}. The relative position of the 4-quinolone in vepridimerine C (9) and D (10) on the left hand side of the molecule has also been unambiguously established by 2D long-range δ_C/δ_H correlation experiments. In these experiments on (9) and (10), only Hc correlates with the 2-quinolone carbonyl thus showing that the 2-quinolone is on the right hand side {21}.

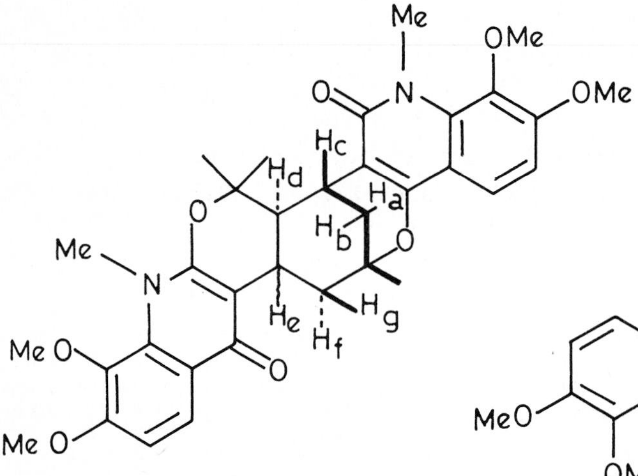

(9) α - He

(10) β - He

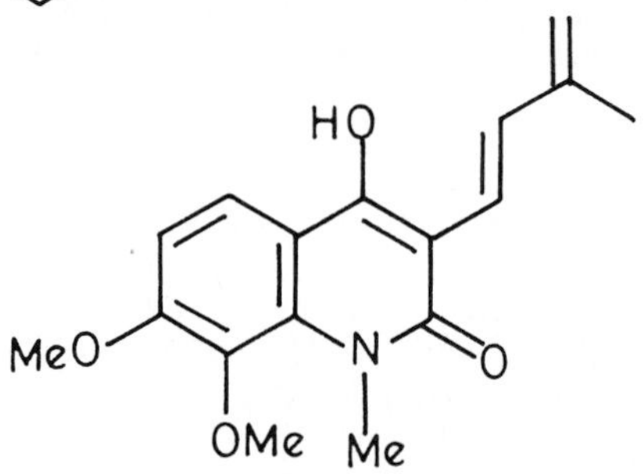

(11)

The biosynthetic derivation of the vepridimerines can be rationalized by Diels-Alder dimerizations of the diene (11) followed by addition of the 2- or 4-hydroxyl groups to the residual double bonds of the adducts. Since they are all racemic, the reaction is not enzyme-controlled. Somehow, the chromene ring of veprisine (6) must be opened to give the requisite diene (11). The stereochemical details of these dimerizations are outlined in schemes 1 and 2.

To obtain veprisimerines B and D two moles of the diene (11) are required. With the orientation shown (scheme 1), the adduct can cyclise directly both ways to give B(8). But here, there is also a possibility to obtain D (10).

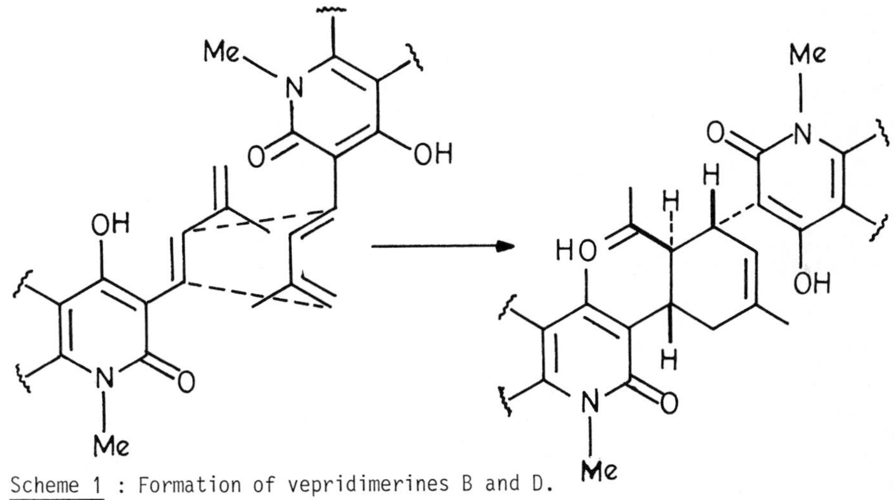

Scheme 1 : Formation of vepridimerines B and D.

In a similar fashion and by changing the orientation of the diene (scheme 2) we can obtain A (7), (C) (9) and the alternative trans, E (vide infra).

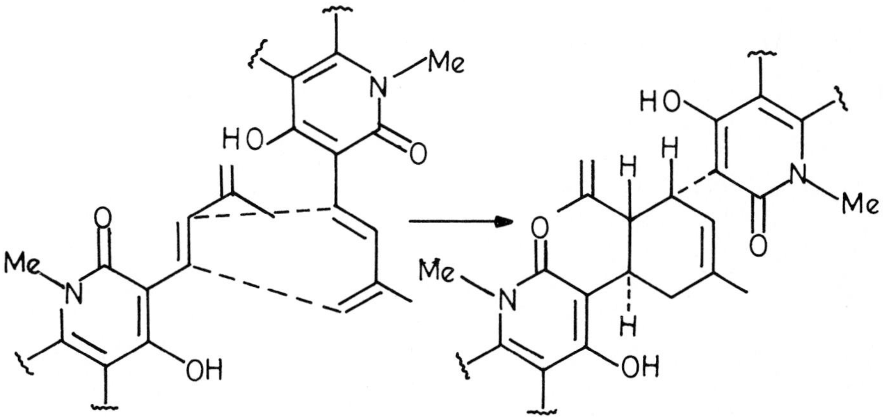

Scheme 2 : Formation of veprisimerines A, C and E.

Grundon {22} has demonstrated that compounds similar to veprisine (6) are excellent dienophiles. The cis compound vepridimerine A (7) could also arise by a Diels-Alder reaction involving veprisine (6) and the diene (11) but the formation of vepridimerine C (9) is not easily rationalized in this route.

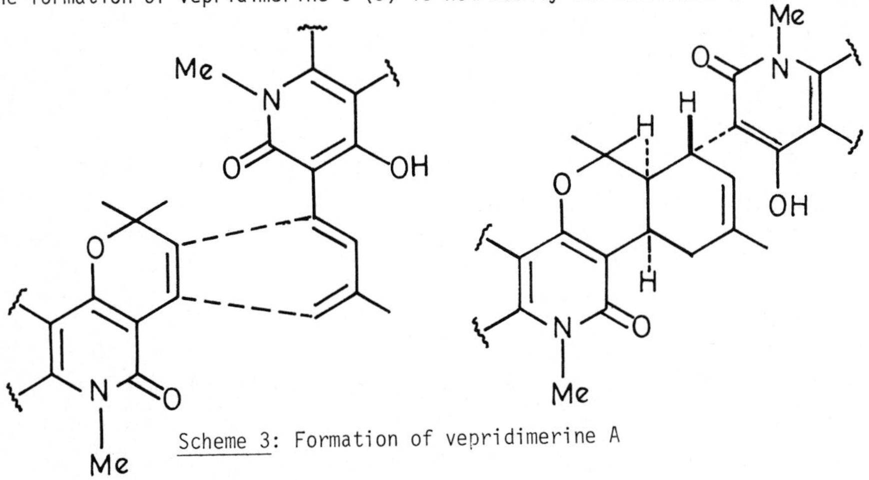

Scheme 3: Formation of vepridimerine A

After our preliminary publication of the structures of vepridimerines A-D {16} we discovered some work by Barnes and his colleagues {23} in Australia with coumarin. Methylmagnesium bromide treatment of coumarin followed by acid treatment of the product furnished a dimer to which they ascribed structure (12). Since this represents the alternative trans ring junction to vepridimerines B and D, we decided to reinvestigate the structure by highfield NMR. The structure was based in 1963 on the 60 MHz ^{1}H NMR spectral data. Even at 360 MHz we found that the ^{1}H NMR spectrum wasn't well-resolved but we could assign most of the couplings. Our results were in accord with the proposed structure (12). The ring junction is trans and as predicted by models the cyclohexane ring is in the chair conformation and the expected ^{4}J coupling (3 Hz) between Ha and Hf is observed.

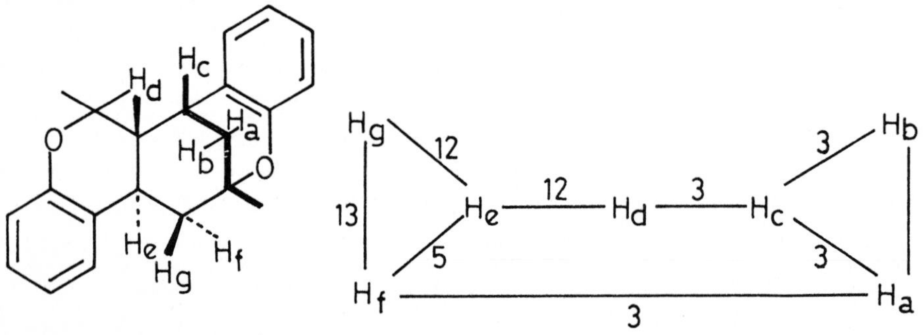

(12) Proton couplings (in Hz) of the Coumarin Dimer. (12a)

Attempted Syntheses of the Vepridimerines.

Encouraged by the successful conversion of the coumarin product into the dimer (12) we embarked on the *in vitro* preparation of vepridimerines A-D. In a first attempt we treated veprisine (6) with acid and obtained as a major product a crystalline dimer, diveprisine, whose structure (13) was assigned on the basis of chemical and spectroscopic evidence {24}.

Diveprisine (13), $C_{34}H_{38}N_2O_8$, m.p. 280-181°, m/z 602 (M^+, 100 %), 302 (84), 301 (40) ; νmax (CCl_4) 1651, 1639 and 1592 cm^{-1}, has resonances for four tertiary methyl groups (δ_H 1.27, 1.49, 1.56 and 1.91 ; δ_C 24.1, 26.6, 27.3 and 29.1), two N-methyl groups (δ_H 3.71, 3.76 ; δ_C 33.4, 33.5), four methoxy groups (δ_H 3.81, 3.38, 3.92, 3.94 ; δ_C 56.17, 56.20, 61.55, 61.63) and two *ortho*-coupled AB systems {δ_H 7.69 and 7.68 and 6.85 (all d, J 9 Hz)}, and obviously contains two veprisine units joined via the chromene rings. The presence of four methyl groups clearly indicates that the carbon skeleton of veprisine is different from that of the vepridimerines. The remaining spectroscopic features of diveprisine include a trisubstituted double bond {δ_H 6.29(s) ; δ_C 141.2(s) and 112.4(d)} and an AMX system comprising a methylene group {δ_H 1.92 (H_A, dd, J 10.0, 13.9 Hz, 2.19 (H_M, dd, J 7.2, 13.9 Hz) ; δ 43.81(t)} and a methine {δ_H 3.49 (H_X, dd, J 7.2, 10.0 Hz) ; δ_C 31.66 (d)}. These data were readily accommodated in structure (13) for veprisine. The three other possible ways of joining the two veprisine units were excluded on the basis of the unsymmetrical nature of the hydrogenation products and on the magnitude of the geminal coupling constant (J 13.9 Hz) of the methylene group (see below).

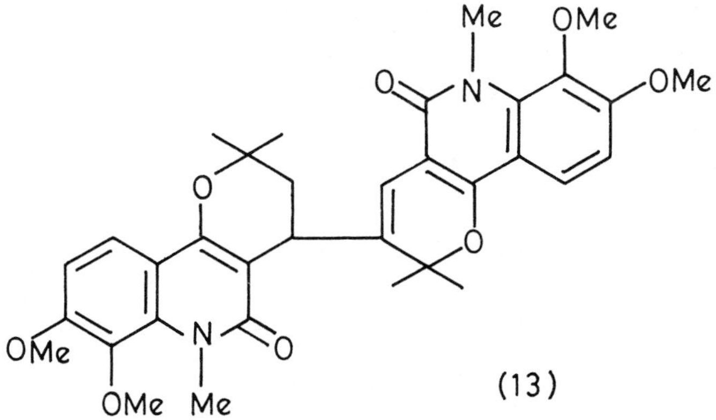

(13)

Hydrogenation of diveprisine afforded two dihydro-derivatives both of which show a small parent ion at 604 (1%) and a base peak at 302. It is clear from the ^{13}C NMR spectrum of these products that they are non-symmetrical. The ^1H and ^{13}C NMR spectra of the major isomer, m.p. 218-220°, are, in part, broad, presumably as a result of restricted rotation. The minor isomer, m.p. 148-150° is unaffected by this phenomenon and has resonances for two methylene groups

(δ_C 19.76 and 34.01) and two methines (δ_C 30.85 and 37.29). The associated proton shifts and coupling constants are shown in (14). The fact that the newly created methylene group has a geminal coupling of 18.2 Hz indicates that it is α to the veprisine nucleus and confirms the structure of diveprisine as (13). The relative configuration of the two dihydroderivatives remains undertermined. It is unclear from an inspection of models which diastereoisomer will suffer from restricted rotation.

The formation of diveprisine can be rationalized in terms of the protonation of the chromene double bond of veprisine (6) to give a resonance-stabilized electrophilic species (15) which then reacts with a second veprisine molecule followed by proton loss. Presumably the presence of a methoxyl group at C-7 favours protonation at the expense of formation of the diene (11).

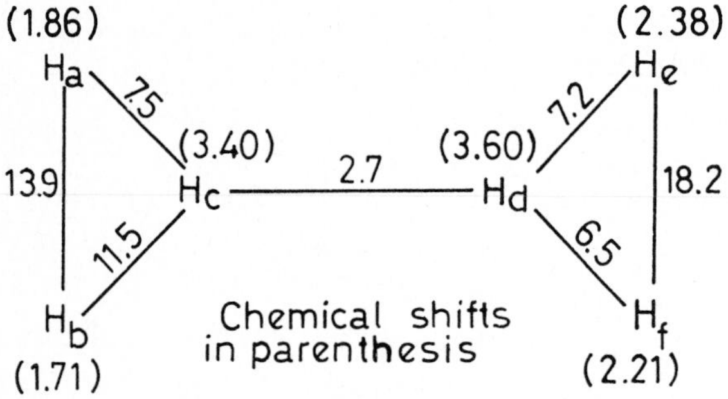

(1.86) Ha, (1.71) Hb, (3.40) Hc, (3.60) Hd, (2.38) He, (2.21) Hf; couplings: 13.9, 7.5, 11.5, 2.7, 7.2, 6.5, 18.2

Chemical shifts in parenthesis

(14)

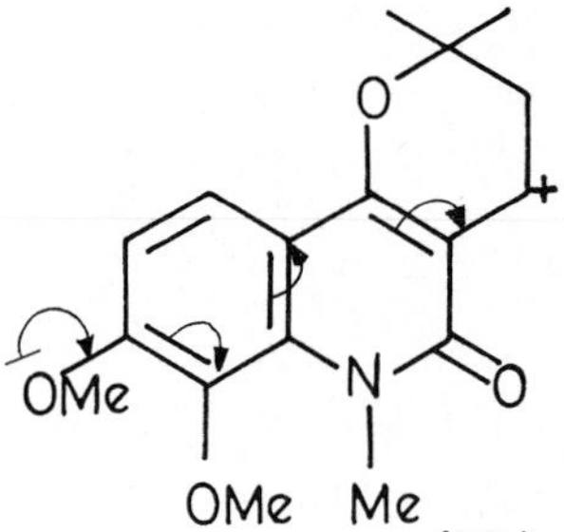

(15)

In order to survey the scope of this novel acid calalysed dimerization reaction of chromenes, the reaction was extended to the antijuvenile hormones precocene II (16) and precocene I (17) {25}. Treatment of precocene II (16) under the same conditions gave the previously reported {26} dimer [as (13)] whose spectroscopic properties are in accord with its structure. The mode of coupling is readily revealed by the size of the geminal coupling constant (J 12.4 Hz) of the methylene group. The structure of the dimer has recently been confirmed by X-ray analysis {26}.

Unexpectedly, the corresponding monomethoxychromene, precocene I (17), did not yield the analogous dimer on exposure to acid but instead afforded a cyclic tetramer (18), $C_{48}H_{56}O_8$, m.p. 242-243°, m/z 760 (M^+, 35%) which has eight tertiary methyl groups (δ_H 0.77, 11.1, 1.22, 1.21, 1.28, 1.40, 1.45, 1.46), four methoxy groups {δ_H 3.84 (2), 3.45, 3.29}, six aromatic proton singlets (δ_H 6.86, 6.42, 6.34, 6.18, 6.16, 5.95) and one ortho-coupled AB system {6.76, 6.27 (both d, J 8.3 Hz)}. It is immediately apparent that the formation of the tetramer must have involved aromatic substitution. The tetramer also has resonances for four methylene groups (δ_C 41.30, 41.29, 38.59, 37.72) and four methines

(δ_C 37.98, 35.57, 32.33, 28.03) which constitute four separate AMX systems {δ_H 4.92 (dd, J 3.8, 12.7 Hz), 2.47 (t, J 12.7 Hz), 1.62 (dd, 12.7 Hz) ; 4.31 (t, J 6.8 Hz), 2.15 (dd, J 6.8, 13.7 Hz), 1.74 (dd, J 6.8, 13.7 Hz) ; 4.22 (dd, J 5.3, 2.6 Hz), 2.37 (dd, J 2.6, 13.3 Hz), 1.82 (dd, J 5.3, 13 Hz) ; 3.68 (dd, J 6.2, 12.8 Hz), 2.41 (t, J 12.8 Hz), 1.63 (dd, J 6.2, 12.8 Hz)}. These must arise from the chromene double bonds of the four monomer units and clearly indicate that the tetramer is cyclic. A possible mode of formation of the tetramer involves attack by electrophilic species (19) on C-6 of a second molecule. The resulting dimer is protonated and reacts with a third monomer molecule. The process is repeated once more to give a linear tetramer which cyclises by electrophilic substitution at the more hindered C-8 position of the starting unit. The relative configuration and the conformation of (18) must however await X-ray analysis.

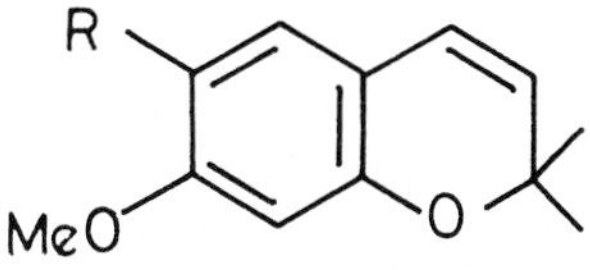

(16) R = OMe
(17) R = H

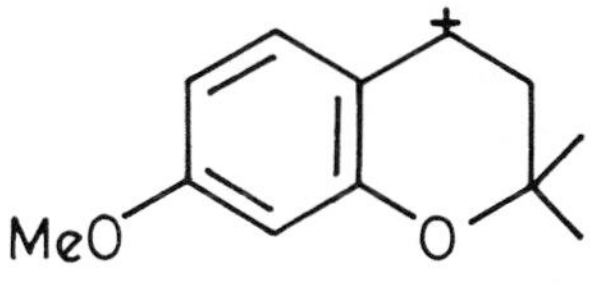

(19)

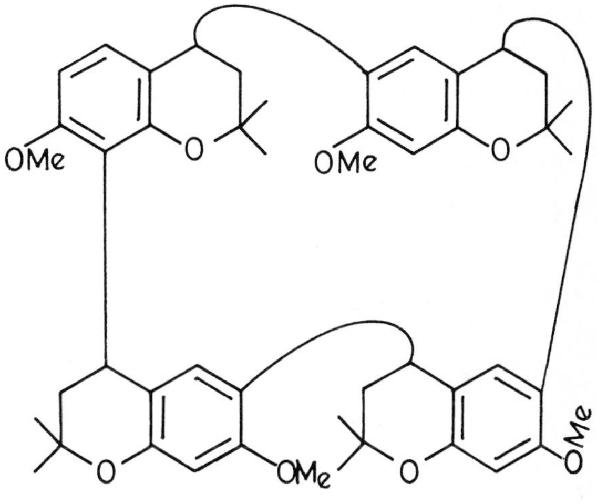

(18)

Recently Cordell and his co-workers have reported {27} that acid treatment of the antitumor alkaloid, acronycine, afforded dimers and trimers whose formation also involved electrophilic substitution of an aromatic ring by a protonated chromene species.

Synthesis of Vepridimerines A-D and a New Synthetic Dimer, Vepridimerine E.

Since acid-catalysed dimerization of veprisine (6) failed to yield the crucial intermediate diene (11) and thus the vepridimerines, we turned our attention to other synthetic methods. First, thermolysis of veprisine (6) was considered. We reasoned that (6)would undergo a thermal heteroelectrocyclic ring opening reaction of the "cyclohexadiene $\rightleftarrows$ hexatriene" type to give the quinoline quinone methide (20) which was expected to yield the requisite diene (11) following a {1,7} sigmatropic shift {28} (Scheme 4). Diels-ALder dimerization of the diene (11) thus generated _in situ_ followed by addition of the free hydroxyl groups to the residual double bonds of the resulting adducts according to schemes 1,2 and 3 above would then give vepridimerines A-D and possibly some new ones. Interestingly, the transformations invisaged in scheme 4 are all thermal pericyclic reactions permissible by the Woodward-Hoffmann rules {29}.

We were gratified to note that on heating veprisine in an evacuated sealed pyrex tube at 200° in an oil-bath for 15 h vepridimerines A-D were obtained in an overall 42% recovery yield after repeated column chromatography together with a novel dimer, vepridimerine E {30}.

Vepridimerine E (21) is the alternative _trans_ isomer of vepridimerine D. Again, models show that with the cyclohexane ring in a chair conformation, Hf and Ha have an ideal W relationship for the observed ^4J coupling (3.0 Hz) and that the position of Hg and Hc relative to the carbonyl groups account for their lowfield chemical shifts (δ_H 3.80 and 3.65 respectively). The 4-quinolone was shown to be on the left hand side by two-dimensional long-range δ_C/δ_H correlation experiments. Thus only Hc correlates with the 2-quinolone carbonyl. Therefore the 2-quinolone is again on the right hand side in E as in D and C. He shows no such correlation {21}.

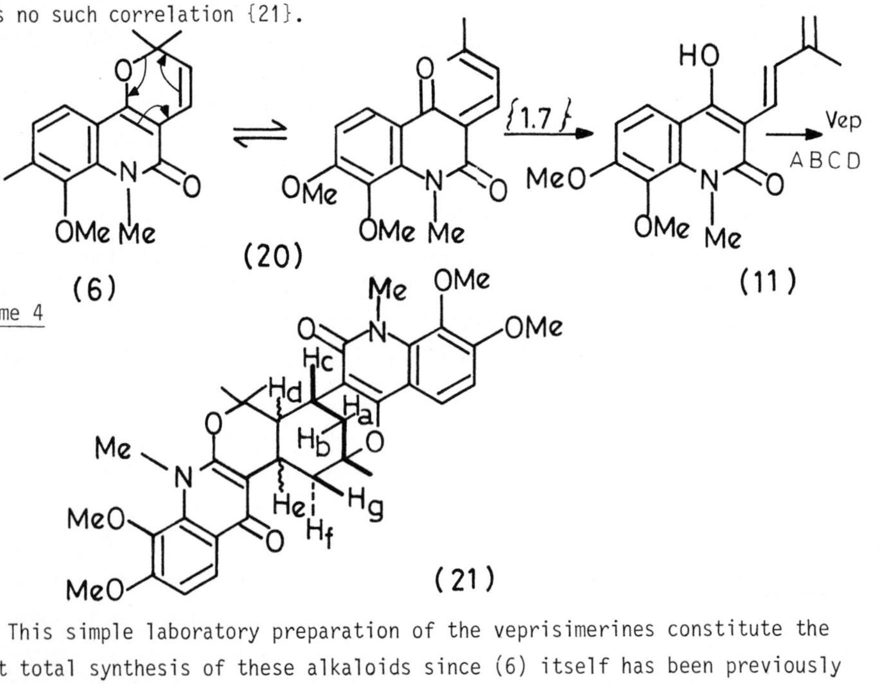

(6) (20) (11) (21)

Scheme 4

This simple laboratory preparation of the veprisimerines constitute the first total synthesis of these alkaloids since (6) itself has been previously synthesized {32,33}.

HOSLUNDIN, ISOHOSLUNDIN, HOSLUNDAL AND HOSLUNDIOL FROM HOSLUNDIA OPPOSITA (LABIATAE)

Hoslundia opposita Vahl is a small shrub which is widely distributed in West Africa. The infusions of its leaves have found wide use in indigenous medicine as an antibiotic, antiseptic, febrifuge, diuretic, cholagogue and purgative. Various parts on the plants are also a popular remedy for snake bite, jaundice, herpes, conjunctivities, vertigo, epilepsy, chest pain, yellow fever, stomach troubles and mental disorders {34}. In the only study carried out on *H. opposita* Kan and co-workers {35} reported strong antibacterial activity for the crude extracts of the entire plant. Our interest in the systematic investigation of the chemical constituent of Cameroonian medicinal plants has led us to a chemical investigation of the leavy twigs of *Hoslundia opposita*. This has resulted in the isolation of four novel flavonoids : hoslundin, isohoslundin, hoslundal and hoslundiol, together with the known compounds oleanolic acid and ursolic acid, and the sterols, stigmasterol, stigmastanol, campesterol, β-sitosterol and their respective glucosides. The antitumor triterpene acid, jacarandic acid {36}, was also obtained.

The first new compound for which we propose the trivial name hoslundin, m.p. 287-288°, analyzed for $C_{23}H_{18}O_7$ (m/z 406). Colour tests with magnesium-concentrated hydrochloric acid together with the uv spectral data {λ_{max} (MeOH) : 250 (ε 18 300), 273 (25 400) and 312 nm (11 000) ; λ_{max} (MeOH + NaOMe) : 253 and 275 nm ; λ_{max} (MeOH + $AlCl_3$) ; 252, 280, 330 and 385 nm ; λ_{max}(MeOH + $AlCl_3$+ HCl) : 252, 280, 328 and 383 nm} suggested that hoslundin was a flavone bearing a hydroxyl group at C-3 or C-5. The uv spectrum particularly resembled that of techtochrysin {37}. Since hoslundin readily gave a strong chelate ferric chloride reaction and showed a hydroxylic proton at low field (δ 12.98 ppm) and the characteristic flavone 3-H signal (δ 6.66), it was clear that it is a 5-hydroxyflavone. Diagnostic fragments at m/z 105 (15%) and m/z 102 (16%) in the mass spectrum further defined the unsubstituted nature of the flavonoid ring-B.

Hoslundin readily formed a monomethyl ether on treatment with diazomethane, m/z 420.1204. The substitution pattern and the nature of the various substituents of O-methylhoslundin viz : a disubstituted 4-pyrone, a methyl and three methoxyl groups, recognised from the detailed analysis of the 200 MHz ^{1}H NMR spectrum, was supported by extensive homonuclear $^{1}H-\{^{1}H\}$ decoupling experiments. The 50.325 MHz ^{13}C NMR chemical shifts, their assignments, the values of the direct and long-range carbon-proton couplings and the observed correlations are listed in Table 2. They were obtained from broad band proton-decoupled and single frequency nuclear Overhauser enchanced (n.O.e) spectra. 2D one-bond δ_C/δ_H correlation permitted the assignment of all the protonated carbons. In the assignment of the ^{13}C resonances, use was also made of chemical shift values and (C,H) coupling constants {1,2}. As in the case of ekeberginine, the structure (23) of O-methylhoslundin was then established by comparing the pattern and size of the couplings with the observed qualitative correlations and considering the carbon resonances in an appropriate sequence. Although the results could be presented in terms of the flavonoid structure, we have decided to use the bond connectivities obtained.

In the 1D proton coupled ^{13}C NMR spectrum of (23) the signal at δ_C 153.2 is a neat doublet (^1J 200 Hz). The 2D long-range δ_C/δ_H correlations show that this resonance is coupled to Ha which in turn shows a ^3J coupling (^3J 6.9) with the carbonyl carbon δ_C 173.6. Further coupling of Ha with three other carbons δ_C 158.4, 122.0 and 113.0 was also observed. The arrangement A_1 fits this information. The alternative arrangement A_2 was discounted at a later stage because of the observed correlation of H_b with δ_C 113.0. The carbon atom with δ_C 145.0 has a ^3J coupling with a methoxyl group (OMe_β) and a ^3J with a methyl group. This information is presen-

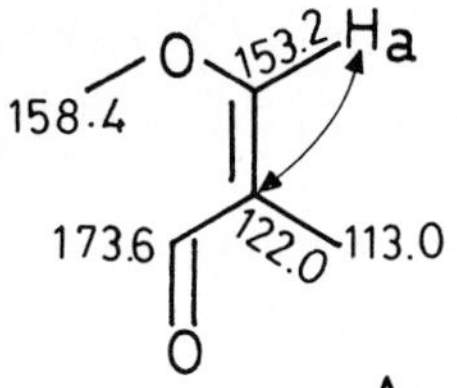

A_1

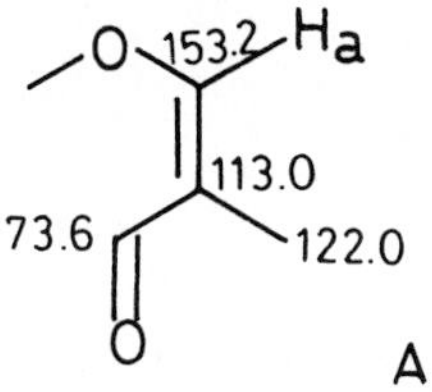

A_2

ted in fragment A_3. A_3 and A_1 on combination afford the fragment I whose overall ^{13}C NMR data are entirely consistent with those reported for 4-pyrone derivatives {38}.

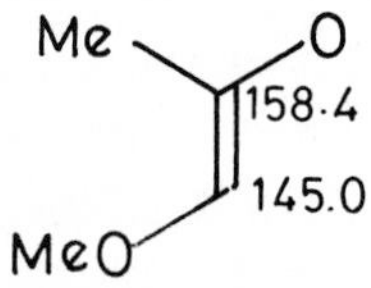

A_3

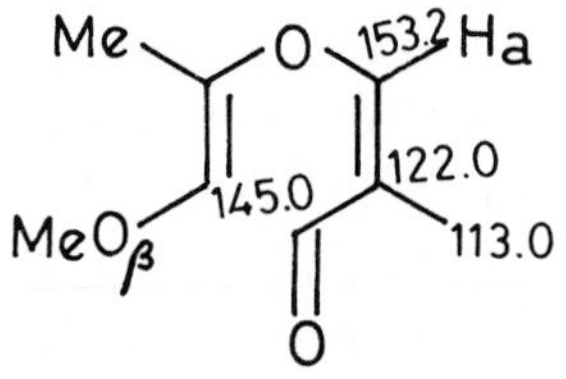

I

TABLE 2

50.325 MHz ^{13}C NMR data of O-methyl hoslundin (23).

Carbon	δ_C	$^1J_{CH}$	Long-range couplings	correlations
4	176.7	-	t (1.5)	H_c, H_b
4"	173.6	-	d (6.9)	H_a
7	162.2	-	dq (3.5, 4.0)	H_b, OMe_γ
2	161.0	-	dt (4.2, 4.9)	H_c, 0
1a	159.6	-	d (5.1)	H_b
5	158.7	-	q (4.0)	OMe_α
2"	158.4	-	dq (8.7)	H_a, Me
6"	153.2	200	-	H_a
3"	145.0	-	Septet (3)	OMe_β, Me
1'	131.33	-	m	H_c
4'	133.27	?	-	-
3',5'	128.9	?	-	-
2',6'	125.9	?	-	-
5"	122.0	-	d (6.8)	H_a
6	113.0	-	dd (5.7, 2.5)	H_b, H_a
4a	112.4	-	t (3.9)	H_b, H_c
3	108.8	?		
8	95.9	?		
OMe_α	62.3	?		
OMe_β	59.9	-		
OMe_γ	56.3	-		
Me	14.7	-		

The immediate surrounding of the already defined monosubstituted benzene ring was determined as follows :Selective irradiation of the aromatic proton δ_{Hc} 6.63 caused a n.O.e of the ortho protons (δ_H 7.86) of the phenyl group which showed a three/bond coupling with δ_C 161.0 2D. Long-range δ_C/δ_H correlation also showed that δ_{Hc} is coupled to δ_C 131.33, 112.4 and 176.7. This evidence leaves no doubt that the orientation of substituents in the immediate neighbourhood of the phenyl group is as defined by part structure B.

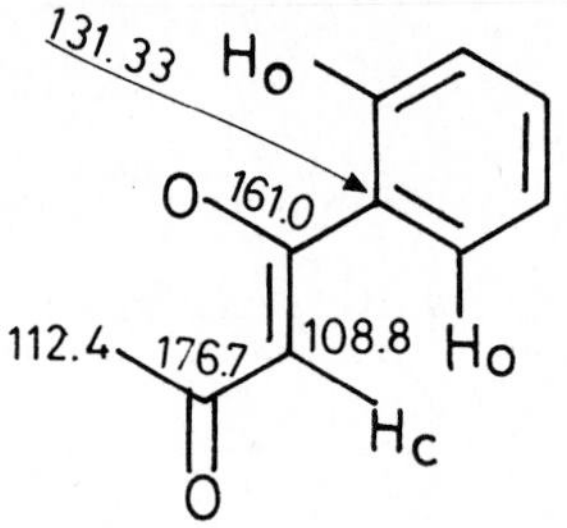

131.33 Ho
O 161.0
112.4 176.7 108.8 Ho
Hc
O
B

Irradiation of the aromatic proton H_b (δ_{Hb} 6.81) gave a n.O.e. with MeO_γ indicating that the latter group, which exhibits a three-bond coupling with δ_C 162.2, must have at least one ortho proton. H_b was further observed to correlate with δ_C 112.4, 113.0, 159.6 and 162.2 thus affording the part structures C_1 or C_2.

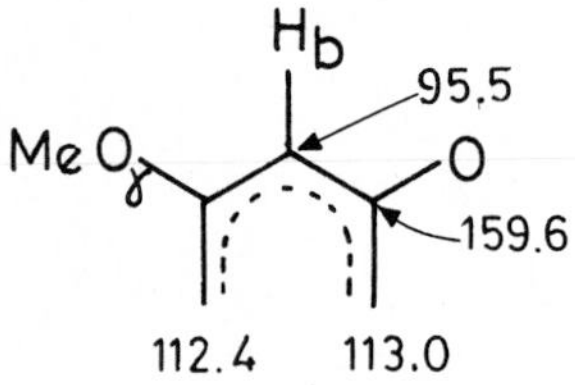

Hb
95.5
MeOγ
O
159.6
112.4 113.0

C_1

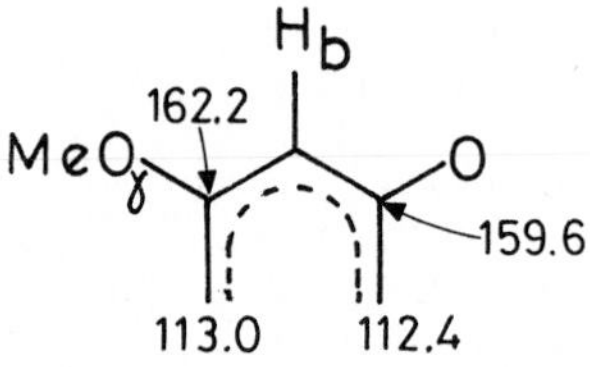

Hb
162.2
MeOγ
O
159.6
113.0 112.4

C_2

From Table 2, only the MeO_α which shows a ^3J coupling with δ_C 158.7 remains to be assigned and is readily accommodated by fragment D.

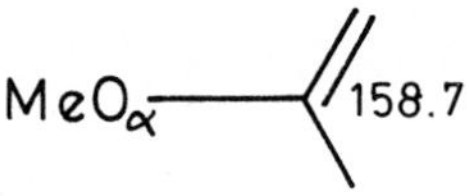

MeOα 158.7

D

If we consider that the free oxygen atom in C_2 and B are the same, a combination of these two fragments with this in mind leads to fragment II.

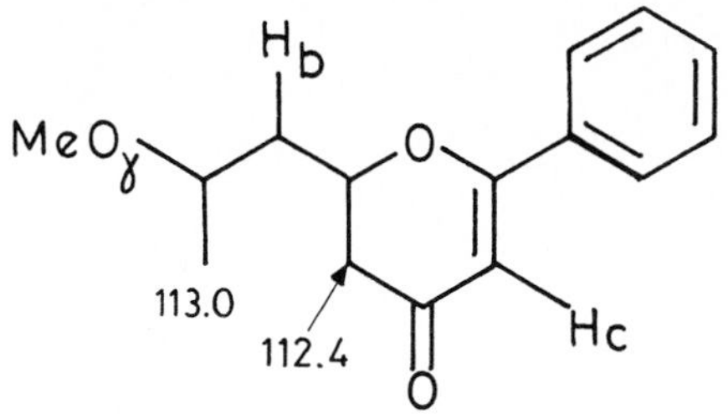

II

Finally, structure (23) was obtained for O-methylhoslundin on putting together part structures I and II and fragment D after due consideration of its stability and biogenetic derivation. Thus the natural product, hoslundin, was assigned structure (22).

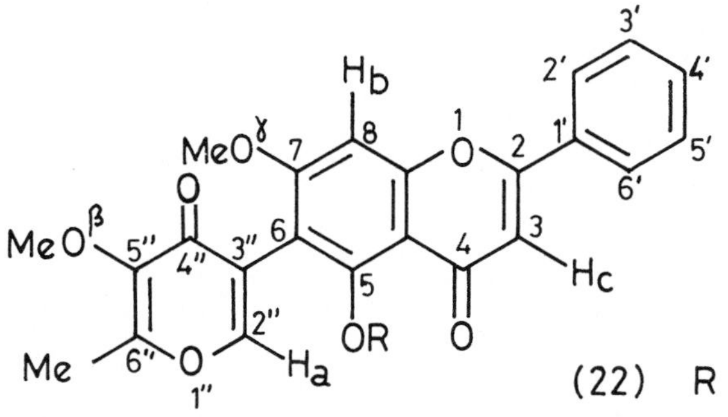

(22) R = H

(23) R = Me

In the course of the isolation of hoslundin (22), it was observed on TLC plate that it co-occurs with a minor product. All attempts to isolate this minor product were unsuccessful as it progressively disappeared increasing the size of the hoslundin spot. We concluded that the minor compound was probably an isomer of hoslundin (22) which was being rapidly converted into hoslundin. A literature survey revealed that 5-hydroxy-6-glycosylflavones actually naturally occur in pairs of the 6- and 8-glycosylisomers {39}. This mainly results from a Wessley-Moser rearrangement involving ring opening of the pyrone ring followed by ring closure on either of the two phenolic hydroxyl groups *ortho* to the carbonyl group. A mixture of the two isomers, the normal and the *iso*-

is thus produced. The minor compound was thus considered to be the naturally occurring isomer of hoslundin, isohoslundin (24).

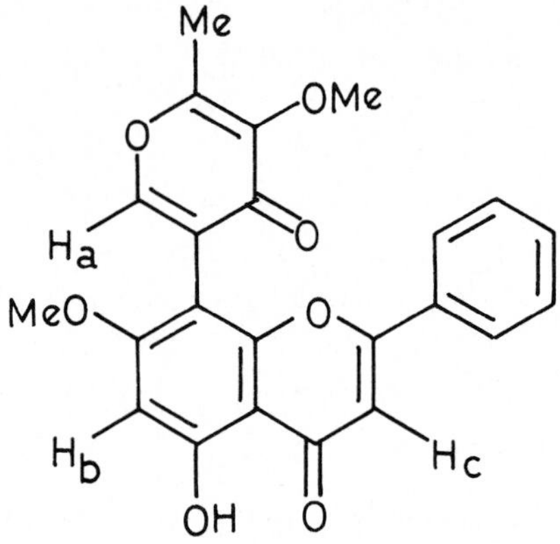

(24)

It is worthy to note that hoslundin (22) and isohoslundin (24) belong to the small group of 4-pyrone derivatives. The members of this group are common flavoring agents and food preservatives and some of them have shown anti-bacterial activity {40}.

The third new flavonoid compound isolated from the more polar fractions of the MeOH extracts of H. opposita, hoslundiol, was assigned structure (25). The substance afforded a u.v. spectrum which was typical of a flavone, showing absorption maxima at λ_{max} 252, 275 and 312 nm. The IR spectrum revealed the presence of a chelated carbonyl at νmax 1642 cm^{-1}, a hydrogen-bonded hydroxyl at 3320 cm^{-1} and free hydroxyls at 3520 cm^{-1}. The mass spectrum was very informative and showed inter alia the existence of an unsubstituted flavone B-ring {m/z 105 (27%) and m/z 103 (22%)}. A high resolution mass measurement on the molecular ion afforded the exact mass m/z 398.1375 in agreement with the formula $C_{22}H_{22}O_7$.

The ^1H NMR spectrum of (25) recorded on the 200 MHz instrument was well resolved and revealed that the methoxymethyl-4-pyrone substituent in (22) and (23) was replaced by a sugar moiety in (25). The chemical shifts of the sugar {δ_H 1.32 (3H, J 6.2 Hz, 3H-6"), 3.83 (dq, J 9.5, 6.2 Hz, H-5"), 3.43 (dd, J 9.5, 2.9 Hz, H-4"), 4.21 (q, J 3.0 Hz, H-3"), 2.93 (ddd, J 14.3, 12.0, 2.5 Hz, H-2"), 1.81 (δt, 14.0, 2.5 Hz, H_{eq}-2"), and 5.45 (dd, J 12.2, 2.0 Hz, H-1")} indicated that it was digitoxose.

The chemical shift δ 5.45 of the anomeric proton H-1" in (25) was very significant. The large coupling constant to H-2" {δ 2.93 (ddd, J 12.0, 2.5 and

14 Hz)} is indicative that H-1" was axially oriented and that the sugar linkage to the flavonoid aglycone was probably β-equatorial.

Hoslundiol was thus defined as 6-C-β-digitoxopyranosyltectochrysin (25).

The structure of hoslundiol, 6-C-β-digitoxopyranosyltectochrysin (25), in which the C-1 carbon atom of the sugar, digitoxose, is directly attached to the aromatic A-ring of tectochrysin is compatible with the biogenetic derivation of these compounds. This is however not the case with hoslundin (22) in which the 4-pyrone moiety is attached at C-2". If one of them has to be the precursor of the other, then an explanation is needed for the attachment of the 4-pyrone substituent to the aromatic A-ring at C-2" rather than at the biogenetically expected C-1" position.

The fourth flavonoid, a yellow oil, $C_{18}H_{14}O_5$ {m/z 282.0890 (70%) (M^+-CO)}, hoslundal, was similarly established to have structure (26).

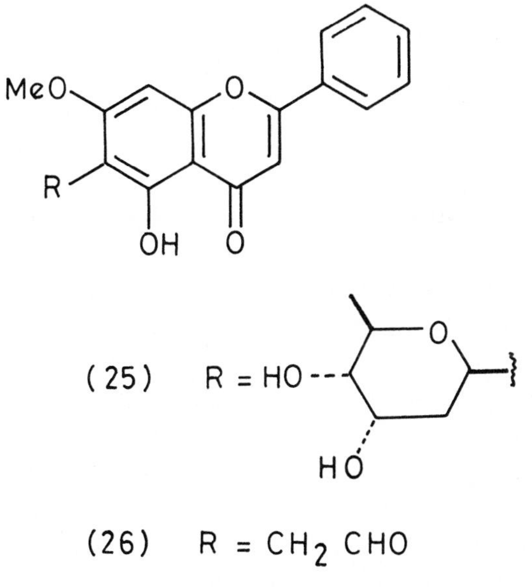

MeO R OH O

(25) R = HO HO

(26) R = CH_2CHO

ACKNOWLEDGMENTS

We thank the International Foundation for Science (IFS) Stockholm and the University of Yaounde Research Grants Commission for support of the work in Yaounde.

REFERENCES

1. H.E. Gottlieb, Israeli J. Chem., 16 (1984) 57.
2. J.L. Marshall, Carbon-carbon and Carbon-Proton NMR Couplings. Applications to Organic Stereochemistry and Conformational Analysis, Verlag Chemie International, Deerfield Beach, Florida, 1983.
3. H. Kessler, C. Griessinger, J. Zarbock, and H.R. Loosli, J. Mag. Reson., 57 (1984) 331.
4. C.C.J. Culvenor, P.A. Cockrum, J.A. Adgar, J.L. Frahm, C.P. Gorst-Allman, A.J. Jones, W.F.O. Marasas, K.E. Murray, L.W. Smith, P.S. Steyn, R. Vleggar, and P.L. Wessels, J. Chem. Soc. Chem. Commun., (1983) 1259.
5. H. Seto, H.K. Furihata, and N. Otake, Tetrahedron Letters (25) (1984) 337 ; M.J. Gidley and S.M. Bociek, J. Chem. Soc., Chem. Commun., (1985) 220.
6. C. Bauer, R. Freeman, and S. Wimperis, J. Mag. Reson., 58 (1984) 526.
7. C.W.L. Bevan, D.E.U. Ekong, and D.A.H. Taylor, Nature, 206 (1965) 1323.
8. J.I. Okogun, V.U. Enyenihi, and D.E.U. Ekong, Tetrahedron, 34 (1978) 1221.
9. D. Lontsi, J.F. Ayafor, B.L. Sondengam, J.D. Connolly, and D.S. Rycroft, Tetrahedron Letters, 26 (1985) 4248.
10. B.S. Joshi and D.H. Gawad, Indian J. Chem. 12 (1974) 437
11. B.S. Joshi, V.N. Kamat, D.H. Gawat, and T.R. Govindachari, Phytochemistry, 11, (1972) 2065.
12. I. Mester, The Alkaloids of the Rutales, in P.G. Waterman and M.F. Grundon (Eds.), Chemistry and Chemical Taxonomy of the Rutales, Academic Press, London, 1983, pp. 31.
13a. R. Freeman and G.A. Morris, J. Chem. Soc., Chem. Commun., (1978) 684.
13b. A. Bax and G.A. Morris, J. Mag. Reson., 42 (1981) 501.
14. J.F. Ayafor, B.T. Ngadjui, and B.L. Sondengam, Submitted for publication.
15. J.F. Ayafor, T.B. Ngadjui, and B.L. Sondengam, Tetrahedron Letters, (1980) 3293.
16. T.B. Ngadjui, J.F. Ayafor, B.L. Sondengam, J.D. Connolly, D.S. Rycroft, S.A. Khalid, P.G. Waterman, N.M.D. Brown, M.F. Grundon, and V.S. Ramachandran, Tetrahedron Letters, 23 (1982) 2041.
17. J. Reisch, I. Mester, K. Szendrei, and J. Körösi, Tetrahedron Letters, (1978) 3687.
18. I. Mester, J. Reisch, and J. Körösi, Liebigs Ann. Chem., (1979) 1785.
19. L. Jurd and M. Benson, J. Chem. Soc. Chem. Commun., (1983) 92
20. L. Jurd, R.Y. Wong, and M. Benson, Aust. J. Chem., 35 (1982) 2505.
21. J.F. Ayafor, B.L. Sondengam, B.T. Ngadjui, J.D. Connolly and D.S. Rycroft, Unpublished Results.
22. M.F. Grundon, Tetrahedron Letters, (1981) 3105.
23. C.S. Barnes, M.I. Strong and J.L. Occolowitz, Tetrahedron 19 (1963) 839
24. J.F. Ayafor, B.T. Ngadjui, B.L. Sondengam, J.D. Connolly and D.S. Rycroft, In Preparation.
25. W.S. Bowers, T. Ohta, J.S. Cleere and P.A. Marsella, Science, 193 (1976) 542.
26. B.M. Fraga, V.P. Garcia, A.G. Gonzaley, M.G. Hernandez, J.R. Hanson and P.T. Hitchcock, J. Chem. Soc., Perkin Trans I, (1983) 2687.
27. S. Fumayana and G.A. Cordell, Planta medica, 48 (1983) 263.
28. G. Desimoni, G. Tacconi, A. Barco, and G.P. Pollini, Natural Products Synthesis through Pericyclic Reactions, ACS Monograph 180, American Chemical Society, Washington D.C. 1983 pp. 361.
29. I. Flemming, Frontier Orbitals and Organic Chemical Reactions, John Wiley and Sons Ltd, Chichester, 1978 pp. 86.
30. J.F. Ayafor, B.L. Sondengam, J.D. Connolly and D.S. Rycroft, Tetrahedron Letters, 26 (1985) 4529.
31. J.F. Ayafor, B.L. Sondengam, D.S. Rycroft, and J.D. Connolly, In Preparation
32. J.F. Ayafor, B.L. Sondengam and B.T. Ngadjui, Phytochemistry 21 (1982) 2733
33. M. Ramesh, P.S. Mohan, and P. Shanmugam, Tetrahedron, 40 (1984) 4041.
34. E.S. Ayensu, Medicinal Plants of West Africa, Reference publications Inc. Michigan 1978, pp. 162.

35. M.R. Khan, G. Ndaalio, M.H.H. Nkunya, H. Wevers, and A.N. Sawhney, Planta Medica. Supplement (1980) 91.
36. M. Ogura, G.A. Cordell, and N.R. Farnsworth, Lloydia, 40 (1977) 157.
37. T.J. Mabry, K.R. Markham, and M.B. Thomas, The Systematic Identification of flavonoids, Springer Verlag, New York, 1970 pp. 69.
38. C.A. Kingsbury, M. Clifton, and J.H. Looker, J. Org. Chem. 41 (16) (1976) 2777.
39. J. Chopin and M.L.Bouillant, C-Glycosylflavonoids, in : J.B. Harborne, T.J. Mabry and H. Mabry (Eds.), The Flavonoids, Chapman and Hall, London 1975 pp. 632.
40. T. Kotani, I. Ichimoto, and C. Tatsumi, Hukko Kogaku Zasshi, 51 (1973) 66.

THE STRUCTURE OF MYCORHODIN

LLOYD M. JACKMAN, KIMBERLY L. COLSON AND TIKAM JAIN

1. ISOLATION AND PRELIMINARY CHARACTERIZATION

In the course of screening a variety of microorganisms for antibiotic activity, a strain, designated AAA566, of Streptomyces produced a broth, the solvent extract of which exhibited high activity against gram positive and gram negative organisms and anaerobes. Crystallization of the crude extract from dichloromethane afforded two weakly active compounds, the structures of which were established as 1 and 2.[1] The major source of activity

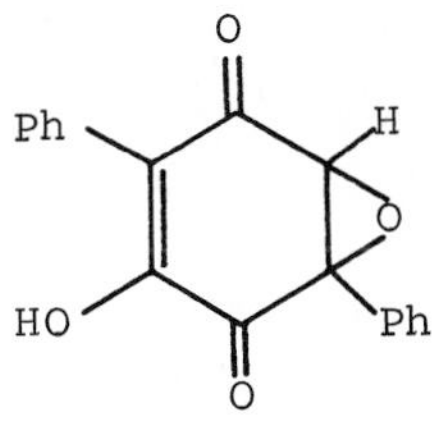

1

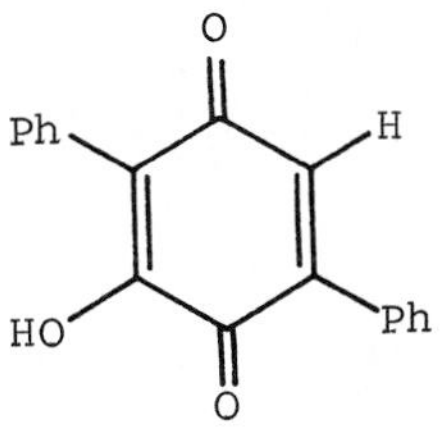

2

was recovered from the mother liquors and, after rigorous chromatographic purification and crystallization, was obtained as brick red crystals, m.p. 198-200°C, from CH_2Cl_2/CH_3OH. This compound showed very high activity against gram positive organisms.

The molecular formula was found to be $C_{34}H_{35}O_{15}$ by microanalysis and FAB mass spectrometry. The uv/visible spectrum (methanol) possesses λ_{max} 2158 (ε 48,000), 250 (ε 25,600), 2750 (ε 19,500) 415 (ε 5,215), and 470 (ε 5,410) nm. On the addition of base, the long wave length band moves to 580 (ε 7,400) nm, indicating that the chromophore bears a phenolic hydroxyl group. Typical color reactions for quinones are positive.

At this point, a literature search revealed that a similar antibiotic, mycorhodin, had been isolated from a strain of *Streptomyces* by Bristol Laboratories.[2] Direct comparison of the two substances established their identity. Since the structure was unknown and probably novel and since the antibiotic is also a potential antitumor lead, a structural investigation was initiated.

During the purification of mycorhodin, an isomer, isomycorhodin, was isolated. Equilibration of the two isomers occurs under acid conditions, and mycorhodin is the major isomer. Both isomers were used in the subsequent structural studies.

Although mycorhodin crystallizes extremely well from a number of solvents, the resulting crystals we obtained all existed in the same space group as α-quartz and their X-ray diffraction data could not be refined presumably because of twinning. Isomycorhodin failed to yield crystals large enough for X-ray diffraction studies. As initial attempts to obtain simple derivatives (acetates, etc.) afforded complex mixtures, it was decided that an exhaustive nmr spectroscopic study using a wide range of 1D and 2D techniques offered the best approach to the structural problem.

2. PROTON AND ^{13}C SPECTROSCOPY

Preliminary inspection of the proton spectrum (Figure 1) of mycorhodin indicates the presence of four methoxyl and two C-methyl

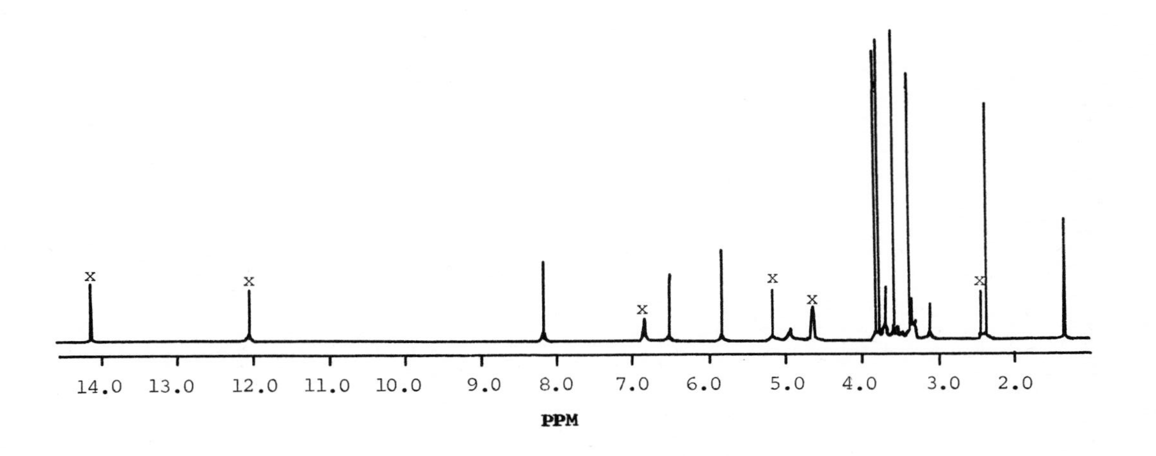

Figure 1: 360 MHz proton spectrum of mycorhodin in $CDCl_3$. "X" denote protons which exchange with D_2O.

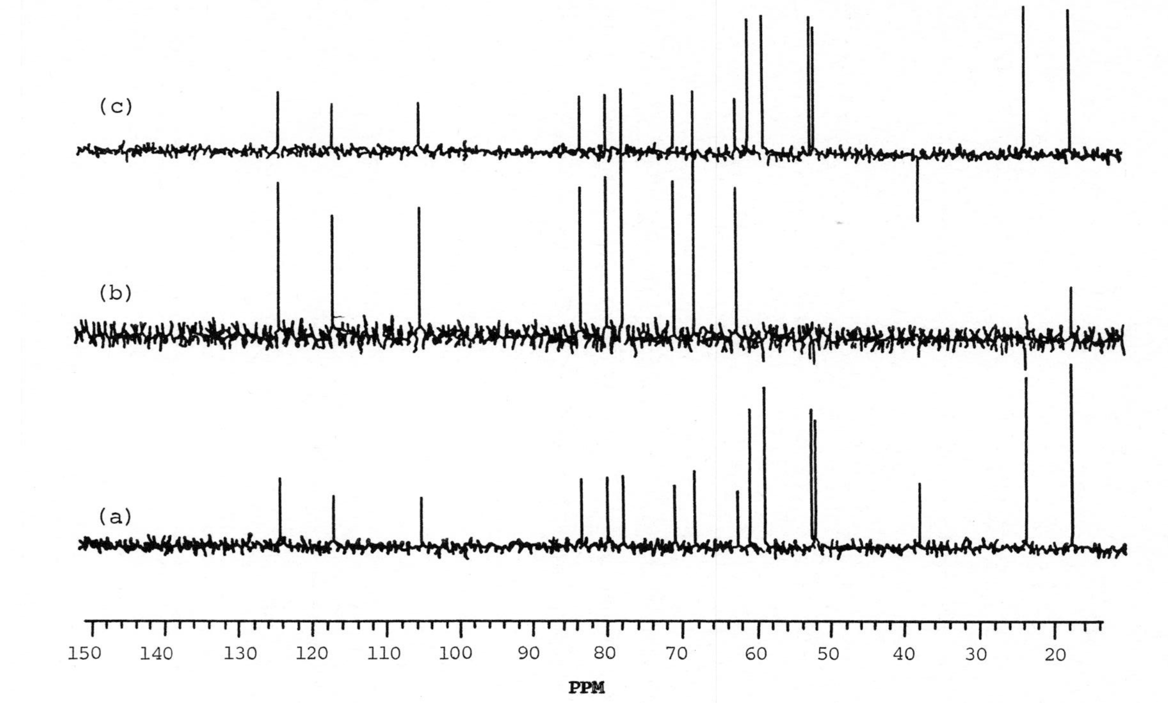

Figure 2: The DEPT spectra of isomycorhodin in $CDCl_3$. A delay τ of 3.7 milliseconds was used. In spectrum (a) an anlge θ of 45° selected for only protonated carbons (all positive). In spectrum (b) only methine carbons appear as a result of $\theta = 90^o$. A pulse angle $\theta = 135^o$ spectrum (c) distinguished methylene carbons (negative intensity) from other protonated carbons.

groups. In addition, the molecule has six readily exchangeable protons. Of the remaining eleven protons, five appear to be associated with a sugar residue. The spectra of mycorhodin and isomycorhodin differ principally in the chemical shifts of the sugar protons.

The DEPT ^{13}C spectra[3] (Figure 2) confirm the presence of six methyl groups and show that, of the remaining eleven non-exchangeable protons, two are present as a methylene group, the remainder being methines. Correlation of the proton and ^{13}C chemical shifts was established by heterocorrelated 2D spectroscopy (Figure 3). All 34 carbon resonances can be resolved (Table 1), and the existence of five signals in the region 170-200 ppm indicates the presence of five carbonyl groups.

3. THE SUGAR RESIDUE

We decided to attack the identification of the sugar residue first, using both 1D and 2D techniques. The COSY spectrum of isomycorhodin allowed the establishment of a six carbon sequence terminating in a C-methyl group (δ 1.23 ppm). By using selective decoupling and deuterium exchange, it was possible to simplify some of the multiplets and to evaluate all of the vicinal coupling constants (Table 2). The values of $^{3}J_{3,4}$ and $^{3}J_{4,5}$ corresponds to diaxial interactions and that of $^{3}J_{2,3}$ to an equatorial-axial coupling. The sugar is thus established as a rhamnose derivative. NOE experiments show the presence of methoxyl substituents at positions 2 and 4. Finally, the anomeric configuration follows from the observation of a strong (diaxial) NOE between H(1) and H(5) in mycorhodin which is absent in isomycorhodin. The ^{13}C chemical shift (78.8 ppm) of the anomeric carbon in mycorhodin is too shielded for an O-glycoside, and since H(1) exhibits a vicinal (8.5 Hz) coupling with an exchangeable proton, mycorhodin is a β-NH-2,4-di-O-methyl-rhamoside.

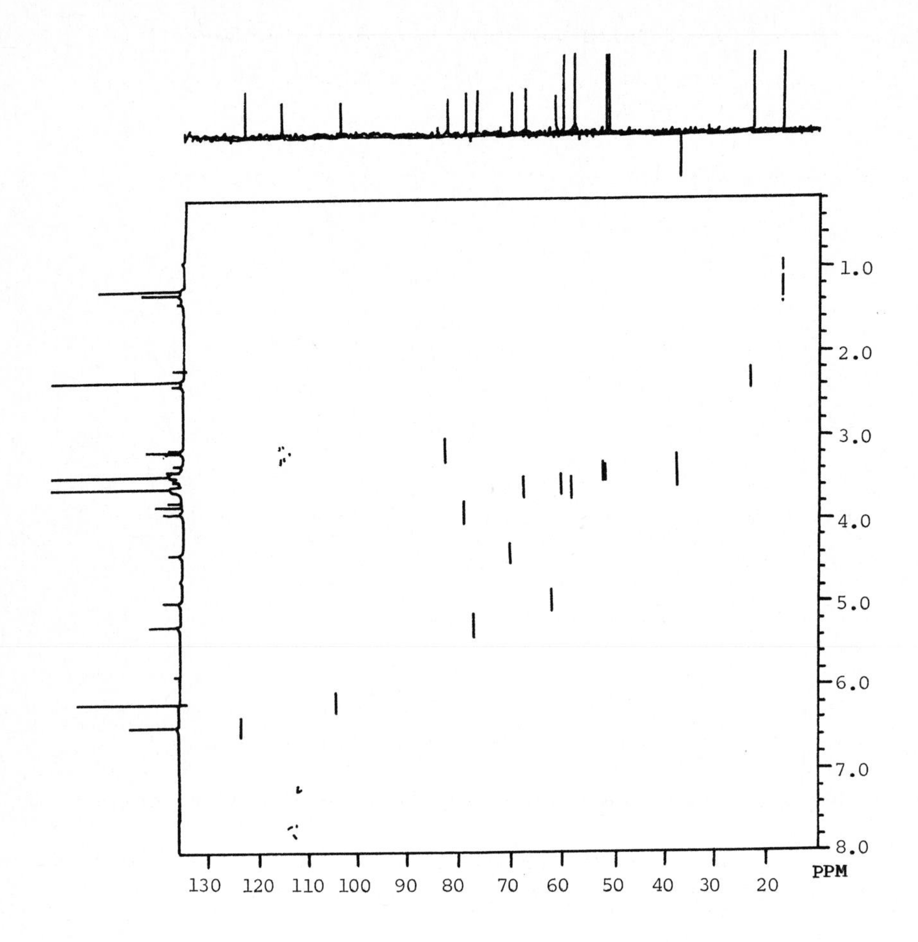

Figure 3: Heteronuclear correlated spectrum of isomycorhodin in $CDCl_3$ acquired using a 256 x 512 data matrix. The pulse sequence used was 90° ^1H-t_1-90° ^1H-τ-180° ^1H, 180° ^{13}C-τ-90° ^1H-t_1-τ-90° ^1H, 90° ^{13}C-τ_m-^1H BB, ^{13}C-acquire in which the delays, τ and τ_m, were set to 3.7 milliseconds and 1.85 milliseconds, respectively. The initial value for the incremented delay was 0.003 microseconds.

TABLE-1a: ^{1}H AND ^{13}C CHEMICAL SHIFT DATA FOR MYCORHODIN IN $CDCl_3$

C(#)	δ(ppm)	#H's	(ppm) attached proton(s)	C(#)	(ppm)	#H's	δ(ppm) attached proton(s)
1	196.3	0		18	109.8	0	
2	190.9	0		19	104.2	1	5.882
3	188.7	0		20	84.5	0	
4	179.3	0		21	82.8	1	3.137
5	172.1	0		22	79.5	1	3.753
6	162.4	0		23	79.2	0	
7	160.1	0		24	78.8	1	4.697
8	147.2	0		25	75.2	1	3.708
9	143.3	0		26	73.4	1	3.354
10	142.8	0		27	62.6	1	4.972
11	140.1	0		28	62.7	3	3.806
12	133.9	0		29	61.3	3	3.618
13	124.3	1	6.55	30	52.6	3	3.856
14	124.3	0		31	52.4	3	3.416
15	119.7	0		32	38.0	2	3.562,3.373
16	118.6	0		33	23.9	3	2.412
17	116.3	1	8.219	34	18.0	3	1.372

TABLE-1b: ^{1}H AND ^{13}C CHEMICAL SHIFT DATA FOR MYCORHODIN IN $CDCl_3$

C(#)	δ(ppm)	#H's	(ppm) attached proton(s)	C(#)	(ppm)	#H's	δ(ppm) attached proton(s)
1	196.6	0		18	109.7	0	
2	190.2	0		19	105.3	1	6.20
3	188.8	0		20	84.6	0	
4	179.6	0		21	83.2	1	3.12
5	171.9	0		22	79.9	1	3.79
6	163.6	0		23	79.2	0	
7	160.1	0		24	77.6	1	5.23
8	147.3	0		25	70.9	1	4.27
9	143.1	0		26	68.6	1	3.59
10	142.7	0		27	62.8	1	4.97
11	140.5	0		28	60.7	3	3.58
12	133.9	0		29	58.9	3	3.59
13	124.1	1	6.51	30	52.6	3	3.37
14	124.0	0		31	52.2	3	3.42
15	120.1	0		32	38.1	2	3.34, 3.57
16	118.7	0		33	23.8	3	2.35
17	116.7	1	8.45	34	17.8	3	1.23

TABLE-2: VICINAL COUPLING CONSTANTS (Hz) IN THE SUGAR RESIDUES OF MYCORHODIN AND ISOMYCORHODIN IN $CDCl_3$

$^3J_{i,j}$	Mycorhodin	Isomycorhodin
1,2	1.5	1.6
2,3	2.4	3.0
3,4	9.2	8.9
4,5	9.2	8.9
5,6	6.1	6.2

4. A PARTIAL STRUCTURE

In addition to the sugar residue, two other structural features, 3 and 4, can be established from COSY and NOE experiments.

	--- CH_2 ------------	CH ------------	OH
δ(H)	3.37 3.56	4.97	4.86 ppm
δ(C)	38.0	62.6 ppm	
^2J	19.0 Hz		
^3J	6.8	8.5 ppm	

3

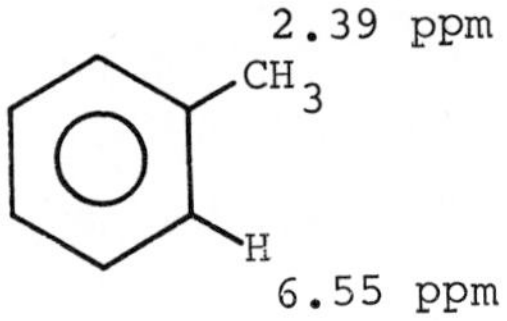

2.39 ppm
CH_3
H
6.55 ppm

4

The partial structure presented in Figure 4 summarizes the structural findings to this point and emphasizes the formidable nature of the remaining problem.

$C_{16}H_3$ —
- ----- ($C_8H_{16}NO_4$)
- ----- OH Chelated Phenolic
- ----- OH Phenolic
- ----- OH tert. Aliphatic
- ----- OCH_3
- ----- OCH_3
- ----- CH_3 Aromatic
- ----- ($-CH_2-CHOH-$)
- ----- (5 x C = O) Based on ^{13}C δ's

$C_8H_{16}NO_4$ = NH-2,4-Di-O-methylrhamnoside

Figure 4. A Partial Structure for Mycorhodin

5. THE CHROMOPHORE

Because of the very heavily substituted nature of the chromophore it has been necessary to exploit fully the information concerning connectivities which can be inferred from observations of long range carbon-hydrogen coupling constants. Three types of experiments are useful in this context. They are selective heteronuclear J-resolved spectroscopy,[4] long range heteronuclear COSY,[5] and, for long range couplings involving OH and NH, deuterium exchange.

Selective heteronuclear J-resolved spectroscopy is particularly effective for a system, such as the chromophore of mycorhodin, in which the few protons present are widely separated in chemical shift and are, therefore, readily subjected to the selective (soft) 180° proton pulse. Table 3 shows the information available from the experiments in which the protons at 6.20 and 8.45 ppm are irradiated. In particular, the two protons share three bond couplings to two carbon atoms, an observation which allows the formulation of the partial structure, 5. Some additional connectivities involving a hydroxyl proton are established by dueterium exchange.

The long range heteronuclear COSY experiment, while it does not readily facilitate the evaluation of magnitudes of coupling constants, has several major advantages. It can establish correlations arising from very small couplings constants and from couplings to protons which, because of overlap or near overlap, cannot be selectively irradiated. In the present instance, it complements the data from the selective heteronuclear J-resolved experiments for the 6.20 and 8.45 ppm protons. As shown in Figure 5 and

TABLE-3: LONG RANGE ^{13}C, ^{1}H COUPLING CONSTANT DATA FOR ISOMYCORHODIN IN $CDCl_3$ FROM SELECTIVE HETERONUCLEAR J-RESOLVED SPECTROSCOPY

$\delta(^{13}C)$ ppm	$\delta(^{1}H)$ ppm)	J (CH) Hz	J (CXH) Hz
179.6	6.20	8.6	4.3
179.6	8.45	4.2	
162.6	8.45	1.6	5.2
124.0	8.45	6.5	
118.7	8.45	6.6	5.5
118.7	6.20	6.1	
116.7	8.45	174.2	
105.3	6.20	169.3	4.9

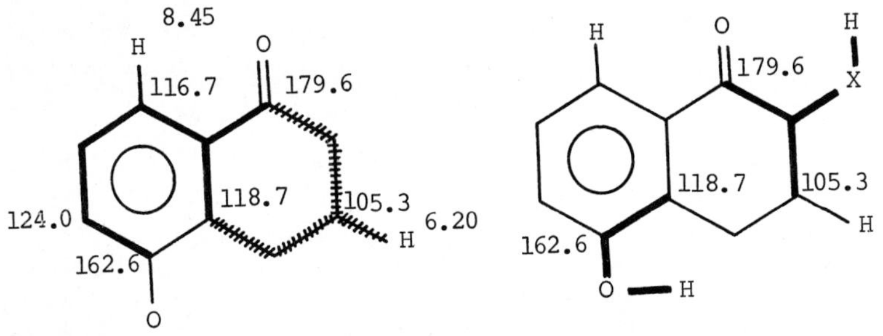

8.45 H O 116.7 179.6 124.0 118.7 105.3 6.20 H 162.6 O

H O H 179.6 X 118.7 105.3 H 162.6 O H

5

Table 4, analysis of the cross peaks leads to the elucidation of the "juglone" moiety 6 and the nature of attachment of the N-glycoside. The nitrogen is thus present as a vinologous amide which is consistent with its lack of basicity.

TABLE-4: CROSS PEAKS OBSERVED IN THE LONG RANGE ^{13}C, ^{1}H HETEROCOSY EXPERIMENTS WITH ISOMYCORHODIN (s=strong, m=medium, w=weak, d=one bond correlations)

	14.3	11.9	8.45	6.70	6.51	6.20	5.23	4.97	3.42	3.37	3.57	3.34	2.35
196.6			s										
190.2													
188.8						w							
179.6			s	w		s							
171.9										s			
162.6	s					w							
160.1		s											
147.3						m	w						
143.1[a]		w											
142.7[a]													s
140.5													
133.9													
124.1					d							s	
124.0	s		s										
120.1		s			s						w	w	
118.7	s		s			m							
116.7			d										
109.7		s			s								s
105.3				s		d							
84.6								s	s				
79.2								s					
62.8								d					
52.6										d			
52.2									d				
38.1					s						d	d	
23.8					s								d

[a]Assignment of the cross peaks to the 142.7 and 143.1 resonances may be ambiguous.

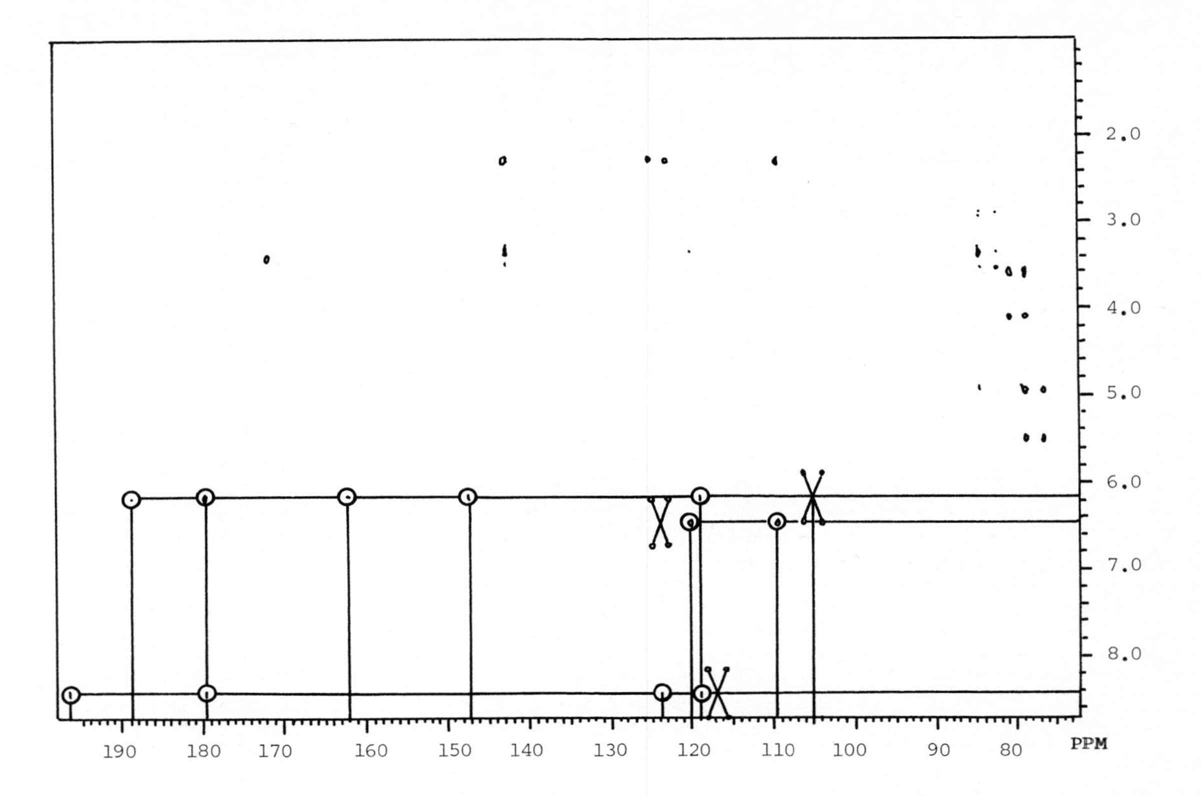

Figure 5: The long range heteronuclear COSY spectrum of isomycorhodin in $CDCl_3$ acquired at 300 MHz on a Bruker AM-300. The initial delay (t_1=0.03 microsec.) was incremented through 256 values. The phase cycling programs used were: $\phi_1=0^{\circ}, 180^{\circ}$, $\phi_2=0^{\circ}, 180^{\circ}, 0^{\circ}, 90^{\circ}, 270^{\circ}, 270^{\circ}, 90^{\circ}$, $\phi_3=0^{\circ}$, and $\phi_4=0^{\circ}, 180^{\circ}, 180^{\circ}, 90^{\circ}, 270^{\circ}, 270^{\circ}$.

The long range heteronuclear COSY spectrum has also been used to establish a partial structure 7 for a ring to which is attached the $-CH_2CH(OH)$-fragment and also to demonstrate the presence of a carbomethoxy group. These correlations are summarized in Table 4.

The final structure requires the linking of a carbonyl group and the two partial structures 6 and 7. Of the limited number of possible structures, most can be eliminated on the basis of the observed carbonyl stretching frequencies or from ^{13}C chemical shift arguments. The two structure 8 and 9 represent the most likely candidates. It should be noted that the final structural proof should follow from the three bond couplings of the tertiary hydroxyl proton to three vicinal carbon atoms. Correlations involving the aliphatic hydroxyl functions have not yet been observed evidently because their exchange lifetimes are short compared with the length of the pulse sequence. We hope to overcome this problem by using dimethyl sulfoxide as a cosolvent and, if necessary, cooling the solution below ambient temperatures, but at the time of writing, the results of these experiments are unavailable.

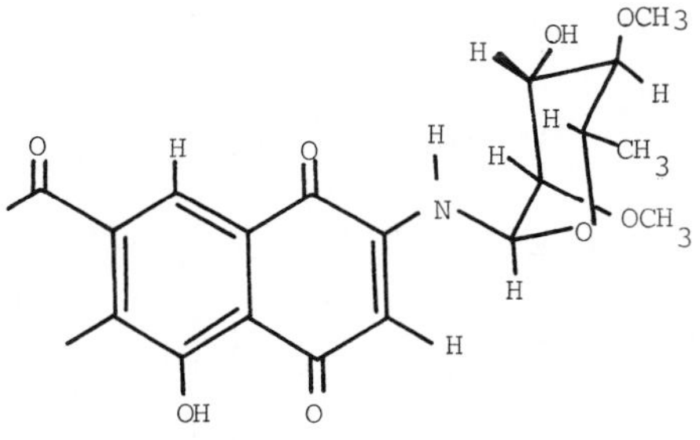

6

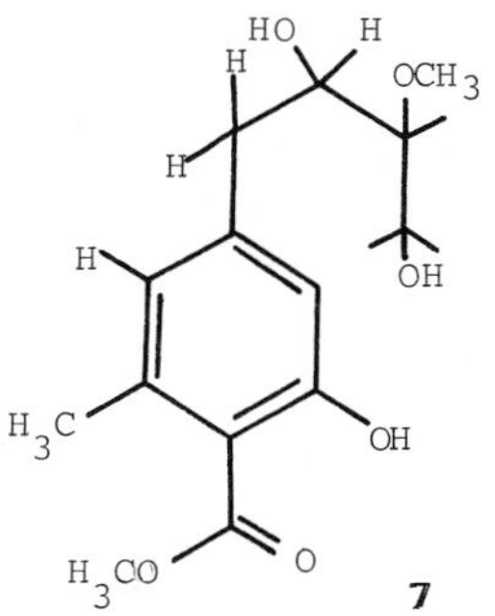

7

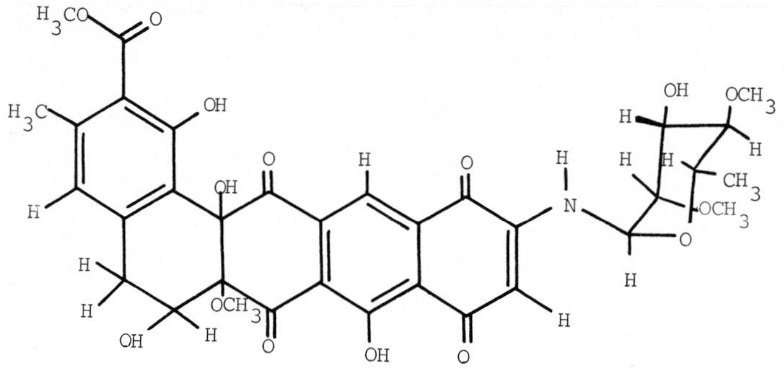

H3CO O
H3C OH
O H O H OH OCH3
H OH H H
N H CH3
H OCH3
O
H H
H OCH3 H
OH H O OH O

8

OH OCH3
H H
HO H O H O H H CH3
H OCH3 H
H N OCH3
O
H
H OH H
O OH O
H3C OH

H3CO O

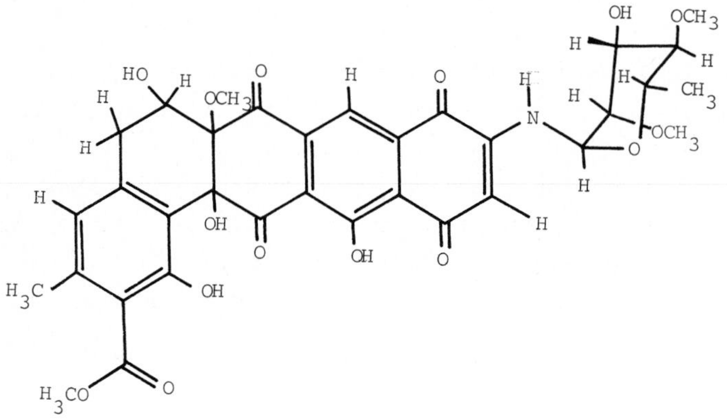

9

REFERENCES

1. K.L.Colson, L.M.Jackman, T.Jain, G.Simolike and J.Keeler, Tetrahedron Lett., 1985, 26, 4579.

2. M.Misiek, A.Gourevitch, B.Heinemann, M.J.Cron, D.F.Whitehead, H.Schmitz, I.R.Hooper, J.Lein, Antibiotics and Chemotherapy, 1959, 9, 280.

3. D.M.Doddrell, P.T.Pegg, and M.R.Bendall, J.Magn.Reson., 1982, 48, 323.

4. A.Bax, J.Magn.Reson., 1983, 53, 517.

5. A.Bax and R.Freeman, J.Am.Chem.Soc., 1982, 104, 1099.

6. Y.Sato, M.Geckle, and S.Gould, Tetrahedron Lett., 1985, 26, 4019.

USE OF CHIROPTICAL PROPERTIES IN STRUCTURE DETERMINATION OF NATURAL PRODUCTS

G. Snatzke

1 INTRODUCTION

"Chiroptical properties" are spectroscopical properties of "optically active" compounds which are given when polarized light is allowed to interact with them. They can be detected in absorption as well as in emission experiments, but only for the first type instrumentation is commercially available. The three best known chiroptical properties are "Circular dichroism" (CD), "Optical rotatory dispersion" (ORD) and "Ellipticity" (no abbreviation in use). To understand these phenomena we have to know about the properties of light and of (chiral) molecules, as well as about the interaction between polarized light and the electron cloud of the molecules.

2 LIGHT

Depending on the conditions of an experiment light can be described by either a stream of photons of energy $E = h\nu$, or as electromagnetic radiation of frequency ν. In the latter way, if z is the direction of propagation of light, an oscillating electric field vector $\underline{E}$ could be measured at time t_o along z which is perpendicular to this direction, and whose magnitude changes like a cosine function. At the same time a magnetic field vector $\underline{H}$ is present, which is perpendicular to both z and $\underline{E}$, so that $\underline{E}$, $\underline{H}$ and z form a righthanded co-ordinate system in this order; $\underline{H}$ is in phase with $\underline{E}$ and proportional to it (if proper units are used, their magnitudes are the same). When observed at a given distance z_o from the light source, both $\underline{E}$ and $\underline{H}$ are found to oscillate with time (again like a cosine function). If for all rays the direction of $\underline{E}$ is the same, then we call this light "linearly polarized", and such can be made from natural light (all directions for the electric field vectors are possible) with the help of a polarizer, as e.g. a Nicol or Rochon prism.

Fresnel has already at the beginning of last century shown that mathematically linearly polarized light can always be thought to be composed of two rays of "circularly polarized" light of opposite senses of helicity, if their refractive indices and their intensities are the same. In order to imagine circularly polarized light we think of an auxiliary cylinder whose axis coincides with the z-axis. Onto its surface we wind a regular helix. If the tips of all electric field vectors $\underline{E}$ along z lie on such a helix then this light is called circularly polarized. Obviously the helix can be righthanded or lefthanded, and so can be the circularly polarized light (definition of sense of helicity: if the direction of the thumb of the right hand follows the direction of translation of the light and the other fingers the rotation, we have a righthanded helix, otherwise a lefthanded). It should be noted that the sense of helicity does not depend on the way how we inspect the helix, along or against the direction of propagation of the light.

Viewing such light at a given distance z_o from the light source <u>against the ray</u> (here the direction is important !) we see the electric field vector $\underline{E}$ of constant length rotating clockwise (for righthanded polarized light) or anticlockwise (for lefthanded) with constant angular frequency.

This splitting of linearly polarized light into two components of circularly polarized one is not only a formal mathematical process but can also be proved by experiment. The most general type of polarized light is the "elliptically polarized" one; in this case we have to think of an auxiliary cylinder of elliptical cross-section, so the tip of $\underline{E}$ (and also $\underline{H}$) describes an ellipse with time (for details and figures see e.g. ref. 1).

The main characteristics of a ray of light are wavelength λ, frequency ν, period T, and velocity c. They are correlated by the well-known equations

$$c = \lambda . \nu \tag{1}$$

and

$$T = 1/\nu \tag{2}$$

It should be noted that only ν (and T) are constant in any medium, whereas the others are not (see below). Nevertheless chemists still today use to draw their UV/VIS-spectra with the wavelength λ as abscissa instead of the more preferable frequency.

3 MOLECULES

3.1 General

Since light of frequency in the UV/VIS-range oscillates too quickly for exciting nuclear motions the description of molecules can make use of the Born-Oppenheimer approximation, in which the positions of the nuclei are considered to be constant during excitation of the electron cloud. The latter is described by molecular orbitals, and to understand (and even to predict) chiroptical properties of simple molecules the use of LCAO-MOs is sufficient. Any molecule can adopt different states of discreet values of energy, and any deformation of either the geometry of the molecule or of the electron cloud leads to a new state. Energy absorption (and emission) is quantized (with the exception of the translations of the whole molecule), the quanta increasing from those for the rotations of the whole molecule, for torsional, bond angle, and bond length vibrations, to those of excitation of the electrons. At room temperature practically all molecules are in their electronic ground state; only a few percent are in an excited vibrational state, but usually several rotations of the whole molecule take place.

3.2 Chirality

As defined by Lord Kelvin last century chirality is the property of a geometric object to be not superposable to (or: not congruent with) its mirror image. This same definition can easily be applied to rigid molecules too; if they are, however, flexible then there in general exists an infinite number of chiral conformations. If at least one conformation is achiral we speak of an "achiral" molecule. Thus any possible conformation of (+)-tartaric acid is chiral, whereas meso-tartaric acid has two achiral conformations (rotations around the bonds to the HO- and the carboxylic groups not counted), one with a plane of symmetry, another one with a center of inversion. Since chirality is a purely geometric property it is bad use to speak of a "chiral substance", and it should be understood as abbreviation for "substance composed of chiral molecules of same sense of chirality".

4 INTERACTION OF LIGHT WITH MOLECULES

4.1 Polarization of achiral molecules

When a molecule is in the electric field of a beam of light both the negatively charged electrons and the positive nuclei would like to oscillate in opposite directions with the frequency of the light. In the UV/VIS-range the nuclei are too heavy to follow such quick oscillations, but electrons can. According to the theory of radiation any such oscillating dipole becomes a source of radiation, and so the absorbed energy will be emitted again. This "elastic kick" of the photon with the electron cloud leads to a time lag and thus to a decrease of the velocity c of the light. Quantitatively this is described by

$$c = c_o/n \tag{3}$$

where n is the refractive index, c the speed of light in the medium, and c_o that in vacuum. Together with eqn. (1) this leads to the important conclusion that the wavelength also follows an analogous law:

$$\lambda = \lambda_o/n \tag{4}$$

$\underline{H}$ also interacts with the electric and magnetic properties of the electrons, as does $\underline{E}$ with the magnetic properties, but these interactions are several orders of magnitude smaller and can thus be neglected.

4.2 Ordinary absorption of light

If the energy of the impinging photon ($h\nu$) is the same as the energy difference between the electronic ground state and one of the electronically excited states then this photon can be absorbed, provided the excitation leads to a deformation of the electron cloud in such a way that it corresponds to a translation of charge during this excitation. By this one part of the electron cloud will become more negative than in the ground state, another less, and this can be interpreted as the generation of a dipole, which is called the "electric transition dipole moment" ($\vec{\mu}$). Once the molecule has reached the excited state these "transition charges" disappear again, the transition dipole "lives" so ιo say only during that short while of the excitation. Whereas thus the frequency of the light determines the position of an absorption band, the transition dipole determines the strength of the absorption.

Quantitatively we describe absorption by an absorption coefficient; if one measures the pathlength in the cuvette (d) in cm, and the concentration of the dissolved substance (c_m) in mol/l then this coefficient is called the "molar decadic absorption coefficient" (ε). Be I_o the intensity of the light entering the absorbing matter, and I that of the light leaving it, then the well-known relation holds (Bouguer-Lambert-Beer law):

$$A = \lg(I_o/I) = \varepsilon . c_m . d \tag{5}$$

(A: "absorbance", is directly recorded by the spectrometer).

One never records a single line but either a series of lines or only a broad band. The reason for this is explained by the Franck-Condon principle: the geometry of a molecule in the electronic and vibrational ground state will in the electronically excited state often correspond to a vibrationally excited state, and several of these may be hit by a "vertical excitation" (i.e. one in which the geometry has not changed). Since these effects cannot be separated we speak of "vibronic coupling". The broadening of the lines is the result of splitting each vibronic niveau into several by the rotations of the whole molecule, and by the nonquantized transfer of translational energy (collisions with solvent molecules). Therefore, a much better quantitative measure of the strength of the absorption is the area under the whole band instead of ε at the maximum, which is normally cited in publications. According to Mulliken one better takes the "wavelength-weighted" absorption coefficient in calculating this area, and for theoretical reasons this area is multiplied by some factor F, leading to a number, which is still an experimental one and which is called "dipole strength" D:

$$D = F \times \int \varepsilon/\lambda \, d\lambda = \mu^2; \; F = 0.918 \times 10^{-38} . \tag{6}$$

Theory shows, that D is equal to the square of the electric transition moment μ^2, and it could easily be calculated if we would know this $\vec{\mu}$. There exists a simple recipe to estimate the direction and magnitude of $\vec{\mu}$ (cf. refs. 1,2): one has to multiply the MO, from which the electron is excited by that, into which it jumps, and the obtained signs have to be inverted (this part of the recipe is not important for the estimation of D, but is necessary when estimating CD-signs). If the formed multipole has the characteristics of a dipole, then this is already $\vec{\mu}$. The magnitude of $\vec{\mu}$ increases with better overlap between the two mentioned MOs. Fig. 1 gives examples. For this procedure it is of no relevance where one starts with the hatching of MO-lobes, since $\underline{E}$ is oscillating back and forth, and in one of the two half-phases it will then always have a component along $\vec{\mu}$.

As one can easily prove, for the $n \rightarrow \pi^*$-transition of a carbonyl chromophore the centers of gravity of both charges coincide, thus

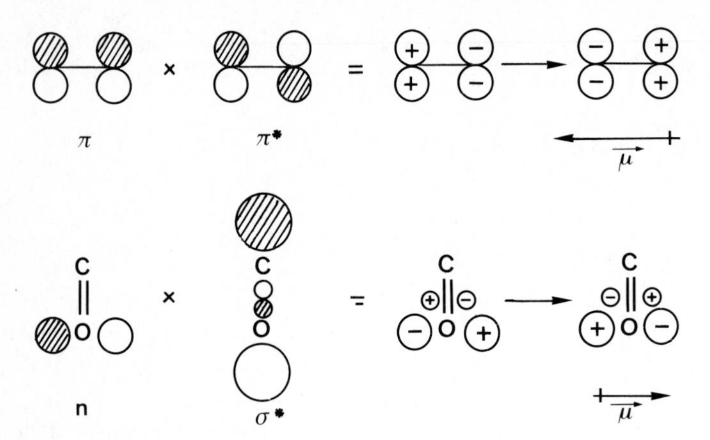

Fig. 1. Recipe for estimating $\vec{\mu}$. Top: $\pi \rightarrow \pi^*$ - transition of an olefin. Since overlap is best we expect a strong $\vec{\mu}$, and, therefore, a large ε (usually in the range of 5 000 to 10 000). $\vec{\mu}$ is directed along the C=C - bond, the "transition is polarized along C=C". Bottom: $n \rightarrow \sigma^*$ - transition of a carbonyl. Overlap is weak, furthermore there is partial compensation, so ε is predicted to be small (appr. 1 000).

$\vec{\mu}$ is zero, the excitation is "electronically forbidden". In fact, ε is around 20, and for oxocompounds other than formaldehyde this weak absorption comes from lowering the symmetry of the chromophore by appropriate vibrations.

4.3 Polarization of chiral molecules

From chemistry is well known that any interaction of molecules which leads to diastereomers depends on the relative senses of chirality of the two partners (D/D or L/L vs. D/L or L/D). The same is true for the interaction of a chiral molecule with a circularly polarized light beam: with a given enantiomer the refractive indices for left (n_L) and right circularly polarized light (n_R) will differ, and so will then the wavelengths and the speeds of the two rays:

$$\Delta n = n_L - n_R \neq 0, \ \Delta\lambda \neq 0, \ \Delta c \neq 0 . \qquad (7)$$

If we look against a beam of linearly polarized light at a distance z_o the two vectors $\underline{E}_L$ and $\underline{E}_R$ will not be any more symmetrical with respect to their starting direction, so their sum-vector will be rotated ,too. The dependence on time at z_o is, however, still the same, so we observe linearly polarized light whose plane of polarization has been rotated: we get "optical rotation". As n, and therefore also Δn, depends on the wavelength

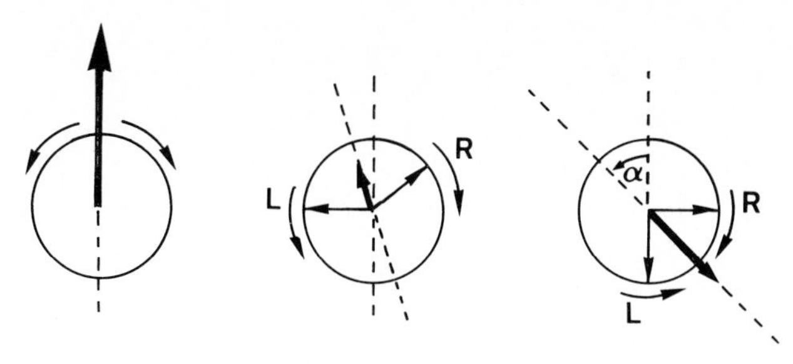

Fig. 2. Generation of rotation of plane of polarization. It is assumed that Δn is negative. Left: Starting positions of both electric field vectors. Middle: at a distance z_o $\underline{E}_R$ has been rotated less than $\underline{E}_L$, so their sumvector has been rotated to the left when looking against the light beam. Right: At the distance $2z_o$ the phase difference between $\underline{E}_L$ and $\underline{E}_R$ has doubled, and so has also the angle of rotation.

the angle of rotation α shows "Optical rotatory dispersion" (ORD). This angle is proportional to the length of the cuvette (Fig. 2) and in order to become independent on the conditions of measurement, a specific rotation $[\alpha]$ has been defined (eqn. 8). It should be noted that the unit of this specific rotation is not degree! In order to compare better such rotations for homologues the "molar rotation" $[\Phi]$ can be used (eqn. 9).

$$[\alpha] = \alpha/(c'.l) = 100\alpha/(c''.l) \tag{8}$$

$$[\Phi] = [\alpha].M/100 \tag{9}$$

where l is the length of the cell measured in dm, c' is the concentration measured in g/ml, c" the concentration given in % (still kept for historical reasons), and M is the molar mass.

Outside ranges of absorption the rotation follows a "plain curve"; the magnitude $|\alpha|$ rises with decreasing wavelength. ORD-curves are rarely used nowadays.

4.4 Absorption of polarized light by chiral molecules

When circularly polarized light is absorbed by chiral molecules one has again enantiomeric interactions and so ε_L and ε_R will differ from each other. This behaviour is called "circular dichroism", and in the stricter sense CD is the difference

$$\Delta\varepsilon = \varepsilon_L - \varepsilon_R \tag{10}$$

As a result of such differential absorption of the two circularly polarized components of linearly polarized light the intensities I_L and I_R will become different. As intensity I is the square of the amplidude also $\underline{E}_L$ and $\underline{E}_R$ will differ, and their sum vector describes then not anymore a circle but an ellipse: the originally linearly polarized light became elliptically polarized after having traversed the absorbing medium. The ratio of the minor (b) to the major axis (a) of the ellipse (Fig. 3) characterizes its shape, and an angle "ellipticity" (ψ) is defined as

$$\psi = \arctan(b/a). \tag{11}$$

In analogy to the specific and molar rotation a "specific" ($[\psi]$) and a "molar ellipticity" ($[\theta]$) are given by

$$[\psi] = \psi/(c'.l) = 100\psi/(c''.l) \quad \text{and} \tag{12}$$

$$[\theta] = [\psi].M/100, \tag{13}$$

and molar ellipticity and CD are proportional to each other:

$$[\theta] = 3\ 300\ \Delta\varepsilon. \tag{14}$$

In ranges of absorption the ORD-curve has a sigmoidal part superposed onto the plain curve (called now "background rotation") (Fig. 3). Such a curve is called "anomalous ORD", and by coming from the long-wavelength side one reaches first a peak and then a trough, when the CD is positive; for a negative CD it is the other way round. All three effects, anomalous ORD, ellipticity and CD, are known under the common name "Cotton effect". As amplitude a of the anomalous ORD we understand

$$a = ([\Phi_1] - [\Phi_2])/100 \tag{15}$$

and it is correlated with $\Delta\varepsilon_{max}$ (for Gaussian CD-curves) approximately by

$$a \approx 40\ \Delta\varepsilon_{max}. \tag{16}$$

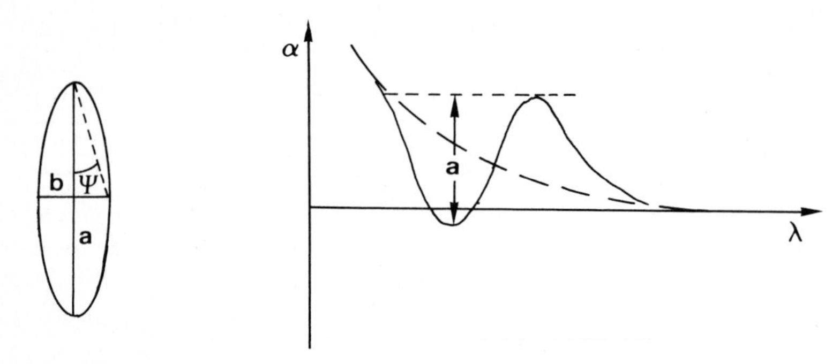

Fig. 3. Left: Definition of ellipticity ψ. Right: Anomalous ORD-curve for a negative Cotton effect, superposed onto a positive background rotation (- - - - -).

If finestructure appears in the CD-curve then it is found in the ORD-curve as well. There is no doubt about the sign of a Cotton effect from inspection of a CD-curve, but from the anomalous ORD it is then impossible to obtain its magnitude and often even its sign. This is one of the main reasons why CD is scarcely used any more.

CD-curves resemble UV-absorption curves, with one major difference: whereas ordinary absorption curves are always positive, CD-curves may be positive or negative, and it is obvious, that the two enantiomers must give enantiomorphous Cotton effects. Theory shows that CD is only generated when the excitation is associated with not only a translation of charge ($\vec{\mu}$), but also with a rotation of charge within the electron cloud. This is caused by the interaction of magnetic with electric properties of light and the electrons (neglected for ordinary absorption) and creates a "magnetic transition moment vector" $\vec{m}$. Here again only the recipe (cf. ref. 1) is given.

If by going from the ground into the excited state MOs are rotated, the radius for such a rotation is not zero, and the angle of rotation is ≠ 180° then one follows this rotation through the smaller angle with the fingers 2-5 of the right hand; the thumb points then into the direction of $\vec{m}$. As follows from Fig. 1 the transition from the π- into the π*-MO of an olefin or ketone does not create such a rotation and, therefore, no $\vec{m}$, but the $n \rightarrow \sigma^*$-transition of a ketone does (applying the right-hand-rule shows that $\vec{m}$ is perpendicular to the plane of the oxo-group and with the signs chosen for the MO-lobes it points to the viewer).

As in ordinary absorption, also for the CD-spectrum the area under the curve is a better quantitative measure of the Cotton effect than $\Delta\varepsilon_{max}$, and in a similar way as was defined the dipole strength D one has defined a "rotational strength" R by eqn. 17

$$R = (F/4) \int \Delta\varepsilon/\lambda \; d\lambda = \mu . m . \cos(\vec{\mu},\vec{m}) ; \; F/4 = 0.229 \; 10^{-38} \qquad (17)$$

(Note that for theoretical reasons one has to use one fourth of the factor used in eqn. 6). The theoretical value of the rotational strength R is given by the product of the magnitudes of the two transition moments, multiplied by the cosine of the angle between them. Thus the determination of the sign of a Cotton effect is achieved when we know this angle. If it is acute (or in the extreme case, when both vectors are parallel) then the CD is positive, if it were oblique (extremum: $\vec{\mu}$ and $\vec{m}$ are antiparallel), the CD is negative.

5 CLASSIFICATION OF CHIRAL MOLECULES

5.1 General remarks

As can easily be proved (e.g. with the help of elementary group theory) both $\vec{\mu}$ and $\vec{m}$ are nonzero only when the chromophore itself is "inherently chiral" (class A of Moscowitz, ref. 3), otherwise we have an "achiral chromophore chirally perturbed by its invironment" (class B). In this latter case the missing transition moment has to be stolen from another transition. The mechanism for this can also often be estimated in elementary way; examples are given e.g. in refs. 1 and 4. For pragmatical reasons we can subdivide class B into such molecules where the (achiral) chromophore is built into a twisted ring (B-I) and those, where this is not the case (B-II). This can be extended to molecules for which the chromophore is in a chain: is the chromophore (syn or anti) periplanar to the neighbouring bonds they belong to class B-II, otherwise to B-I. If a very strong perturber as Cl, Br, NR_2, SR is in non-coplanar arrangement with respect to the chromophore then we are dealing with class A.

5.2 Inherently chiral chromophores

Inherently chiral chromophores are mostly twisted π-systems as non-planar conjugated dienes, enones, or biphenyls. p-Orbitals may also be included, as in twisted disulfides or vinyl ethers:

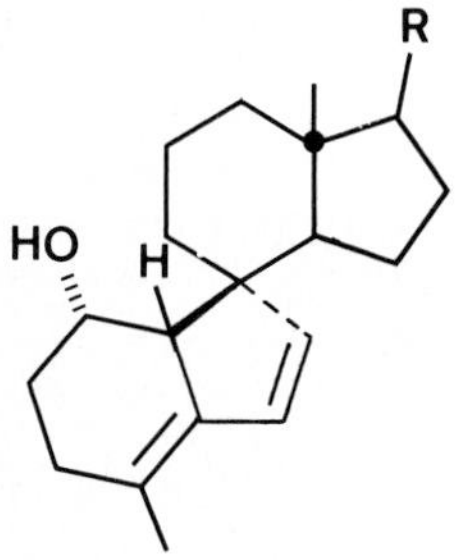

R
HO
H
parasorbic acid 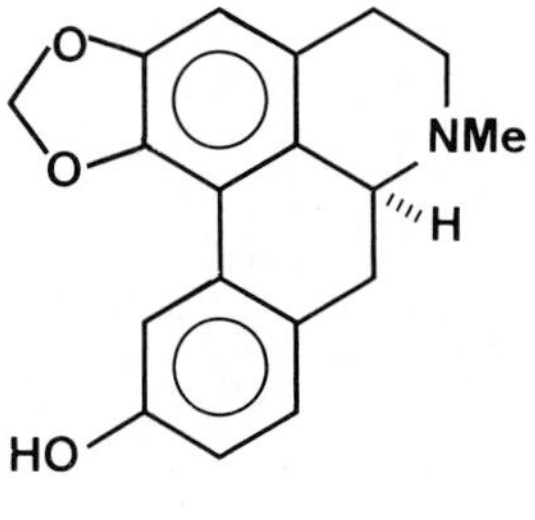
NMe
H
HO
toxisterol-B
mecambrine

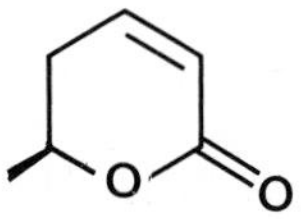

CH_2OH
R – S – S – R'
HO
HO
O

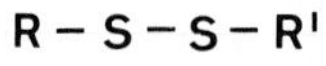

peptides with disulfide bridge

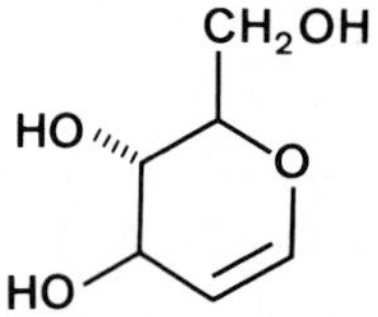

D-glucal

5.3 Achiral chromophores chirally perturbed by adjacent bonds

Some examples of molecules of this class B-I are twisted cycloalkanones, tetrahydroisoquinolines, and lactones. Also 20-ketopregnanes belong to this class!

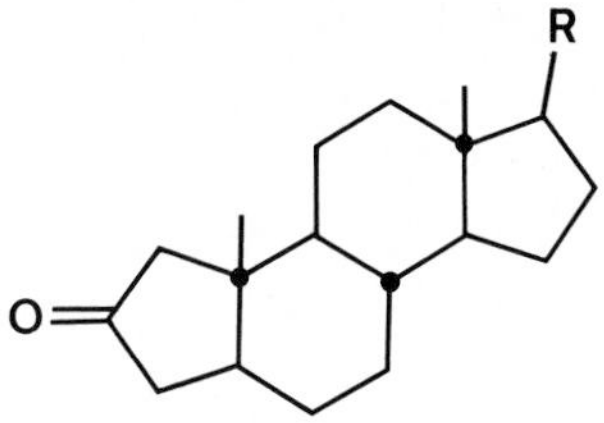

A-nor-2-ketosteroid

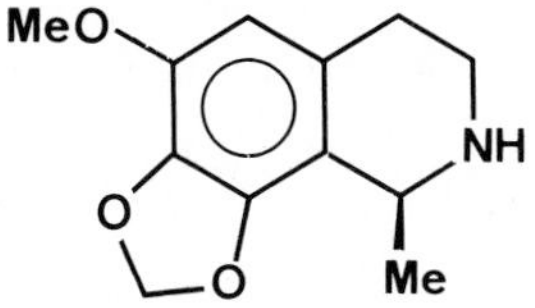

(-)-anhalonine

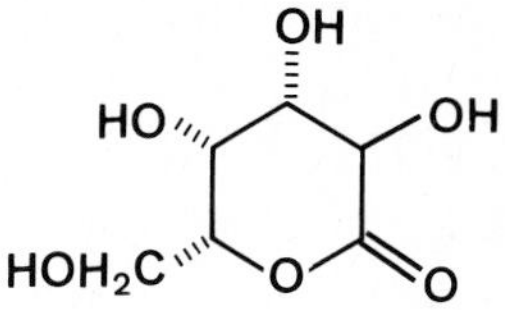

D-galactonic acid δ-lactone

5.4 Achiral chromophores chirally perturbed by other bonds

To this class B-II belong most other molecules, as e.g. cyclohexanones in the chair conformation, crossed-conjugated cyclohexadienones, most aliphatic ketones, acids, etc. With α-amino acids the situation is, however, more complicated: conformers with the amino group periplanar to the carboxylic group belong to class B-I, others to class A!

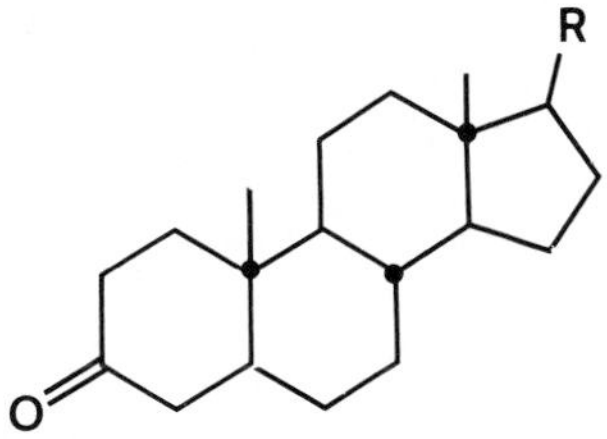

cholestane-3-one

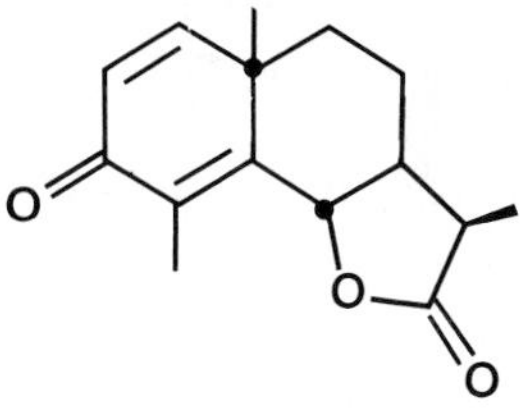

ß-santonin

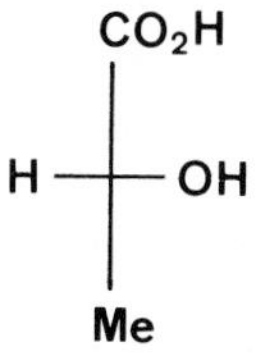

lactic acid

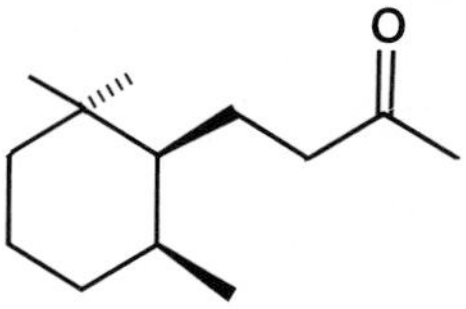

(+)-tetrahydroionone

5.5 Exciton interaction

If two chromophores, associated with strong electric transition moments, are arranged in a molecule in such a way that the two $\vec{\mu}$-s form together a chiral unit then the combined translation of transient charges has at the same time the character of an overall rotation of charge (think of two tangents on a helix!) and generates thus an $\vec{m}$. For the two possible relative combinations of $\vec{\mu}_1$ and $\vec{\mu}_2$ once the resulting electric transition moment sumvector is parallel to $\vec{m}$, once antiparallel. The two combined translations have different energies, which can be estimated with the help of Coulomb's law, so two Cotton effects appear in the CD-spectrum at different wavelength: one speaks of exciton splitting and of a CD-couplet. It is not important whether the two chromophores are identical or not, and whether they are inherently chiral or not. A detailed treatment is given in ref. 5 (cf. also refs. 1 and 4). The best known cases are (substituted) dibenzoates of glycols, but also abscisic acid is a good example. These exciton-Cotton effects are usually very strong, and in the UV/VIS-spectrum a corresponding strong absorption must appear, too.

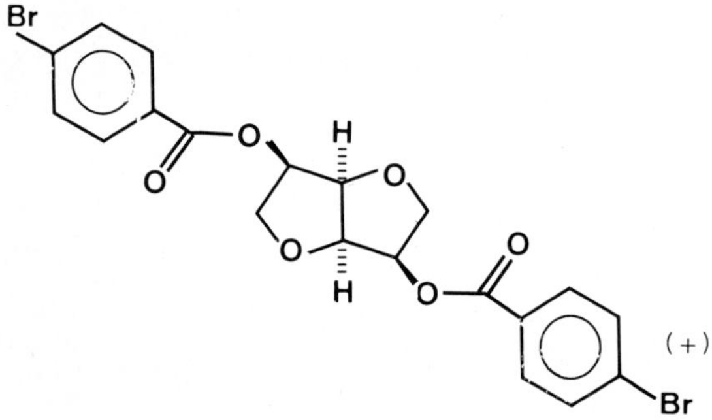

Br O O H O O H O O Br

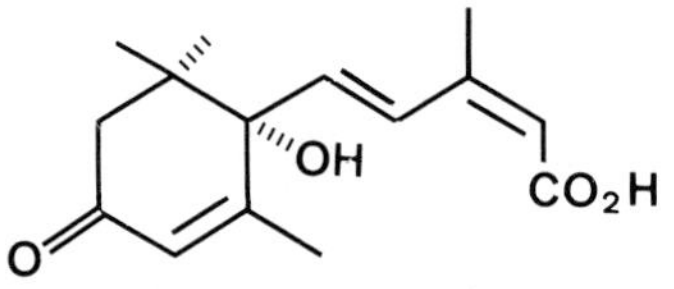

OH CO_2H O

(+)-abscisic acid

1,4:3,6-dianhydro-D-mannitol di-p-bromobenzoate

6 RULES FOR CORRELATION OF STRUCTURE WITH CD

6.1 Helicity rules

To the chiroptical properties contributes the whole molecule, but the closer a chiral perturbation is to the chromophore the stronger will in general be its influence. For molecules belonging to class A the Cotton effects will thus be governed mainly by the inherent chirality of this chromophore itself; it is its sense of helicity which determines to greatest extent the CD, so such chromophores follow a "helicity rule". By the same argumentation for molecules of class B-I the validity of a helicity rule is expected, too, and so it must be for the exciton coupling case.

6.2 Sector rules

In class B-II the chromophore and its immediate surrounding are achiral, the "through-bond" perturbation of more distant groups which is the predominant effect in class B-I will diminish (exception: groups which are connected to the chromophore through an antiperiplanar train of bonds), and the far-reaching Coulomb "through-space" interactions will then determine the CD effects. Such contributions of a substituent (assumed for simplicity as a dimensionless point) to the chiroptical properties depend solely on its co-ordinates, and in such a case we derive at a "sector rule": the space around the chromophore is divided into sectors, and the contributions of alike groups in adjacent sectors will have opposite signs. A group positioned in a nodal sphere separating two sectors has no influence upon the CD.

According to Schellman (ref. 6) all symmetry planes of the chromophore become such nodal planes for the sector rule. If the MOs involved in the transition have additional nodal spheres besides these planes then they have also to be taken into consideration. So for a ketone of class B-II a quadrant rule follows from the symmetry of the C=O - group, since only two symmetry planes are present, but the π*-orbital adds one more nodal surface (for simplification assumed to be also a plane) and leads thus to the well-known octant rule for the $n\rightarrow\pi^*$ - band of ketones. Is only one nodal plane present, as e.g. for the sulfoxide chromophore, then in first approximation the simplest sector rule, a "hemisphere rule" (earlier called a "planar rule") will hold.

6.3 Absolute configuration from CD?

For all these rules there was never mentioned an "asymmetric carbon atom" or any other element of chirality. In fact in general all such rules can give only the sense of helicity of the chromophore or of its immediate invironment, or the overall distribution of substituents in the sectors, but not the absolute configuration! We need one more argument (e.g. conformational analysis, X-ray diffraction data, NMR-spectra, etc.) to correlate the sense of helicity with the absolute configuration of the whole molecule. Chiroptical methods can thus either determine the conformation or absolute configuration, but never both, at least when one uses only the sign of the Cotton effects for interpretations. In a few rare cases the magnitude of a CD-band may give such a second argument, as e.g. in case of axial α -bromoketones (see below), and only then both, conformation and absolute configuration, can be determined from CD alone.

7 COTTONOGENIC DERIVATIVES

Since CD is observed only in the range of absorption of light compounds like saturated hydrocarbons, alcohols etc. can not be investigated by this method. A plain ORD-curve could be recorded, but the information from this is not very conclusive. Often absorbing derivatives can be prepared, and if the newly introduced chromophore is close to the chiral part of the molecule then its CD could be used for correlation with stereochemistry. Such derivatives are, e.g. benzoates, hemiphthalates, or xanthates of alcohols, salicylidene derivatives or phthalimides of primary, and N-chloro, N-nitroso or N-dithiocarbamate derivatives of secondary amines, N,N-dimethyl-thioamides of acids, etc. Another possibility is the _in situ_ generation of complexes with transition metals. The advantage of the latter method is, that one has not to isolate and purify any derivative and can often work with very small amounts of substance directly in the cell. By this one gets, of course, no quantitative values; for the purpose of determination of absolute configurations from the sign of the Cotton effects this does not matter. A prerequisite for this method is, that such complexes are thermodynamically relatively stable, but kinetically quite labile. For alcohols copper hexafluoroacetylacetonate can be used, for glycols Cu-tetrammine complexes or $Mo_2(OAc)_4$, etc. Such derivatives are called "cottonogenic" ones.

8 A FEW EXAMPLES

8.1. _Ketones_

Since the oxo-chromophore is the best investigated one it deserves more detailed discussion. Depending on its immediate invironment this chromophore can belong to any of the classes A, B-I or B-II, and for each of these types the appropriate rule has to be chosen. Experience shows that for ketones of class B-II the magnitude of Δε does not exceed an upper limit of appr. 3. Any compound with an isolated keto group which gives a CD of 3.5 to appr. 6 should then be classified into B-I; is the carbonyl in a ring, then this is most probably twisted, is it in a chain, then one of the torsional angles (O=)C-C(-C) deviates strongly from 180° (or 0°). It is general good practice to draw a standard projection from O towards C of the carbonyl. In Fig. 4 this is done for ring A of A-nor-5α-androstan-2-one, and this sense of helicity of the twisted ring gives rise to a positive CD for the CD around 300 nm (Δε ≈ +6.0, with pronounced finestructure). For the "molecule torso" shown also in Fig. 4 the CD-contribution is thus appr. +3. For 20-ketopregnanes usually Δε -values of appr. +3.5 can be recorded, so one can conclude that the C-acetyl group is more or less fixed in its conformation with a similar torsional angle as shown, and indeed from X-ray studies in the crystalline state and from calculations this conformation has been proved to be the most preferred one.

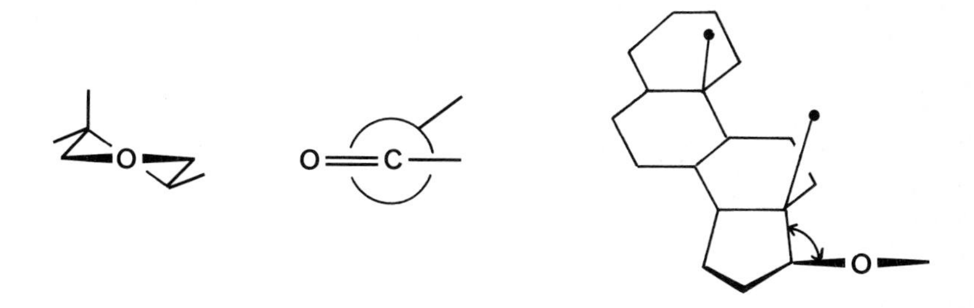

Fig. 4. Left: Standard projection from O to C of the oxo group of an A-nor-5α-androstan-2-one. Middle: Newman projection along the bond from (O=)C to C_α , which leads to a strong positive contribution to the Cotton effect around 300 nm. Right: Standard projection for a 20-ketopregnane in its preferred conformation.

Two typical examples for ketones of class B-II are 5α-cholestan-3-one and (+)-citronellal. To them the famous octant rule can be applied, and this is done in Fig. 5. If in the standard projection only the back (or rear) octants are considered, then the contributions of any group besides fluorine have the sign given in Fig. 5 (for nitrogen the sign of the contribution may depend on the direction of the lone pair). In the cholestanone case only atoms C(6) and C(7) will give (positive) contributions, because C(15) and C(16) are too far away from the chromophore, and all other C-atoms lie in nodal planes or appear in pairs whose contributions cancel each other (e.g. C(8) and C(11)). The CD is thus predicted to be weakly positive, and the experimental value is +1.3. From the projection of citronellal is obvious that indeed an octant and not a quadrant rule is valid.

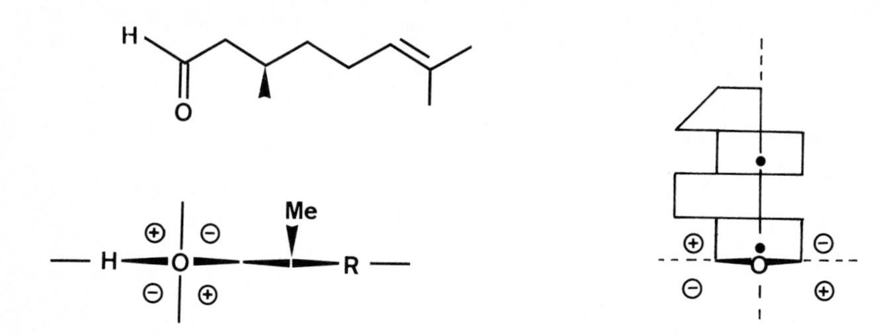

Fig. 5. Left: Standard projection for (+)-citronellal; the longest chain is anti-periplanar in the most preferred conformation, and the methyl group lies then in a front octant. Since the signs given are for the rear octants the contribution of the methyl is positive! Right: Standard projection of 5α-cholestan-3-one. Signs are again for back octants. Note that in the standard projections the C-C(=O)-C plane has to be projected horizontally; whether the rest of the projection is then drawn above or below this line is, however, of no importance.

For conjugated ketones in which the C=C-C=O system is transoid and coplanar another rule holds, in which the signs are opposite to those of the original octant rule. For non-coplanar enones in nearly all cases it is the sign of the torsional angle (C=)C-C(=O) which determines the sign of the $n \rightarrow \pi^*$-band CD according to Fig. 6 (exception: cyclopentenones). In ecdysones this same conformation is present, so they have positive CD-maxima around 340 nm. Usually the values for 5ß-derivatives are double as large (+2.6) as those of the 5α-isomers (+1.4). For the $\pi \rightarrow \pi^*$ Cotton effect the correlation of the sign with the geometry is more complicated: it is not only this mentioned helicity of the enone unit but also the number, position, and direction of bonds arranged axial-allylic with respect to the C=C which govern all together sign and magnitude of this Cotton effect.

Like non-coplanar conjugated enones also several cyclopropyl and oxido ketones belong to class A: it is the helicity of this moiety which determines the Cotton effect (Fig. 7). The same is true for axial α-halogeno ketones (Cl, Br, I), which belong to those rare cases where one can determine conformation <u>and</u> configuration from CD alone, because only in axial conformation this halogen exerts a very strong bathochromic shift (e.g. with Br about 25 nm), furthermore the Cotton effects are unusually large.

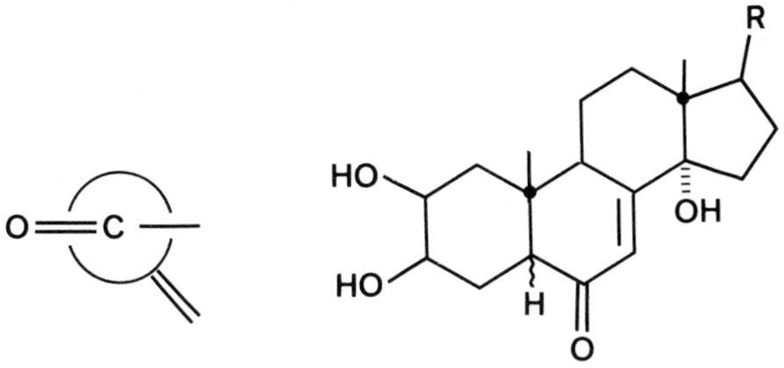

Fig. 6. Newman projection for the C=C-C=O moiety of the ecdysones; this torsional angle leads to positive CD.

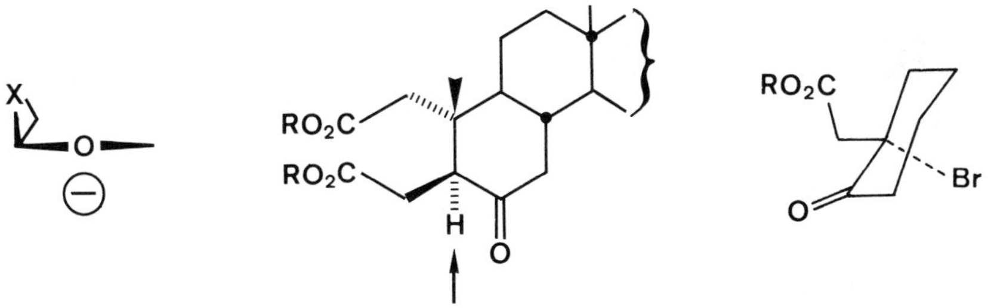

Fig. 7. Left: Standard projection of a cyclopropyl (X = C) or oxido(X = O) ketone giving rise to a negative CD around 300 nm. Middle: An A-seco-6-ketosteroid which had been brominated in the 5-position. The Cotton effect of the bromoketone was strongly negative and showed a large bathochromic shift. The Br must, therefore, adopt the axial conformation, and from the sign follows the stereoformula drawn to the right. The absolute configuration as read from this projection is thus (5S).

Both these characteristic features give thus information about the conformation, and the absolute configuration follows from the sign, which is the same as would be expected from the usual octant rule. Fig. 7 shows a 6-ketone which was brominated on the tertiary C next to the carbonyl. The resulting Cotton effect around 320 nm was strongly negative, so the configuration must be that shown on the right side, i.e. (5S). Substituents in the same position like -SR or $-NR_2$ give similar results.

The acetophenone chromophore is much more complicated, but still the carbonyl $n \rightarrow \pi^*$-Cotton effect around 325 nm can be used for structure determination: that conformation which is given in Fig. 8 leads to a negative CD. Application of this rule to flavanones explains the well established rule for these compounds.

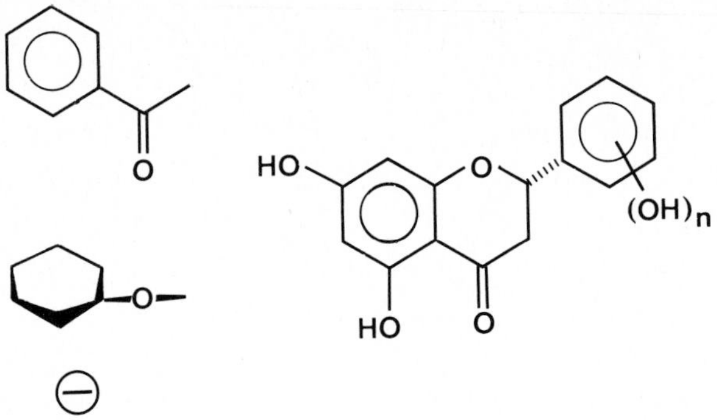

Fig. 8. Left: Acetophenone helicity which leads to a negative CD around 325 nm. Right: Rule for the same Cotton effect of flavanones: independent of the substitution pattern of the aromatic ring a positive Cotton effect is given for a (2S)-configuration.

Still larger effects ($|\Delta\varepsilon|$ up to appr. 35) are given by certain ß,γ-unsaturated (homoconjugated) oxo compounds, if the C=C-C-C=O system can adopt such a conformation as given in Fig. 9 (or its mirror image). Highly characteristic for this type of chromophore is the also quite strong ordinary absorption (ε up to 1000). This rule of Fig. 9 is together with the axial α -bromoketone rule and that for exciton couplets one of the most reliable ones. Even for apparently very flexible molecules like homoconjugated aldehydes the $\Delta\varepsilon$ -values can nevertheless be very high. Measuring the CD at lower temperature may often even lead to increased Cotton effects.

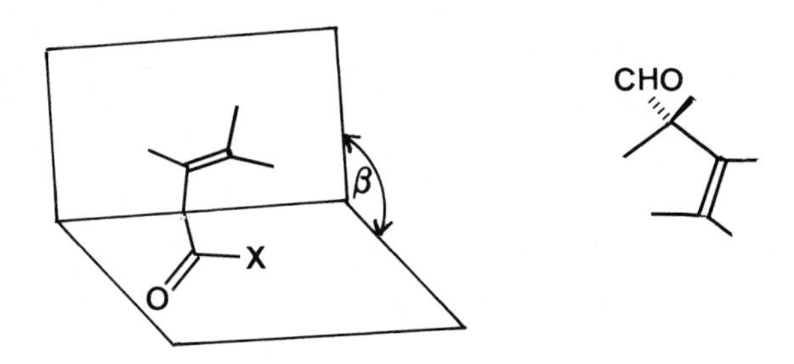

Fig. 9. Left: Necessary conformation (ß appr. 100 - 120°) of a homoconjugated oxo-compound for interaction to give very large ellipticities. The helicity indicated leads to a positive Cotton effect. Right: Laurolenal, for which $\Delta\varepsilon$ = +10.3 was found.

8.2 Acids and their derivatives

The n→π*- and π→π*-bands of acids, esters, lactones and amides lie much closer to each other than for ketones (appr. 220 and 200 nm, resp.). Saturated acids are best investigated via a cottonogenic derivative, the effects of simple esters are usually small and difficult to interprete. For lactones in which the C-C(=O)-O-C- moiety is forced into a cisoid conformation the torsional angle along the (O=)C-C_α - bond can, however be correlated with high probability with the CD around 220 nm (Fig. 10). It is actually the same type of correlation as that for ketones given in Fig. 4, and it is again a helicity and not a sector rule, which applies.

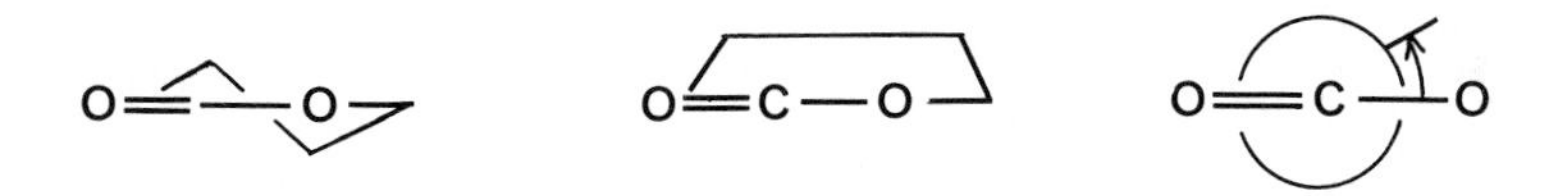

Fig. 10. Left: Halfchair - Middle: Boat conformation of a δ-lactone leading to a positive CD around 220 nm. Right: Newman projection along the (O=)C-C(-C) - bond for both these conformations (cf. Fig. 4).

For conjugated and homoconjugated unsaturated acids and their derivatives as well as for the ones with a threemembered ring next to the CO_2 - moiety the same rules hold as for the corresponding ketones.

8.3 Aromatic chromophores

The typical benzene UV-spectrum can easily be understood in an elementary way, too, but this needs involvement of electron correlation and configurational interaction, which cannot be discussed here. The situation with most natural products is complicated by the presence of heterosubstituents, which even then strongly influence magnitudes, band positions, and signs of the Cotton effects when they are achiral, like e.g. a methoxy group. So it has e.g. been found that the helicity rule for tetralins (Fig. 11) is not changed by 5-mono-, 6,7-di- or 5,6,7-trisubstitution with OR - groups on the aromatic ring, whereas a 6-monosubstitution can lead to sign inversion within the first band around 270 nm. The CD is also very sensitive to the type of substitution, not only to the pattern: the first band of estradiol and its 3-methyl ether is negative, but positive for the acetate!

<u>Tetrahydroisoquinolines</u> follow a similar helicity rule as the tetralins when 6,7-di- or 5,6,7-trisubstituted with OR-groups. If more rings are present, as e.g. in morphinanes, "local rules" can be developed; each new system needs careful study with model compounds in order to find appropriate rules. If the generalizations mentioned are taken into account then with only a few characteristic model compounds such new rules can be put forward.

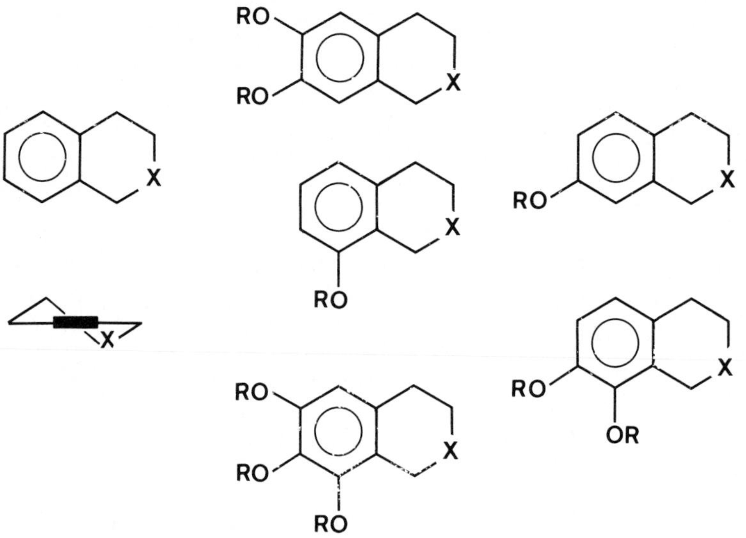

Fig. 11. Helicity rule for tetralins and tetrahydroisoquinolines. Left: Projection along the C_2-axis of the twisted system from the aromatic ring towards the cyclohexene (piperideine) which leads to a positive CD within the first band around 270 nm. Middle: Substitution patterns which do not change the sign. Right: Substitution patterns changing the sign. For tetrahydroisoquinolines without such group on the aromatic ring the CD is strongly influenced by other substituents (especially in 1- and 4-position).

The CD-spectra of many <u>indole</u> alkaloids have been published. With this chromophore the second benzene band is strongly bathochromically shifted and appears very close to the first one. The two degenerate bands of benzene (called ß,ß') show up between 235 and 210 nm and often have opposite signs. If the pyrrole ring of the indole system is incorporated into a ring, which is mostly the case, then the helicity of this ring will in first place determine the signs of the Cotton effects. So e.g. for diastereomeric yohimbanes the helicity rule of Fig. 12 has been found. The correlation between the sign of the first CD-band and the sense of helicity of the piperideine ring is the same as that for 6-monosubstituted tetralins: obviously the pyrrole - Nacts in a similar unsymmetric way as strong perturber as the OR-group in the tetralin system, when in 6-position.

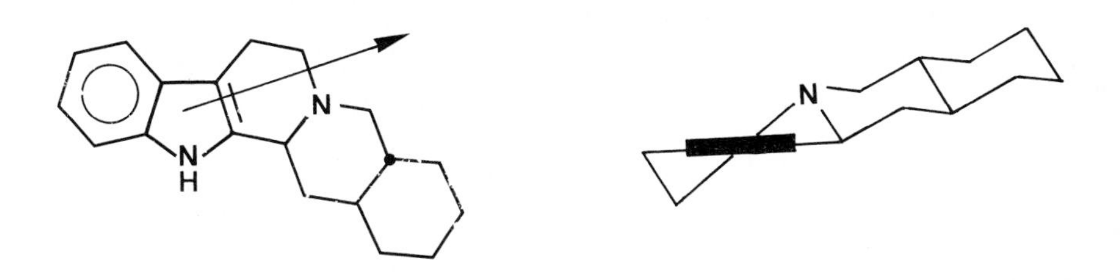

Fig. 12. Projection of yohimbane from the pyrrole towards the piperideine ring along the C_2-axis. The first band (first line at 297 nm) is positive ($\Delta\varepsilon$ = +0.72), the second (broader, maximum around 274 nm) is also positive (+0.83), and so is the third (+4.96 at 235 nm); the fourth around 217 nm is negative (-12.3).

8.4 Exciton coupling

Whereas it is in general not difficult to predict from the positions and directions of the two individual electric transition moments the signs of the two branches of the CD-couplet it is not always obvious which of the two transitions is at higher and which at lower energy. One can make use for this of Coulomb's law to estimate whether a combined transition has more attractive or repulsive interactions. If the two chromophores are quite close to each other, then one has to use for this all individual transient charges and not only the two poles of each over-all moment vector. This is especially difficult for many aromatic systems, but can in general be done quite easily for dibenzoates and related systems, as has been practized by Harada and Nakanishi (ref. 5) so successfully. For dibenzoates of vic-glycols with a torsional angle of ±60° the rule holds that the sign of the CD-couplet (defined as the sign of the first branch at longer wavelengths) is the same as the sign of the (HO-)C-C(OH) - torsional angle. If carefully applied then this rule can be extended to dibenzoates of more distant diols, too. By using other, stronger absorbing acids, a method has been worked out recently by the same authors for the determination of the individual sugars and their branching in oligosaccharides on the basis of this exciton coupling in microscale.

REFERENCES

1 G. Snatzke, Chem. Unserer Zeit **15** (1981) 78, - **16** (1982) 160.
2 H. Labhart, Einführung in die Physikalische Chemie, Teil V: Molekülspektroskopie, Springer, Berlin, 1975.
3 A. Moscowitz, Tetrahedron **13** (1961) 48.
4 G. Snatzke, Angew. Chem. **91** (1979) 380 - Angew. Chem. Int. Ed. Engl. **18** (1979) 363.
5 N. Harada and K. Nakanishi, Circular Dichroism Spectroscopy - Exciton Coupling in Organic Stereochemistry, University Science Books, Mill Valley, 1983.
6 J. A. Schellman, J. Chem. Phys. **44** (1966) 55.

The most modern introduction to the whole field of the application of chiroptical methods in organic chemistry is given in

7 M. Legrand and M. J. Rougier, Application of the Optical Activity to Stereochemical Determinations, in: H. B. Kagan (Ed.) Stereochemistry, Vol. 2, p. 33, Thieme, Stuttgart, 1977.

A collection of definitions and rules can be found in

8 G. Snatzke and F. Snatzke, Chiroptische Methoden, in: H. Kienitz et. al (Ed), Analytiker-Taschenbuch, Bd. 1, p. 217, Springer, Berlin, 1980.

A review on some applications in natural products chemistry:

9. P. M. Scopes, Applications of the Chiroptical Techniques to the Studies of Natural Products, Fortschr. Chem. Organ. Naturst. **32** (1975) 167.

STRUCTURAL STUDIES ON NEW STEROIDAL ALKALOIDS OF BUXUS PAPILOSA - SOME GENERALIZED SPECTRAL DEDUCTIONS

ATTA-UR-RAHMAN* AND M.IQBAL CHOUDHARY

1.0 INTRODUCTION

Buxus papilosa C.K.Schneider is a member of plant family Buxaceae. The isolation and structure elucidation of thirty new steroidal alkaloids from the leaves of B.papilosa is discussed. Some generalizations have been drawn which relate spectral data (^{1}H-NMR, MS, specific rotations) to structural features of Buxus alkaloids.

2.0 NEW STEROIDAL ALKALOIDS

The ethanolic extract of Buxus papilosa leaves collected from the northern regions of Pakistan, was evaporated to a gum, and partial separation of the alkaloids was carried out by extraction into $CHCl_3$ at different pH values. These fractions were subjected to column and thin layer chromatography to afford the following new alkaloids.

2.1 (+)-MOENJODARAMINE (1) (ref. 1)

The UV spectrum of compound **1** exhibited absorptions characteristic of 9(10 → 19) abeo-diene system with λ_{max} at 237 and 245 nm and shoulders at 207 and 254 nm (ref. 2). The IR spectrum showed bands at 2940 (C-H), 1360 (C-N) and 1595 (C=C) cm^{-1}.

The ^{1}H-NMR spectrum ($CDCl_3$) showed three singlets, corresponding to the three tertiary methyl groups, at δ0.71, 0.75 and 1.03. The C-21 secondary methyl group resonated as a doublet at δ0.88 ($J_{21,20}$= 6.0 Hz). A 3-H singlet at δ 2.10 was assigned to the N_b-CH_3 group, while another 6-H singlet resonating at δ 2.20 was due to $N(CH_3)_2$ group. A set of AB doublets centered at δ 3.24 and 3.82 ($J_{31\alpha,31\beta}$= 10.6 Hz) was assigned to the C-31 methylene protons. Another set of AB doublets at δ 3.56 and 4.42 ($J_{33\alpha,33\beta}$= 7.5 Hz) was attributed to the -N-CH_2-O group. A singlet at δ5.98 was ascribed to the C-19 isolated olefinic proton , while a multiplet centered at δ 5.55 was assigned to the C-11 olefinic proton.

The ^{13}C-NMR spectrum ($CDCl_3$) showed four signals at δ9.44, 77.94, 13.95 and 17.14 which were assigned to 21,31,18 and 32 methyl carbons respectively. The olefinic carbon atoms C-11, C-19, C-10 and C-9 appeared at δ 129.92, 130.13, 133.46 and 137.99, respectively. The N_a-methyl carbon resonated at δ 39.80 and the N_b-methyl carbon appeared at δ 36.35. The assignments to the various carbons are summarized around expression **1.**

The mass spectrum of the compound afforded the molecular ion at m/z 426.3609 corresponding to the molecular formula $C_{28}H_{46}N_2O$, indicating the presence of seven double bond equivalents. The peak at m/z 127 was due to the cleavage of ring A along with the tetrahydrooxazine ring. The substance showed a base peak at m/z 58.0650 corresponding to the composition C_3H_8N due to N,N-dimethyliminium ion, CH_2=N $(CH_3)_2$, resulting from the cleavage of ring A accompanied by intramoleclar protons transfer. Another peak at m/z 85.0883 was in accordance with the composition $C_5H_{11}N$ formed by the cleavage of ring A. A peak at m/z 72.0810 having the composition $C_4H_{10}N$ corresponded to trimethyliminium, ion $CH_3\text{-}CH\text{=}\overset{+}{N}(CH_3)_2$, commonly encountered in alkaloids bearing a nitrogen-containing side chain on ring D. Another peak at m/z 71.0734 having the formula C_4H_9N was formed by cleavage of ring A.

On the basis of these spectral data the alkaloid was assigned structure **1** (ref. 1). It is the first member of a new class of pentacyclic natural products bearing both a tetrahydrooxazine ring and a 9(10 → 19) abeo-diene system. The substance has not been isolated previously but has been prepared synthetically from desoxy-16-buxidienine (ref. 3).

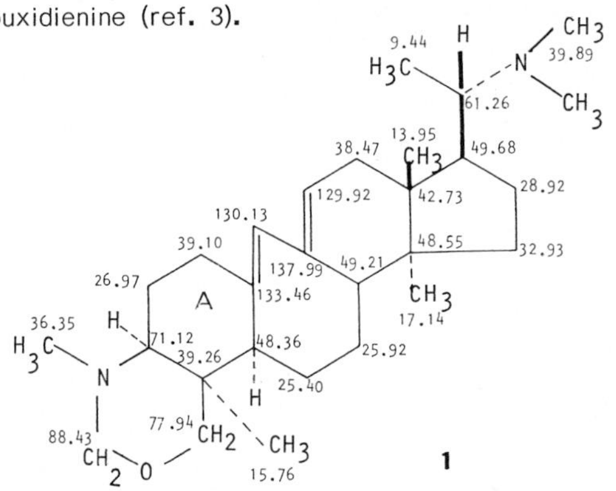

1

2.2 (+)-HARAPPAMINE (**2**) (ref. 4)

Another closely related alkaloid, (+)-harappamine, was isolated from the leaves of B.papilosa showing UV absorptions at 237 and 246 nm with shoulders at 207 and 258 nm, again due to the presence of 9(10→ 19) abeo-diene system. The IR spectrum includes bands at 3400 (N-H), 2840 (C-H) and 1650 (C=C) cm^{-1}.

The ^{1}H-NMR spectrum ($CDCl_3$) showed three singlets at δ 0.68, 0.76 and 1.03 due to three tertiary methyl groups. A 3-H doublet at δ 0.72 was assigned to C-21 secondary methyl group. Two singlets at δ 2.10 and 2.41 were assigned to the two $N(CH_3)_2$ and N_b-CH_3 groups respectively. A set of AB doublets resonating at δ 3.24 and 3.82 were assigned to C-31 methylene protons ($J_{31\alpha,31\beta}$=

10.6 Hz), while another set of AB doublets centered at δ 3.57 and 4.42 ($J_{33\alpha,33\beta}$ =7.4 Hz) were attributed to the methylene protons α-to the C-3 nitrogen. A singlet at δ 5.98 was assigned to C-19H, while a multiplet at δ 5.56 was attributed to the C-11 olefinic proton respectively. The ^{13}C-NMR ($CDCl_3$) spectrum is summarized around structure **2** (ref. 5).

The mass spectrum of compound **2** included a molecular ion at m/z 412.3454 in agreement with the formula $C_{27}H_{44}N_2O$. A considerably large peak at m/z 127 was due to the cleavage of ring A along with the attached tetrahydrooxazine ring. The base peak at m/z 58 was again due to the cleavage of ring A with the nitrogen substituents. Another fragment at m/z 72 was due to trimethyl-iminium side chain.

On the basis of the above spectroscopic studies structure **2** was assigned to this new alkaloid, which is actually a nor-derivative (+)-moenjodaramine (**1**) (ref. 1).

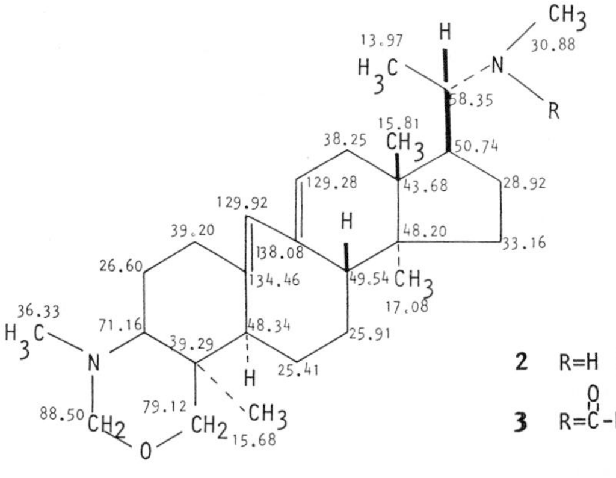

2 R=H (^{13}C-NMR)

3 R=C(=O)-H

2.3 (+)-N-FORMYLHARAPPAMINE (**3**) (ref. 6)

The N-formyl derivative of (+)-harappamine (**2**) was isolated from B.papilosa. The UV spectrum of the compound was identical to that of (+)-harappamine (**2**). The IR spectrum incorporated an intense amide carbonyl absorption at 1653 cm^{-1}, in addition to the N-H and C=C absorption.

The ^{1}H-NMR spectrum of **3** bore a distinct similarity to that of (+)-harapp-amine (**2**) and included three singlets at δ0.70, 1.09 and 1.20. The secondary methyl absorbed as a doublet at δ 0.83 (J=6.6 Hz). A 3-H singlet located at 2.20 was assigned to N_a-CH_3 group. A singlet at δ8.11 accompanied by a much smaller singlet at δ 7.98, represented the N-formyl proton. Similarly, the N_b-methyl group was indicated by a singlet at δ 2.74, followed by a smaller singlet at δ 2.80, due to geometrical isomerism. The rest of ^{1}H-NMR spectrum was identical to that of (+)-harappamine (ref. 4).

The mass spectrum of N-formylharappamine (**3**) displayed the molecular ion at m/z 440.3423 in agreement with the formula $C_{28}H_{44}N_2O_2$. The base peak at m/z 86 represented the N-formyldimethyliminium cation, $CH_3\text{-}CH{=}\overset{+}{N}(CH_3)CHO$. The fragment at m/z 127 arose through cleavage of ring A.

These studies led to structure **3** for N-formylharappamine (ref. 6).

2.4 (+)-PAPILININE (4) (ref. 7)

It is the fourth member of a new group of alkaloids bearing both a tetrahydrooxazine ring and 9(10→ 19) abeo-diene system. The IR spectrum showed absorptions at 1380 (C-N) and 1600 (C=C) cm^{-1}.

The ^1H-NMR spectrum ($CDCl_3$) exhibited close resemblance to those of (+)-moenjodaramine (**1**) and (+)-harappamine (**2**) bearing a tetrahydrooxazine ring attached to ring A. Three singlets at δ 0.70, 0.74 and δ1.02 were attributed to the three tertiary methyl groups. A doublet at δ 0.72 ($J_{21,20}$= 6.0 Hz) was assigned to the secondary methyl group. A 3-H singlet resonating at δ 2.10 was assigned to the N-CH_3 group attached to C-20. A set of AB doublets resonating at δ 3.23 and 3.79 ($J_{31\alpha,31\beta}$= 9.0 Hz) was due to the C-31 methylene protons, while another AB doublet at δ3.56 and 4.42 ($J_{33\alpha,33\beta}$= 7.0 Hz) was ascribed to the C-33 methylene protons flanked between nitrogen and oxygen of the tetrahydrooxazine ring. The vinylic protons at C-19 and C-11 resonated at δ5.97 and 5.51, respectively.

The mass spectrum afforded the molecular ion peak at m/z 398.3307, corresponding to the molecular formula $C_{26}H_{42}N_2O$, while the base peak appeared at m/z 58.9659 corresponding to the composition C_3H_8N. The latter probably arises by the cleavage of ring D side chain. Other major peaks were found at m/z 71, 127 and 100. The presence of a peak at δ 127.0999 was formed by the cleavage of ring A.

In the light of the above spectral studies structure **4** was assigned to this new alkaloid, named "(+)-papilinine"(**4**) (ref. 7) . (+)-Papilinine may be the biosynthetic precursor of (+)-harappamine (**2**) (ref. 4).

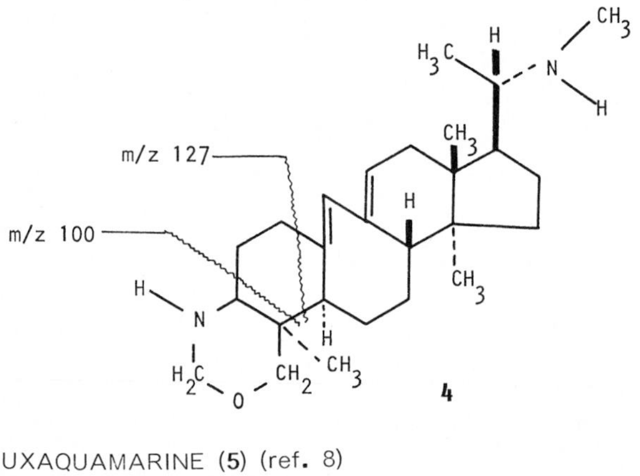

2.5 (+)-BUXAQUAMARINE (5) (ref. 8)

This new steroidal alkaloid showed UV absorptions at 238 and 245 and shoulders at 205 and 253 nm, characteristic of the presence of a 9(10 → 19) abeo diene system. The IR spectrum included an intense band at 1685 cm^{-1} which represents ketonic absorption.

The ^1H-NMR spectrum ($CDCl_3$) showed three tertiary methyl singlets at δ0.68, 0.75 and 1.06, while the N-CH_3 and 21-methyl protons (α to the carbonyl function) appeared as singlets at δ 2.11 and 2.17. A set of AB doublets at δ 3.27 and 3.84 ($J_{31\alpha,31\beta}$= 10.6 Hz) were assigned to C-31α and β-protons, while another set of AB doublets at δ3.61 and 4.45 ($J_{33\alpha,33\beta}$= 7.5 Hz) was attributed to the C-33 methylenic protons. A multiplet at δ 5.60 and a singlet at δ6.00 were ascribed to C-11 and C-19 olefinic protons. Two dimensional spectra (2D J-resolved, COSY 45°, NOESY) (ref. 8a) fully agreed with above mentioned assignments. The ^{13}C-NMR assignments ($CDCl_3$) are presented around expression **5**

The mass spectrum of the compound showed M^+ 397.2980 corresponding to the molecular formula $C_{26}H_{39}NO_2$ (calcd. 397.2981). The peak at m/z 382 was due to the loss of methyl group from the molecular ion, while m/z 354.2796 resulted from the loss of acetyl group. An important ion m/z 127.0997 arose by the cleavage of ring A along with the attached tetrahydrooxazine ring (ref. 7). The peak at m/z 58.0651 (C_3H_8N) was attributed to the loss of fragment ion CH_2=$\overset{+}{N}(CH_3)_2$. Structure **5** was assigned to (+)-buxaquamarine on the basis of the above data (ref. 8).

Biogenetically compound **5** may arise from (+)-moenjodaramine (ref. 1) or (+)-harappamine (ref. 2) by oxidation of the N-bearing side chain to the corresponding ketimine, followed by its hydrolytic removal.

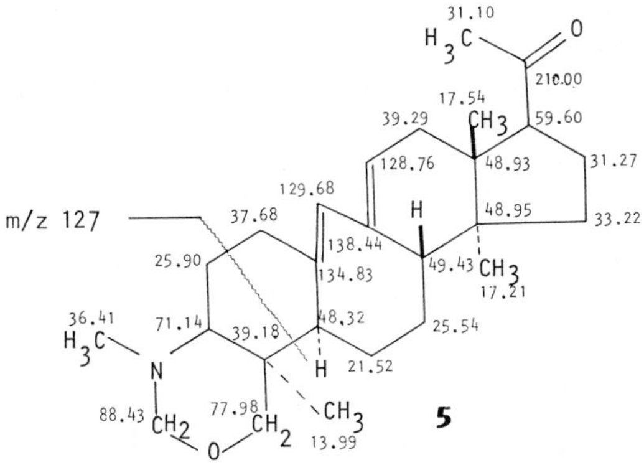

2.6 (+)-HOMOBUXAQUAMARINE (6) (ref. 9)

Our sixth steroidal alkaloid, (+)-homobuxaquamarine, showed a UV spectrum with maxima at 236 and 245 nm, and shoulders at 225 and 255 nm, characteristic of 9(10 → 19) abeo-diene bases. Similar absorptions have been encountered in the case of (+)-buxaquamarine (**5**) (ref. 8). The IR spectrum displayed a strong absorption at 1695 cm^{-1} indicating the presence of a carbonyl group.

The ^{1}H-NMR spectrum ($CDCl_3$) of **6** bore a distinct similarity to that of (+)-buxaquamarine (**5**) (ref. 8). It incorporated signals due to the three tertiary methyl groups at δ0.68, 0.75 and 1.01. A three-proton singlet at δ2.12 was assigned to the methyl group adjacent to the carbonyl function. Another three-proton singlet at δ 2.15 could be assigned to the N-methyl group. A set of AB doublets resonating at δ 3.30 and 3.84 represented the C-31 methylene protons (J=10.0 Hz). However, unlike the spectrum of (+)-buxaquamarine (**5**), the C-33 proton appeared as a quartet at δ3.63. Additionally, a three-proton doublet at δ1.33 indicated that this methyl group was attached to C-33 in the tetrahydrooxazine ring. The H-11 vinylic proton appeared as a doublet of doublets at δ5.52 (J_1=2.5 Hz, J_2=1.6 Hz) while H-19 resonated in the form of a singlet at δ5.98.

The mass spectrum of (+)-homobuxaquamarine (**6**) displayed the molecular ion at m/z 411.3147, in agreement with the molecular formula $C_{27}H_{41}NO_2$. The base peak at m/z 396 corresponded to the loss of a methyl group, while another peak at m/z 368 resulted from the loss of acetyl group from the molecular ion. An important fragment at m/z 141 could arise by the cleavage of ring A along with the tetrahydrooxazine ring. Peaks at m/z 71 and 58 arose by the cleavage of ring A along with the nitrogen substituents.

On the basis of the above studies, structure **6** was assigned to (+)-homobuxaquamarine (ref. 9).

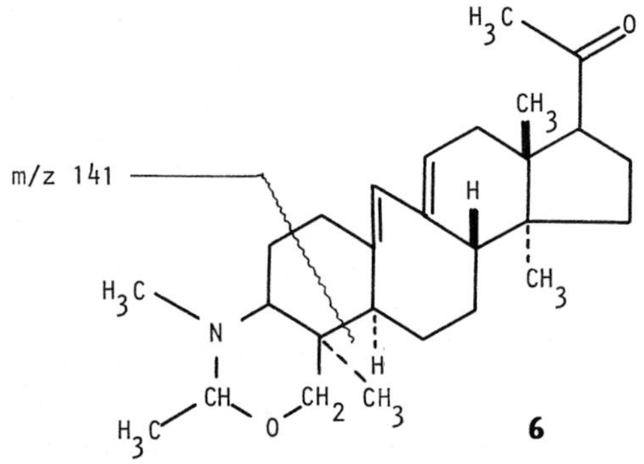

2.7 KARACHICINE (**7**) (ref. 10)

The UV spectrum of this minor alkaloid showed maxima at 241 and 248 nm and shoulders at 210 and 257 nm, characteristic of the presence of a conjugated 9(10 $\rightarrow$ 19) abeo-diene system. The UV spectrum also included weak absorptions at 283 and 291 nm. The IR spectrum of the substance displayed a band at 1680 cm^{-1}, indicating the presence of a six-membered $\alpha\beta$-unsaturated ketone.

The ^{1}H-NMR spectrum ($CDCl_3$) of the alkaloid showed four tertiary methyl singlets at δ 0.58, 0.68, 1.13 and 1.21. A doublet at δ 0.75 (J=6.6 Hz) was assigned to the secondary methyl group at C-21. Two singlets at δ 2.03 and 2.09 were ascribed to the $N(CH_3)_2$ protons at C-3, while another singlet at δ 2.41 was assigned to the $N\text{-}CH_3$ group at C-20. A singlet at δ5.16 was assigned to the olefinic proton at C-19 while another singlet at δ 5.93 was ascribed to the C-11 olefinic proton.

The mass spectrum displayed the molecular ion at m/z 412.3446 in agreement with the formula $C_{27}H_{44}N_2O$. The base peak at m/z 58 was attributed to the loss of fragment $CH_3\text{-}CH=\overset{+}{N}HCH_3$ from ring D. Other major peaks were present at m/z 85 and 71, these fragments being similar to those of (+)-moenjodaramine (**1**). The only two positions of the carbonyl group which would explain its absorption at 1680 cm^{-1} are C-1 and C-12 (assuming that the position of the olefinic bond is the same as that normally encountered in the Buxus alkaloids). It was apparent from the mass spectral fragmentation pattern that the carbonyl group was not located in ring A. Hence its position at C-12 appears reasonable and is consistent with the downfield shift of the C-11 proton (δ5.93) as compared to the corresponding proton found in (+)-moenjodaramine (**1**) (δ 5.55) (ref. 1) and harappamine (**2**) (δ5.56) (ref. 4).

In the light of above studies we proposed structure **7** for this new alkaloid (ref. 10).

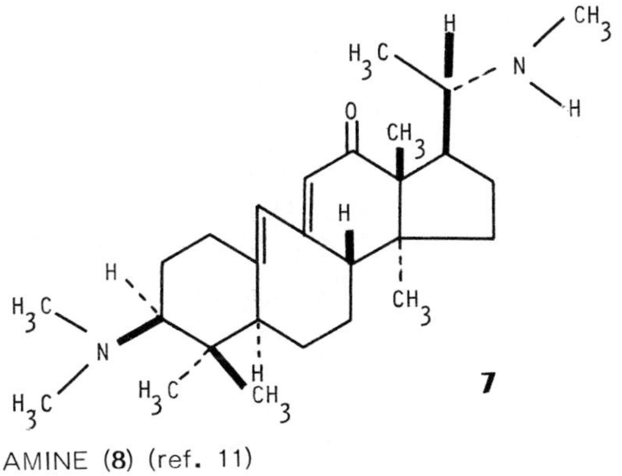

7

2.8 (+)-PAPILAMINE (**8**) (ref. 11)

The UV spectrum of (+)-papilamine (**8**) was similar to that of (+)-moenjodaramine (**1**) and harappamine (**2**), showing the presence of a 9(10 → 19) abeo-diene system. The IR spectrum (KBr) showed absorptions at 3400 (N-H) and 1585 (C=C) cm^{-1}.

The ^1H-NMR spectrum ($CDCl_3$) showed four singlets at δ 0.72, 0.97 and 1.25 and 1.38 which were assigned to the four tertiary methyl groups. A doublet at δ 0.97 (J= 6.5 Hz) was attributed to the C-21 secondary methyl group. Two singlets at δ 2.43 and 2.59 (δ 2.67 and 2.76 in CD_3OD) each integrating for 3-H were assigned to the $N\text{-}CH_3$ groups at C-3 and C-20 respectively. The two olefinic protons at C-19 and C-11 resonated at δ 5.90 and 5.56 respectively.

An accurate mass measurement on the molecular ion peak showed the exact mass to be 384.3504 which was in agreement with the formula $C_{26}H_{44}N_2$. A peak at m/z 341 was attributed to the loss of gem-C-dimethyl group from ring A. The base peak appeared at m/z 58 indicating the cleavage of CH_3-CH=$\overset{+}{N}$HCH_3 grouping from ring D. A peak at m/z 85 was due to the cleavage of ring D along with the nitrogen-containing side chain.

The above studies led to structure **8** for (+)-papilamine (ref. 11).

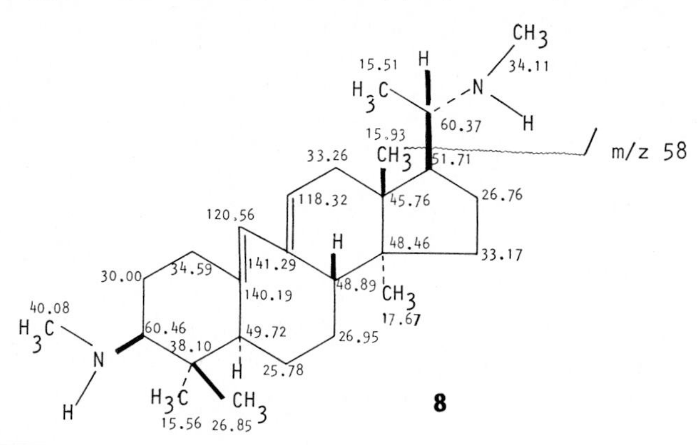

2.9 (+)-PAPILICINE (**9**) (ref. 12)

The UV spectrum of (+)-papilicine (**9**) exhibited characteristic absorptions typical of the 9(10 → 19) abeo-diene system, λ (MeOH): 210, 241, 248 and 257 nm. The IR spectrum showed peaks at 1080 (C-N) and 1600 cm^{-1} (C=C).

The ^{1}H-NMR spectrum ($CDCl_3$) showed four singlets corresponding to the four tertiary methyls at δ 0.69, 0.71, 0.75 and 1.01. A doublet appeared at δ 0.97 (J=7.0 Hz), and was assigned to the C-21 methyl group. A 3-H singlet at δ 2.46 was due to the -NCH_3 group at C-20, while another 6-H singlet at δ 2.28 was assigned to the -$N(CH_3)_2$ group. A singlet at δ 5.91 was attributed to the vinylic proton at C-19, while a multiplet centered at δ 5.49 was due to the C-11 vinylic proton.

The mass spectrum of **9** afforded M^+ at m/z 398.3665 in agreement with the composition, $C_{27}H_{46}N_2$. The base peak at m/z 58 represented the dimethyl-iminium cation, CH_3-CH=$\overset{+}{N}$HCH_3. Other major peaks were present at m/z 85

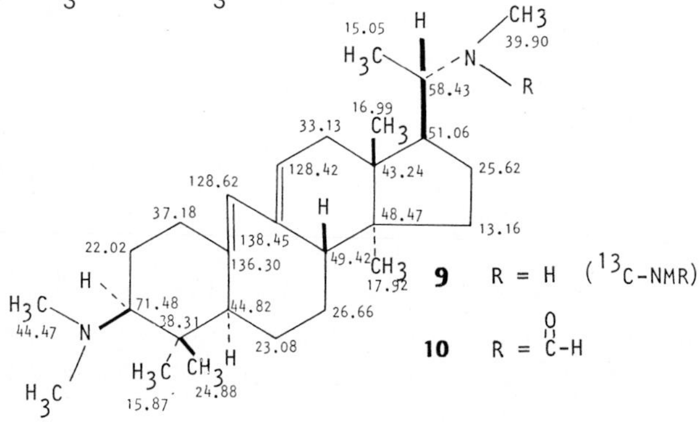

and 71 indicating the presence of $-N(CH_3)_2$ grouping on ring A.

The above studies led to structure **9** for (+)-papilicine (ref. 12).

2.10 (+)-N-FORMYLPAPILICINE (10) (ref. 6)

This compound showed a UV spectrum nearly identical with that of (**9**). The IR spectrum showed intense bands at 1658 (amide carbonyl) and 1599 cm^{-1} (C=C).

The ^1H-NMR spectrum ($CDCl_3$) was closely related to that of the known (+)-papilicine (**9**) (ref. 12). It included four three-proton singlets at δ0.70, 0.80, 0.81 and 1.14, representing the four tertiary methyl groups present. The 21-secondary methyl group resonated as a doublet at δ 1.20. A six-proton singlet at δ 2.43 was assigned to the $N(CH_3)_2$ group attached to C-3. The C-11 vinylic proton appeared as a doublet of doublets at δ 5.53, while the C-19 olefinic proton resonated as a singlet at δ 5.93. A singlet at δ 8.11 accompanied by a much smaller singlet at δ 7.98, represented the N_b- formyl proton. Similarly, the N_b- methyl group was indicated by a singlet at δ 2.74 followed by a smaller singlet at δ 2.80 due to geometrical isomerism.

The mass spectrum of (+)-N-formylpapilicine (**10**) showed the molecular ion at m/z 426 in agreement with the molecular formula $C_{28}H_{46}N_2O$. The peak at m/z 383 resulted from the loss of the methyliminium moiety from ring A. A fairly large ion at m/z 86 represented the $[CH_3-CH=N(CH_3)(CHO)]^+$ fragment which arose by the cleavage of ring D side chain. Finally, the base peak at m/z 71 arose by cleavage of ring A along with the dimethylamino substituent.

On the basis of the above spectroscopic studies, structure **10** was assigned to (+)-N-formylpapilicine (ref. 6).

2.11 (+)-BUXAMINOL-G (11) (ref. 13)

(+)-Buxaminol-G (**11**) is another member of the 9(10 → 19) abeo-diene series. Its IR spectrum displayed the presence of hydroxyl group (3550 cm^{-1}) and C=C (1550 cm^{-1}).

The ^1H-NMR spectrum ($CDCl_3$) showed a set of AB doublets at δ 3.53 and 3.84 (J=10.0 Hz), which were assigned to the C-31 methylene protons α -to the hydroxyl group. The presence of the hydroxyl group at C-31 in α-disposition was indicated by comparison of chemical shift and coupling constants with other similarly substituted alkaloids. A multiplet at δ4.21 was ascribed to the C-16 proton. A 6-H singlet at δ 2.30 was assigned to the $N(CH_3)_2$ group, while another singlet at δ2.33 was due to the $N(CH_3)_2$ group at C-3.

The mass spectrum of (+)-buxaminol-G showed the molecular ion at m/z 444.3709 corresponding to the molecular composition $C_{28}H_{48}N_2O_2$. Other major peaks were present at m/z 84, 71 and 58 indicating the presence of dimethylamino grouping at C-3, whereas a base peak at m/z 72 was indicative of a dimethylamino substituent at C-20. Peaks at m/z 115 and 129 suggested the presence of hydroxyl group on ring D.

Structure **11** was assigned to (+)-buxaminol-G on the basis of the above spectroscopic studies (ref. 13).

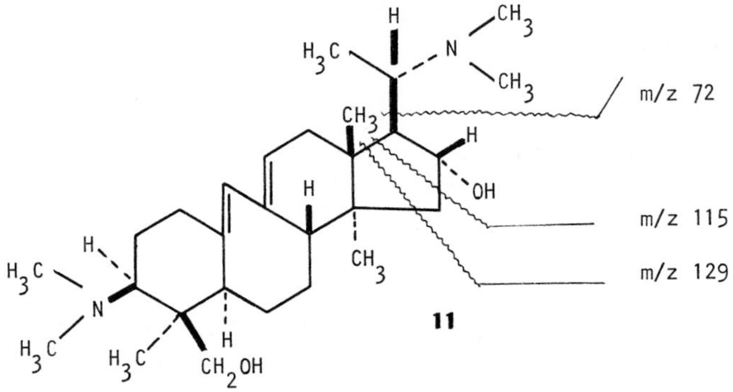

2.12 (+)-31-ACETOXY-N_a-BENZOYLBUXIDIENINE (**12**) (ref. 14)

The substance showed maxima at 238 and 245 nm with shoulders at 205 and 253 nm in its UV spectrum, indicating the presence of a 9(10→19) abeo-diene system. The IR spectrum included intense absorptions at 3350 (OH), 1716 (ester carbonyl) and 1662 cm^{-1} (α, β-unsaturated amide carbonyl).

The ^{1}H-NMR spectrum ($CDCl_3$) showed three tertiary methyl groups at δ 0.76, 0.77 and 0.94 while a secondary methyl group resonated as a doublet at δ 0.93 (J=6.4 Hz). A six-proton singlet at δ 2.61 was assigned to the $-N(CH_3)_2$ group. A three-proton singlet at δ 2.12 was due to the acetyl methyl group. The C-31 methylene protons appeared as a set of AB doublets at δ 3.82 and 4.02 (J=10.9 Hz). A multiplet centered at δ 3.94 was assigned to the C-16 proton geminal to the hydroxy group. A close multiplet at δ 5.52 and a singlet at δ 6.07 were assigned to C-11 and C-19 vinylic protons. The aromatic protons appeared as two groups of 3H and 2H multiplets at δ 7.43 and 7.11 respectively. The NOESY spectrum of the compound served to establish the relative stereochemistry at several key points in the molecule. Strong cross peaks were observed corresponding to the C-11 and C-19 olefinic protons. The N_b-CH_3 signal showed NOE interaction with the aromatic region (phenyl 4'-H). This interaction requires ring A to be in a twist-boat conformation. The NOE interaction between N_b-methyl protons and C-18 methyl group established the β-orientation of the C-18 methyl group. Similarly NOE interaction between the C-30 methyl protons at δ 0.94 could be seen with the C-6 methylene protons at δ 1.91, which established the α-orientation of the C-30 methyl group. This showed that the acetate group was attached to the β-oriented C-31 methylene group. The NOESY interactions are shown in Fig.

12. Other 2D ^{1}H-NMR experiments and ^{13}C-NMR spectra (DEPT, GASPE etc.) also favour the proposed structure **12**.

The mass spectrum showed the M^+ at m/z 562.3801 corresponding to the formula $C_{35}H_{50}N_2O_4$. A peak at m/z 547 was due to the loss of methyl group, while m/z 503 (M-59) was due to the loss of acetate group. The loss of $OCOCH_3$ (59 m.u) rather than CH_3COOH (60 m.u) suggested that the acetate was attached to a carbon with no α-hydrogen. A peak at m/z 115 resulted from the cleavage of ring D along with the nitrogen-bearing side chain, and indicated the presence of a hydroxy group on ring D. The base peak at m/z 72 was attributed to the trimethyliminium side chain. Acetylation of **12** produced the diacetate, $C_{37}H_{52}N_2O_6$ (12a).

On the basis of the above studies, structure **12** was assigned to this new alkaloid (ref. 14).

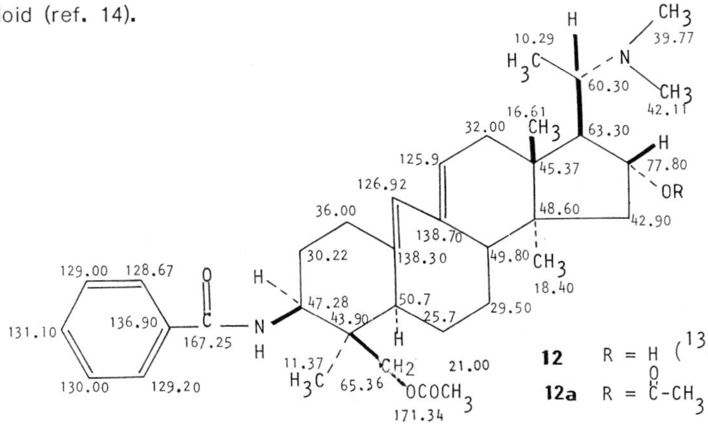

12 R = H (^{13}C-NMR)

12a R = $\overset{O}{\parallel}$C-CH_3

2.13 (+)-BUXABENZAMIDIENINE (**13**) (ref. 15)

The compound showed a UV spectrum characteristic of the 9(10 → 19) abeo-diene bases. The IR spectrum displayed intense bands at 3680 (NH) and 1652 cm^{-1} (α,β-unsaturated amide).

The ^{1}H-NMR spectrum ($CDCl_3$) of **13** exhibited four tertiary methyl singlets at δ 0.73, 0.77, 0.80 and 1.01. A doublet at δ 0.93 is due to the C-21 methyl group. A six-proton singlet at δ 2.30 was due to $-N(CH_3)_2$ group. Another singlet at δ 6.00 represented the isolated vinylic H-19, while a doublet of doublets centered at δ 5.56 was due to the vinylic H-11. The -NH doublet appeared at δ 5.91 (J=9.8 Hz), while multiplets at δ 7.44 and 7.88 were assigned to aromatic protons.

The mass spectrum of (+)-buxabenzamidienine (**13**) included the molecular ion at m/z 488.371, corresponding to the molecular formula $C_{33}H_{48}N_2O$. A peak at m/z 473 indicated the loss of a methyl, while a peak at m/z 383 reflected the loss of a benzoyl substituent. The base peak at m/z 72 was attributed to the trimethylinimium side chain fragment, $CH_3-CH=\overset{+}{N}(CH_3)_2$.

On the basis of the above studies, structure **13** was assigned to (+)-buxabenzamidienine (ref. 15).

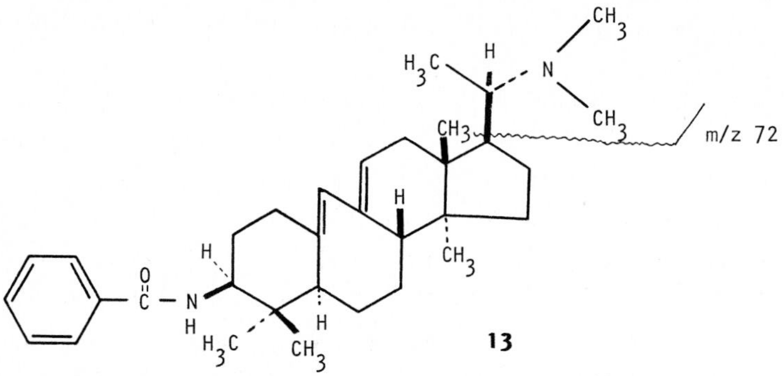

2.14 (+)-16α-ACETOXYBUXABENZAMIDIENINE (**14**) (ref. 15)

(+)-Buxabenzamidienine (**13**) was accompanied in the plant by its 16α-acetoxy derivative (**14**). This compound showed a UV spectrum nearly identical with that of **13**. The IR spectrum was also somewhat similar, but with an additional band at 1732 cm^{-1} because of the ester carbonyl function.

The ^{1}H-NMR spectrum ($CDCl_3$) showed four tertiary methyl singlets at δ 0.77, 0.86, 0.87 and 1.11. The C-21 methyl doublet at δ 1.26 was appreciably further downfield than in the case of **13** which lacks the C-16 acetoxyl function. A singlet at δ 1.80 represented the acetyl methyl group, while a multiplet at δ 5.00 was due to H-16. The rest of the ^{1}H-NMR spectrum was similar to compound **13**.

The mass spectrum included the molecular ion peak at m/z 546.3729, corresponding to the formula $C_{35}H_{50}N_2O_3$ (calcd. 546.3720). The compound showed a peak at m/z 531 due to the loss of methyl group from the molecular ion. A peak at m/z 441 represented the cleavage of benzoyl substituent, while the base peak at m/z 72 was again due to the trimethyliminium cation. Other significant ions were at m/z 60 representing acetic acid, and m/z 157 and 171 which were formed by cleavage of ring A alongwith the side chain. These ions indicated that the acetate function must be attached to the C-16 position of ring D.

The above mentioned studies led to structure **14** for 16α-acetoxybuxabenzamidienine (ref. 15).

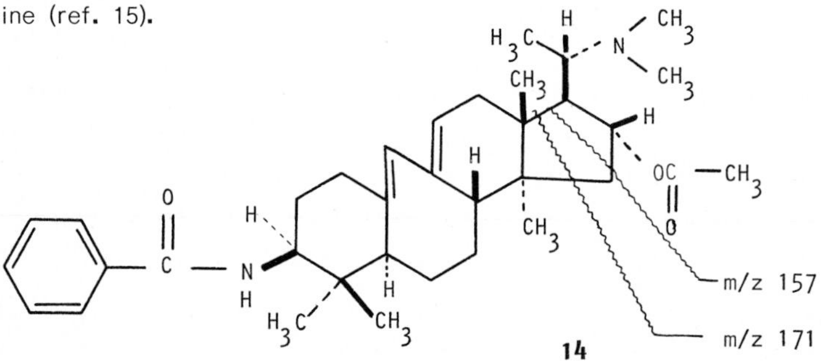

2.15 (-)-BUXANOLDINE (**15**) (ref. 15)

The UV spectrum of compound **15** showed a benzamide chromophore at 228 nm. The IR spectrum included peaks at 3580 (NH), 3360 (O-H), 1640 (α,β-unsaturated amide) and 1600 cm^{-1} (C=C).

The ^1H-NMR spectrum ($CDCl_3$) of (-)-buxanoldine was particularly informative. Sufficient quantities were on hand to allow for detailed NOE difference studies, in addition to the usual 2D ^1H-NMR experiments. It featured three tertiary methyl groups as singlets at δ 0.69, 0.72 and 0.90, while a secondary methyl group resonated as a doublet at δ 0.90 (J= 6.5 Hz). Since a hydroxy group was present at C-31, the C-31 methylene protons appeared as two doublets at δ 3.23 and 3.49 (J=12.4 Hz). A multiplet at δ4.11 was due to H-16 which is geminal to a hydroxyl group. A doublet of doublets at δ 5.26 (J_1= 2.8 Hz, J_2=1.6 Hz) and another at 5.43 (J=4.0 Hz, J=1.8 Hz) represented the vinylic hydrogens at C-11 and C-1 respectively. The two double bonds are thus not conjugated, and this finding is in accord with the UV absorption discussed above. A doublet at δ 6.05 (J= 9.0 Hz) was due to -NH coupled with 3-H. The aromatic protons appeared as two multiplets centered at δ 7.43 and 7.76. Some of the more salient NMR NOE results are shown in diagram **15.1**. Irradiation of the H-17 doublet of doublets (δ1.90) led to a 12.5% enhancement of the C-32 methyl singlet (δ 0.90). Irradiation of the H-16 multiplet (δ4.11) resulted in no enhancement of the H-17 signal (δ 1.90), but instead led to an increase in the area of the H-15 absorption (δ1.95), thus furnishing an insight into the stereochemistry of the ring D substituents. Specifically, the C-16 hydroxyl is α, as in all other Buxus alkaloids hydroxylated at that site. Another significant NOED result reflected on the spatial proximity of the C-19 methylene protons (δ 2.68 and 2.86) to the vinylic H-1 and H-11 protons (δ5.43 and 5.26) since these absorptions also showed reciprocal NOED's.

The mass spectrum of (-)-buxanoldine (**15**) showed the molecular ion m/z 520.3677 corresponding to the molecular formula $C_{33}H_{48}N_2O_3$ (calcd. 520.3644). A peak at m/z 183 resulted from the retro-Diels Alder cleavage of ring C. The peaks at m/z 129 and m/z 115 were formed by ring D cleavage, and indicated the presence of -OH group at the C-16 position of ring D. The base peak at m/z 72 represented the trimethyliminium ion.

Acetylation of (-)-buxanoldine (**15**) using acetic anhydride in pyridine afforded (-)-buxanoldine diacetate, $C_{37}H_{52}N_2O_5$ (**15a**), the IR spectrum of which incorporated an intense ester carbonyl absorption at 1720 cm^{-1}, in addition to the amidic carbonyl band at 1650 cm^{-1}. The ^1H-NMR spectrum of the diacetate displayed downfield shifts of H-16 from δ4.11 to 5.15, and of the C-31 methylene protons from δ 3.23 and 3.49 to δ 3.67 and 3.85 respectively.

Hydrolysis of (-)-buxanoldine (**15**) using ethanolic KOH yielded (-)-desbenzoylbuxanoldine (**15b**), $C_{26}H_{44}N_2O_2$. The UV spectrum of this compound showed only a terminal absorption, whereas the IR spectrum showed lack of amidic carbonyl absorption band. The ^{1}H-NMR spectrum was devoid of any proton in the aromatic region. The relatively facile hydrolysis of the benzamide contrasted with the previously-observed resistance to hydrolysis of benzamides of Buxus alkaloids, which was attributed to neighbouring -OH group participation (ref. 16,17).

On the basis of above studies, structure **15** was assigned to (-)-buxanoldine (ref. 15).

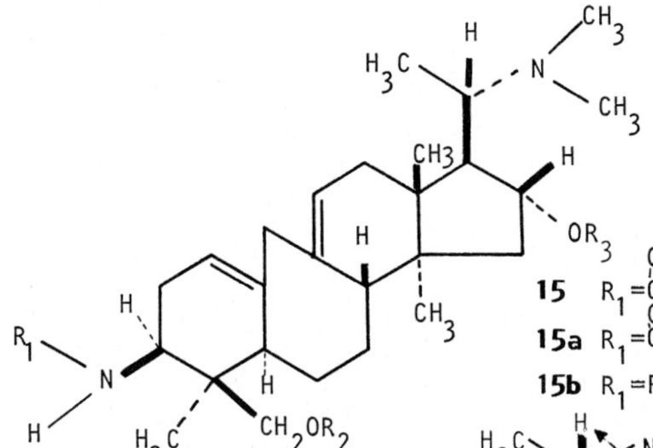

15 $R_1=\overset{O}{\overset{\|}{C}}-Ph, R_2=R_3=H$

15a $R_1=\overset{O}{\overset{\|}{C}}-Ph, R_2=R_3=\overset{O}{\overset{\|}{C}}-CH_3$

15b $R_1=R_2=R_3=H$

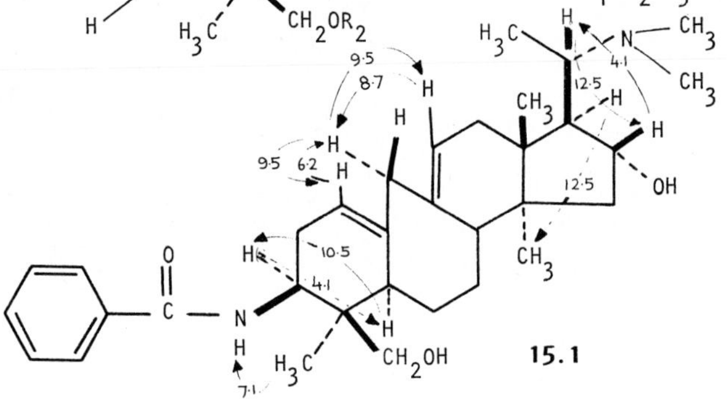

15.1

2.16 (+)-BUXANALDININE (16) (ref. 15)

(+)-Buxanaldinine (**16**) is the first Buxus alkaloid to incorporate a C-31 aldehyde group in lieu of the more common hydroxyl or ester functions. The UV spectrum displayed a maximum at 228 nm characteristic of the benzamide substituent. The IR spectrum showed absorptions at 3360 (NH), 1735 (ester carbonyl), 1722 (aldehyde carbonyl) and 1656 cm^{-1} (α,β-unsaturated amide carbonyl).

The ^{1}H-NMR spectrum ($CDCl_3$) exhibited singlets at δ0.88, δ0.883 and δ0.89 for the three tertiary methyl groups. Another singlet at δ 1.91 can be assigned to the acetate methyl group. The doublet at δ0.88 (J=5.8 Hz) belongs to the secondary methyl group. A multiplet centered at δ4.71 represented H-16 which is alpha to the acetate group. The C-1 olefinic proton appeared as a doublet of doublets at δ 5.69 (J=4.0 Hz, J=9.0 Hz), while the two multiplets at δ7.40 and 7.66 were due to aromatic protons. Finally, the aldehydic proton absorbed as a singlet at 9.50.

The mass spectrum of (+)-buxanaldinine (**16**) has molecular ion at m/z 562.3766 corresponding to the molecular formula, $C_{35}H_{50}N_2O_4$. Sodium borohydride in methanol reduction of (+)-buxanaldinine (**16**) supplied the corresponding alcohol (+)-buxanoldinine, $C_{35}H_{52}N_2O_4$ whose ^{1}H-NMR spectrum significantly included two one-proton doublets at δ 3.60 and 3.76 representing the C-31 methylene hydrogens, but was devoid of the aldehydic singlet at δ 9.50.

Structure **16** was assigned for (+)-buxanaldinine on the basis of above studies (ref. 15).

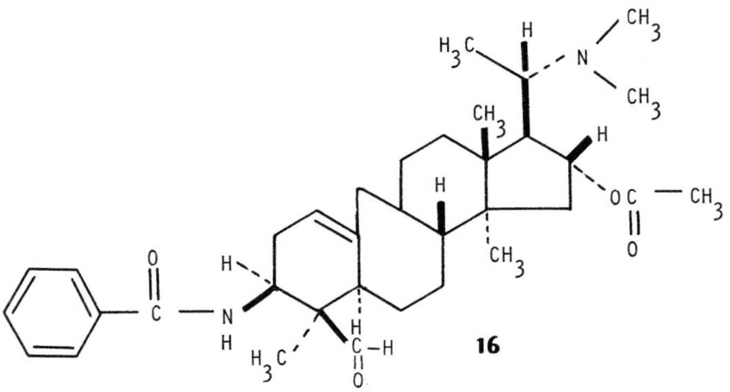

16

2.17 (+)-BUXUPAPINE (**17**) (ref. 9)

The UV spectrum of (+)-buxupapine (**17**) showed only terminal absorption, and the IR spectrum included bands at 3670 (N-H) and 1600 cm^{-1} (C=C).

The ^{1}H-NMR spectrum ($CDCl_3$) of (+)-buxupapine (**17**) included four singlets for the four tertiary methyl groups at δ 0.70, 0.73, 0.95 and 0.96. The C-21 secondary methyl group is represented by a doublet at δ 0.88. A six-proton singlet at δ2.25 was diagnostic of a dimethylamino group, and another singlet at δ 2.40 of an N_b-methyl group. A doublet of doublets centered at δ 5.26 can be ascribed to H-11 which is split by the C-12 methylene protons. Another broad multiplet at δ 5.33 was due to H-1. The two double bonds are not conjugated, and this finding is in accord with the lack of a distinct UV absorption.

The mass spectrum of **17** exhibited the molecular ion at m/z 398.3621 corresponding to the molecular formula $C_{27}H_{46}N_2$. The peak at m/z 355 reflects the loss of $HC{=}NHCH_3$ from ring A. Another peak at m/z 85 resulted from retro-Diels Alder cleavage of ring A (ref. 18). The compound showed the base peak at m/z 72 representing the trimethyliminium chain. These studies led to structure **17** for buxupapine (ref. 9).

2.18 (+)-N_b-NORBUXUPAPINE (**18**) (ref. 9)

(+)-Buxupapine (**17**) was accompanied in the plant by its N-demethyl analog, (+)-N_b-norbuxupapine (**18**). It has an IR spectrum very close to that of **17**, and its UV spectrum again showed only terminal absorption.

The ^{1}H-NMR spectrum ($CDCl_3$) of (+)-N_b-norbuxupapine (**18**) incorporated signals due to four tertiary methyl groups at δ0.70, 0.71, 0.93 and 0.94 while the secondary methyl group resonated as a doublet at δ0.97 (J=6.0 Hz). A 3-H singlet at δ 2.39 was assigned to N_a-CH_3 group, while another singlet at δ 2.40 was due to N_b-CH_3 group. The C-1 vinylic proton appeared as a doublet of doublet at δ5.26 (J_1= 2.5 Hz, J_2= 1.4 Hz). The other vinylic proton resonated at δ 5.33 as a broad multiplet. The C-19α and β methylenic protons appeared as AB doublets at δ2.77 and 2.93 (J=12.0 Hz). The mass spectrum of the compound included the M^+ ion at m/z 384.3517 in agreement with the composition, $C_{26}H_{44}N_2$. A peak at m/z 341 was again due to the loss of $CH=\overset{+}{N}HCH_3$ from ring A (ref. 18). Another peak at m/z 85 reflected the retro-Diels Alder cleavage of ring A (ref. 18). Finally, the base peak at m/z 58 was characteristic of the dimethyliminium cation, CH_3-$CH=\overset{+}{N}HCH_3$, produced by fission of the ring D side chain.

These studies established that compound **18** was the N-norderivative of buxupapine (ref. 9).

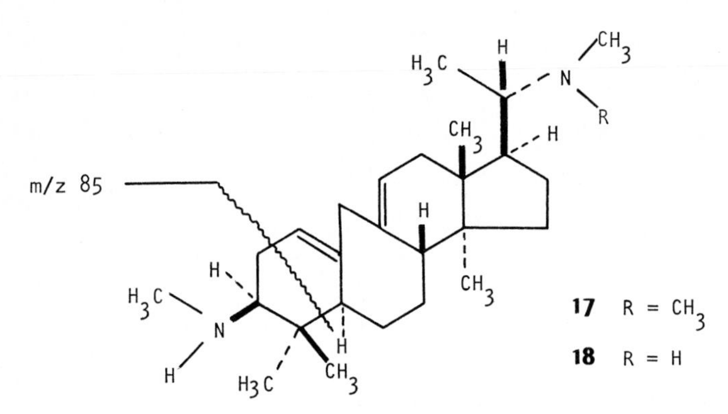

2.19 (-)-BUXAPAPINOLAMINE (**19**) (ref. 19)

The alkaloid possessed a UV spectrum with a maximum at 227 nm, characteristic of a benzamidic chromophore. The IR spectrum displayed intense bands at 3670 (N-H), 3310 (O-H), 1732 (ester carbonyl), 1722 (aldehydic carbonyl) and 1658 cm^{-1} (α,β-unsaturated amide).

The ^{1}H-NMR spectrum ($CDCl_3$) of **19** manifested signals representing three tertiary methyl groups at δ0.73, 0.88 and 1.01. A 3-H doublet at δ0.92 was assigned to the secondary methyl group of the side chain. A relatively downfield singlet at δ1.92 was diagnostic of an acetate methyl group. The dimethylamino function appeared as a singlet at δ2.27. A multiplet at δ 4.71 was due to H-16 which is geminal to the acetate function. A doublet of doublets at δ5.31 and another at δ5.66 represented the vinylic hydrogens at C-11 and C-1, respectively. The two double bonds are not conjugated, a finding which is in accord with the UV absorption. The aromatic protons appeared as two multiplets centered at δ7.39 and 7.65.

The aldehydic proton absorbed as a singlet at δ9.51. An interesting feature of the ^{1}H-NMR spectrum was the presence of a multiplet at δ4.20 denoting H-6 which is geminal to the hydroxyl group.

In accord with all other related Buxus alkaloids, the C-3 aminated substituent in (-)-buxapapinolamine (**19**) has been placed in a beta configuration. Furthermore, whenever biogenetic oxidation of the gemdimethyl substituent of ring A occurs, it is always the C-31 site that is affected.

It is relevant to point out at this stage that more than a dozen 6α-hydroxy-cycloartenol triterpenoids are known which possess an alcohol function in ring B at exactly the same site as in (-)-buxapapinolamine (**19**).

The ^{1}H-NMR spectrum of species **19** was also re-run in pyridine-d_5. It is known that under these conditions protons adjacent to the hydroxyl group will suffer a pronounced paramagnetic shifts in relation to the $CDCl_3$ spectrum. Indeed, the downfield shifts experienced by the C-31 aldehydic proton (δ 9.51 to 9.94), the C-6 proton (δ 4.20 to 4.91) and the C-30 methyl protons (δ 0.88 to 1.24) argue convincingly in favour of the proposed C-6 position for the hydroxyl function (ref. 21). This finding was also confirmed by partial NOED experiments.

The mass spectrum of **19** evidenced molecular ion at m/z 576.3562 corresponding to the molecular formula $C_{35}H_{48}N_2O_5$. Loss of trimethyliminium side chain accounted for the base peak m/z 72. Other significant ions were m/z 157 and 171 which were formed by the cleavage of ring D along the lines indicated. It follows that the acetoxyl group must be attached to ring D. More specifically, this function is linked to C-16. Finally, the peak at m/z 105 was due to the benzoyl substituent, while the relatively large m/z 28 peak may be accounted for through loss of carbon monoxide. Acetylation of (-)-buxapapinolamine (**19**) using acetic anhydride in pyridine afforded, as expected, (-)-6α-acetylbuxapapinolamine, (**19a**), $C_{37}H_{50}N_2O_6$. The ^{1}H-NMR spectrum of this acetate showed a downfield shift of H-6 from δ 4.20 to 5.18 and a new singlet at δ1.90 representing the acetyl protons. Alternatively, sodium borohydride in methanol reduction of (-)-buxapapinolamine (**19**) supplied the corresponding C-31 alcohol, $C_{35}H_{50}N_2O_5$, whose IR spectrum showed the lack of aldehydic absorption near 1722 cm^{-1}. The ^{1}H-NMR spectrum of this alcohol included two one-proton doublets at δ 3.68 and 3.78 representing the C-31 methylene hydrogens.

On the basis of the above spectral and chemical studies, structure **19** was assigned to (-)-buxapapinolamine (ref. 19).

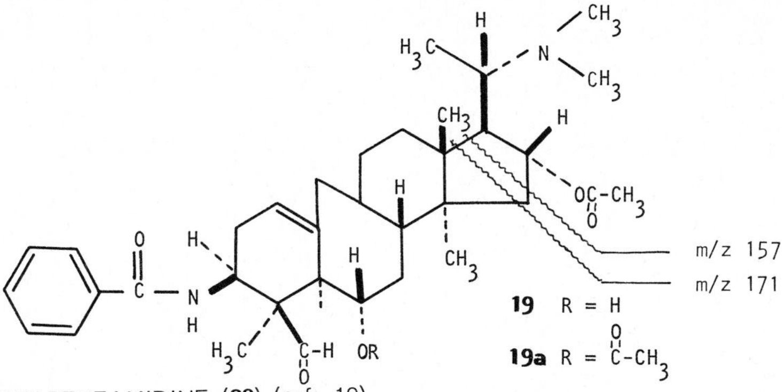

19 R = H

19a R = $\overset{O}{\overset{\|}{C}}$-$CH_3$

2.20 (+)-BUXABENZAMIDINE (**20**) (ref. 19)

(+)-Buxabenzamidine (**20**) showed a UV spectrum with a maximum at 226 nm, characteristic of a benzamidic chromophore. The IR spectrum in chloroform solution displayed intense bands at 3670 (N-H) and 1650 cm^{-1} (α,β-unsaturated amide).

The ^{1}H-NMR spectrum ($CDCl_3$) of (+)-buxabenzamidine featured four singlets, for the tertiary methyl groups at δ0.69, 0.77, 1.00 and 1.01, while a doublet centered at δ 0.91 was assigned to the C-21 secondary methyl group. A 6-H singlet at δ2.27 was ascribed to the $N(CH_3)_2$ group. A multiplet centered at δ 4.12 represented H-3. A doublet of doublets at δ5.34 was related to the vinylic H-11, whose signal is split by the C-12 methylene protons.

The mass spectrum of (+)-buxabenzamidine (**20**) included molecular ion m/z 490.3934 corresponding to the molecular formula $C_{33}H_{50}N_2O$. A peak at m/z 323 was the result of retro-Diels Alder cleavage of ring ring C (ref. 18). The base peak at m/z 72 represented the trimethyliminium cation, $H_3C\text{-}CH\text{=}\overset{+}{N}(CH_3)_2$.

In the light of the above spectral studies structure **20** was proposed for (+)-buxabenzamidine (ref. 19).

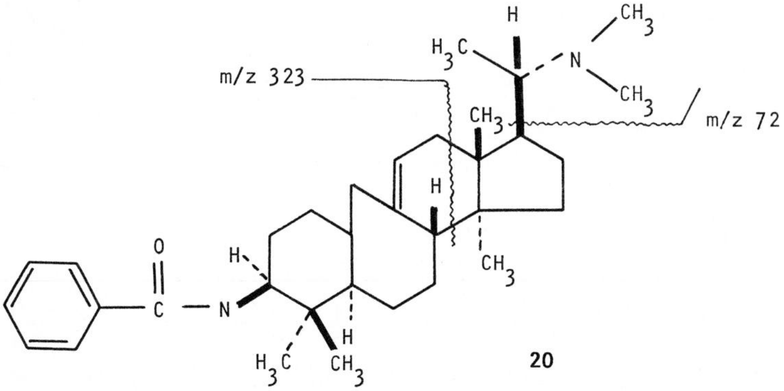

20

2.21 (-)-CYCLOBUXOVIRICINE (**21**) (ref. 22)

This new mono-amino steroidal alkaloid showed UV absorption at 268 nm, indicative of the presence of α,β-unsaturated ketonic group conjugated with the cyclopropane ring. The IR absorption band at 1647 cm^{-1}, further confirmed it.

The ^{1}H-NMR spectrum ($CDCl_3$) showed four three-proton singlets at δ 0.90, 0.95, 0.97 and 1.09 representing four tertiary methyl groups. A set of AB doublets at δ 0.75 and 0.81 ($J_{19\alpha,19\beta}$= 5.0 Hz) was assigned to the cyclopropyl protons. The downfield shift from the usual values is due to the deshielding effect of the neighbouring olefinic-functionality (ref. 23). Two doublets at δ 6.76 and 5.95 were ascribed to the C-1 and C-2 olefinic protons respectively (J=10.1 Hz). The ^{13}C-NMR ($CDCl_3$) of compound **21** are sumarized around expression **21**.

The mass spectrum of the compound afforded a molecular ion at m/z 369.3025, corresponding to the formula $C_{25}H_{39}NO$ (calcd. 369.3031) indicating seven double bond equivalents. A peak at m/z 354 resulted from the loss of methyl group from M^+. The base peak at m/z 58, corresponding to CH_3-CH=$\overset{+}{N}HCH_3$, resulted from the cleavage of ring D side chain. N-Methylation by formic acid/ formaldehyde yielded the known (-)-cyclobuxoviridine (**21a**). These studies led to structure **21** for cyclobuxoviricine (ref. 22).

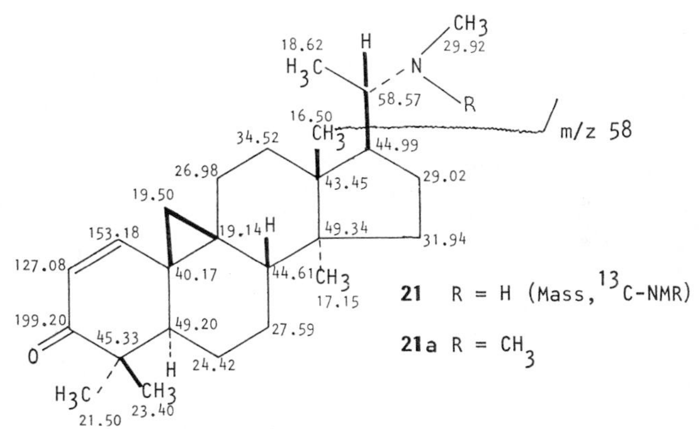

2.22 (-)-CYCLOXOBUXOVIRICINE (**22**) (ref. 24)

The 16α-hydroxy derivative of cyclobuxoviricine (**22**) has been isolated from the weakly basic alkaloidal fractions of <u>B.papilosa</u> leaves. Compound **22** showed UV absorption at 265 nm, characteristic of a conjugated ketonic group. The IR spectrum displayed bands at 1661 (α,β-unsaturated carbonyl), 3100 (cyclopropyl) and 3400 cm^{-1} (OH).

The ^{1}H-NMR spectrum ($CDCl_3$) showed four singlets for the four tertiary methyl groups at δ 0.95, 0.97, 1.10 and 1.13. A doublet at δ 1.20 (J=6.09 Hz) was assigned to the 21-secondary methyl group. The presence of cyclopropyl group was shown by the presence of half of an AB double doublet at δ 0.70 (J= 4.59 Hz), the other half being embedded in the methyl/methylenic region of the spectrum. A multiplet centered at δ 4.28 was assigned to C-16 proton geminal to the hydroxy group. The three-proton singlet at δ 2.59 was assigned to the N-CH_3 group. Doublets at δ 5.94 and 6.73 (J= 10.0 Hz) were assigned to C-2 and C-1 olefinic protons, conjugated with the carbonyl function. Two dimensional ^{1}H-NMR measurements (ref. 25,26) and homo-decoupling experiments also favour the above mentioned assignments. The NOE difference measurements were also carried out to establish the relative stereochemistry of the various protons.Irradiation of the C-16 proton at δ 4.28 resulted in 1.68% NOE of C-20 H, while irradiation at C-20H resulted in a reciprocal NOE effect of C-16H. This suggested the β-orientation of C-16H and C-20H. The absence of any NOE interactions with C-17H indicated the α-orientation of geminal OH group at C-16. Irradiation at C-20H resulted in 1.92% NOE of the N-CH_3 and 2.64% NOE of the 18-CH_3. This showed the α-orientation of C-20H. Irradiation at C-17H resulted in 4.5% NOE of the 32-CH_3 which established the α-orientation of both C-17H and 32-CH_3.

The mass spectrum of **22** showed the molecular ion at m/z 385.2955 in agreement with the formula $C_{25}H_{39}NO_2$ (calcd. 385.2955). The substance readily lost a methyl group to give the ion m/z 370. The fragment at m/z 115 was formed by the cleavage of ring D along with the nitrogen-bearing side chain. The peak at m/z 100 ($C_5H_{10}NO$) is particularly informative, as it established that the OH group is present on a five carbon fragment which resulted from the cleavage of ring D along with the N-side chain; the lack of oxygen in the ion of m/z 71 (C_4H_9N) confirmed the position of the -OH group at C-16. The compound showed a base peak at m/z 58 attributed to the loss of dimethyliminium (CH_3-CH=$\overset{+}{N}$HCH_3) side chain.

In the light of the above studies, structure **22** was assigned to (-)-cycloxobuxoviricine (ref. 24).

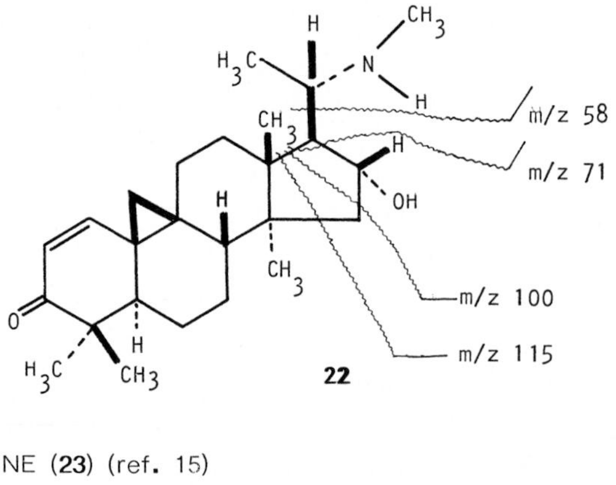

22

2.23 (+)-CYCLOBUXOTRIENE (**23**) (ref. 15)

The somewhat unusual nature of (+)-cyclobuxotriene was initimated by its yellow colour, which could be clearly distinguished on silica gel TLC plates. The

UV spectrum was characterized by an intense absorption maximum at 324 nm, denoting extensive conjugation. The IR spectrum included peaks at 1652 (α,β-unsaturated ketone) and 1600 cm^{-1} (C=C).

The ^{1}H-NMR spectrum ($CDCl_3$) was very informative. Four singlets at δ 0.72, 0.78, 0.96 and 1.17 corresponded to the four tertiary methyl groups. The C-21 secondary methyl group resonated as a doublet at δ 0.93 (J= 6.00 Hz). The $-N(CH_3)_2$ group appeared as a singlet at δ 2.30. Two doublets at δ 6.00 and 6.71 were ascribed to the vinylic C-2 and C-11 protons, respectively, whereas the vinylic H-19 appeared as a close doublet at δ 7.42 (J= 2.2 Hz). Additionally the vinylic H-11 appeared as a doublet of doublets centered at δ 6.24. The chemical shift of H-11 clearly indicated that conjugation only extended until C-19, while the double bond between C-9/C-11 behaves as an isolated double bond due to non-planarity of the molecule.

The mass spectrum evidenced molecular ion at m/z 381.3037 in agreement with the molecular formula $C_{26}H_{39}NO$ (calcd. 381.3031), corresponding to the presence of eight double bond equivalents. A peak at m/z 366 was due to the loss of a methyl group. Another interesting peak was at m/z 167, representing ring D with its substituents, and resulting from retro-Diels Alder cleavage of ring C along the line indicated. Loss of the trimethyliminium side chain accounted for the base peak at m/z 72.

On the basis of above spectral evidence, structure **23** was assigned for (+)-cyclobuxotriene (ref. 15).

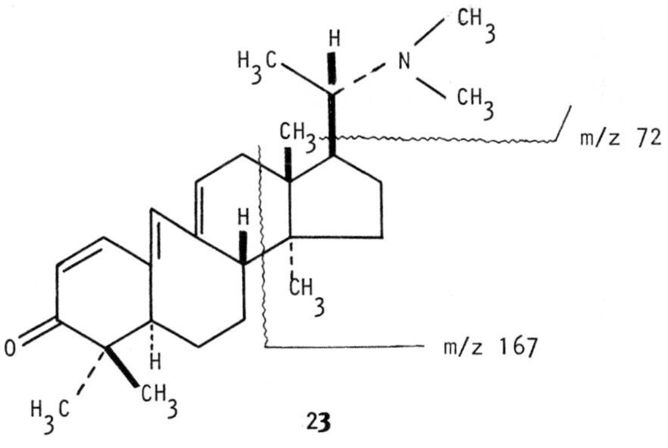

CH_3 H H_3C N CH_3 CH_3 m/z 72 H CH_3 m/z 167 O H CH_3 H_3C

23

2.24 (+)-NORCYCLOMICROBUXEINE (**24**) (ref. 9)

(+)-Norcyclomicrobuxeine (**24**), showed a UV spectrum with a maximum at 240 nm, indicating the presence of an α,β-unsaturated ketone. The IR spectrum incorporated bands at 3680 (NH), 1657 (α, β-unsaturated carbonyl) and 1588 cm^{-1} (C=C).

The ^{1}H-NMR spectrum ($CDCl_3$) of (+)-norcyclomicrobuxeine (**24**) showed close resemblance with that of cyclomicrobuxeine (**24a**) (ref. 23). It exhibited two AB doublets at δ 0.10 and 0.39 ($J_{19\alpha,19\beta}$ 4.4 Hz) accounting for the cyclopropyl methylenic protons. Signals due to two tertiary methyl groups were at δ 0.99 and 1.20. A three-proton singlet at δ 2.28 was assigned to a methyl group next to the carbonyl function. Another three-proton singlet at δ 2.50 represented the N-methyl group. The terminal methylidine protons appeared as singlets at δ 4.64 and 4.88. The C-16 olefinic proton absorbed as a doublet of doublet at δ 6.67 (J= 3.4 Hz, J= 2.0 Hz). The overall spectral image thus resembled that of (+)-cyclomicrobuxeine (**24a**), and in fact species **24** corresponds to the N-nor derivative of (+)-cyclomicrobuxeine.

Some NMR NOE difference measurements were obtained in order to clarify structural features and confirm some of the assignments. Irradiation of the C-21 methyl group (δ2.28) led to an 11.1% enhancement of H-16 (δ6.67), while irradiation of H-16 resulted in a 7.8% increase of the C-21 methyl signal. These data supported a transoid geometry for the enone system. Similarly, reciprocating NOED's were found between the exocyclic vinylic proton signal at δ4.88 and the N-methyl at δ 2.50.

The mass spectrum of the compound included molecular ion at m/z 353.2721 corresponding to the molecular formula $C_{24}H_{35}NO$. The peak at m/z 310 reflected the loss of $HC{=}\overset{+}{N}HCH_3$ from ring A. The base peak at m/z 44 derives from cleavage of ring A along with intramolecular proton transfer. Structure **24** was assigned to the substance on the basis of these studies (ref. 9).

Biogenetically compound **24** is intermedate between cholesterol and lanosterol.

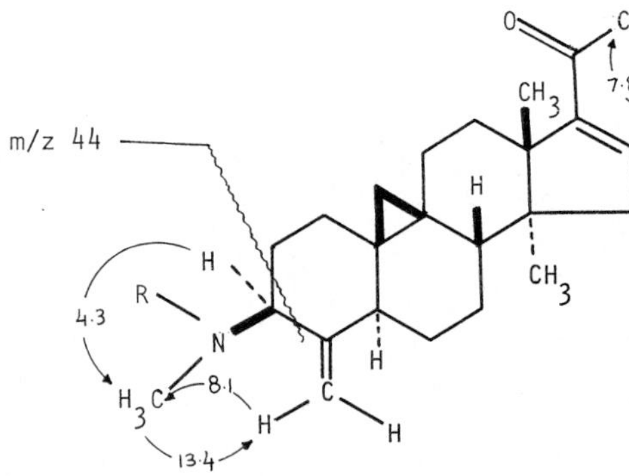

2.25 (+)-N-FORMYLCYCLOMICROBUXEINE (**25**) (ref. 27)

(+)-N-Formylcyclomicrobuxeine (**25**) showed UV maximum at 239 nm, characteristic of an α,β-unsaturated ketone. The IR spectrum displayed bands at 1670 (amide carbonyl), 1660 (α,β-unsaturated carbonyl) and 1600 (C=C) cm^{-1}.

The ^1H-NMR spectrum ($CDCl_3$) of **25** bore a distinct similarity to that of (+)-norcyclomicrobuxeine (**24**) (ref. 9) and (+)-cyclomicrobuxeine (**24a**) (ref. 23). It included two AB doublets at δ0.1 and 0.41 (J=4.2 Hz) accounting for the cyclopropyl methylenic protons. Two 3-H singlets at δ 1.00 and 1.20 represent the C-18 and C-32 methyl groups, respectively. Another 3-H singlet located at δ 2.27 was assigned to a methyl group α-to the carbonyl group. The terminal methylidene protons absorbed as singlets at δ4.59 and 4.67. The C-16 olefinic proton appeared as a doublet of doublets at δ 6.67 (J_1=3.4 Hz, J_2=2.0 Hz). A singlet at δ 8.12, accompanied by a much smaller singlet at δ8.18, represented the N_a-formyl proton. Similarly, the N_a-methyl group was indicated by a singlet at δ 2.80 along with a smaller singlet at δ 2.93, due to non-separable geometrical isomers. The NOESY spectrum showed strong cross peaks between the C-21 methyl group and the C-16 olefinic proton. This observation supported an s-transoid (or E) geometry for the enone system. Similarly NOE interaction was seen between the exocyclic vinylic proton signal at δ4.67 and the N_a-methyl group at δ 2.88 in the NOESY spectrum. A strong cross peak between the N_a-formyl proton and the C-3 proton was also observed.

The high resolution mass spectrum showed the molecular ion at m/z381.2648 corresponding to the molecular formula $C_{25}H_{35}NO_2$. The peak at m/z 366.2420 ($C_{24}H_{32}NO_2$) resulted from the loss of methyl group from the molecular ion. A considerably large peak at m/z 322 was due to the cleavage of the nitrogen containing side chain. The substance afforded a base peak at m/z 58.0292, corresponding to the composition C_2H_4NO, a fragment resulting from the cleavage of the ring A side chain. Structure **25** was assigned to N-formylcyclomicrobuxeine on the basis of these studies (ref. 27).

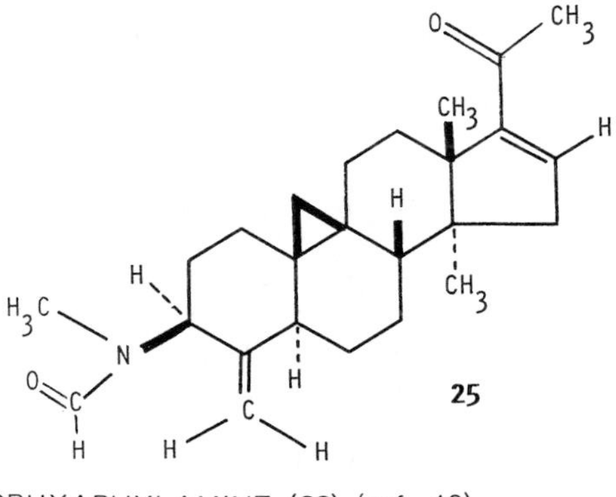

2.26 (-)-E-CYCLOBUXAPHYLAMINE (**26**) (ref. 19)

The compound has a UV absorption with a maximum at 224 nm, characteristic of an α,β-unsaturated ketone. The IR spectrum displayed peaks at 1712 and 1636 (α,β-unsaturated cyclopentanone) cm^{-1}.

The ^{1}H-NMR spectrum ($CDCl_3$) featured four singlets at δ1.14, 1.17, 1.27 and 1.33, corresponding to the four tertiary methyl groups. Only half of the cyclopropyl AB quartet could be observed at δ 0.45, while the other half had shifted to the methyl-methylene region due to the deshielding influence of the neighbouring olefinic functionality. A doublet at δ 1.84 was due to the 21-methyl group. The C-20 vinylic proton appeared as a quartet at δ 6.55 (J=7.5 Hz)(ref.28,29). Singlet at δ2.58 was readily assigned to the N-CH_3 residue. The C-7 vinylic proton resonated as a doublet of doublets at δ5.06 (J_1=5.4 Hz, J_2=2.0 Hz) (ref.30).

The mass spectrum of (-)-E-cyclobuxaphylamine was particularly informative. It included molecular ion m/z 367.2873 corresponding to the formula $C_{25}H_{37}NO$. A peak at m/z 138 was the result of allylic cleavage of ring B along the lines indicated. This ion was important in locating the position of the double bond at C-7(8). The base peak at m/z 71 resulted from ring A cleavage as shown in expression **26**.

These studies led to structure **26** for E-cyclobuxaphylamine (ref. 19).

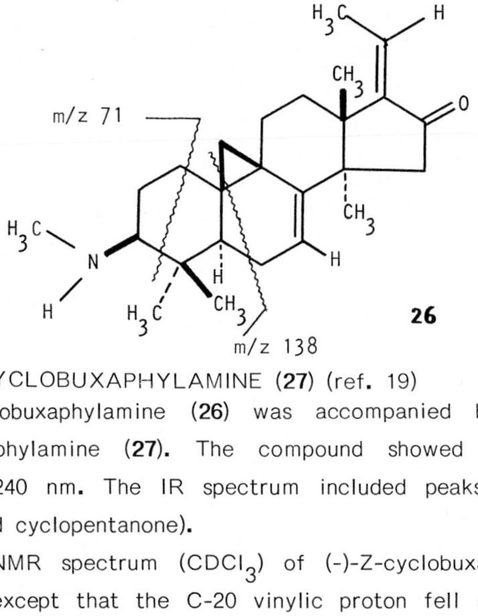

2.27 (-)-Z-CYCLOBUXAPHYLAMINE (**27**) (ref. 19)

(-)-E-Cyclobuxaphylamine (**26**) was accompanied by its geometric isomer, (-)-Z-cyclobuxaphylamine (**27**). The compound showed a UV spectrum with a maximum at 240 nm. The IR spectrum included peaks at 1709 and 1637 cm^{-1} (α, β-unsaturated cyclopentanone).

The ^{1}H-NMR spectrum ($CDCl_3$) of (-)-Z-cyclobuxaphylamine (**27**) resembled that of **26**, except that the C-20 vinylic proton fell relatively upfield at δ5.78. In contrast, the 21-methyl doublet at δ 2.12 was further downfield than in the case of **26** since this methyl is now syn to the C-16 carbonyl oxygen (ref.28,29).

The mass spectrum exhibited the molecular ion at m/z 367.2871 corresponding to the molecular formula, $C_{25}H_{37}NO$. The base peak at m/z 71 was again due to the cleavage of ring A. Another fragment at m/z 138 resulted from fission of ring B along the lines indicated.

It has been justifiably pointed in the literature that species of type **26** and **27** could be true alkaloids or else they could be artifacts of isolation produced through β-elimination of the corresponding C-20 dimethylamino group (ref. 28).

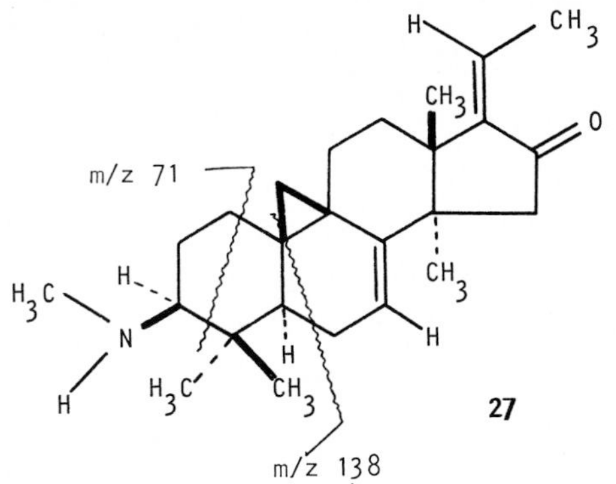

27

2.28 (+)-BUXAPROGESTINE (**28**) (ref. 19)

A very interesting alkaloid (+)-buxaprogestine (**28**) was isolated. It has a simple pregnane skeleton which lacks substituents at C-4 and C-14. The compound exhibited a UV spectrum characteristic of a six-membered α,β-unsaturated carbonyl group showing λ_{max} at 239 nm. The IR spectrum showed bands at 1660 (α,β-unsaturated carbonyl) and 1610 cm^{-1} (C=C).

The ^{1}H-NMR spectrum ($CDCl_3$) included two three-proton singlets at δ 0.72 and 1.18 corresponding to C-18 and C-19 methyl groups, respectively. The secondary 21-methyl group resonated as a doublet at δ0.97 ($J_{21,20}$= 6.0 Hz). A six-proton singlet at δ2.30 was assigned to the $N(CH_3)_2$ group. A close doublet centered at δ5.73 (J=1.1 Hz) can be ascribed to the C-4 olefinic proton which showed long range coupling with H-6.

The mass spectrum of (+)-buxaprogestine (**28**) featured the molecular ion at m/z 343.2869 corresponding to the formula $C_{23}H_{37}NO$. The substance showed a peak at m/z 328 resulting from the loss of methyl group from the M^{+}. The base peak at m/z 72 represented the trimethyliminium cation fragment through the cleavage of ring D side chain.

Structure **28** was assigned to buxaprogestine on the basis of these spectroscopic studies (ref.19). (+)-Buxaprogestine (**28**) represented the second example of the occurrence of a simple pregnane derivative within the Buxaceae and Apocynaceae families. The first example, (-)-irehine, was found in *Buxus sempervirens* (ref. 31), and later isolated from *Buxus papilosa* (ref. 32). Both examples provide valuable information about the biogenetic routes to *Buxus* alkaloids.

Due to close similarity of (+)-buxaprogestine (**28**) with that of the human pregnancy hormone progesterone (**28a**), compound **28** can act as a potential source for this hormone. Oxidation of N-bearing side chain to the corresponding ketimine, followed by its hydrolytic removal would result in the formation of progesterone

(**28a**).

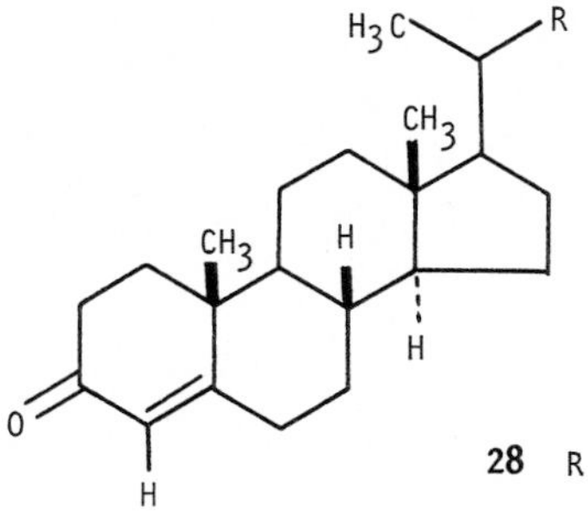

28 R = $N(CH_3)_2$

28a R = =O

2.29 (+)-N-BENZOYL-16-ACETOXYCYCLOXOBUXIDINE-F (**29**) (ref. 33)

The alkaloid showed intense UV absorption at 227 nm indicating the presence of a secondary benzamide and carbonyl group adjacent to the cyclopropane ring. The IR spectrum exhibited intense absorptions at 3400 cm^{-1} (OH), 1640 (carbonyl α-to cyclopropyl group, amide carbonyl) , 1730 (ester carbonyl) and 1595 (C=C) cm^{-1}.

The ^1H-NMR spectrum ($CDCl_3$) showed three tertiary methyl groups at δ0.80, 0.85 and 1.19 while the secondary methyl group resonated as a doublet at δ0.81 (J=7.0 Hz). The cyclopropyl protons normally resonate in the region of δ 0.2-0.5 but the electron-withdrawing effect of the α-carbonyl group shifts these signals considerably downfield, so that they become embedded in the methyl/methylene region. A three-proton singlet at δ 2.06 was assigned to the acetate group. Another six-proton singlet at δ2.17 was assigned to the $N(CH_3)_2$ group. The C-31α and β methylene protons appeared as a set of AB doublets at δ 3.92 and 4.12 ($J_{31\alpha,31\beta}$=9.6 Hz). A multiplet centered at δ4.80 was assigned to 16 H, which is geminal to the acetoxy group. The aromatic protons appeared as two groups of 3H and 2H multiplets centered at δ7.45 and 7.78, respectively.

Two dimensional ^1H-NMR measurements fully agreed with the proposed structure **29.** The multiplicities of overlapping proton signals could be determined from the 2D-J resolved spectrum while the COSY 45° spectrum established the coupling interactions among vicinal protons. The C-21 methyl protons showed cross peaks in the COSY spectrum with the C-20 proton at δ 2.41, which, in turn, showed cross peaks corresponding to the coupling with the C-17 proton at δ 1.81. The COSY 45° spectrum also showed coupling interaction of the C-17 proton with the C-16 methine proton at δ4.80. The C-16 proton on the other hand showed coupling not only with the C-17 proton but also with C-15 methylene protons. These relationships firmly established the position of the acetoxy group at C-16.

The ^{13}C-NMR spectrum ($CDCl_3$) showed four signals at δ 9.10, 10.50, 17.60

and 18.00, which were assigned to C-21, C-30, C-18 and C-32 methyl carbons, respectively. The methyl carbon of the acetyl group resonated at δ 21.71 while the C-31 methylenic carbon resonated at δ61.00. The amide, acetyl and the C-11 carbonyl carbons appeared at δ 170.00, 168.20 and 204.00. The assignments to the various carbons are presented around expression **29.**

The high resolution mass spectrum of the alkaloid showed the molecular ion at m/z 578.3722, corresponding to the formula $C_{35}H_{50}N_2O_5$, indicating the presence of twelve double bond equivalents in the molecule. The peak at m/z 563 resulted from the loss of methyl group from the molecular ion. Another peak at m/z 518 was due to the loss of acetic acid from the molecular ion. A peak at m/z 171.1242 having the formula $C_9H_{17}NO_2$ resulted from the cleavage of ring D. The composition of this ion indicated that the acetate ion was located on ring D. Another important fragment at m/z 157.1101 with the composition $C_8H_{15}NO_2$ resulted from the cleavage of ring D along with the attached substituents. This ion indicated that C-16 was the probable position of the acetate substituent on ring D. The peak m/z 105 was due to the benzoyl ion. The substance showed a base peak at m/z 72, representing dimethyliminium cation.

Structure **29** was proposed for (+) N-benzoyl-16-acetoxycycloxobuxidine-F on the basis of above studies (ref. 33). It may arise in the plant by the acetylation of N-benzoylcyclobuxidine-F (**29**a) (ref. 17).

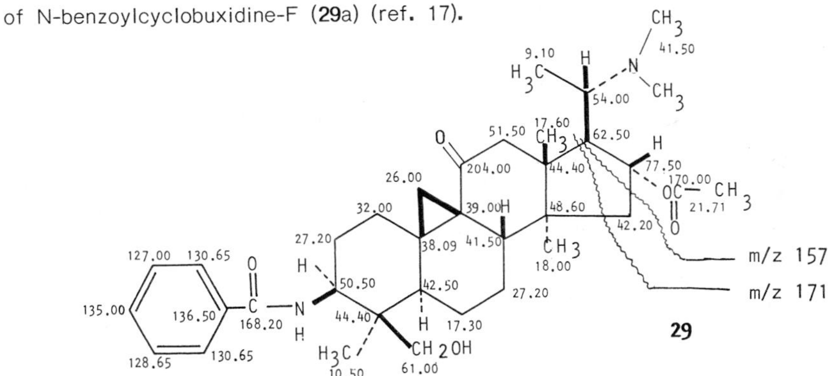

2.30 (+)-BUXAMINONE (**30**) (ref. 34)

Buxaminone (**30**) was obtained as colourless amorphous solid. The UV spectrum showed maxima at 238 and 248 nm with shoulders at 225 and 254 nm, characteristic of a 9(10 → 19) <u>abeo</u>-diene system. The IR spectrum of the compound displayed absorptions at 1690 (ketonic carbonyl) (ref. 8) and 1596 (C=C) cm^{-1}.

The ^{1}H-NMR spectrum included four 3-H singlets at δ 0.66, 0.76, 0.77 and 1.07, assigned to the four tertiary methyl groups. Another 3-H singlet at δ 2.10 was due to the methyl group α-to the 20-carbonyl function. A 6-H singlet at δ 2.35 was assigned to the $N(CH_3)_2$ group attached to C-3 of ring A. Vinylic protons at C-11 and C-19 appeared as a doublet of doublet at δ5.55 (J_1=2.3 Hz, J_2= 1.8 Hz) and as a singlet at δ 5.93, respectively.

The mass spectrum of the compound included a molecular ion at m/z 383.3186 corresponding to the molecular formula, $C_{26}H_{41}NO$ (calcd. 383.3187). A peak at m/z 340 was due to the loss of the C-17 carbonyl-containing side chain from the molecular ion. Another peak at m/z 338 resulted from the loss of $N(CH_3)_2$ group. The compound showed the base peak at m/z 71, which was due to the ion $CH_2\text{-}CH\text{=}\overset{+}{N}(CH_3)_2$, common in Buxus alkaloids containing dimethylamino substituent at C-3 of ring A. A considerably large peak at m/z 57 also resulted by the cleavage of ring A along with the nitrogen-containing substituents.

On the basis of the above spectroscopic studies structure **30** was assigned to (+)-buxaminone (ref. 34)

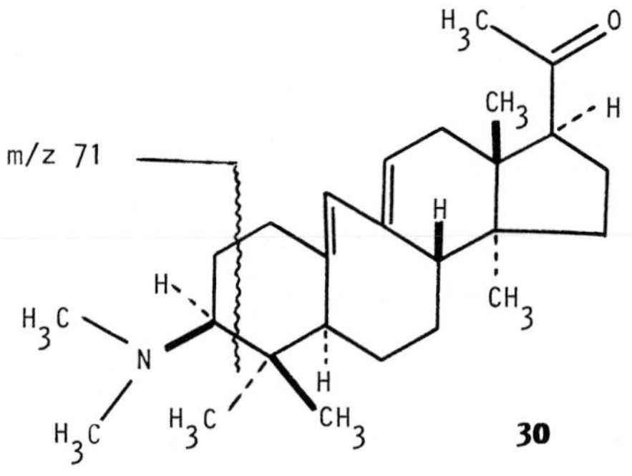

30

3.0 SPECTRAL BEHAVIOUR OF BUXUS ALKALOIDS

The Buxus alkaloids show characteristic spectral behaviour. Some generalizations have been drawn on the basis of this spectral behaviour which are very useful in structure elucidation of new steroidal alkaloids.

3.1 PROTON NMR

Buxus alkaloids belong to the rare group of natural products which often contain a cyclopropyl ring in their structures. This confers characteristic spectral and chemical properties to them. The ^{1}H-NMR chemical shifts of the cyclopropyl methylenic protons are very diagnostic about structural changes in the neighbouring rings A,B and C.

Normally, in the case of non-substituted cycloartenol triterpenoid skeletons, the cyclopropyl methylenic protons appear in the region of δ 0.1-0.5 as AB doublets (J=4.0 Hz) (ref. 28). Compounds which contain a C-11 carbonyl function in rings exhibit the absence of cyclopropyl methylenic proton signals in this region of the ^{1}H-NMR spectrum. This is due to the electron-withdrawing effect of the conjugated carbonyl function which deshields the methylenic protons, and the AB doublets are consequently shifted to the methyl/methylenic region of the spectrum (ref. 33,35,36).

The ^{1}H-NMR spectrum of compounds which contain a hydroxyl function at C-11 display only one half of the AB double doublet relatively downfield from its normal position at δ 0.5-0.8, while the other half is shifted to the methyl/methylenic region due to the deshielding effect of the vicinal hydroxyl group (at C-11) (ref. 35).

Similarly, compounds of type **31** and **32** bearing a double bond at 1-2 or 11-12 positions also show only one half of the AB doublets for the cyclopropyl methylene protons at δ 0.5-0.7, while the other half of the pattern is embedded in the methyl/methylenic region. This is due to the deshielding effect of the adjacent double bond. Both cases can be distinguished on the basis of chemical shift and coupling patterns of the olefinic protons, which in the case of **31** will be relatively downfield, while in **32** they will resonate upfield as clear AB doublets (ref. 36).

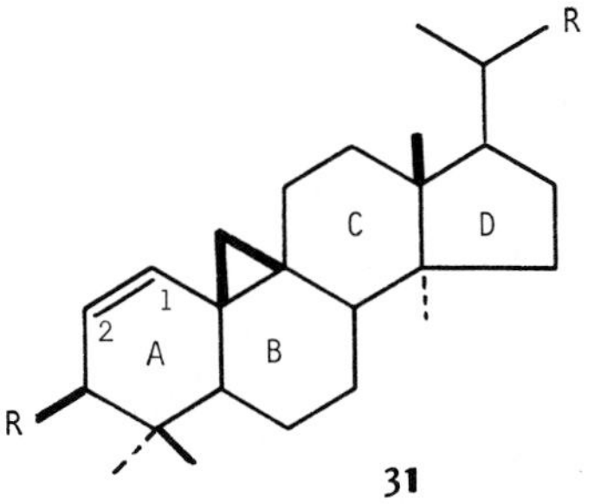

31 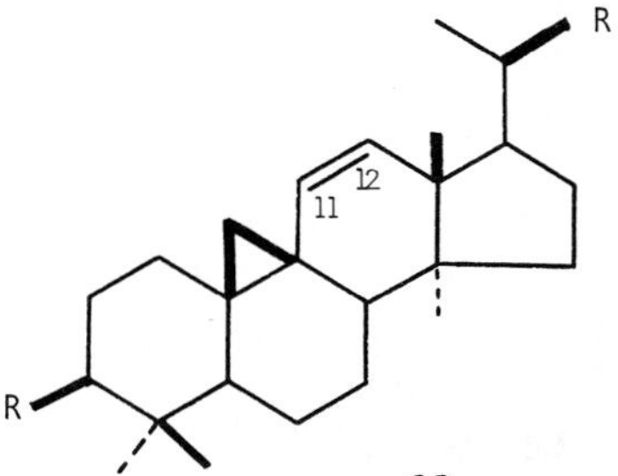 **32**

Compounds such as **33** bearing a 6-7 double bond showed pronounced upfield shifts of cyclopropyl methylenic doublets which then resonate in the region of δ0.1-0.4. This is due to the fact that the β-oriented cyclopropyl methylene proton lies in the shielding region of 6-7 double bond (ref. 37,38). Compounds having 7-8 or 5-6 exocyclic double bonds also show downfield shifts of one of the cyclopropyl AB doublets into the methyl/methylenic region (ref. 19,30).

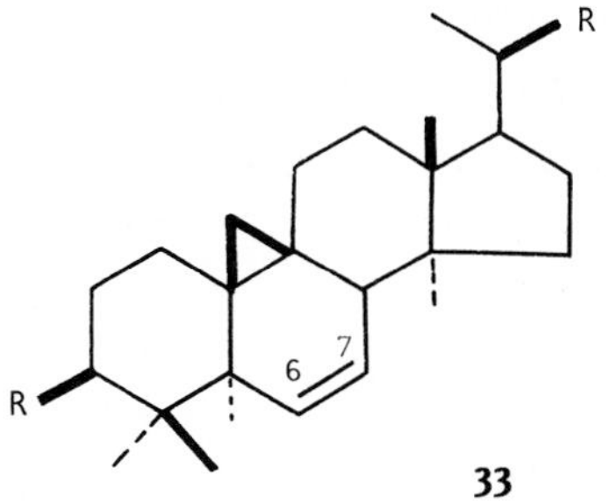

33

3.2 MASS SPECTROSCOPY

Mass fragmentation patterns of Buxus alkaloids are highly informative. They provide useful information about individual structural types, positions of functional group and the position and extent of unsaturation in different groups. The lower mass region of the spectra are of considerable importance in this connection (ref. 39).

Systematic mass spectroscopic studies of a large number of Buxus alkaloids using computer monitored high resolution mass measurements, FAB, FD, CI and linked scan measurements have supplied valuable generalizations summarized below:

Compounds bearing oxygenated functions on ring D exhibit characteristic peaks in their mass spectra and the position of such group(s) can often be deduced from the mass fragmentation pattern. This is exemplified by buxanoldine (**15**) (ref. 15). The considerably large fragment ion at m/z 129 ($C_7H_{15}NO$) in the mass spectrum of (-)-buxanoldine (**15**) is formed by the cleavage of -OH attached to ring D along with the nitrogen-bearing side chain. This hydroxy group may be attached to C-15, C-16 or C-17 of ring D. Another peak at m/z 115 ($C_6H_{13}NO$) established that the -OH group was present on a six-carbon fragment. The probable points of attachment of the -OH group were therefore concluded to be at C-16 or C-17. Another important fragment at m/z 85 ($C_5H_{11}N$) arose by the cleavage of ring D along with nitrogen bearing side chain and an intramolecular proton transfer. The absence of oxygen in this five-carbon fragment in m/z 85 established that the hydroxyl group was located at C-16 of ring D (ref. 15,24).

Compounds bearing acetate group instead of -OH show characteristic mass fragmentation patterns. The diagnostic ions at m/z 157 and 171 indicate the presence of acetate group at the C-16 position of ring D (e.g. (+)-buxanaldinine (**16**)) (ref. 15).

The fragment ions resulting from cleavage of ring D side chain predominate in the mass spectrum of Buxus alkaloids. Compounds bearing monomethylamino substituents at C-20 (e.g. (-)-cyclobuxoviricine (**21**)) show the base peak at m/z 58 due to the fragment, $CH_3\text{-}CH\text{=}\overset{+}{N}H\text{-}CH_3$, (ref. 40) while compounds containing dimethylamino side chain (e.g. (+)-buxabenzamidienine (**13**)) exhibit a fragment at m/z 72 as the base peak (ref. 15). This ion represents the trimethyliminium cation ($CH_3\text{-}CH\text{=}\overset{+}{N}(CH_3)_2$) (ref. 41). It is important to note that fragment ions arising by cleavage of the nitrogen-containing side chain in ring D are more abundant as compared to fragments arising from ring A when the latter contains a nitrogen bearing side chain (ref. 42). Compounds which contain nitrogen at C-3 give rise to the base peak at m/z 57 or m/z 71. The former is formed from compounds containing a dimethylamine $N(CH_3)_2$ group at C-3 and result from the cleavage of ring A (e.g. (+)-N-formylpapilicine (**10**), (+)-buxaminone (**30**) while the latter ion is present in compounds containing monomethylamino group at C-3 (ref. 37,42).

The mode of unsaturation in various rings of Buxus alkaloids have a pronounced effect on the fragmentation pattern. Compounds having isolated double bonds show fragment ions resulting from the retro-Diels Alder cleavage of rings containing such unsaturated linkages. For example the mass spectrum of (+)-buxabenzamidine (**20**) shows an ion at m/z 323 resulting from the retro-Diels Alder cleavage of ring C while the ion at m/z 85 is formed by cleavage of ring A by a similar process in the case of (+)-buxupapine (**17**) (ref. 9,15,18).

Compounds such as (+)-buxabenzamidienine (**13**) which contain a 9(10 → 19) abeo diene system, would also be expected to produce similar fragments but this is not observed in practice (ref. 8,15).

Buxus alkaloids containing an unsubstituted tetrahydrooxazine ring e.g. (+)-buxaquamarine (**5**) show considerably large m/z 127 ions in their mass spectra (ref. 6,8), while alkaloids containing methyl-substituted tetrahydrooxazine rings (e.g. (+)-homobuxaquamarine (**6**)) exhibit large fragments ions at m/z 141, formed by the cleavage of ring A along the lines indicated in structure **6** (ref. 9).

3.3 CORRELATION BETWEEN STRUCTURE AND SPECIFIC ROTATION

An aspect of Buxus alkaloids that had never previously been considered is the relationship between structure on the one hand, and specific rotation on the other. Fortunately, the specific rotations for most Buxus alkaloids, have been measured in chloroform, so that the values recorded for different alkaloids may be compared. Indeed, a simple measurement of the specific rotation can immediately throw light on some of the structural features of Buxus bases. The following rules relate certain molecular features to specific rotations, with specific rotations quoted between brackets. These rules sometimes provide a last chance to check the structure before publication.

(a) 9(10 → 19) abeo-dienes of type **34** are dextrorotatory; examples are buxamine-E [+42°] (ref. 2), papilamine [+23.3^{o}] (ref. 11), buxenine-G [+30] (ref. 43) buxamine-C [+24°] (ref. 38), desoxy-16-buxidienine [+33°] (ref. 38), buxaquamarine [+24°] (ref. 8), papilicine [+47°] (ref. 12), buxamine-A [+40°] (ref. 3), papilinine [+29.4^{o}] (ref. 7), buxaminol-B [+20°] (ref. 43a), buxaminol-E [+40°] (ref. 3) and moenjodaramine [+33.3^{o}] (ref. 1). The one exception is N-benzoylbuxidienine [-36°] (ref. 35), but unfortunately this N-benzoylated compound was not available to us for a redetermination of its optical activity.

(b) 9β,19-Cyclo-11-oxo alkaloids of type **35** are also dextrorotatory, but the magnitude of the specific rotation is usually larger than for dienes of type A. Examples include baleabuxoxazine-C [+116°] (ref. 3), N-benzoylbaleabuxidine-F [+52°] (ref. 3), N-isobutyroylcycloxobuxidine-H [+76°] (ref. 44), baleabuxidine [+127°] (ref. 3), N-isobutyroylcycloxobuxine [+115] (ref. 45), buxarine [+98°] (ref. 42), buxatine [+108°] (ref. 46) etc.

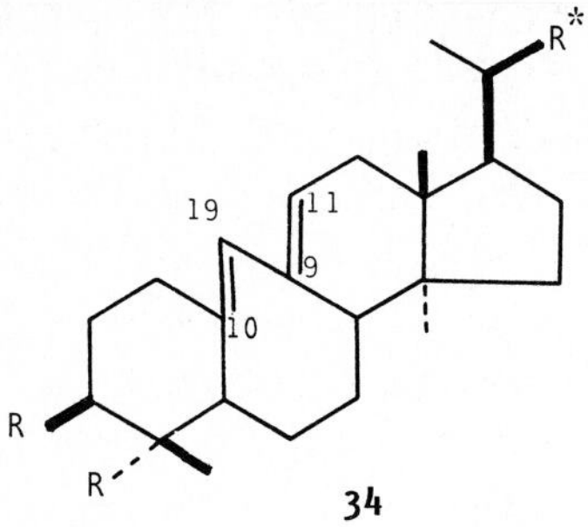

34

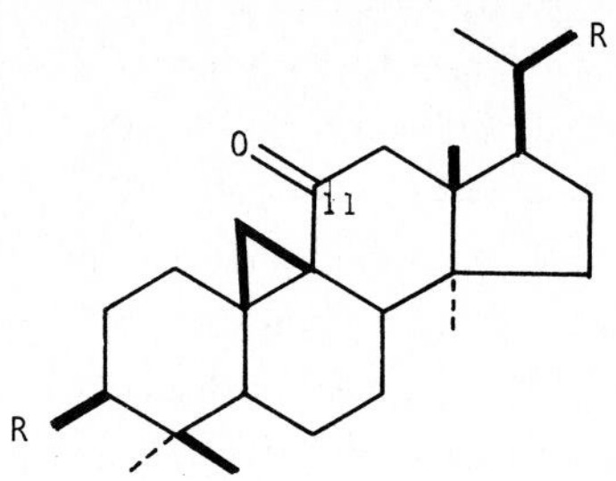

35

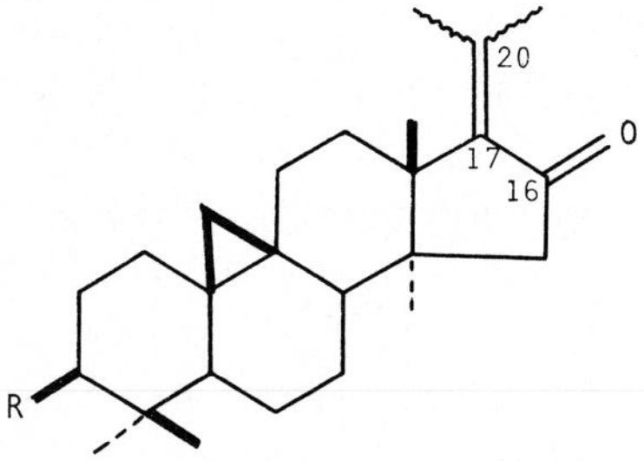

36

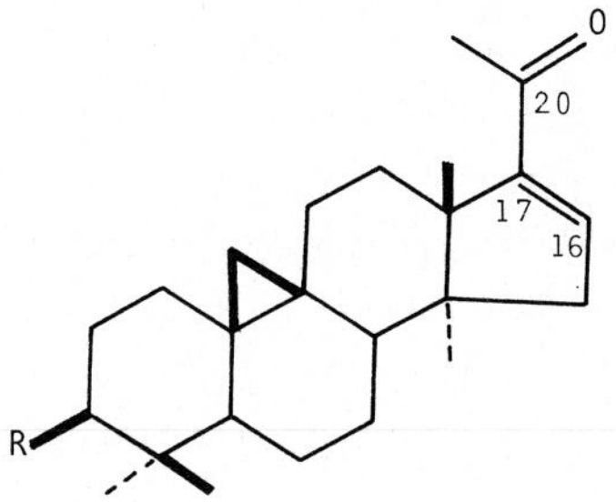

37

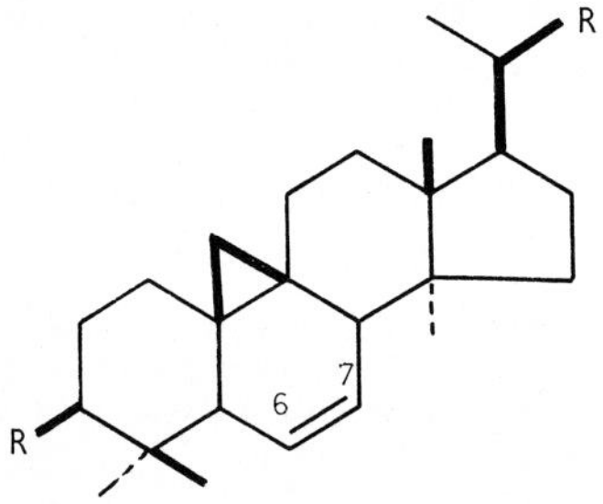

38

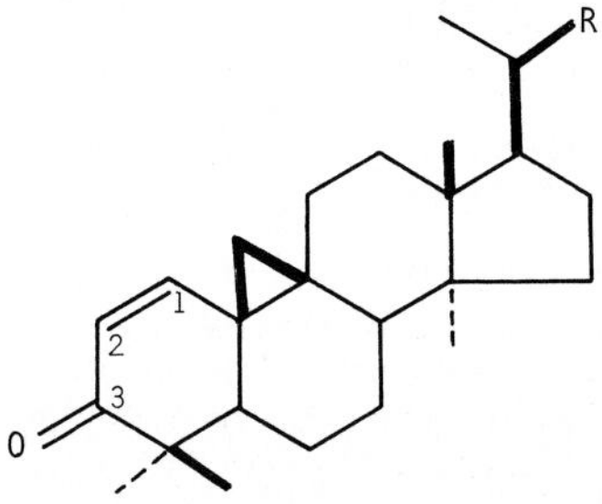

39

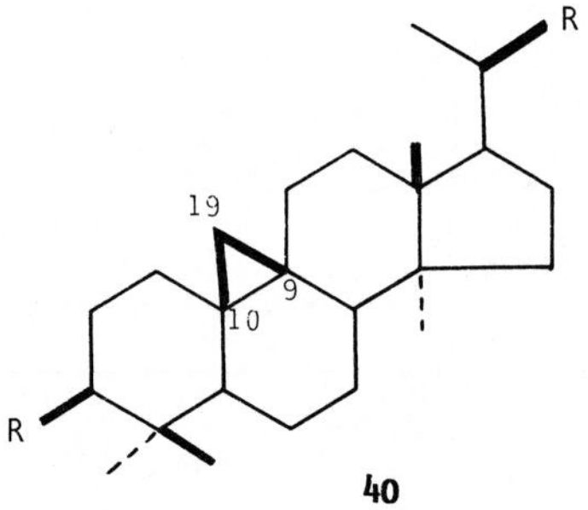

40

*R = Amino or carbonyl group

(c) 9β ,19-Cyclo-16-oxo- Δ(17 → 20) alkaloids of type **36** are levorotatory, regardless of the geometrical isomerisn about the C-17 (20) double bond. Some examples are cyclobuxophylline-O [-61°] (ref. 47), buxenone [-48°] (ref. 42), cyclobuxophylline [-72°] (ref. 28), methylbuxene [-104°] (ref. 48) and cyclobuxosuffrine [-62°] (ref. 28).

(d) 9β,19-Cyclo-20-oxo-Δ (16 → 17) compounds of type **37** are dextrorotatory. Only two examples are available, namely cyclomicrobuxeine [+126] (ref. 28), and cyclobuxomicreine [+37°] (ref. 28). The former of these also incorporated an exocyclic methylene at C-4 instead of the usual gem dimethyl.

(e) 9β,19-Cyclo-Δ(6 →7) compounds of type **38** are levorotatory as exemplified by cyclobuxopaline-C [-37°] (ref. 38), cyclobullatine-A [-99°] (ref. 43a), cyclovirobuxeine-B [-80°] (ref. 17), cyclovirobuxeine-A [-87°] (ref. 17), N-benzoyldihydrocyclomicrophylline-F [-20°] (ref. 49), cyclomicrosine [-33°] (ref. 28), and cyclomalyanine-B [-61°] (ref. 17).

(f)The enone containing alkaloids of type **39** are levorotatory. Examples are cyclobuxoviricine [-54°] (ref. 22), cyclobuxoviridine [-20] (ref. 9) and cycloxobuxoviricine [-42°] (ref. 24).

(g) Simple Buxus alkaloids with no unsaturation, but with a 9β,19-cyclo system of type **40** are dextrorotatory. Relevant examples are cycloprotobuxine-F [+42°] (ref. 3), cycloprotobuxine-C [+68°] (ref. 43a), buxocyclamine-A [+87°] (ref. 41), cyclovirobuxine-D [+63°] (ref. 17), cyclopapilosine-D [+54°] (ref. 38), buxozine-C [+65°] (ref. 50) etc.

REFERENCES

1. Atta-ur-Rahman, M.Nisa and S.Farhi, Planta Medica, **49**, 1983, 126
2. F.Khuong-Huu, D.H. Gaulier, M.M.Q.K. Huu, E.Stanislas and R.Goutarel, Tetrahedron, **22**, 1966, 3321 -
3. F.Khuong-Huu, R.Paris, R.Razafindrambao, A.Cave and R.Goutarel, Compt.Rend.Acad.Sci.Paris, **273** (c), 1971, 558 - 560.
4. Atta-ur-Rahman and M.Nisa, Heterocycles, **20**, 1983, 69 - 70.
5. Atta-ur-Rahman, M.Nisa and S.Farhi, Z.Naturforsch, **39**, 1984, 524- 527.
6. M.I.Choudhary, Atta-ur-Rahman and M.Shamma, Phytochemistry (accepted).
7. Atta-ur-Rahman, M.Nisa and T.Zamir, Z.Naturforsch., **40**, 1984, 565-566.
8. Atta-ur-Rahman, M.I.Choudhary and M.Nisa, Heterocycles, **23**, 1985, 1951-1953
9. M.I. Choudhary, Atta-ur-Rahman, A.J.Freyer and M.Shamma, J.Nat.Prod., **50**, 1987, 84-88.

10. Atta-ur-Rahman and M.Nisa, Z.Naturforsch., **39**, 1984, 839.

11. Atta-ur-Rahman, S.Farhi, G.A.Miana, M.Nisa and W.Voelter, Z.Naturforsch., **40**, 1985, 567-568.

12. Atta-ur-Rahman, M.Nisa and T.Zamir, Z.Naturforsch., **39**, 1984, 127-128.

13. Atta-ur-Rahman, M.Nisa and K.Jahan, Phytochemistry, **24**, 1985, 1398-1399.

14. Atta-ur-Rahman and M.I. Choudhary, J.Chem.Soc., Perkin Trans. I, 1986, 919-921.

15. M.I.Choudhary, Atta-ur-Rahman, A.J.Freyer and M.Shamma, Tetrahedron, **42**, 1986, 5747 - 5752.

16. K.S.Brown, Jr., and S.M.Kupchan, J.Am.Chem.Soc., **86**, 1964, 4424 - 4430.

17. F.Khuong-Huu-Laine, M.Magdeleine, N.Bisset and R.Goutarel, Bull.Soc.-Chim.France, 1966, 657.

18. F.Khuong-Huu, D.Herlem and M.Benechie, Bull.Soc.Chim.France, 1972, 1092.

19. M.I. Choudhary, Atta-ur-Rahman and M.Shamma, Phytochemistry (in press)

20. M.I.Isaev, M.B.Gorovits and N.K. Abubakirov, Khim.Prir.Soedin, **4**, 1985, 431

21. P.V. Demarco, E.Farkas, D.Doddrell, B.L. Mylari and E.Wenkert, J.Am.Chem.Soc., **90**, 1968, 5480 - 5486.

22. Atta-ur-Rahman, M.I.Choudhary and M.Nisa, Phytochemistry, **24**, 1985, 3082 - 3083.

23. T.Nakano and S.Terao, J.Chem.Soc.(C), 1965, 4512

24. Atta-ur-Rahman, M.I.Choudhary, I.Ali and Habib-ur-Rehman, J.Nat.Prod., **49**, 1986, 106 - 110.

25. G.A.Morris, Mag.Res.Chem., **24**, 1986, 371-403.

26. Atta-ur-Rahman, "Nuclear Magnetic Resonance Spectroscopy", Springer--Verlag, New York, 1986, **20**.

27. Atta-ur-Rahman, M.Alam, M.I.Choudhary and S.Firdous, Planta Medica **53**, 1987, 496-497.

28. T.Nakano, S.Terao and Y. Saeki, J.Chem.Soc.(C), 1966, 1412 - 1421.

29. K.S.Brown Jr. and S.M.Kupchan, J.Am.Chem.Soc., **84**, 1964, 4414 - 4424.

30. T.Takemoto, G.Kusano and N.Yamamoto, Yakugaku Zasshi, **90**, 1970, 68-72.

31. Z.Voticky and J.Tomko, Coll.Czech.Chem.Commun., **30**, 1965, 348 - 350.

32. Atta-ur-Rahman, M.Nisa, J.Nat.Prod., (in press).

33. Atta-ur-Rahman, M.I.Choudhary and M.Nisa, Planta Medica, **53**, 1987, 75-77.

34. Atta-ur-Rahman, M.Alam and M.I.Choudhary, J.Nat.Prod., (accepted).

35. S.M.Kupchan, R.M. Kennedy, W.R. Schleigh and G.Ohta, Tetrahedron, **23**, 1967, 4563 - 4586.

36. T.Nakano, S.Terao, Y.Saeki and K.D.Jin, J.Chem.Soc.(C), 1966, 1805-1810.

37. F.Khuong-Huu and M.J.Magdeleine, Ann.Pharm.Fr., **23**, 1970, 211

38. M.Shamma, V.S.Greogiev, G.A.Miana and F.S.Khan, Phytochemistry, **12**, 1973, 2051- 2054.

39. G.R.Waller and O.C.Dermes, "Biochemical Application of Mass Spectrometry", John Wiley and Sons, New York, 1980, 783- 784.

40. J.Dolejs, V.Hanus, Z.Voticky and J.Tomko, Coll.Czech.Chem.Commun., **30**, 1965, 2869

41. W.Dopke, B.Mueller and P.W.Jeffs, Pharmazie, **23**, 1969, 37- 38.

42. W.Dopke, B.Muller and P.W.Jeffs, Pharmazie, **21**, 1966, 643

43. S.M.Kupchan and W.L.Asbun, Tetrahedron Lett., 1964, 3145 - 3150.

43a. Z.Voticky, O.Bauerova and V.Paulik, Coll.Czech.Chem.Commun., **40**, 1975, 3055 -3059.

44. D.H.Gaulier, F.Khuong-Huu-Laine and R.Goutarel, Bull.Soc.Chim.France, 1968, 763 - 773.

45. D.H.Gaulier, F.Khuong-Huu-Laine and R.Goutarel, Bull.Soc.Chim.France, 1966, 3478 - 3486.

46. W.Dopke and B.Muller, Naturewiss., **54**, 1967, 249 .

47. L.T.Huong, Z.Voticky and V.Paulik, Coll.Czech.Chem.Commun., **46**, 1981, 1425- 1431.

48. W.Dopke, R.Hartel and H.W.Fahlhaber, Tetrahedron Lett., 1969, 4423 - 4424.

49. W.Dopke, B.Muller, Pharmazie, **22**, 1967, 666 .

50. Z.Voticky, L.Dolejs, O.Bauerova and V.Paulik, Coll.Czech.Chem.Commun., **42**, 1977, 2549 -2554.

ISOLATION AND STRUCTURAL STUDIES OF SOME NOVEL QUINONES AND ALKALOIDS OF TWO SRI LANKAN PLANTS

A.A. LESLIE GUNATILAKA

1. GENERAL INTRODUCTION

Sri Lanka being an island possesses a unique and varied flora. There are about 3 500 recorded flowering plant species in about 1 300 genera and 172 families (ref. 1,2). Of this, total, about 830 species are reported to be endemic to the country (ref. 3) and around 750 species are claimed to have uses in the indigenous system of medicine (ref. 4) which provides primary health care for over 70% of the Sri Lankan population (ref. 5). A good number of local plants are also used as sources of drugs in Western medical practice.

For the past decade or so we have been engaged in the chemical studies of medicinal and related plants of Sri Lanka which were undertaken with the hope of harnessing natural product resources for development of Sri Lanka (ref. 6). Some main objectives of these studies were : (1) isolation and characterization of compounds responsible for the claimed biological effects of medicinal plants; (2) search for the presence of these bio-active compounds in their related species; (3) investigation of these plants for toxic effects, if any; and (4) study of 'biologically inactive' but chemically and structurally interesting compounds present in these plants. Thus far we have investigated in detail over two dozen Sri Lankan plants results of which have been reviewed elsewhere (ref. 7-10). Our studies on two Sri Lankan plant species namely, *Plumbago zeylanica* and *Broussonetia zeylanica*, are presented in the following pages.

2. QUINONES OF *PLUMBAGO ZEYLANICA*

2.1 Introduction

Plumbago zeylanica (Family: Plumbaginaceae) is a perennial herb with long succulent roots growing especially in the warm dry districts of Sri Lanka. Roots of *P.zeylanica* have been well known in many Asian countries for a long time for its medicinal properties (ref. 11-13). Some important ethnomedical uses of *P.zeylanica* root are summarized in Table 1.

TABLE 1

Ethnomedical claims of *Plumbago zeylanica* root.

Plant part (type extract/ method of adminis- tration)	Country	Ethnomedical claims
Root (Hot water)	India	Antifertility agent
	India, Philippines and Taiwan	Abortifacient, coughs, worms, ulcers, leprosy, astringent and digestive
	Japan	Diuretic, ecbolic, emmenagogue, intermittant fever, syphilis, scabies, carcinoma and bold headness
	Malaysia	Abortifacient
Root (Insertion to the os uterus)	India Philippines	Abortions

Previous chemical investigations of *P.zeylanica* root have led to the isolation and characterization of plumbagin [5-hydroxy-2-methyl-1,4-naphthoquinone] (1), 3,3'-biplumbagin (2), 3-chloroplumbagin (3), droserone (4), elliptinone (5), chitranone (6), zeylanone (7) and isozeylanone (8) (ref.14-16). In continuing our studies on the use of inedigenous plants for fertility regulation, we have investigated the root extractives of *P. zeylanica* and have isolated and characterized plumbagin (1), droserone (4), chitranone (6), zeylanone (7), quinones which have been reported from this plant previously; 2-methylnaphthazarin (9), maritinone (10), which are natural quinones but not known previously from *P.zeylanica*; isoshinanolone (11), a naphthalenone previously encountered in *Diospyros maritima* (Ebenaceae) (ref. 17); a new naphthalenone, 1,2(3)-tetrahydro-3,3'-biplumbagin (15) (ref. 18); a new naphthoquinone, methylene-3,3'-biplumbagin (16) (ref. 19); and a trimer of plumbagin, plumbazeylanone (ref. 20).† In addition to root extracts, plumbagin and isoshinanolone, leaf and stem extractives were also tested in the anti-fertility assay.

2.2 Chemical Studies

2.2.1 Isolation of Sitosterol, Quinones and Naphthalenones

The hot light petroleum extract of *P. zeylanica* roots were separated into phenolic (sodium hydroxide-soluble) and neutral fractions. The neutral fraction constituted chiefly of sitosterol. The phenolic fraction on column chromatography over acidic silica gel and elution with varying percentages of ethyl acetate in light petroleum afforded ten quinones and naphthalenones (Fig. 1). The major constituent of the phenolic fraction was identified as plumbagin (1). Other known quinones, 2-methylnaphthazarin (9), droserone (4), chitranone (6), zeylanone (7) and maritinone (10) were identified from the spectral data and direct comparison with authentic samples.

2.2.2 Structure of methylene-3,3'-biplumbagin

The least polar compound [$C_{23}H_{16}O_6$; m.p. 208-210°C] in the mixture of quinones eluted with 7.5% ethyl acetate in light petroleum was found to be a new quinone and its structure was elcudiated as methylene-3,3'-biplumbagin (16) from spectral data and partial synthesis of its dimethyl ether (17) starting with plumbagin (1).

The carbonyl stretching region of the IR spectrum of this quinone [ν_{max}(KBr) 1655, 1632 and 1597] showed close resemblance to that of plumbagin [ν_{max}(KBr) 1660, 1645 and 1602] suggesting the presence of similar type of quinone system. The ^{1}H NMR showed a D_2O exchangeable singlet at δ 11.93, a multiplet at δ 7.63-7.13, a broad singlet at δ 3.96 and a sharp singlet at δ 2.30 in the ratio of 1:3:1:3. Appearence of the signal due to methyl group as a sharp singlet coupled with the absence of signals due to vinylic protons suggested that the

† The original tentative structure (18) proposed for plumbazeylanone (ref. 20) has now been revised and a probable structure (19) has been proposed.

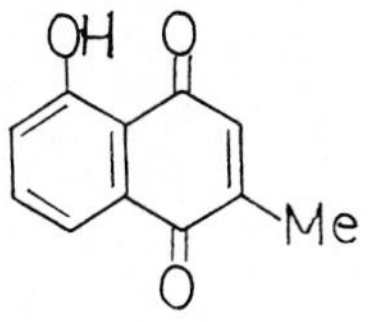

(1) (2)

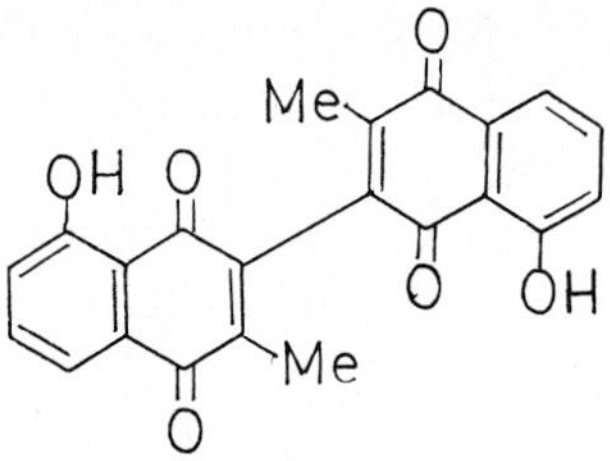

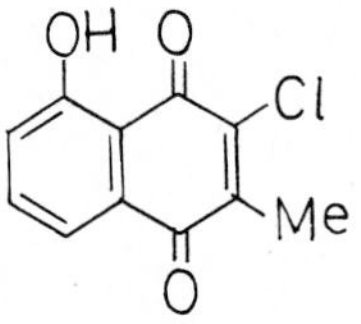

(3) (4)

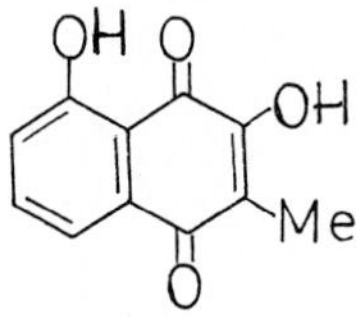

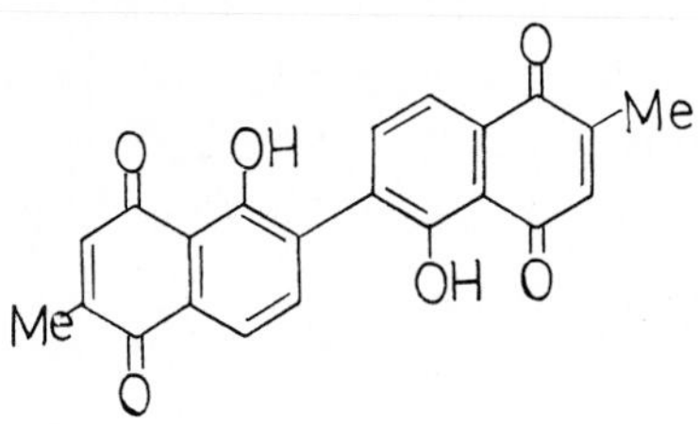

(5) (6)

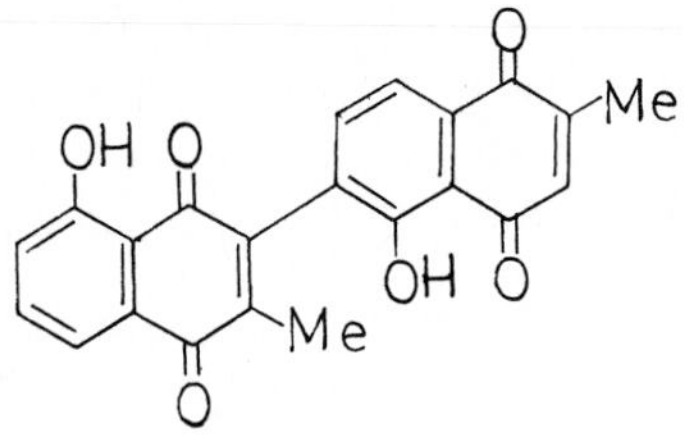

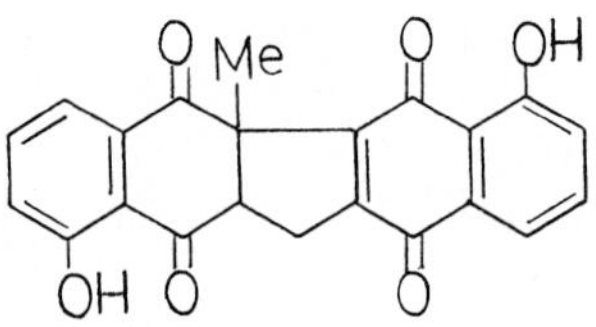

(7) (8)

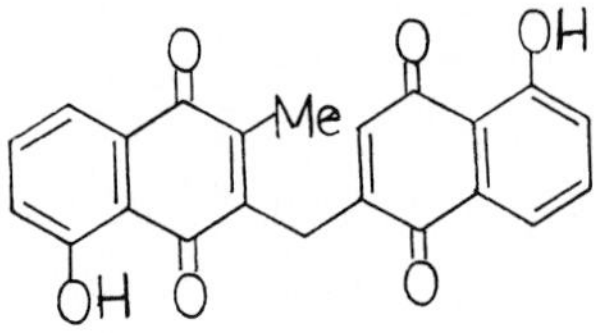

C-3 was occupied by a substituent. Molecular weight (M^+, 388) of the pigment along with its ^{1}H NMR data suggested it to contain two plumbagin units connected by a methylene bridge. Protons of the methylene bridge appeared as a broad

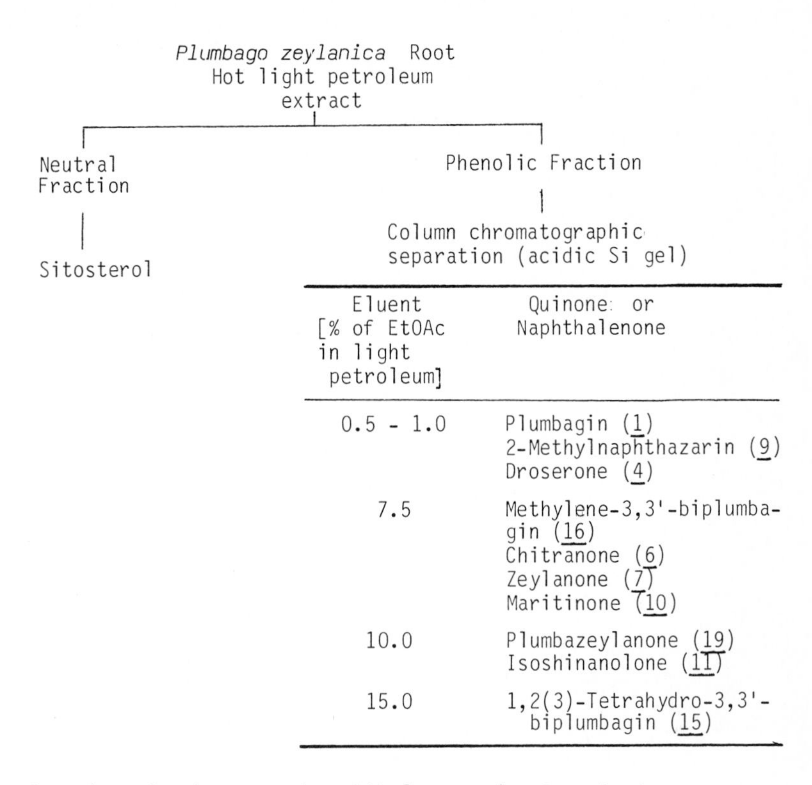

Eluent [% of EtOAc in light petroleum]	Quinone or Naphthalenone
0.5 - 1.0	Plumbagin (1) 2-Methylnaphthazarin (9) Droserone (4)
7.5	Methylene-3,3'-biplumbagin (16) Chitranone (6) Zeylanone (7) Maritinone (10)
10.0	Plumbazeylanone (19) Isoshinanolone (11)
15.0	1,2(3)-Tetrahydro-3,3'-biplumbagin (15)

Fig. 1. Fractionation of quinones and naphthalenones in *P.zeylanica* root extractive.

singlet at δ 3.96. All these data suggested the structure (16) for this new quinone. This was further confirmed by comparison of its dimethyl ether (17) with a synthetic sample prepared from plumbagin (1) as described below.

Plumbagin was converted into plumbagin methyl ether (20) by treatment with iodomethane in chloroform in the presence of silver oxide (ref. 21). This was treated with diazomethane at low temperature under anhydrous conditions in ether to yield the indazole dione (21) (ref. 22). When decomposed with aqueous sodium hydroxide in the presence of a slight excess of plumbagin methyl ether (20), the indazole dione (21) afforded the methylene bridged dimer of plumbagin (17) in 41% yield. In the presence of alkali alone, the indazole dione (21) gave ethylene bridged diquinone (22) and thermal decomposition

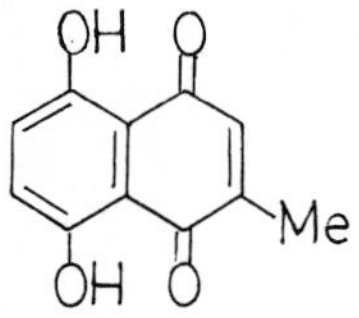

(9)

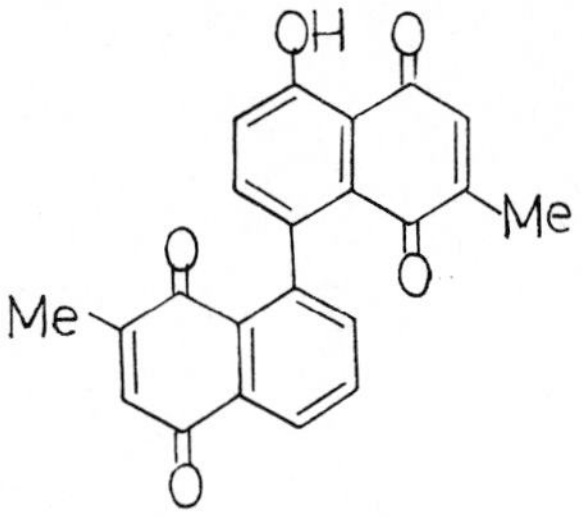

(10)

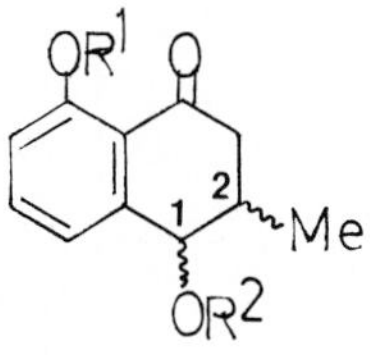

(15)

(11) $R^1 = R^2 = H$
(12) $R^1 = Me$; $R^2 = H$
(13) $R^1 = R^2 = COMe$
(14) $R^1 = R^2 = COPh$

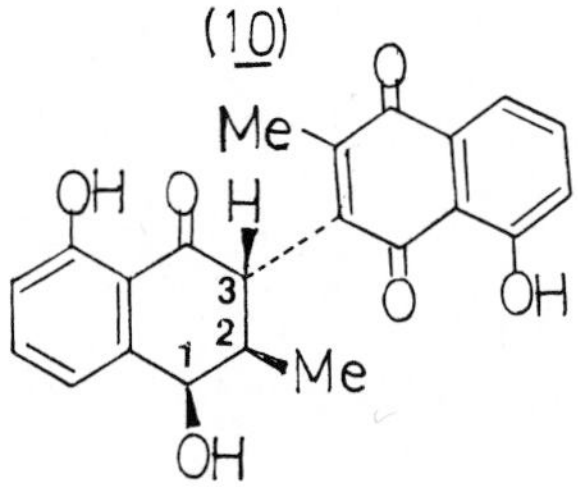

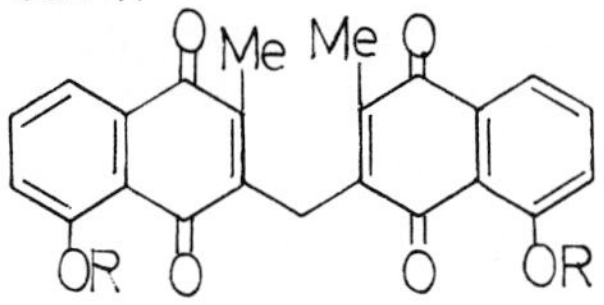

(16) R = H
(17) R = Me

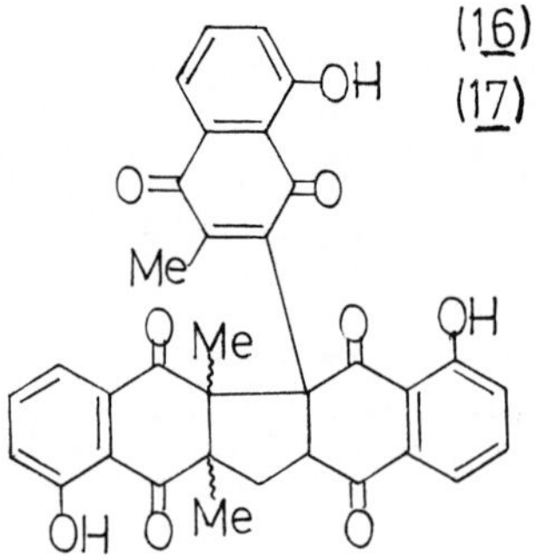

(18)

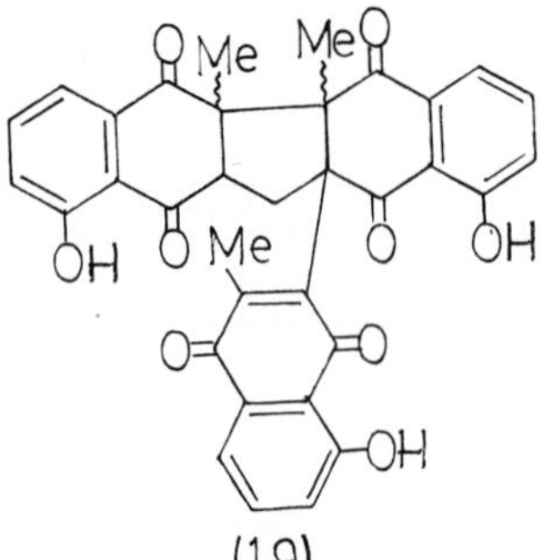

(19)

yielded exclusively 2,3-dimethyl-5-methoxy-1,4-naphthalenedione (23) (Scheme 1).

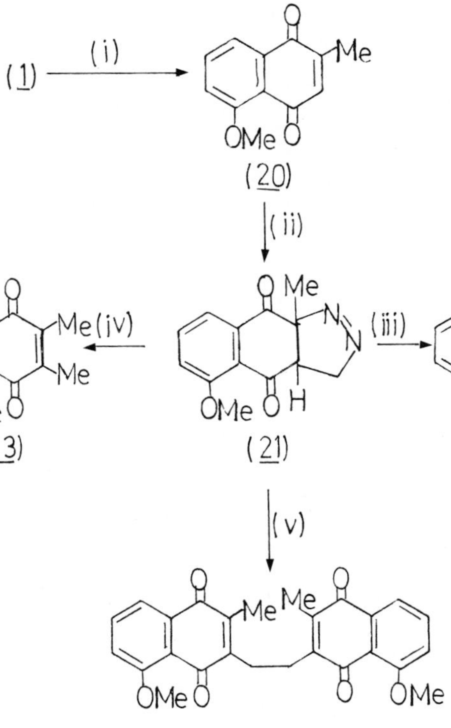

(1) —(i)→ (20)
(20) —(ii)→ (21)
(21) —(iv)→ (23)
(21) —(iii)→ (17)
(21) —(v)→ (22)

Reagents : (i) $MeI-Ag_2O-CHCl_3-27^oC$, (ii) CH_2N_2-ether-0^oC - 3h (iii) Slight excess of (20) - $CHCl_3$-MeOH-NaOH-20 min, (iv) toluene-reflux-20 min., (v) $CHCl_3$-MeOH-NaOH-20 min.

Scheme 1. Partial synthesis of methyl-3,3'-biplumbagin methyl ether from plumbagin and some reactions of indazoline dione (21) derived from plumbagin methyl ether.

2.2.3 Structure of plumbazeylanone

The column fraction obtained with 10% ethyl acetate in light petroleum (Fig. 1) contained two compounds and these were separated by PLC into less polar bright orange crystalline compound and a more polar pale yellow semi-solid.

The bright orange pigment [m.p. 246-248^{0}C, $(\alpha)_D$ = 0^{0}] had the molecular formula $C_{34}H_{24}O_9$ indicating it to be a trimer of plumbagin. It gave a reddish purple solution in aqueous sodium hydroxide and was reversibly reduced by dithionite. The UV-VIS spectrum and IR carbonyl absorptions corresponded approximately to a combination of UV-VIS and IR spectra, respectively of plumbagin (1) and zeylanone (7). Thus the new quinone was thought to consist of these two part structures and was named plumbazeylanone.

The ^{1}H NMR (360 MHz) showed signals for peri-hydroxy protons (D_2O exchangeable 1H singlets at δ 11.55, 11.44 and 11.41), nine aromatic protons (multiplet, δ 7.71, 7.07), three methyl groups (3H singlets at δ 2.49, 2.32 and 1.82), and an ABX ($-CH_2-CH<$) system (1H double doublets at δ 4.30, 3.25 and 2.74; J_{AB} 13Hz, J_{AX} 9Hz and J_{BX} 3Hz) in a cyclopentane ring. Olefinic protons were absent but evidently a methylene bridge was present as in zeylanone (7). Thus plumbazeylanone appeared to be formed from three plumbagin molecules, two of which are probably linked as in zeylanone (7) or methylene-3,3'-biplumbagin (16) suggesting two possible structures (18) and (19) for plumbazeylanone.

Majority of the data available favoured structure (18) for plumbazeylanone (ref. 20). After publication of this structure for plumbazeylanone we performed a crucial experiment to differentiate between the two possibilities. Sublimation of plumbazeylanone *in vacuo* afforded among other products, plumbagin and methylene-3,3'-biplumbagin (16) as the only identifiable products. This is possible only if plumbazeylanone has the structure (19) (see Scheme 2) (ref. 23).

The proposed structure (19) for plumbazeylanone was further supported by ^{13}C NMR data (Table 2). Off-resonance decoupling technique was employed to distinguish between the carbon atoms. The proton noise decoupled ^{13}C NMR spectrum showed 34 signals for all 34 carbon atoms in the molecule. Assignments were made by comparison with reported data for quinones (ref. 24-27). The nine down-field singlets were assigned to six carbonyl and three aromatic carbons bearing hydroxy groups. The nine down-field doublets were attributed to the unsubstituted aromatic carbons. The doublet at δ 55.7 and the triplet at 24.1 were assigned, respectively, to $>CH-$ and $>CH_2$ of the cyclopentane ring system. The quartet at δ 21.1 and two overlapping quartets centred at δ 13.7 and 13.5 were assigned to three methyl carbons.

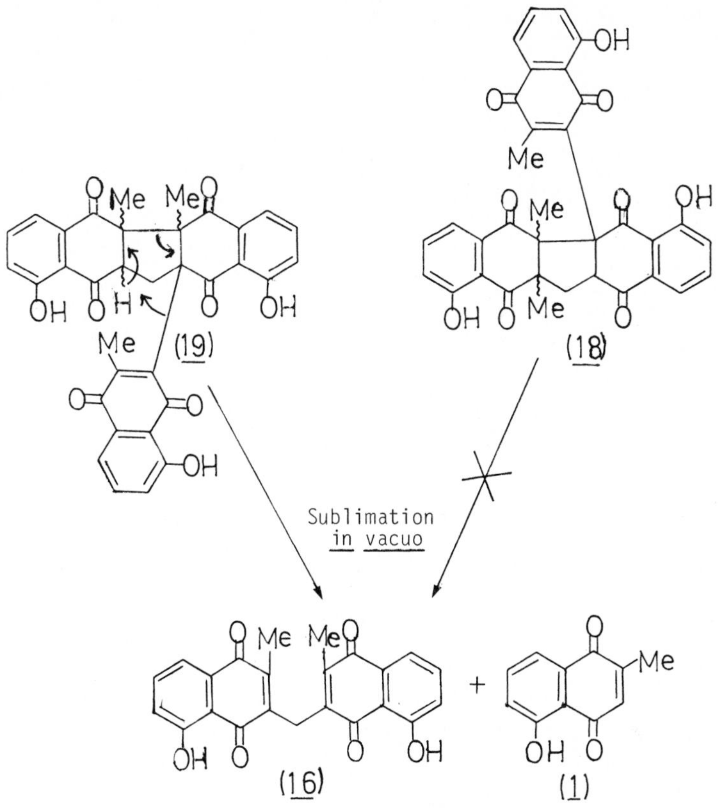

Scheme 2. Sublimation of plumbazeylanone giving plumbagin (1) and methylene-3,3'-biplumbagin (16)

The arrangement of substituents around cyclopentane ring of plumbazeylanone can be deduced from the MS which showed a molecular ion at *m/z* 576 and was dominated by three intense peaks at *m/z* 388(88%), 202(94%) and 188(100%). Other major peaks occurred at *m/z* 120(81%) and 92(73%) as expected for juglone derivatives substituted in the benzenoid ring, and at 374(50%) and 373(62%) (Scheme 3). Further work is in progress in order to confirm the structure (19) proposed for plumbazeylanone and to determine the stereochemical arrangement of groups around the cyclopentane ring.

TABLE 2

^{13}C NMR chemical shifts (δ ; 90.56 MHz; $CDCl_3$) of plumbazeylanone (19).

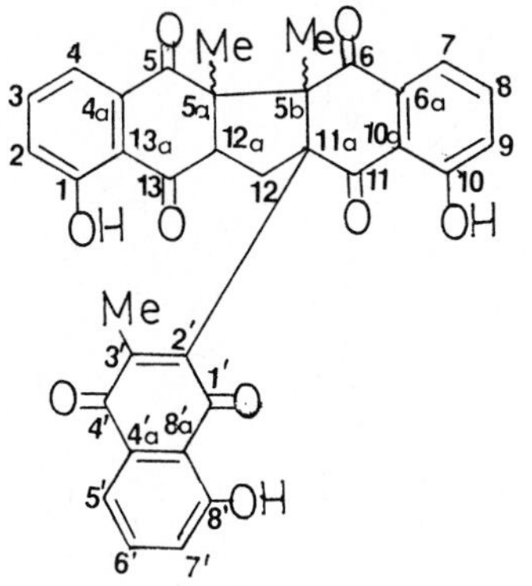

Chemical shift(s) (multiplicity)[a]	Assignment(s)
202.8, 196.3 (s)	11, 13
190.0 (s)	1'
189.7, 183.6 (s)	5,6
184.1 (s)	4'
161.2, 161.0, 160.4 (s)	1, 10, 8'
148.7, 147.2, 147.0, 143.5 (s)	5a, 11a, 2', 3'
137.1, 136.4, 136.1 (d)	3, 8, 6'
132.8, 131.8, 131.4 (s)	4a, 6a, 4'a
124.1 (d)	7'
123.8, 123.4 (d)	2, 9
119.7, 119.2, 119.0 (d)	4, 7, 5'
116.4, 115.1, 114.5 (s)	10a, 13a, 8'a
57.9 (s)	5b
55.7 (d)	12a
24.1 (t)	12
21.1 (q)	3'-Me
13.7, 13.5 (q)	5a-Me, 5b-Me

[a] s, singlet; d, doublet; t, triplet; q, quartet

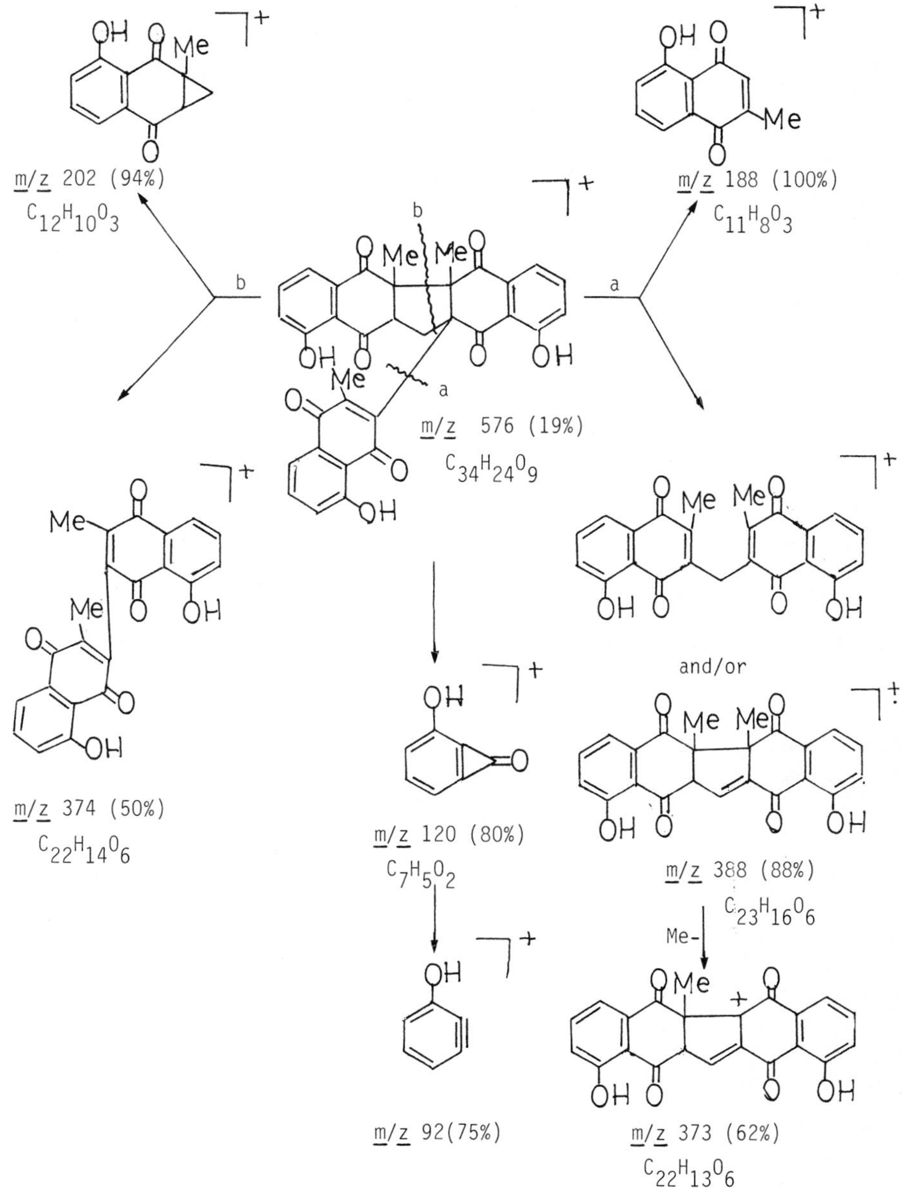

m/z 202 (94%)
$C_{12}H_{10}O_3$

m/z 188 (100%)
$C_{11}H_8O_3$

m/z 576 (19%)
$C_{34}H_{24}O_9$

m/z 374 (50%)
$C_{22}H_{14}O_6$

and/or

m/z 120 (80%)
$C_7H_5O_2$

m/z 388 (88%)
$C_{23}H_{16}O_6$

Me-

m/z 92(75%)

m/z 373 (62%)
$C_{22}H_{13}O_6$

Scheme 3. Possible mass spectral fragmentation of plumbazeylanone.

2.2.4 Structure and stereochemistry of 'isoshinanolone'

The more polar compound obtained from PLC of the column fraction eluted with 10% ethyl acetate in light petroleum was a pale yellow semi-solid [$C_{11}H_{12}O_3$; $(\alpha)_D$ + 24.2^o]. This was identified as an isomer of isoshinanolone (11) from its spectral data and chemical interconversions (Scheme 4).

Oxidation of the natural product with DDQ in dioxan afforded plumbagin (1), revealing it to be isoshinanolone or a stereoisomer. The ^{1}H NMR spectrum of the natural product was identical with that reported for isoshinanolone obtained from *Diospyros maritima* (ref. 17). It yielded a crystalline monomethyl ether (12) (m.p. 99-101°C), a crystalline diacetate (13) (m.p. 91-93° C) and a non-crystalline dibenzoate (14) (Scheme 4). The evidence for the stereochemistry at C-1 and C-2 came from its ^{1}H NMR spectrum (see above) and the Circular Dichroism (CD) curve of the dibenzoate (14) (Fig. 2) which were almost super-imposable with those reported (ref. 17). Although certain physical characteristics [m.p. and $(\alpha)_D$ of our sample differed from those of isoshinanolone isolated by Natori et al. (ref. 17) all the remaining physical data were in full agreement.

It would be of interest to review some stereochemical problems associated with natural 'isoshinanolones'. 'Isoshinanolone' [1,2(3)-tetrahydroplumbagin], an isomer of shinanolone (24), was first isolated by Natori et al. from *Diospyros maritima* (Ebenaceae) (ref. 17). There are four theoretically possible stereoisomers for isoshinanolone (11), viz. (11a) - (11d);

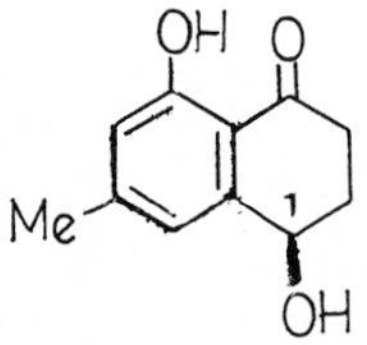

(24)

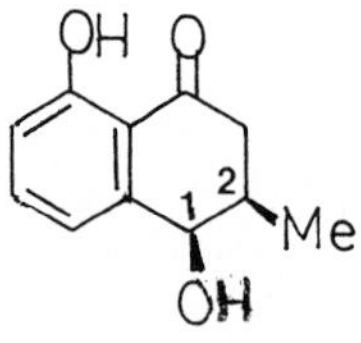

(11a) 1R,2R

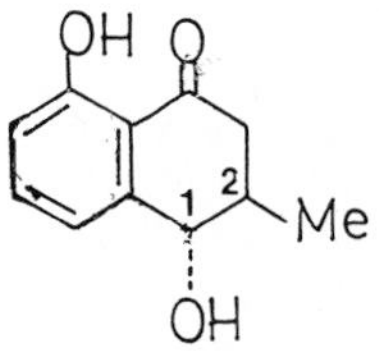

(11b) 1S,2R

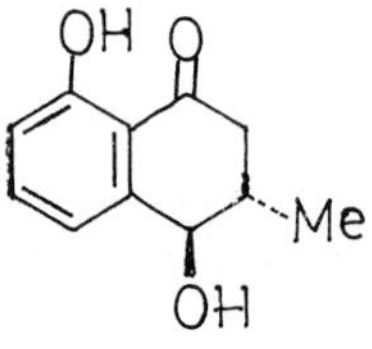

(11c) 1R,2S

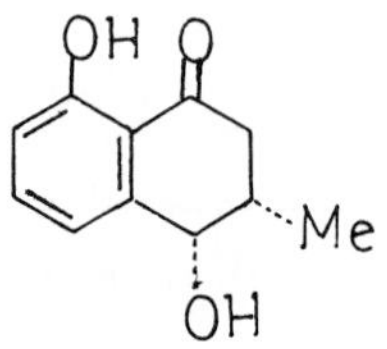

(11d) 1S,2S

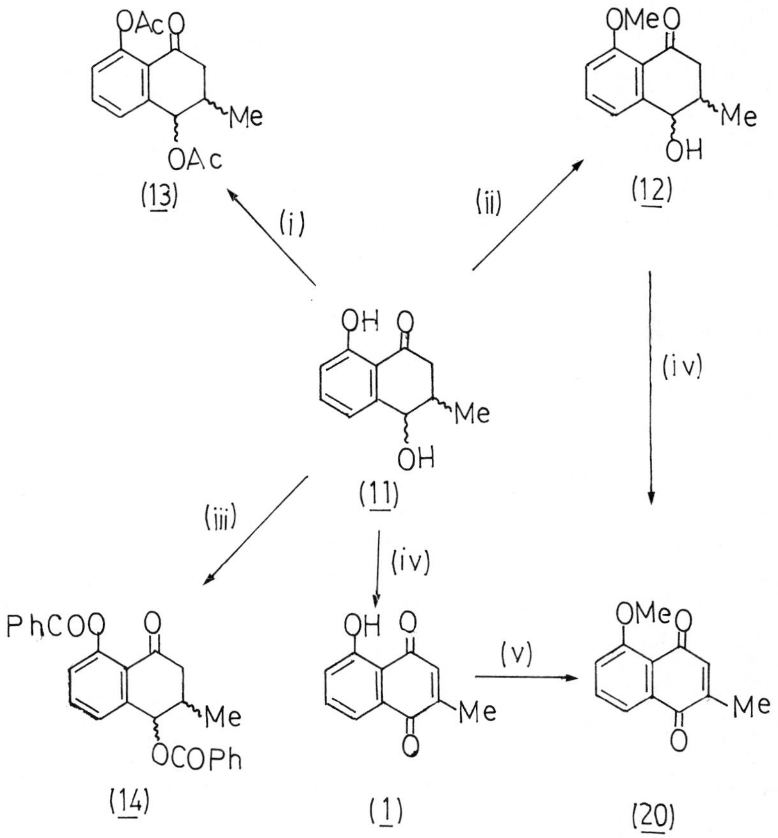

OAc O
Me
OAc
(13)

(i)

(ii)

OMe O
Me
OH
(12)

(iv)

OH O
Me
OH
(11)

(iii)

(iv)

PhCOO O
Me
OCOPh
(14)

OH O
Me
O
(1)

(v)

OMe O
Me
O
(20)

Reagents : (i) Ac_2O-pyridine-27°C, (ii) Me_2SO_4-K_2CO_3-acetone (anhyd.)-reflux-12h, (iii) PhCOCl-pyridine-27°C, (iv) DDQ-dioxan-reflux, (v) MeI-Ag_2O-$CHCl_3$-27°C.

Scheme 4. Chemical interconversions of isoshinanolone (11).

Isoshinanolone isolated by Natori and co-workers was determined to have 1R, 2R configuration (11a) by comparison of the CD curve with that of shinanolone (24) and by ^{1}H NMR coupling constant ($J_{1,2}$ = 2.5 Hz) of H-1 and H-2. However, in (24) the configuration of OH at C-1 was determined by application of the aromatic chirality method (ref. 28) on the dibenzoate, though there exist some limitation in the presence of the benzophenone chromophore.† The CD curve of the dibenzoate of Natori's isoshinanolone showed strong Cotton effects ($\Delta\epsilon_{240}$ + 33; $\Delta\epsilon_{225}$ - 11) quite similar to the dibenzoate of (24), the first positive chirality in the dibenzoate as far as the application of the method is tenable. If so, the absolute configuration at C-1 is assigned as R and that at C-2, R and the absolute configuration can be expressed by (11a).

Waterman and co-workers isolated a mixture of isoshinanolone (11a) and epi-isoshinanolone (11b) from *Diospyros cananiculata* (Ebenaceae) (ref. 30). It was suggested to be an equimolar mixture of (11a) and (11b) by ^{1}H NMR and by optical rotation calculations. Kumar et al. isolated neo-isoshinanolone (11d) from *Aristea ecklonii* (Iridaceae) (ref. 31). ^{1}H NMR spectrum and $(\alpha)_D$ of the natural product and some derivatives are almost identical with isoshinanolone isolated by us. Kumar et al. argue "The $J_{1,2}$ = 2.5 Hz suggests that of the methyl hydroxyl groups, one has an equatorial or a pseudo-equatorial orientation while the other has an axial of a pseudo-axial orientation. Natori et al. (ref. 17) suggested the stereochemisitry of isoshinanolone to be (11a) on the basis of aromatic chirality method, although they express some doubt on the validity of the method. Waterman and co-workers isolated a mixture of isoshinanolone (11a) and epi-isoshinanolone (11b) (ref. 30). The coupling constants from ^{1}H NMR suggest H-1 and H-2 in the latter (11b) to exist in pseudo-axial and axial orientations. However, the assignment of stereochemistry on the basis of comparison of the optical rotation of the mixture with that of shinanolone (24), which has additional methyl substitution in the aromatic ring, is not entirely unequivocal. The CD curve of the dibenzoate of 'neo-isoshinanolone' isolated from *P. zeylanica* (ref. 18) was found to be similar to that of isoshinanolone. The ^{1}H NMR chemical shifts of neo-isoshinanolone and isoshinanolone were similar." Kumar et al. beleives that "neo-isoshinanolone (11d) and isoshinanolone (11a) are enantiomeric and one or both may be

---

† It has been suggested recently that the use of p-substituted benzoate with a chromophore shifting the Cotton effect to wavelengths beyond the tetralone absorbing region would have overcome this problem (ref. 29).

partly racemized. The use of the aromatic chirality method in assigning the stereochemistry of isoshinanolone (11a) appears to be questionable. The higher optical rotation and the isolation of the identical material from two very different sources, *A.ecklonii* and *P.zeylanica* suggest that neo-isoshinanolone (11d) contains a greater amount of the enantiomer of negative rotation. However, the assignment of its stereochemistry will require further work (ref. 31)."

Recently Bhattacharyya and Carvalho reported isolation of a 4:1 mixture ['substance A'] of isoshinanolone (11a)[†] and epi-isoshinanolone (11b) from *Plumbago scandens* (Plumbaginaceae) (ref. 32). Surprisingly they had not seen two earlier reports (ref. 29, 30) and had stated "This is the first report of the existence of epi-isoshinanolone in nature, although, it appears likely that isoshinanolone reported from a related species, *P.zeylanica* (ref. 18), is also a mixture of the two diastereoisomers, probably in a slightly different ratio from that found in 'Substance A' (ref. 32)." Physical data reported for various stereoisomers, stereoisomeric mixtures of isoshinanolones and their derivatives are depicted in Table 3.

TABLE 3

Physical data of stereoisomers, stereoisomeric mixtures of isoshinanolone and their derivatives.

Compound	Appearence	m.p. (^{o}C)	$(\alpha)_D$	Ref.
(11a)	crystalline	160 (dec.) 230 (subl.) 255 (dec.)	-7°	17
	semi-solid	-	$+24.2^{o}$	18
Diacetate of (11a)	crystalline	$91-93^{o}$	-	18
Dibenzoate of (11a)	oil	-	$+12.2^{o}$	17
	oil	-	$+135.8^{o}$	18
(11a) + (11b)[a]	oil	-	$+16^{o}$	30
(11a)[c]+(11b)[b,c]	semi-solid	-	$+19.7^{o}$	32
(11d)	semi-solid	-	$+24^{o}$	31
Diacetate of (11d)	crystalline	$91-93^{o}$	$+128.3^{o}$	31

[a] Ratio of (11a):(11b), 1:1.
[b] Ratio of (11a):(11b), 4:1.
[c] Bhattacharyya and Carvalho assigns structure (11c) for isoshinolone (11a), and (11d) for epi-isoshinanolone (11b) (ref. 32).

[†] However, see footnote c in Table 3.

2.2.5 Structure of 1,2(3)-tetrahydro-3,3'-biplumbagin

Further purification of the column fraction eluted with 15% ethyl acetate in light petroleum afforded an orange yellow crystalline compound, $C_{22}H_{18}O_3$, m.p. 109-110^{0}C, $(\alpha)_D$ + 69.8^0, whose IR spectrum indicated the presence of hydroxy, $\alpha\beta$-unsaturated carbonyl and quinone carbonyl groups. Comparison of the UV (Table 4) and ^{1}H NMR (Table 5) spectra of this quinone with plumbagin (1) and isoshinanolone (11) suggested it to be made up of these two units.

TABLE 4

UV absorption maxima of plumbagin (1), isoshinanolone (11) and 1,2(3)-tetrahydro-3,3'-biplumbagin (15).

Compound	λ_{max}^{EtOH} nm (log ϵ)			
(1)	211(4.39)	266(3.99)	-	404(3.52)
(11)	217(4.52)	259(4.09)	335(3.76)	-
(15)	212(4.70)	263(4.32)	333(3.77)	418(3.63)

In the ^{1}H NMR spectrum of the natural product (Table 5) the absence of the vinyl proton at C-3 of the plumbagin unit indicated the possible linkage to be through C-3 carbon atom of the two units. This was confirmed by oxidation of the natural product with DDQ yielding 3,3'-biplumbagin (2). Therefore, the natural quinone should be 1,2(3)-tetrahydro-3,3'-biplumbagin (15). It remained then to determine the stereochemical arrangement of groups at C-1, C-2 and C-3 [see (15)] . The ^{1}H NMR coupling constant of the H-1 with H-2 is 2.5 Hz and this suggested that the methyl and hydroxyl groups are cis (the methyl is equatorial and hydroxyl is pseudo-axial) as in isoshinanolone (11a). Furthermore, the similarity of the CD curves of (15) and isoshinanolone (11) (Fig. 2) aided to assign the absolute configuration at C-1 as R (ref. 18). The coupling constant of the H-2 and H-3 ($J_{2,3}$) is 11 Hz which suggested the trans arrangement of groups at these two positions (ref. 33,34).

2.3 Biosynthetic aspects

Several biosynthetic pathways have been postulated for the origin of naphthoquinones in higher plants (ref. 35-38). The structural features of the naphthoquinones encountered in *P.zeylanica* clearly suggest that they are derived from acetate-polymalonate units. Occurrence of naphthalenones (11) and (15) suggest an alternate pathway to quinones from the cyclised product (25) and/or the presence of a reducing enzyme in *P.zeylanica* (Scheme 5). Natural

TABLE 5

^{1}H NMR data (60 MHz; CCl_4) of plumbagin (1), 3,3'-biplumbagin (2), isoshinanolone (11) and 1,2(3)-tetrahydro-3,3'-biplumbagin (15).

Compound	Aromatic H and OH		Quinone H and Me		Reduced quinone H and Me			
	H-6 – H-8	OH-5*	Me-2	H-3	H-1	H-2	H-3	Me-2
(1)	7.06 - 7.60 m	11.77 brs	2.15 d (J= 1.6 Hz)	6.73 q (J =1.6 Hz)	-	-	-	-
(2)($CDCl_3$)	7.20 - 7.90 m	11.80 brs	2.06 s	-	-	-	-	-
(11)	6.66 - 7.37 m	12.17 brs	-	-	4.57 d (J =2.5 Hz)	2.10-3.10 m		1.11 d (J = 6 Hz)
(15)	6.66 - 7.66 m	11.61 brs 11.90 brs	2.30 s	-	4.66 d (J =2.5 Hz)	2.83 m	4.20 d (J =11 Hz)	1.13 d (J = 7 Hz)

* Exchangeable with D_2O

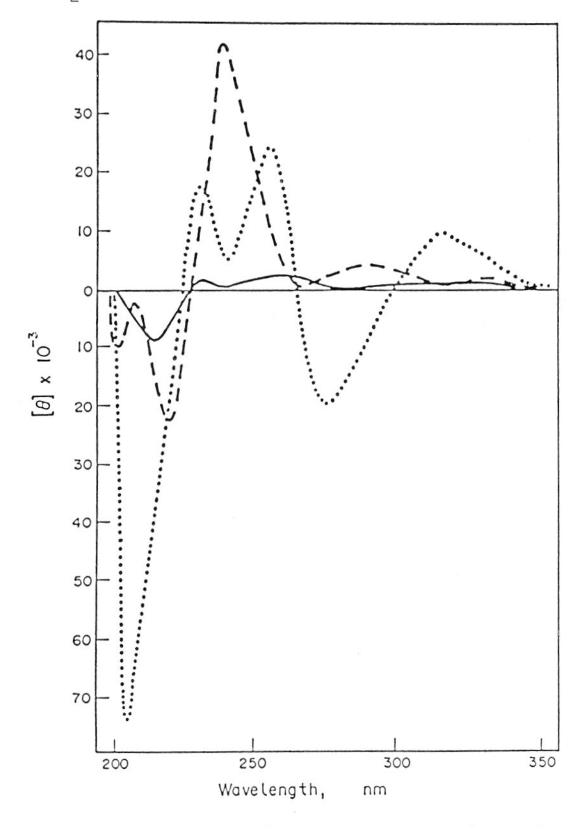

Fig. 2. CD curves (in MeOH) of (—) isoshinanolone (11), (----), isoshinanolone dibenzoate (14) and (....) 1,2(3)-tetrahydro-3,3'-biplumbagin (15).

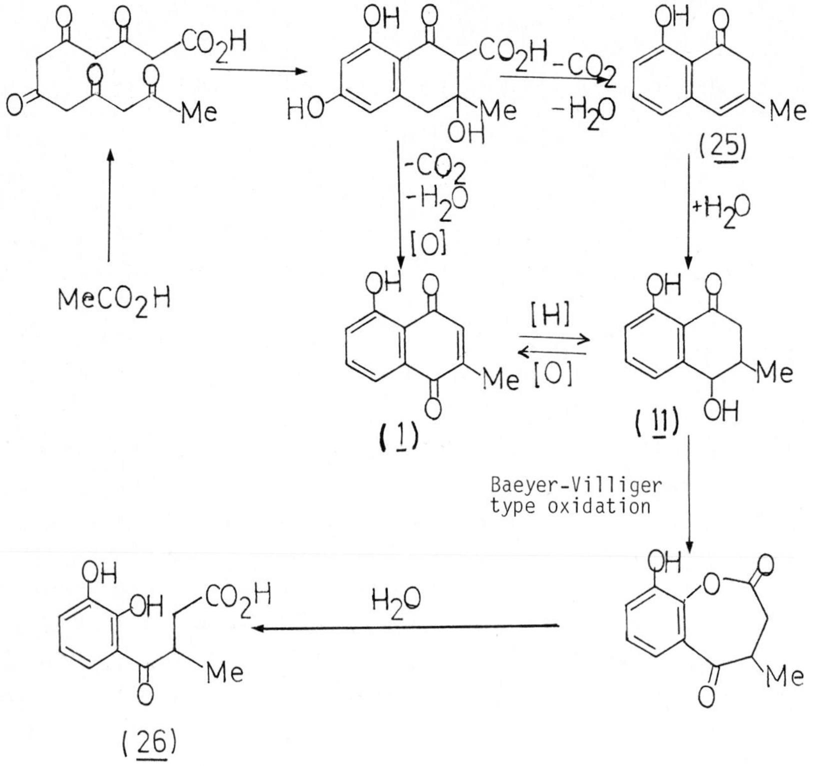

Scheme 5. Possible biosynthetic pathway to isoshinanolone (11) and its intermediacy in the biosynthesis of plumbagic acid (26).

occurrence of isoshinanolone (11) is significant as it may be the immediate biosynthetic precursor of recently encountered plumbagic acid (26) (ref. 39).

Possible biosynthetic route to methylene-3,3'-biplumbagin (16) and plumbazeylanone (19) is depicted in Scheme 6. It is noteworthy that (16) has been obtained together with several other dimers of plumbagin during a biomimetic type reaction of plumbagin (1) with its hydroquinone (27) in buffered methanolic solution (ref. 40). Possible participation of methanol (as formaldehyde) in the formation of methylene-3,3'-biplumbagin (16) in this reaction has been suggested and it is possible that the biosynthesis of (16) in *P. zeylanica* would involve an analogous reaction, involving intermediacy of (29).

Biosynthetically, plumbazeylanone (19) may arise from hemi-hydroquinone (28) of 3,3'-biplumbagin (2) by the reaction of (29) followed by oxidation and cyclisation or from methylene-3,3'-biplumbagin by oxidation, cyclisation and coupling with plumbagin, these reactions taking place in any sequence (Scheme 6).

2.4 Antifertility activity of *P.zeylanica* , plumbagin and isoshinanolone.

Antifertility effect of *P.zeylanica* have been investigated in a lmited number of studies which have been confined mainly to the root extractives. In continuing our studies on the use of indigenous plants for fertility regulation (ref. 41), we have investigated the antifertility activity of root, leaf and stem extractives of this plant in addition to the pure compounds, plumbagin (1) and isoshinanolone (11). In most instances the hot water extracts were tested since this has been the standard method of preparation of extracts according to ethnomedical information available. The results obtained in our studies are summarized in Table 6.

Our results confirm the earlier report (ref. 42) that the root of *P.zeylanica* show uterine stimulant activity *in vitro* . The leaf and stem bark also appear to possess this property, with activity in the leaf surpassing the activity exhibited by the root and stem bark. The two major compounds of the root extractive, plumbagin (1) and isoshinanolone (11) did not show any response, suggesting that the uterine stimulant activity is due to some other constituent(s) in these extracts. The results obtained in the anti-implantation *in vivo* assay employing female rats with the root extractives were contrary to the observations of earlier workers who noted that a 50% alcoholic extract resulted in interference with implantation in 100% of experimental animals (ref. 43), but are in conformity with those of Saksena *et al*. (ref. 44)

It has been reported that the antifertility effect of *P. zeylanica* is due to plumbagin (1), the major constituent of this plant (ref. 45). Significant anti-implantation and abortifacient activity at a dose level of 10 mg/kg body weight was reported for plumbagin isolated from the whole plant extract of

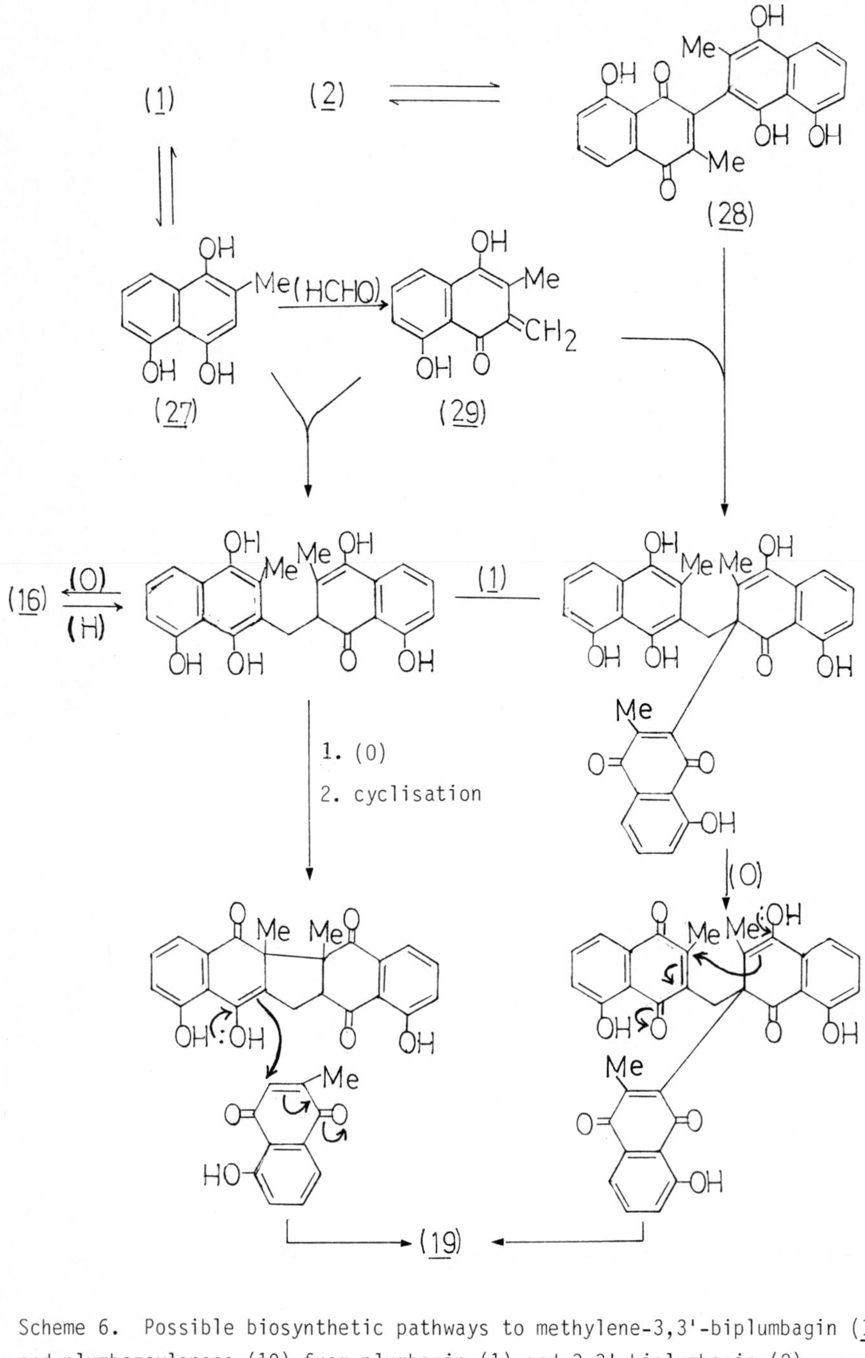

(1) (2) ⇌ (28)

(27) —(HCHO)→ (29)

(16) ⇌ (O)/(H) ← (1)

1. (O)
2. cyclisation

(O)

→ (19) ←

Scheme 6. Possible biosynthetic pathways to methylene-3,3'-biplumbagin (16) and plumbazeylanone (19) from plumbagin (1) and 3,3'-biplumbagin (2).

TABLE 6

Results of antifertility assay of *P. zeylanica* extractives, plumbagin (1) and isoshinanolone (11).

Antifertility assay	Plant part	Extract/ compound	Dose	Result
Uterotonic (*in vitro*)	root	cold water	20 mg/ml	active
	leaf	cold water	20 mg/ml	active
	stem bark	cold water	200 mg/ml	active
		plumbagin	0.2 mg/ml	inactive
		isoshinanolone	0.2 mg/ml	inactive
Anti-implantation (*in vivo*)	root	cold water	0.41 g/kg	inactive
		hot water	1.59 g/kg	inactive
		methanol	1.67 g/kg	inactive
	leaf	cold water	0.27 g/kg	inactive
		hot water	1.20 g/kg	inactive
	stem bark	cold water	0.59 g/kg	inactive
		plumbagin	0.02 g/kg	inactive

P. zeylanica. In our studies we were unable to confirm this activity even at a higher dose level of 20 mg/kg body weight (Table 6). A careful study of the properties of plumbagin used by Premakumari *et al*. in their investigation (ref. 45) reveals the possibility that the activity reported by them may be due to the impurities present in their test sample.

3. ALKALOIDS OF *BROUSSONETIA ZEYLANICA*

3.1 Introduction

The family Moraceae (mulberry family) consists of about 53 genera and over 1400 species, mostly of pantropical distribution (ref. 46). In Sri Lanka, there are 33 species belonging to 8 genera of which 5 are endemic species. They are *Artocarpus nobilis, Broussonetia zeylanica, Ficus costata, F. fergusoni* and *F. diversiformis*. Only seven species have been hitherto reported for the genus *Broussonetia* (ref. 46). The only *Broussonetia* species occurring in Sri Lanka is *B. zeylanica* (Thw.) Corner (= *Allaeanthus zeylanicus* Thw.) which is endemic to the country. It is a graceful tree, the very tough bark-fibres of which are used for string. This species is found in the intermediate zone forests and appears to be dwindling and may become extinct very soon.

HO OH OH

(30)

HO OH OH

(31)

HO OH

(32)

CHO N OH

(33)

OH N OH

(34)

O 2 H 3 4 H 4′ 4″ N 2′ 2″ N OH OH

(35)

At the time we initiated our work on *B. zeylanica* no chemical studies have been reported from any *Broussonetia* species. However, recent investigation of *B.papyrifera* and *B. kazinoki* which are employed in Japan as raw material for paper and rayon manufacture had led to the isolation of three catechins (30), (31), and (32) (ref. 47, 48). During our investigation of the timber of *B. zeylanica* we have encountered one major and two minor alkaloids which have been identified as 4-formyl-8-hydroxyquinoline (33), 3,4'-dihydroxy-2,3'-bipyridine (34) and 3,4-bis(8-hydroxyquinolin-4-yl)-γ-butyrolactone (35) (ref.49-51). The alkaloid (35) has a novel carbon skeleton and was named as broussonetine.

3.2 Chemical studies

3.2.1 Isolation of alkaloids

Timber of *Broussonetia zeylanica* was extracted successively and exhaustively with hot light petroleum and hot benzene. The hot light petroleum extract on column chromatographic purification gave the major alkaloid, 4-formyl-8-hydroxyquinoline (33). The hot benzene extract on column chromatography followed by PLC yielded a further quantity of (33), and two minor alkaloids, 3,4'-dihydroxy-2,3'-bipyridine (34) and broussonetine (35).

3.2.2 Structure of 4-formyl-8-hydroxyquinoline

The major alkaloid ($C_{10}H_7NO_2$, m.p. 155-156^oC) was soluble in dilute hydrochloric acid, and gave positive tests with ferric chloride and Gibb's reagent (ref. 52). it showed bands at ν_{max} 3240 (OH) and 1700 (CHO) cm^{-1} in its IR spectrum and a NMR singlet (1H) at δ 10.5. The remainder of the ^1H NMR spectrum comprised only aromatic signals overlapping a broad hydroxy band, two doublets (1H each) at δ 8.97 and 7.78 (*J*= 4 Hz), two double doublets (1H each) at δ 8.37 and 7.23 (*J* = 8 and 2 Hz), and a one-proton triplet centred at δ 7.60 (*J* = 8 Hz). Irradiation experiments suggested the presence of two and three adjacent protons in two separate aromatic (or heteroaromatic) rings. Addition of [$^2H_4$] methanol to the deuteriochloroform solution reduced the intensity of the aldehyde proton signal, and a corresponding hemi-ketal proton [-C*H*(OCD_3)OD] signal appeared at δ 6.10. The very similar behaviour of 4-formyl pyridine and 4-formyl quinoline (see Section 3.2.3), together with the above NMR spectrum and decoupling experiments suggested a quinoline or isoquinoline skeleton for the alkaloid with the aldehyde function at C-2 or C-4 (quinoline) or C-1 (isoquinoline). As there are three adjacent protons in the benzenoid ring the hydroxy group must be at C-5 or C-8 [see (36a) and (36b)].

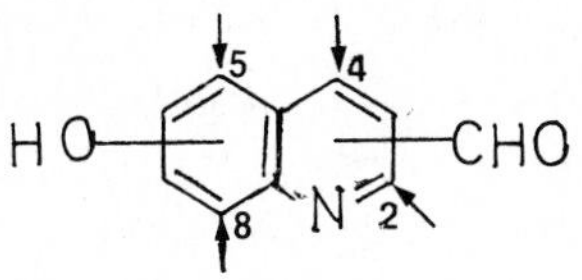

HO — CHO (36a) [positions 2, 4, 5, 8 marked]

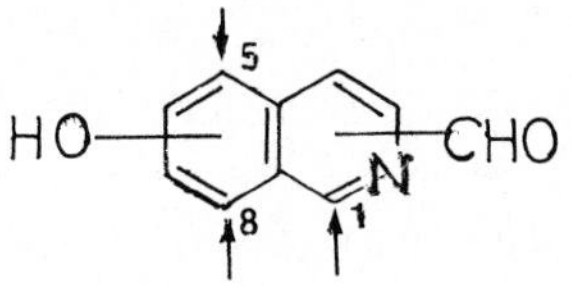

HO — CHO (36b) [positions 1, 5, 8 marked]

CHO (33) OH → (i) → CHO (37) OAc

(33) → (ii) → CH=NOH (41) HO → (i) → CH=NOAc (42) OAc

(33) → (iii) → CH=N-NH-C(=S)NH$_2$ (39) OH

(33) → (iv) → Me (38) OH

(33) → (v) → CO_2Me (40) OH

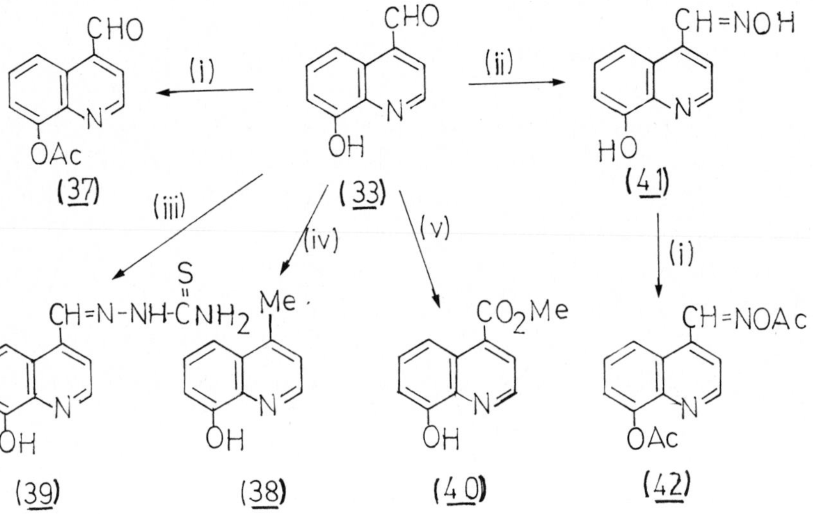

Reagents : (i) Ac_2O-pyridine-25°C-2h, (ii) $NH_2OH.HCl$-pyridine-100°C-3h, (iii) $NH_2NHC(=S)NH_2$-EtOH-H_2O-100°C-0.5h, (iv) $NH_2.NH_2.H_2O$-pyridine-reflux-3h, (v) MnO_2-NaCN-MeOH-CH_3CO_2H-25°C-2h.

Scheme 7. Chemical interconversions of 4-formyl-8-hydroxyquinoline (33).

The orange colour [λ_{max} (EtOH) 325 nm, no max., in Vis. region] of the alkaloid is indicative of a chromophore in which the phenolic group is conjugated with the carbonyl function. This excludes 2,5-disubstitution in the quinoline case but by analogy not 4,5-disubstitution, since 8-hydroxy-1-naphthaldehyde is also orange [λ_{max} (MeOH-H_2O) 350 nm] (ref. 53). On treatment with acetic anhydride and pyridine the alkaloid afforded a leuco O-acetyl derivative (37) (Scheme 7). Modified Wolff-Kishner reduction of the parent alkaloid with excess hydrazine hydrate and a drop of pyridine under reflux yielded the corresponding methylphenol. The ^1H NMR spectrum of this compound showed no low field aldehyde proton signal but a new methyl signal at δ 2.65, while in the aromatic region the double doublet at δ 8.37 had shifted upfield to δ 7.21. This, and the evidence presented above, showed that the aldehyde group is para to the N atom and the adjacent peri position is unsubstituted (ref. 54). Therefore, the methylphenol has to be either 8-hydroxy-4-methylquinoline (38) or 5-hydroxy-1-methylisoquinoline. The methylphenol, although not the parent alkaloid, gave a Mg^{2+} complex which showed a bluish-green fluorescence under UV thus supporting the 8-hydroxyquinoline structure (ref. 55). The physical properties of the methylphenol corresponded to (38), and comparison with an authentic sample (ref. 56) established its identity. The alkaloid is therefore 4-formyl-8-hydroxyquinoline (33).

Further evidence for the structure (33) proposed is provided by the ^{13}C NMR spectrum, the signals in which have been assigned (Table 7). Although 4-formyl-8-hydroxyquinoline (33) has been synthesized (ref. 57), the m.p. of our sample (155-156°) was considerably higher than that recorded (144-145°) and the thiosemicabazone (39) was also found to have a m.p. higher than that reported. Oxidation of (33) under Corey conditions (ref. 58) gave methyl-8-hydroxyquinoline-4-carboxylate (40). The parent alkaloid also gave an oxime (41). Treatment of (41) with acetic anhydride failed to give a 1,2-oxazine, again confirming that the two substituents were not in a peri relationship (ref. 59). However, the new product isolated was the diacetate (42), the MS of which is of some interest (Scheme 8). It showed prominent peaks at m/z 273 (M + 1), 230, 212, 199, 170 (100%), 144 and 142 but the molecular ion was absent. The base peak at m/z 170 ($C_{10}H_6N_2O$) possibly arises via loss of keten and acetic acid from the molecular ion (Scheme 8). An alternative fragmentation pathway through the oxime is ruled out as the oxime (41) itself did not show a prominent peak at m/z 170.

TABLE 7

Observed and calculated ^{13}C NMR chemical shifts of 4-formyl-8-hydroxyquinoline (33).

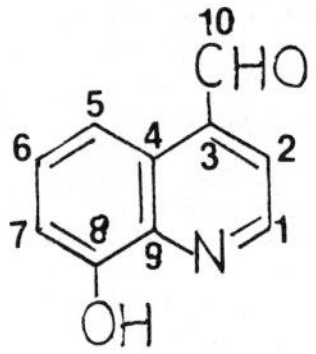

Carbon	^{13}C Chemical Shifts		Relative Intensity
	Observed[a]	Calculated[b]	
C(1)	148.08	151.60	80
C(2)	125.90	122.30	110
C(3)	153.46	144.60	21
C(4)	124.68	129.70	26
C(5)	114.36	120.70	114
C(6)	130.50	127.40	104
C(7)	111.70	117.00	91
C(8)	157.16	157.00	32
C(9)	137.09	136.00	28
C(10)	193.02	-	53

[a] values in ppm downfield from Me_4Si [(δMe_4Si) = $(CD_3)_2SO$ + 39.57 ppm.]

[b] calculated= [for corresponding quinoline carbon (ref. 60) + the effect due to substituents)(ref. 61, 62)].

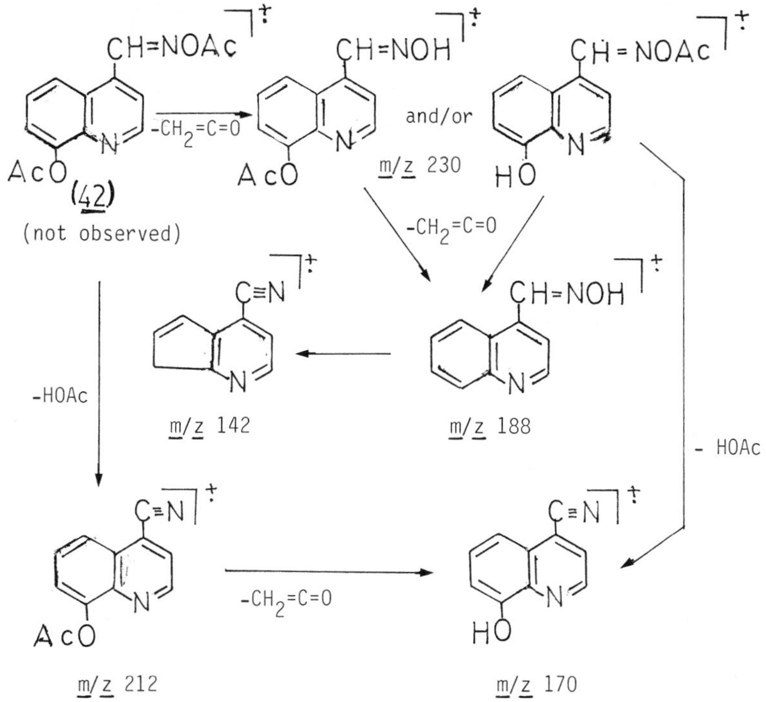

CH=NOAc ·+ (42) (not observed) —$-CH_2=C=O$→ CH=NOH ·+ and/or CH=NOAc ·+ m/z 230

$-CH_2=C=O$ → CH=NOH ·+ m/z 188 → C≡N ·+ m/z 142

-HOAc → C≡N ·+ AcO m/z 212 —$-CH_2=C=O$→ C≡N ·+ HO m/z 170 ← - HOAc

Scheme 8. Mass spectral fragmentation of the oxime diacetate (42).

3.2.3 Hemi-ketal Formation of some Aromatic and Heteroaromatic Aldehydes

As mentioned above (Section 3.2.2), the ^1H NMR spectrum of our new alkaloid, 4-formyl-8-hydroxyquinoline (33), underwent a marked change when a drop of $[^2H_4]$ methanol was added to a deuteriochloroform solution, and the solution then allowed to stand for an hour. The aldehyde proton signal diminished and a new singlet (*ca.* 0.5 H) appeared at δ 6.10 which may be attributed to the ketal proton of (43) (Scheme 9). Simultaneously, the whole spectrum became complex as two compounds (33) and (43) were present. All the signals due to (33) were reduced in intensity whilst most of the signals due to (43) were shifted upfield, relative to (33). Addition of further $[^2H_4]$ methanol had only a little effect while on warming to $45^\circ C$ the ketal signal slightly diminished. Evaporation of the solution led to the recovery of the original aldehyde. However, a MS of

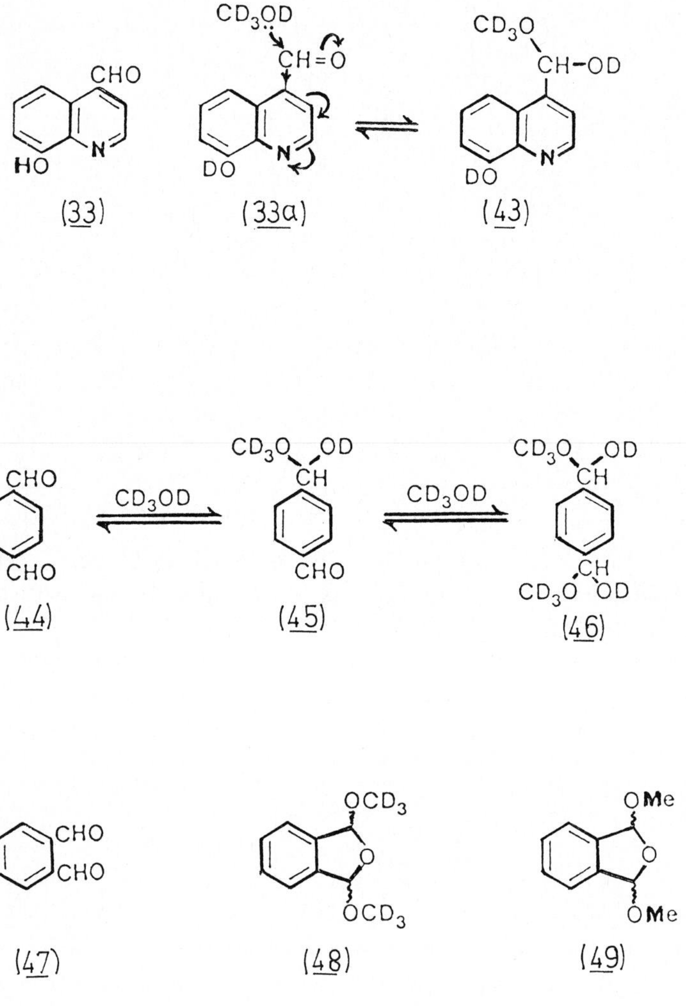

Scheme 9. Hemi-ketal formation of (33), terephthalaldehyde (44), and phthalaldehyde (47).

the NMR solution showed, in addition to the normal fragmentation pattern for (33), very weak peaks at m/z 210, 209 and 208 consistent with (43) and analogues containing less deuterium. Ready formation of the hemi-ketal may be attributed to the enhanced electrophilicity of the aldehyde group arising due to the electron withdrawing power of the nitrogen [see (33a)]. 4-Formyl quinoline and 4-formyl pyridine also gave the corresponding ketals, but in neat methanol (ref. 49).

Same behaviour was exhibited by benzaldehydes with strong -M ortho or para-substituents. Benzaldehyde itself equilibrates very slowly in deuteriochloroform and [$^2H_4$] methanol (9:1) increasing to about 30% hemi-ketal formation after 3 days. p-Nitrobenzaldehyde reacts to the same extent (ca. 30%) in 1 h. and in neat [$^2H_4$] methanol 70% of the aldehyde is rapidly converted to the ketal. The ^{1}H NMR spectra of terephthalaldehyde (44) and phthalaldehyde (47) were found to be interesting although they equilibrated slowly. In the former case, after two days in a 9:1 solution of deuteriochloroform and [$^2H_4$] methanol, (44) was just detectable; the equilibrium mixture consisted essentially (45) (65%) $\rightleftharpoons$ (46) (35%).

The ^{1}H NMR spectrum of phthalaldehyde (47) consisted of a singlet at δ 10.49 and a multiplet centred at δ 7.85 due to aromatic protons. After leaving in a 9:1 solution of deuteriochloroform and [$^2H_4$] methanol the spectrum changed into three singlets at δ 7.39(ArH), 6.31 and 6.04. The last two signals may be attributed to the cis and trans isomers of (48) (ref. 63). The aldehyde proton signal had completely disappeared and the ^{1}H NMR spectrum below δ 6.0 was superimposable on that of (49) prepared by heating phthalaldehyde (47) in methanol containing hydrochloric acid. The above observation made us to conclude that [$^2H_4$] methanol should be used with caution for the NMR spectral determinations of aldehydes. In certain circumstances it can be of diagnostic value; it can also be very misleading (ref. 49).

3.2.4 Structure of 3,4'-dihydroxy-2,3'-bipyridine

The less polar minor yellow alkaloid, m.p. 223-224°C, isolated from *B.zeylanica* (see Section 3.2.1) analysed for $C_{10}H_8N_2O_2$ and gave a green colouration with ferric chloride. The presence of a phenolic hydroxyl was apparent from a broad IR band (ν_{max} 3200 cm^{-1}) and a bathochromic shift in the UV λ_{max} on addition of sodium hydroxide and aluminium chloride. This was further confirmed by acetylation to give a colourless diacetate (50) which did not respond to the ferric chloride test. The UV spectrum was unchanged on addition of sodium hydroxide-boric acid thus ruling out an ortho-dihydroxy system.

Based on the spectral data, we suspected the alkaloid to have a bipyridyl ring skeleton. Since the alkaloid failed to give a red complex with ferrous sulphate (no λ_{max} in the visible region around 500 nm) a 2,2'-bipyridyl ring system was ruled out (ref. 64,65), thus leaving 2,3', 3,3' and 2,4' as possible ring attachments. The latter two were eliminated on the following spectral evidence. The ^{1}H NMR (360 MHz) spectrum with double irradiation experiments revealed three adjacent protons at δ 7.14(d), 7.51(t) and 8.01(d)(J = 7.6 Hz) in one ring and two adjacent protons at δ 7.79(d) and 8.87(d) (J = 4.5 Hz) in the other ring. In addition two D_2O exchangeable protons were also seen, one as a sharp singlet at δ 12.0 and the other as a broad signal at δ 9.90, probably due to a hydrogen bonded OH and NH, respectively. The broad signal at δ 9.90 strongly suggested the presence of a α- or a γ-pyridone type structure which are the preferred tautomers of α- and γ-hydroxypyridines in solution. This was further confirmed by IR which showed a weak stretching at 1620 cm^{-1} in keeping with the γ-pyridone rather than α-pyridone structure [see (34b)] (ref. 66).

The UV, IR, ^{1}H NMR and ^{13}C NMR (Table 8) spectra closely resembled those of a 2,3'-bipyridine ring skeleton. Having ascertained one ring to be a γ-pyridone leaves only one possible position for the second hydroxy group and that is at C-3. The upfield shifts observed for C-4, C-6 and C-5 in ^{13}C NMR spectrum of the alkaloid are compatible with the shifts observed for orelline [3,3',4,4'-tetrahydroxy-2,2'-bipyridine (51)] (Table 8) (ref. 67). All these evidence suggested this alkaloid to be probably 3,4-dihydroxy-2,3'-bipyridine (34). However, the final proof of the structure has to be based on synthesis.

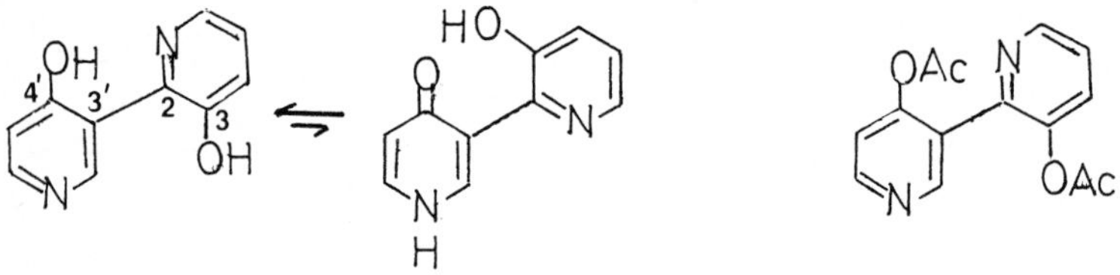

(34a) (34b) (50)

TABLE 8

^{13}C NMR chemical shift(s) of 3,4'-dihydroxy-2,3'-bipyridine (<u>34</u>) and some other bipyridines.

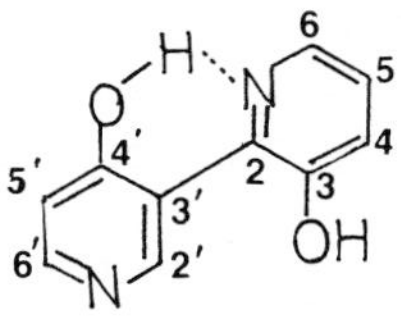

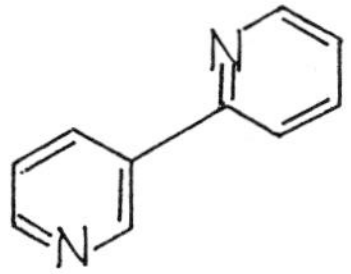

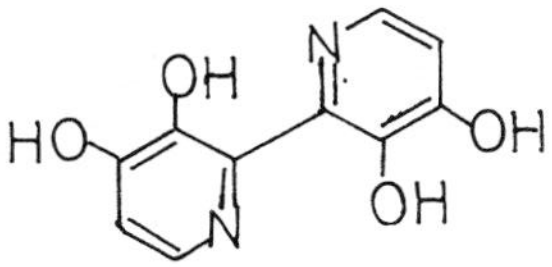

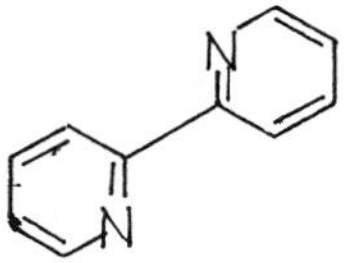

Carbon No.	3,4'-Dihydroxy-2,3'-bipyridine (<u>34</u>)	2,3'-Bipyridine	Δ*	3,3'-4',4'-Tetrahydroxy-2,2'-bipyridine (<u>51</u>)	2,2'-Bipyridine	Δ*
2	153.3 (s)	153.0	- 0.5	-	155.4	-
3	125.5 (s)	121.0	- 4.5	135.0	120.5	+ 15.5‡
4	114.4 (d)	135.5	+ 21.1†	137.0	137.2	- 0.2
5	119.5 (d)	119.0	- 0.5	112.3	124.1	+ 11.8‡
6	128.2 (d)	148.5	- 20.3†	123.7	149.3	+ 25.6†
2'	146.3 (d)	146.9	+ 0.6	-	-	-
3'	136.8 (s)	133.2	- 3.6	-	-	-
4'	139.1 (s)	132.0	- 7.1	-	-	-
5'	114.4 (d)	122.0	+ 10.6‡	-	-	-
6'	147.6 (d)	148.5	- 0.9	-	-	-

*Δ, δ parent bipyridine - δ hydroxylated bypyridine.

† Shift due to <u>ortho</u>/<u>para</u>-N and -OH.

‡ Shift due to <u>ortho</u>-OH and <u>meta</u>-N.

3.2.5 Structure of broussonetine

The more polar minor quinoline alkaloid, m.p. 238-239°C, obtained from *B.zeylanica* (see Section 3.2.1), was named broussonetine. It analyzed for $C_{22}H_{16}N_2O_4$ and was soluble in dilute hydrochloric acid and dilute sodium hydroxide. The alkaloid gave a fluorescent complex with Mg^{2+} ions (ref. 55) and showed UV λ_{max} at 252 and 233 nm. The presence of two 8-hydroxyiso-quinoline moieties in the alkaloid was evident from its ^{13}C NMR spectrum which revealed the presence of eighteen aromatic carbons which can be assigned as given in Table 9 and bear a close resemblance to those of 4-formyl-8-hydroxy-quinoline (33) (see Table 7).

The ^1H NMR (360 MHz) spectrum (Fig. 3) had a two proton D_2O exchangeable singlet at δ 9.71 and in the aromatic region two overlapping doublets at δ 8.89(H-2' and H-2"; J = 4.6 Hz) coupled to doublets at 7.97(H-3", J = 4.6 Hz) and 7.70(H-3', J = 4.6 Hz), respectively, double doublets at δ 7.09(H-5' or H-5", J = 1.0 and 8.60 Hz) and complex signals between δ 7.22 and 7.31(H-5' or H-5", H-6' and H-6"). Extensive homonuclear decoupling of the ^1H NMR spectrum established the presence of two sets of two and three adjacent protons showing that the 8-hydroxyquinoline moieties are substituted at either C-2, C-4, C-5 or C-7. That the substituents are located at C-4' and C-4" in (35) follows from the H-2', H-2" and H-3', H-3" assignments above, supported by the presence of the downfield doublet at δ 147.8(2C) in the off-resonance proton decoupled ^{13}C NMR spectrum which clearly arises from C-2' and C-2", and from the observation that in the ^1H NMR spectrum of broussonetine diacetate (52) the signals from the protons ortho- and para- to the acetoxyls are all shifted downfield. The remainder of the molecule was a fragment $C_4H_6O_2$ which must be a γ-butyrolactone unit (ν_{CO} 1775 cm^{-1}). In the ^1H NMR spectrum it showed signals at δ 6.71(H-4, d, J = 7.4 Hz), 4.84(H-3, dt, J = 7.4 and 8.6 Hz), 3.33 (H-2a, dd, J = 8.6 and 17.3 Hz) and 3.00 (H-2b, dd, J = 8.6 and 17.3 Hz) consistent with the structure 3,4-bis(8-hydroxyquinolin-4-yl)-γ-butyrolactone. The $J_{3,4}$ value (7.4 Hz) and the downfield shift of H-4 indicated that the quinoline rings were cis to each other [cf. 3,4-diphenyl-γ-butyrolactone (ref. 68)] as shown in (35). The presence of a major peak at m/z 171 (89%) further supported the γ-butyrolactone structure (Scheme 10).

3.3 Biosynthetic aspects

This is the first natural occurrence of alkaloids bearing 8-hydroxyquinoline moiety and thus it prompts us to postulate biosynthetic pathways to 4-formyl-8-hydroxyquinoline (33) and broussonetine (35). Possible biosynthetic pathways to these two alkaloids are depicted in Scheme 11. Biosynthetically, (33) could arise from the amino acid tryptophan (54) by the pathway shown for which

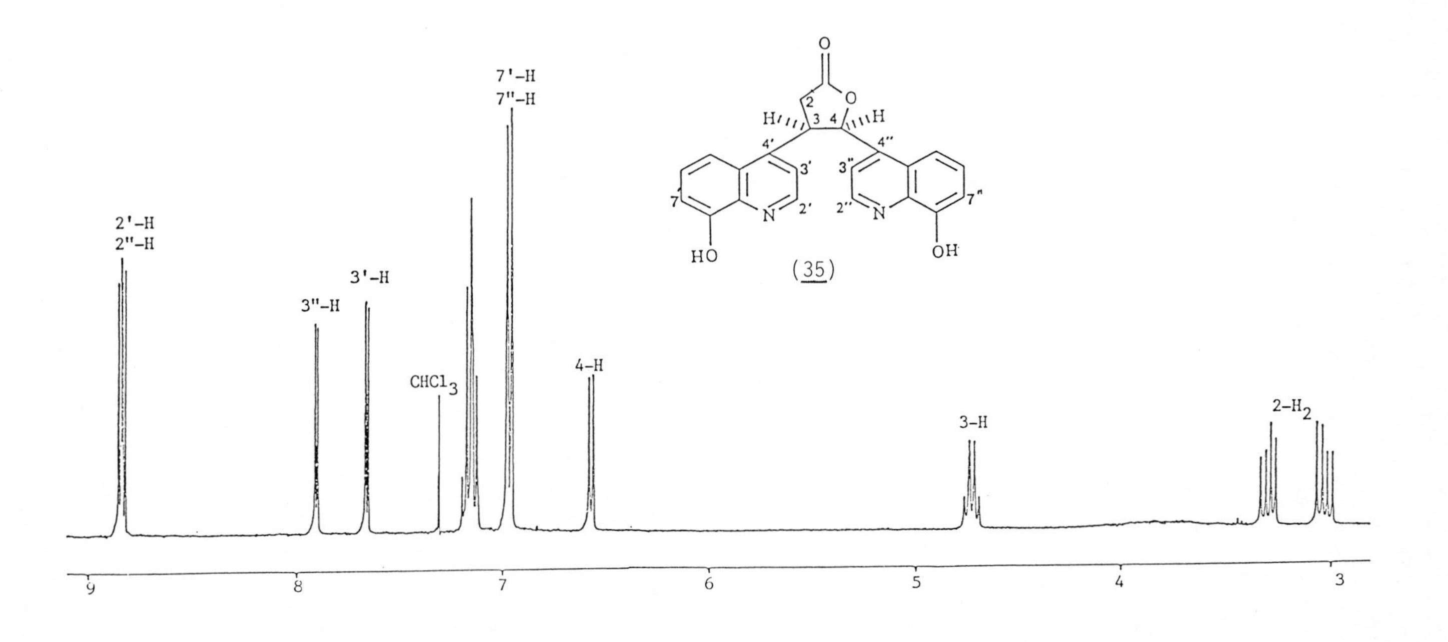

Fig. 3. ^{1}H NMR spectrum (360 MHz; $CDCl_3 + D_2O$) of broussonetine (35)

TABLE 9

^{13}C NMR chemical shifts (90.56 MHz, DMSO-d_6) and multiplicities for broussonetine (35), 4-formyl-8-hydroxyquinoline (33) and 8-hydroxyquinoline (53).

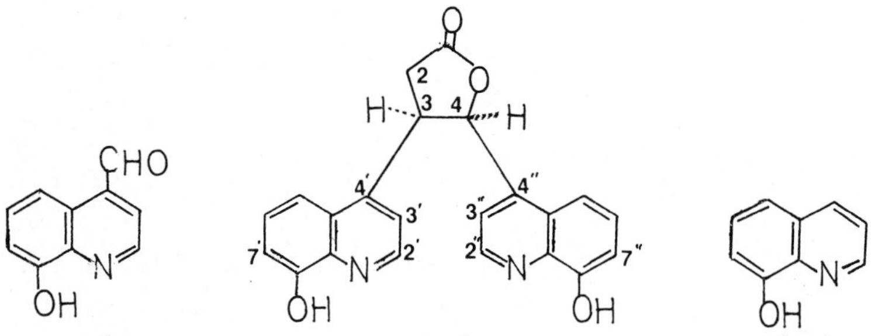

(33) (35) (53)

(33)		(35)		(53)	
C-2	148.08 d	C-2' and C-2"	147.81 d	C-2	148.13 d
C-3	125.90 d	C-3' and C-3"	112.76 d	C-3	121.80 d
C-4	153.46 s	C-4' and C-4"	143.13 s, 145.04 s	C-4	136.03 d
C-4a	124.68 s	C-4'a and C-4"a	125.86 s, 127.26 s	C-4a	128.83 s
C-5	114.36 d	C-5' and C-5"	118.89 d, 119.06 d	C-5	117.77 d
C-6	130.50 d	C-6' and c-6"	127.64 d, 127.85 d	C-6	127.52 d
C-7	111.70 d	C-7' and C-7"	111.06 d, 111.21 d	C-7	111.30 d
C-8	157.16 s	C-8' and C-8"	153.60 s, 153.66 s	C-8	153.32 s
C-8a	137.09 s	C-8'a and C-8"a	138.57 s, 136.60 s	C-8a	139.49 s
C-9	193.02 d				
		C-1	157.90 s		
		C-2	36.79 t		
		C-3	42.43 d		
		C-4	80.45 d		

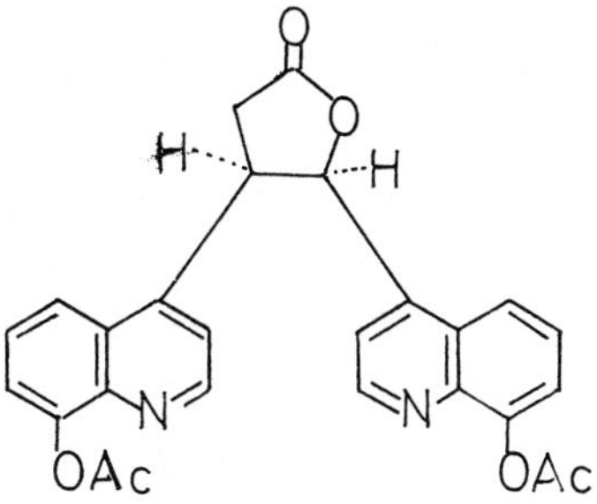

(52)

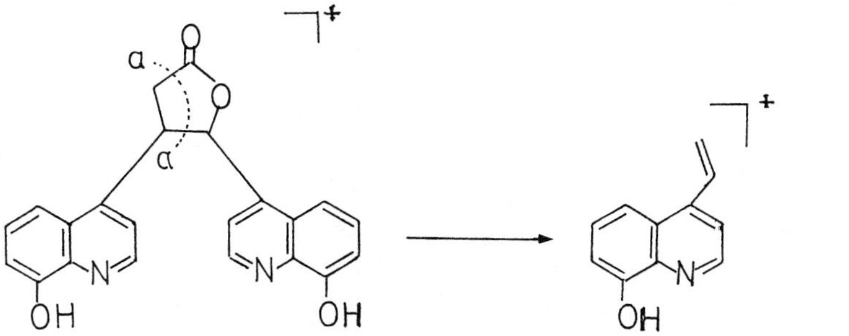

(35)

m/z 171 (89%)
$C_{11}H_9NO$

Scheme 10. Mass spectral fragmentation of broussonetine.

chemical analogues are known (ref. 69). Biosynthesis of broussonetine (35) would involve condensation of two molecules of 4-formyl-8-hydroxyquinoline with a molecule of acetyl-CoA followed by cyclisation as shown in Scheme 11.

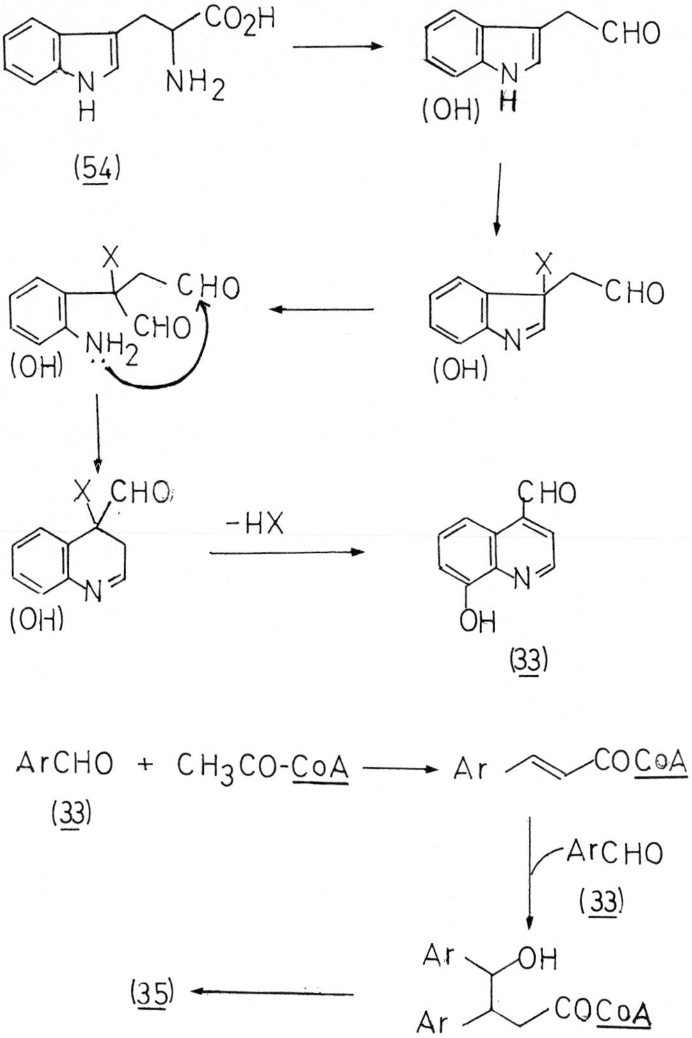

CO_2H
NH_2
N
H
(54)
CHO
N
(OH) H
X
CHO
(OH)
N
X
CHO
CHO
(OH) NH_2
X CHO
(OH)
N
-HX
CHO
N
OH
(33)

ArCHO + CH_3CO-CoA ⟶ Ar ⁄⁄ COCoA
(33)
ArCHO
(33)
Ar OH
Ar COCoA
(35) ⟵

Scheme 11. Possible biosynthetic pathways to 4-formyl-8-hydroxyquinoline (33) and broussonetine (35).

ACKNOWLEDGEMENTS

I am grateful to my coworkers Professors M.U.S. Sultanbawa, R.H. Thomson, S. Balasubramaniam, K. Jayasena, V. Gunawardena and Drs. B.M.R. Bandara, R. Somanathan, G.M.K.B. Gunaherath and S. Surendrakumar for their valuable contributions to the work presented herein. Mrs. S.C.Weerasekera and Mr. W.R. Wimalasiri are thanked for careful preparation of this manuscript. Financial assistance from Special Programme of Research Development and Research Training in Human Reproduction, W.H.O. (Geneva), Natural Resources, Energy and Science Authority (Sri Lanka), International Foundation for Science (Sweden) and University of Peradeniya is gratefully acknowledged.

REFERENCES

1 B.A. Abeywickrema, Cey. J. Sci. (Bio. Sci.), 2 (1959) 187.
2 W.M. Bandaranayake, M.U.S. Sultanbawa, S.C. Weerasekera and S. Balasubramaniam, The Sri Lanka forester, XI (1974) 67.
3 W.M. Bandaranayke and M.U.S. Sultanbawa, A List of Endemic Plants of Ceylon, University of Sri Lanka, Peradeniya, 1969, 60 pp.
4 B.A. Abeywickrema, Proc. Workshop on Natural Products, Colombo, Sri Lanka, 1975, p. 11.
5 Anon., World Health Organization Technical Report No. 622 on the Promotion and Development of Traditional Medicine, Report of a Meeting, Geneva, 1978, p. 10.
6 A.A.L. Gunatilaka, Harnessing Natural Products for Development (Presidential Address : Chemical & Physical Sciences Section), Proc. Sri Lanka Assoc. Advmt. Sci. 42 (1986) in press.
7 A.A.L. Gunatilaka, Chemistry in Sri Lanka, 1 (1984) 24.
8 A.A.L. Gunatilaka, Proc. The Fifth Asian Symposium on Medicinal Plants and Spices, Invited Lectures (ed. B.H. Han, D.S. Han, Y.N. Han and W.S. Woo) Kyung Moon Printing Co., Seoul, 1984, pp 605 - 614.
9 A.A.L. Gunatilaka, Proc. Sri Lanka Assoc. Advmt. Sci., 40 (1984) 141.
10 G.R.C.B. Gamlath, G.M.K.B. Gunaherath and A.A.L. Gunatilaka in Atta-ur-Rahman and P.W. Le Quesne (Eds.), New Trends in Natural Products Chemistry, Vol 26, Elsevier Science Publishers, Amsterdam, 1986, pp 109 - 121.
11 R.N. Chopra, K.L. Handa and L.D. Kapur, Indigenous Drugs of India, U.N. Dhur & Sons Ltd., Culcutta, 1958, 816 pp.
12 J.P.C. Chandrasena, The Chemistry and Pharmacology of Ceylon and Indian Medicinal Plants, H & C Press, Colombo, 1935, 167 pp.
13 J. Attygalle, Sinhalese Materia Medica, M.D. Gunasena & Co. Ltd., Colombo, 1917.
14 G.S. Sidhu and A.V.B. Sankaram, Tetrahedron Letters (1971) 2385.
15 A.V.B. Sankaram, A.S. Rao and G.S. Sidhu, Phytochemistry, 15 (1976) 237.
16 A.V.B. Sankaram, A.S. Rao and J.N. Shoolery, Tetrahedron, 35 (1979) 1777.
17 M. Tezuka, C. Takahashi, M. Kuroyanagi, M. Satake, K. Yoshihira and S. Natori, Phytochemistry 12 (1973) 175.
18 G.M.K.B. Gunaherath, A.A.L. Gunatilaka, M.U.S. Sultanbawa and S. Balasubramaniam, Phytochemistry, 22 (1983) 1245.
19 G.M.K.B. Gunaherath, A.A.L. Gunatilaka and R.H. Thomson, J. Chem. Soc. Perkin Trans. I (1986) in press.
20 G.M.K.B. Gunaherath, A.A.L. Gunatilaka and R.H. Thomson, Tetrahedron Letters 25(1984) 4801.

21 J.F. Garden and R.H. Thomson, J. Chem. Soc. (1957) 2483.
22 F.M. Dean, P.G. Jones, R.B. Morton and P. Sidisunthron, J. Chem. Soc. (1963) 5336.
23 G.M.K.B. Gunaherath, A.A.L. Gunatilaka and R.H. Thomson, unpublished results.
24 I.A. McDonald, T.J. Simpson and A.F. Sierakowski, Aust. J. Chem. 30 (1977) 1727.
25 B.F. Bowen, D.W. Cameron, M.J. Crossley, G.I. Feutrill, P.G. Griffiths and D.P. Kelly, Aust. J. Chem. 32 (1979) 769.
26 H. Seto, L.W. Cary and M. Tanabe, J. Chem. Soc. Chem. Comm. (1973) 867.
27 A.V.B. Sankaram, V.V.N. Reddy and M. Marthandamurthi, Phytochemistry, 25 (1986) 2867.
28 M. Kuroyanagi, K. Yoshihira and S. Natori, Chem. Pharm. Bull., 19 (1971) 2314.
29 N. Harada and K. Nakanishi, Circular Dichroic Spectroscopy - Exciton Coupling in Organic Stereochemistry, University Science Book Publishers, 1983.
30 S.M. Zhong, P.G. Waterman and J.A.D. Jeffreys, Phytochemistry, 23 (1984) 1067.
31 V. Kumar, K.M. Meepagala and S. Balasubramaniam, Phytochemistry, 24 (1985) 1118.
32 J. Bhattacharyya and V.R. De Carvalho, Phytochemistry, 25 (1986) 764.
33 J.T. Pinhey and S. Sternhell, Aust. J. Chem. , 18 (1965) 543.
34 J.W. Clark-Lewis, L.M. Jackmann and L.R. Williams, J. Chem. Soc. (1962) 3858.
35 J.H. Richards and J.B. Hendrickson, The biosynthesis of steroids, terpenes and acetogenins, W.A. Benjamin Inc., 1964, p. 75.
36 R. Durand and M.H. Zenk, Tetrahedron Letters, (1971) 3009.
37 H.V. Schmid and M.H. Zenk, Tetrahedron Letters, (1971) 4151.
38 H. Seto, L.W. Cary and M. Tanabe, J. Chem. Soc. Chem. Comm. (1973) 867.
39 X. Qian, X. Liang and P. Cong, Hua Hsueh Hsueh Pao, 38 (1980) 377 [Chem. Abs. 93 (1980) 21795 g].
40 A.V.B. Sankaram, A.S. Rao and G.S. Sidhu, Tetrahedron Letters, (1975) 3627.
41 S. Balasubramaniam, B.M.R. Bandara, A.A.L. Gunatilaka, V. Gunawardena, K. Jayasena, S. Sotheeswaran, M.U.S. Sultanbawa *et al.*, unpublished results.
42 R.B. Bhatia and S. Lala, Indian J. Med. Res. 23 (1983) 777.
43 M.L. Gupta, T.K. Gupta and K.P. Bhargava, J. Res. Ind. Med., 6 (1971) 112.
44 S.K. Saksena, S.K. Garg and R.R. Chaudhury, Indian J. Med. Res., 58 (1970) 253.
45 P. Premakumari, K. Rathinam and G. Santhakumari, Indian J. Med. Res., 65 (1977) 829.
46 E.J.H. Corner, A revised handbook to the Flora of Ceylon (ed. M.D. Dassanayake and F.R. Fosberg) Amerind Publishing Co. Pvt. Ltd., New Delhi, 1981, pp 213 - 292.
47 M. Takaguri, N. Niino, S. Nagao, M. Anetai, T. Masamune, A. Shirata and K. Takahashi, Chemistry Letters (1984) 689.
48 J. Ikuta, Y. Hano and T. Nomura, Heterocycles, 23 (1985) 2835.
49 A.A.L. Gunatilaka, J.S.H.Q. Perera, M.U.S. Sultanbawa, P.M. Brown and R.H. Thomson, J. Chem. Research (M) (1979) 779.
50 A.A.L. Gunatilaka, M.U.S. Sultanbawa, S. Surendrakumar and R. Somanathan, Phytochemistry, 22 (1983) 2847.
51 A.A.L. Gunatilaka, S. Surendrakumar and R.H. Thomson, Phytochemistry, 23 (1984) 929.
52 F.E. King, T.J. King and L.C. Manning, J. Chem. Soc. (1957) 563.
53 D. Berry and D.C.C. Smith, J. Chem. Soc. Perkin Trans I (1972) 699.

54 A. Godard, P. Duballet, G. Queguiner and P. Pastour, Bull. Soc. Chem. France (1976) 789.
55 F. Feigl, Spot tests in Organic Analysis, 6th Ed., Elsevier, Amsterdam, 1960, p 205.
56 A. Corsini, W.J. Louch and M. Thompson, Talanta, 21 (1976) 252.
57 J. Buchi, A. Aebi, A. Deflorin and H. Hurni, Helv. Chim. Acta, 39 (1956) 1676.
58 E.J. Corey, N.W. Gilman abd B.E. Ganem, J. Amer. Chem. Soc. 90 (1968) 5626.
59 R. Adams and D.E. Burney, J. Amer. Chem. Soc., 63 (1941) 1103.
60 R.J. Pugmire, D.M. Grant, M.J. Tobins and R.K. Robins, J. Amer. Chem. Soc. 91 (1969) 6381.
61 G.C. Levy, G.L. Nelson and J.D. Cargioli, J. Chem. Soc. Chem. Comm. (1971) 506.
62 G.L. Nelson, G.C. Levy and J.D. Cargioli, J. Amer. Chem. Soc., 94 (1972) 3089.
63 T. DoMinh, A.L. Johnson, J.E. Jones and P.P. Senise, J. Org. Chem., 42 (1977) 4217.
64 A.G. McInnes, D.G. Smith, J.L.C. Wright and L.C. Vining, Can. J. Chem., 55 (1977) 4159.
65 W.W. Brandt, F.P. Dwyer and E.C. Gyarfas, Chem. Rev., 54 (1954) 959.
66 A.R. Katritzky and R.A. Jones, J. Chem. Soc., (1960) 2947.
67 W.Z. Antkowiak and W.P. Gessner, Tetrahedron Letters, (1979) 1931.
68 T.V. Mandelšhtam, S.V. Kolesova, T.V. Polina, V.V. Solomentsev and N.S. Osmolovskaya, J. Org. Chem. USSR 16 (1980) 1024.
69 E.E. van Tamelen, V.B. Haarstad and R.L. Orvis, Tetrahedron, 24 (1968). 687.

THE STEREOCHEMISTRY OF PROAPORPHINES

ALAN J. FREYER, HELENE GUINAUDEAU and MAURICE SHAMMA

The proaporphines constitute a group of approximately 70 alkaloids. Their existence in nature was first predicted by Barton in 1957, in a historic review on the role of phenolic oxidative coupling in plant metabolism. They were postulated as tetracyclic molecules incorporating a cross-conjugated dienone system, and they were assumed to be formed from intramolecular oxidative coupling of tetrahydrobenzylisoquinoline precursors, such as N-methylcoclaurine. Their acid catalyzed dienone-phenol rearrangement would then lead to aporphines oxygenated at C-10.

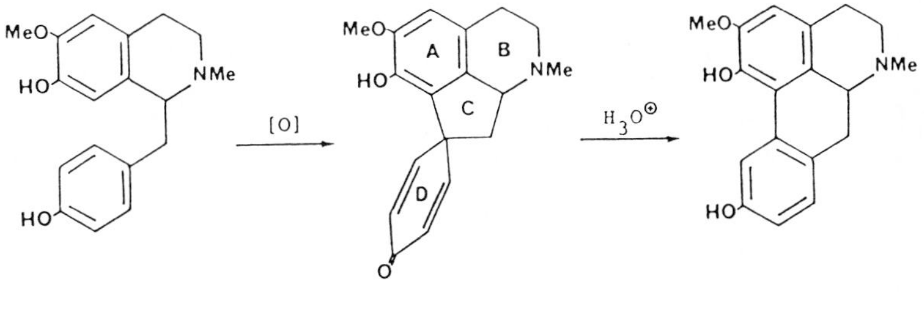

MeO HO NMe HO —[O]→ MeO HO A B C D NMe O —$H_3O^{\oplus}$→ MeO HO NMe HO

That proaporphines may indeed act as biogenetic precursors to some of the aporphines has been confirmed by *in vivo* studies using labeled compounds.

We have adopted the numbering system shown below for the proaporphine alkaloids:

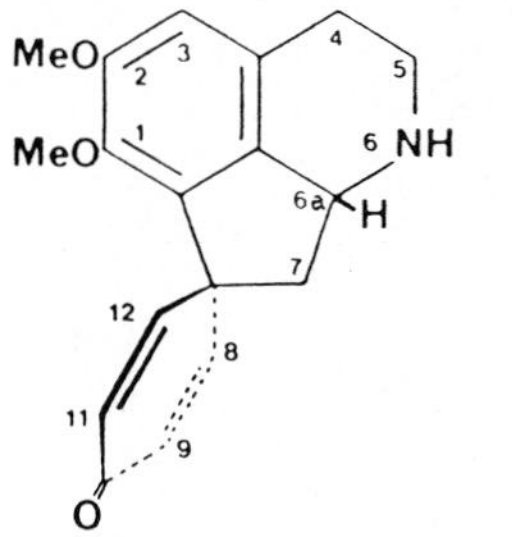

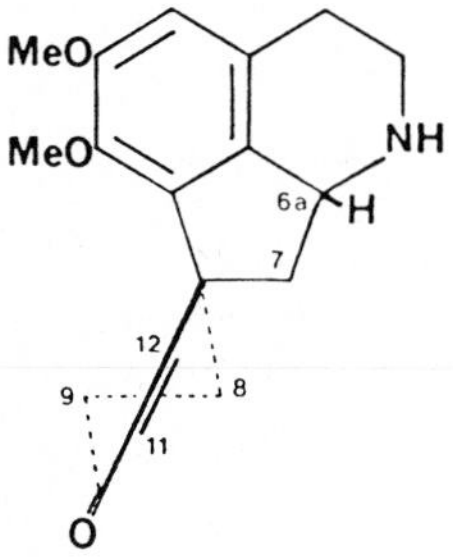

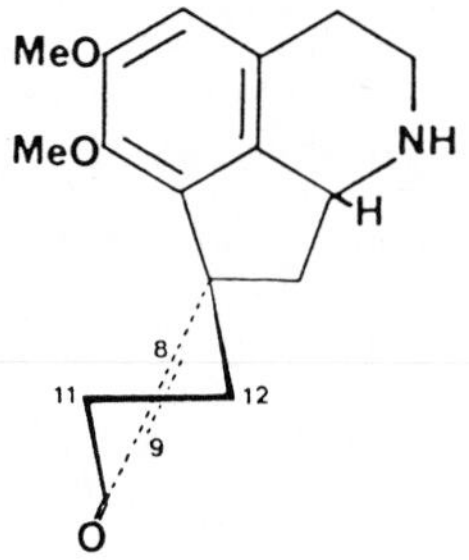

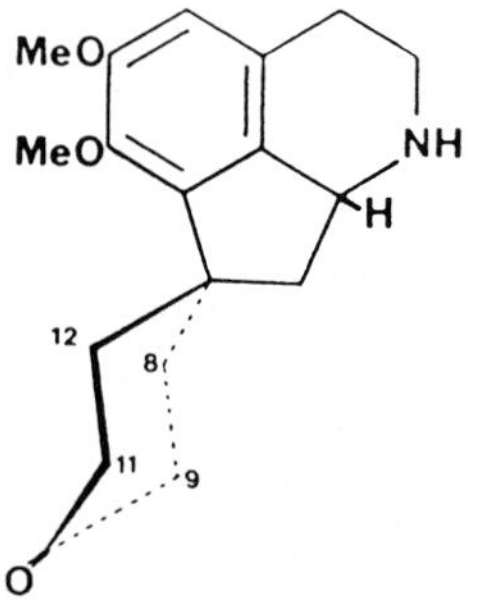

The diagrams have been drawn with C-8 and C-9 below the mean plane of the molecules, and C-11 and C-12 above.

The absolute configuration of a proaporphine may be determined by the sign of the $[\alpha]_D$. It is known that compounds with the C-6a S configuration possess a negative rotation, while the C-6a R proaporphines show a positive $[\alpha]_D$. Proaporphines readily decompose to colored products which make the optical measurements unreliable. One way of avoiding this difficulty consists in the acid catalyzed dienone-phenol rearrangement to the corresponding aporphine. The latter is usually stable. In the aporphine case, the C-6a R configuration leads to a negative $[\alpha]_D$, while the opposite S configuration displays a positive rotation.

In the course of a recent investigation of the alkaloids of Turkish Roemeria species (Papaveraceae), we obtained a series of proaporphine and reduced proaporphine alkaloids. While several reduced proaporphines have been reported in the literature, no complete NMR spectral data were available. Fortunately, we had on hand a supply of the known proaporphine (+)-stepharine, which we had obtained from the roots of Thai Stephania venosa (Menispermaceae). Our objective was to carry out a study of (+)-stepharine itself, as well as its various reduction products, with the aim of setting up reliable spectral and optical standards by which to assign the structures of any new proaporphinoid.

By means of $[\alpha]_D$ measurements of the proaporphine (+)-stepharine, as well as the levorotatory aporphine resulting from its dienone-phenol rearrangement, the C-6a R configuration was confirmed. The assignment of the NMR chemical shifts indicated around the following structure is based upon proton spin decoupling and NOE difference techniques.

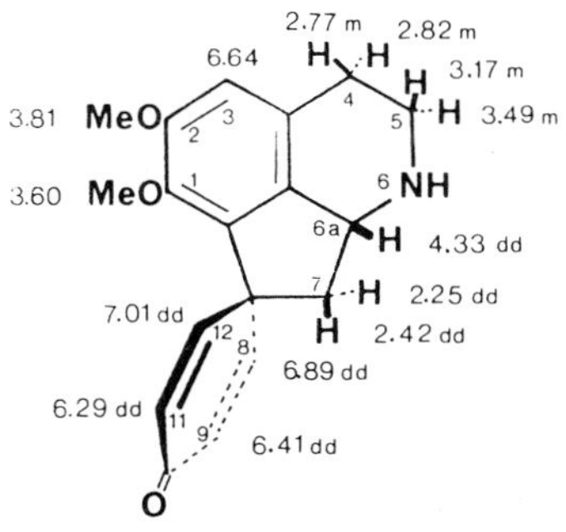

(+)-stepharine

$J_{6a,7\alpha}$ 10.5; $J_{6a,7\beta}$ 6.3; $J_{7\alpha,7\beta}$ 12.0;
$J_{8,9}$ 9.9; $J_{11,12}$ 9.9; $J_{8,12}$ 2.9; $J_{9,11}$ 1.9;

To obtain the two dihydrostepharine diastereomers (+)-stepharine was subjected to controlled catalytic hydrogenation. Following separation of the resulting mixture of diastereomers spin decoupling and NOE difference techniques were used to locate the reduced double bond between C-11 and 12 in one case, and C-8 and 9 in the other.

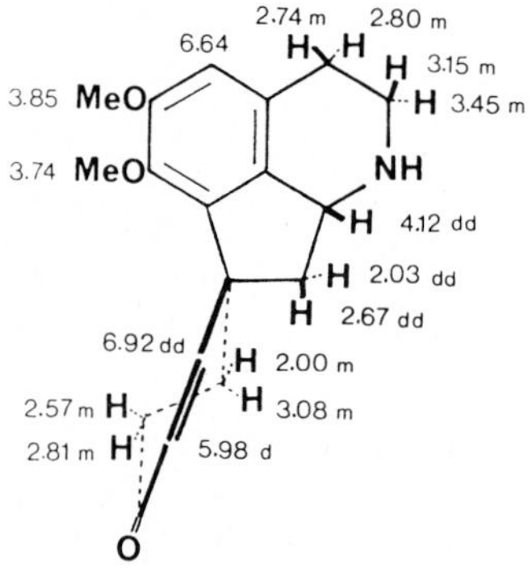

$J_{6a,7\alpha}$ 10.5; $J_{6a,7\beta}$ 6.6; $J_{7\alpha,7\beta}$ 11.6; $J_{11,12}$ 10.1; $J_{8\leftrightarrow eq,12}$ 1.3;

(+)-8,9-dihydrostepharine

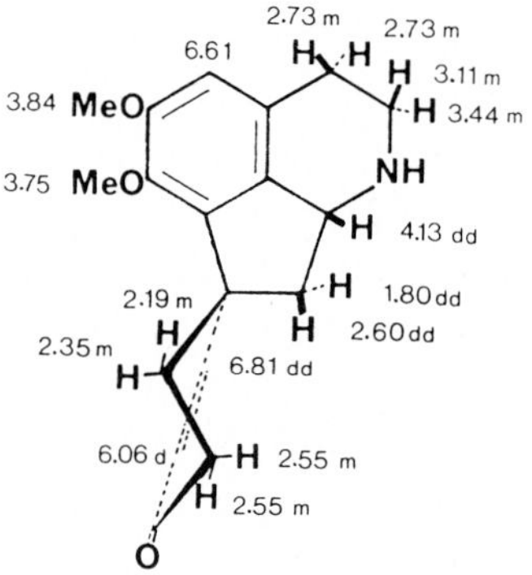

$J_{6a,7\alpha}$ 11.0; $J_{6a,7\beta}$ 6.3; $J_{7\alpha,7\beta}$ 11.4; $J_{8,9}$ 10.1; $J_{8,11\leftrightarrow eq}$ 1.3;

(-)-11,12-dihydrostepharine

For (-)-11,12-dihydrostepharine, spin decoupling identified proton 6a and the vicinal C-7 protons. In the same fashion, the protons at C-11 and C-12 were located. Two important NOE enhancements should be pointed out. Irradiation of the 7β proton affected protons 7α and 6a. Furthermore, irradiation of proton 7α led to enhancement of protons 7β and 8. Thus, the double bond is located anti to proton 6a. An additional long range W-coupling between H-8 and H-12 indicated these two protons to be in the same plane.

A similar experimental approach was employed to elucidate the position of the double bond in ring D of (+)-8,9-dihydrostepharine. The significant NOE in this case consisted of the mutual enhancement between protons 6a and 12.

As far as optical activity is concerned, a telling observation is that the two dihydro derivatives show *opposite* specific rotations.

Allowing catalytic hydrogenation to proceed further, (+)-stepharine may be fully converted to (+)-tetrahydrostepharine. The aliphatic region of the NMR spectrum for this saturated ketone is considerably more complicated. Nevertheless, spin decoupling of H-6a clearly identified the two protons at C-7. Complete NOE and decoupling studies then led to the following assignments:

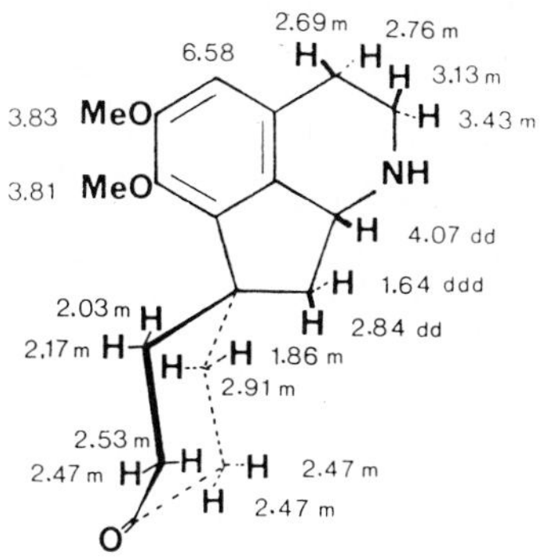

(+)-tetrahydrostepharine

$J_{6a,7\alpha}$ 10.5; $J_{6a,7\beta}$ 6.6; $J_{7\alpha,7\beta}$ 11.7; $J_{7\alpha,12\psi ax}$ 1.2; $J_{8\psi eq,12\psi eq}$ 3.0;

The conformation of ring D, which incorporates the ketonic function, was elucidated by means of two long range W-couplings. The first of these was between the pseudo-equatorial H-8 and H-12, and the second between H-7α and the pseudo-axial H-12. Such an arrangement is possible only when ring D exists in a chair arrangement as indicated.

Since naturally occurring hexahydroproaporphines are known, (+)-tetrahydrostepharine was reduced with sodium borohydride. The resulting mixture of alcohols was separated, and each isomer was fully characterized again by spin decoupling and NOE difference experiments.

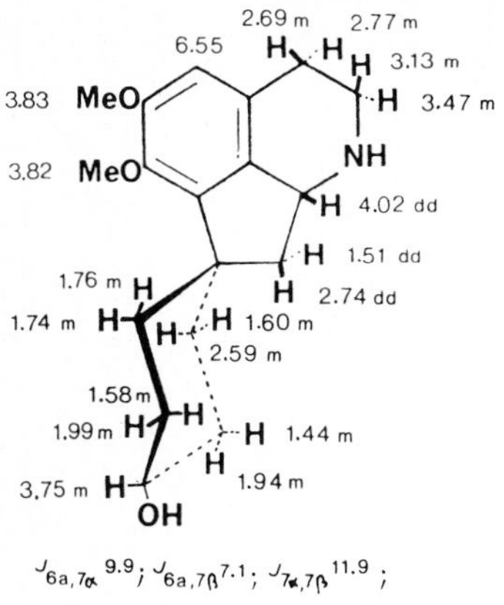

$J_{6a,7\alpha}$ 9.9; $J_{6a,7\beta}$ 7.1; $J_{7\alpha,7\beta}$ 11.9;

(+)-α-hexahydrostepharine

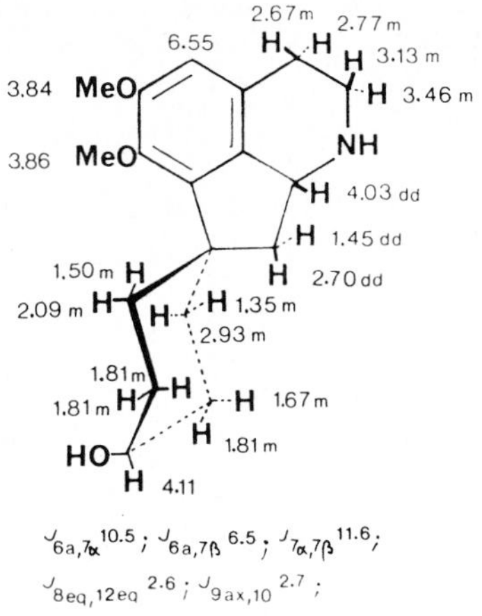

$J_{6a,7\alpha}$ 10.5; $J_{6a,7\beta}$ 6.5; $J_{7\alpha,7\beta}$ 11.6;
$J_{8eq,12eq}$ 2.6; $J_{9ax,10}$ 2.7;

(+)-β-hexahydrostepharine

In the major alcohol, (+)-α-hexahydrostepharine, the proton assignment of ring D was of prime concern. The large coupling constants found in the H-10 multiplet denoted an axial configuration. NOE irradiation of the equatorial H-9 and H-11 showed mutual effect with H-10. Additionally, irradiation of the axial H-8 and H-12 resulted in significant NOE's of H-10. Finally, irradiation of the equatorial H-12 led to enhancement of the H-6a multiplet, thus establishing the configuration around the spiro carbon atom.

Turning now to the minor alcohol (+)-β-hexahydrostepharine, the narrow width of the H-10 signal indicated small coupling constants with the vicinal protons due to its equatorial configuration. Successive decoupling of axial H-8 and H-12 then allowed assignment of the ring D protons. A W-coupling was observed between the planar and equatorial H-8 and H-12.

It should be noted that in the NMR spectra of proaporphine dienones, the difference in chemical shifts, Δδ, between adjacent vinylic protons on ring D depends upon their stereochemical relationship to H-6a. The signals arising from vinylic protons syn to H-6a show a Δδ of about 0.70 ppm, while those that are anti display a Δδ of nearly 0.50 ppm. This observation may prove useful in the case of unknown proaporphine dienones bearing an extra substitutent on ring D, as is the case with (-)-roemeria-linone and (-)-isoroemerialinone.

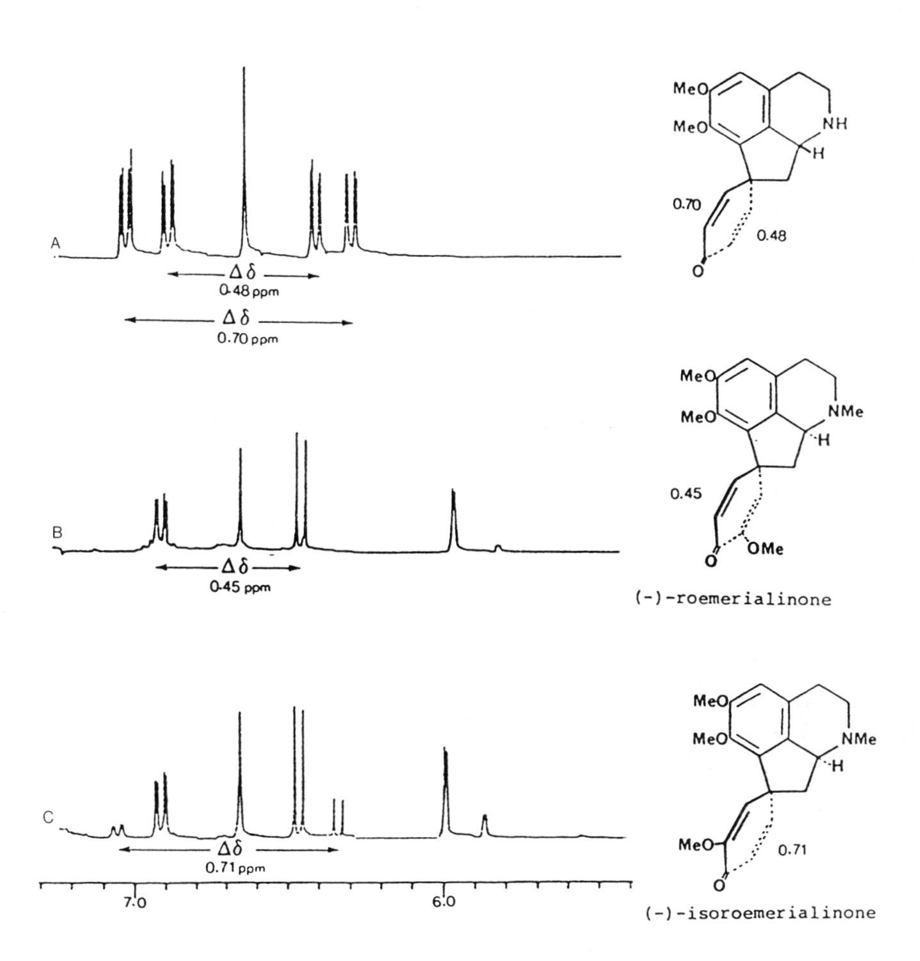

Aromatic region of: (A) (+)-stepharine,
(B) (-)-roemerialinone, and
(C) mixture of (-)-roemerialinone and (-)-isoroemerialinone.

A similar trend can be observed for the vinylic protons of dihydroproaporphines. The Δδ of the vinylic protons is significant and indicative of the relative configuration. When the double bond is anti to H-6a, Δδ is about 0.75 ppm, while in the syn series this value approaches 0.95 ppm.

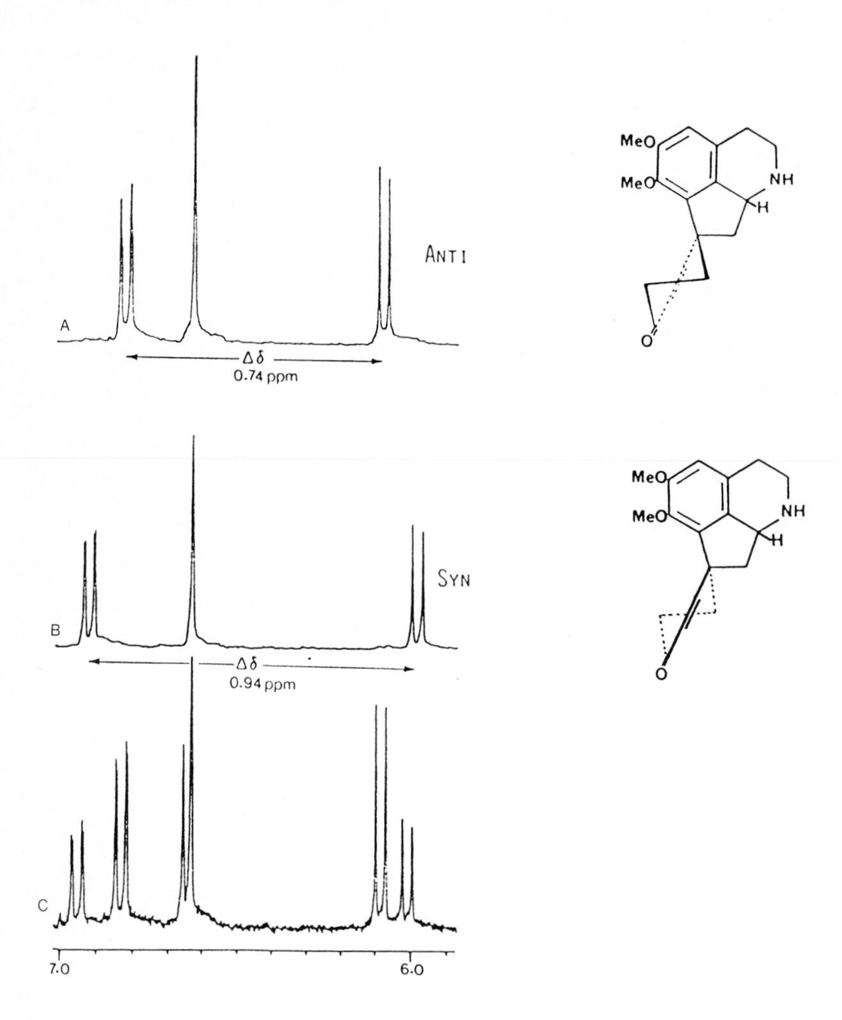

Aromatic region for (A) (-)-11,12-dihydrostepharine,
(B) (+)-8,9-dihydrostepharine, and
(C) mixture of the two diastereomers.

The following scheme thus summarizes the correlations between specific rotations and absolute configurations for the proaporphine dienones and their reduced derivatives.

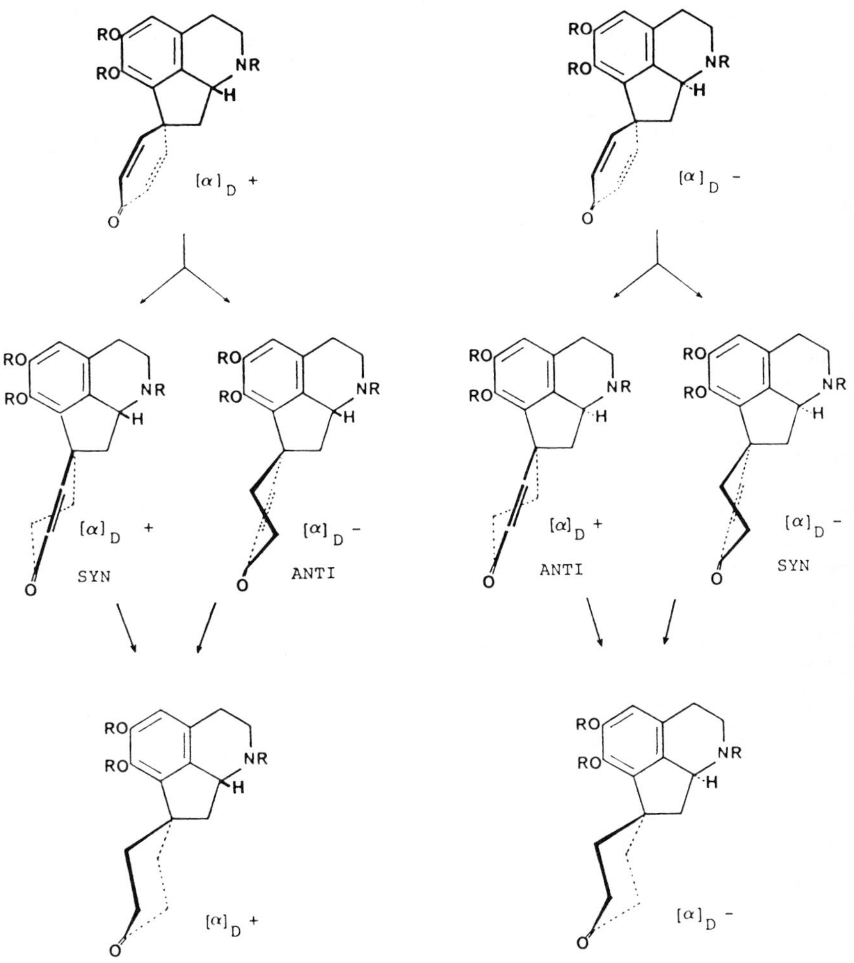

NEW COMPOUNDS FROM THE EUPHORBIACEAE, THE MELIACEAE AND THE HEPATICAE

J.D. Connolly

I wish to discuss our use of modern n.m.r. methods in the structural elucidation of compounds from three families, the Euphorbiaceae, the Meliaceae and the Hepaticae. My principal theme will be the usefulness of 2D long-range δ_C/δ_H correlation to the natural product chemist. This technique is invaluable for the assignment of the ^{13}C resonances of non-protonated carbons and for determination of the position of attachment of esters and may also provide structural connectivity.

The Euphorbiaceae family is well known (ref. 1) as a source of ingol, ingenol, phorbol and related diterpenoids, which often occur as complex mixtures of esters which are difficult to purify. In addition the irritant nature of some of these compounds makes them difficult to handle. It is often relatively straightforward to identify the skeletal type of diterpenoid and the structural problem is reduced to determining the position of attachment of the various esters, usually acetates together with one or more different esters e.g. tiglate, angelate, benzoate, phenylacetate etc. Traditionally the position of the "odd" ester was assigned by partial hydrolysis, a method fraught with danger since intramolecular ester transfer can occur.

Recently we were faced with this type of problem during work (ref. 2) on the constituents of _Euphorbia poisonii_ in collaboration with Dr. C.O. Fakunle, University of Ife, Nigeria. We isolated a 19-hydroxy-ingol derivative (1) which had four acetates and a phenylacetate. Partial hydrolysis was attempted but gave no meaningful result. It seemed to us that the most satisfactory approach was to use a direct n.m.r. method based on the coupling of the skeletal protons and the ester protons to the ester carbonyl carbons [see (2)]. These are of the two-bond and three-bond variety and are small e.g. for theyl acetate $^2J_{CH}$ = 6.8 Hz and $^3J_{CH}$ = 3.2 Hz. Such interactions may be observed directly in a 2D long-range δ_C/δ_H correlation experiment, using appropriate delays. In the case of the 19-hydroxy-ingol derivative (1) the resultant contour plot (ref. 2) shows clearly that the phenylacetate is attached to C-7. Under the conditions of the experiment a complete set of correlations from skeletal protons to the acetate carbonyls was not obtained.

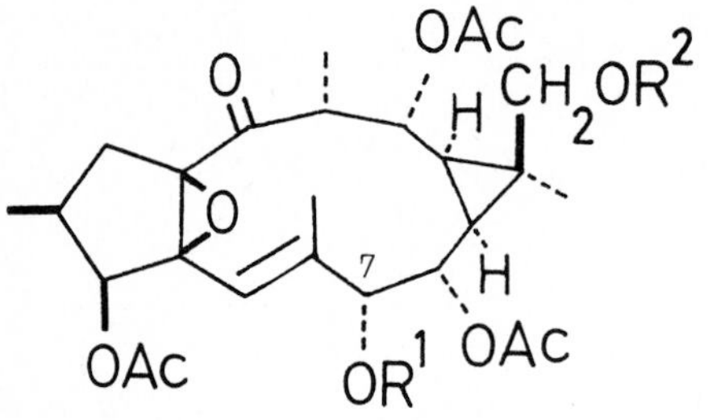

(1) R^1 = phenylacetyl; R^2 = Ac
(3) R^1 = p-hydroxyphenylacetyl; R^2 = Ac
(4) R^1 = R^2 = phenylacetyl

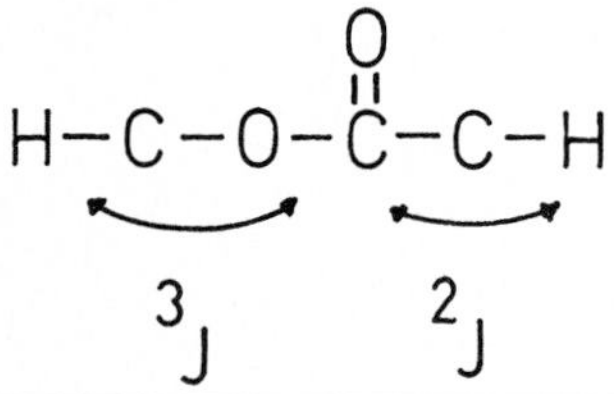

(2)

Two further 19-hydroxy-ingol derivatives, the p-hydroxyphenylacetate (3) and the bis-phenylacetate (4), have since been isolated from E. poisonii. In the case of (3) the attachment of the p-hydroxyphenylacetate to C-7 was confirmed using the above method. In the case of (4), however, no polarisation transfer was observed from skeletal protons to carbonyl carbons, probably because of lack of material. The virtual identity of the ^1H and ^{13}C n.m.r. spectra of (1) and (4), apart from small shifts in 2H-19 [δ_H 3.54, 4.25 in (1), 3.46, 4.19 in (4), both ABq, J 12 Hz] and C-19 [δ_C 64.7 in (1), 64.2 in (4)], suggest that the phenylacetates are attached to C-7 and C-19.

In a continuation of our work on the Euphorbiaceae we have examined E. lateriflora and have isolated a crystalline diterpenoid, enukokurin $C_{34}H_{44}O_{10}$. It has spectroscopic properties (Tables 1 and 2) consistent with the presence of two secondary and two tertiary methyls, two ketonic carbonyls, two hydroxyl groups, one tertiary and one secondary, a trans-disubstituted double bond, an exomethylene and three secondary esters, a benzoate, a propanoate and a 2-methylpropanoate. Thus enukokurin is a derivative of a bicarbocyclic diterpenoid, probably, in the absence of cyclopropyl resonances, a jatrophane. The chemical shift and coupling information revealed the part-structures (5) and (6) which can be readily incorporated, by analogy with kansuinine B (7) (ref. 3), into a jatrophane to give (8) as a plausible working structure. As with kansuinine B (7) H-7 is a singlet and H-8 a doublet

(J 9.6 Hz) coupled to the secondary hydroxyl proton.

TABLE 1.

^{1}H N.m.r. Shifts of Enukokurin

H-1α)	1.8 (m)	H-11	5.96 (d, J 16.1 Hz)
H-1β)	2.3 (m)	H-12	5.67 (dd, J 16.1, 9.6 Hz)
H-2	2.3 (m)	H-13	3.57 (dq, J 9.6, 6.5 Hz)
H-3	5.70 (t, J 3.9 Hz)	15-OH	4.17 (s)
H-4	2.95 (dd, J 10.5, 4.1 Hz)	Me-16	0.98 (d, J 6.3 Hz)
H-5	5.87 (d, J 10.5 Hz)	H-17E	5.75 (s)
H-7	5.64 (s)	H-17Z	6.02 (s)
H-8	4.58 (d, J 9.6 Hz)	Me-18	1.25 (s)
8-OH	3.10 (d, J 9.6 Hz)	Me-19	1.14 (s)
		Me-20	1.38 (d, J 6.5 Hz)
2H-2'''	2.20 (q, J 7.6 Hz)	H-2''	1.53 (septet, J 6.9 Hz)
Me-3'''	0.85 (t, J 7.6 Hz)	Me-3'')	0.52) (each d, J 6.9 Hz)
		Me-4'')	0.78)

TABLE 2.

^{13}C N.m.r. Shifts of Enukokurin.

1	46.4	11	135.5	1'''	173.4	1''	176.3
2	38.5	12	132.8	2'''	27.2	2''	33.2
3	77.2	13	43.5	3'''	8.9	3''	17.5
4	50.1	14	212.1			4''	18.9
5	73.9	15	84.2	1'α	164.7		
6	137.8	16	13.8	1'	129.6		
7	63.9	17	125.4	2', 6'	129.5		
8	71.5	18	23.3	3', 5'	128.2		
9	210.7	19	20.9	4'	132.9		
10	48.1	20					

Following the complete assignment of all the ^{1}H resonances the multiplicities of the ^{13}C resonances and the assignment of the protonated carbons were determined using DEPT and 2D δ_C/δ_H correlation respectively. Evidence for the position of attachment of the three esters was readily obtained from a 2D long-range δ_C/δ_H correlation experiment which also provided strong support for structure (8) for enukokurin. The cross sections through the ester carbonyl carbons are shown in Fig. 1. The propanoate carbonyl at δ_C 173.4 correlates as expected with its methyl and methylene protons and also with H-3 indicating that it is attached to the oxygen at C-3. Similarly the benzoate carbonyl at δ_C 164.6 correlates with its *ortho* protons and with H-5 while the 2-methylpropanoate carbonyl at δ_C 176.3 correlates with its methyl protons and with H-7. Thus the potentially difficult task of assigning the positions of attachment of the three esters was completed unambiguously in a simple manner.

CH_3-CH-CH-CH-CH-C— (OR, OR) (5)

—C-CH(CH_3)-CH=CH-C— (t) (6)

(7)

(8)

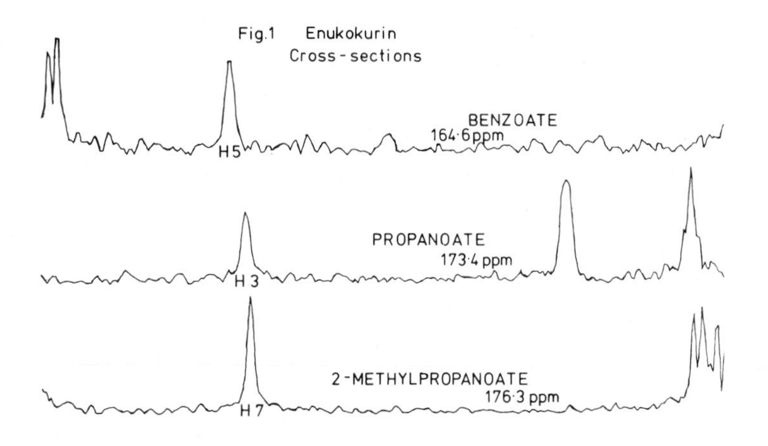

Fig.1 Enukokurin Cross-sections

The information available from the 2D long-range δ_C/δ_H correlation experiment can be used to confirm the proposed structure (8) of enukokurin. For convenience the correlations, over two or three bonds, are shown diagrammatically. Diagram A starts at C-10. The geminal methyl protons correlate with the ketonic carbonyl at δ_C 210.7. Less intense correlations of this carbonyl with 8-OH, H-8 and H-7 continue the sequence. The correlations of 2H-17 with C-7 and C-5 and of H-5 with C-7 complete diagram A. The H-17 at δ_H 5.72 correlates more strongly with C-7 than the other H-17 at δ_H 6.02. The reverse situation pertains for C-5. Thus H-17Z, syn to C-7, resonates at δ_H 6.02 and H-17E, anti to C-7, at δ_H 5.75. The same conclusions were reached on the basis of NOE difference experiments (see later).

Diagram B also starts at C-10. The geminal methyl protons correlate with C-11 while the secondary methyl protons 3H-20 correlate with C-12. The additional correlation of 3H-20 to the second ketonic carbonyl carbon (δ_C 212.1) permits extension of the chain. The ring junction and the beginnings of the five-membered ring are evident from the further correlations of this carbonyl (C-14) with 2H-1, H-4 and the 15-OH. The correlations of H-3 with C-15, 2H-1 with C-2 and 3H-16 with C-1, taken in conjunction with the vicinal couplings $J_{H-2,H-3}$ and $J_{H-3,H-4}$ complete the cyclopentane. Since H-4 has a vicinal coupling to H-5 diagrams A and B can be combined, incorporating the information about ester attachment, to give structure (8) for enukokurin (without stereochemistry).

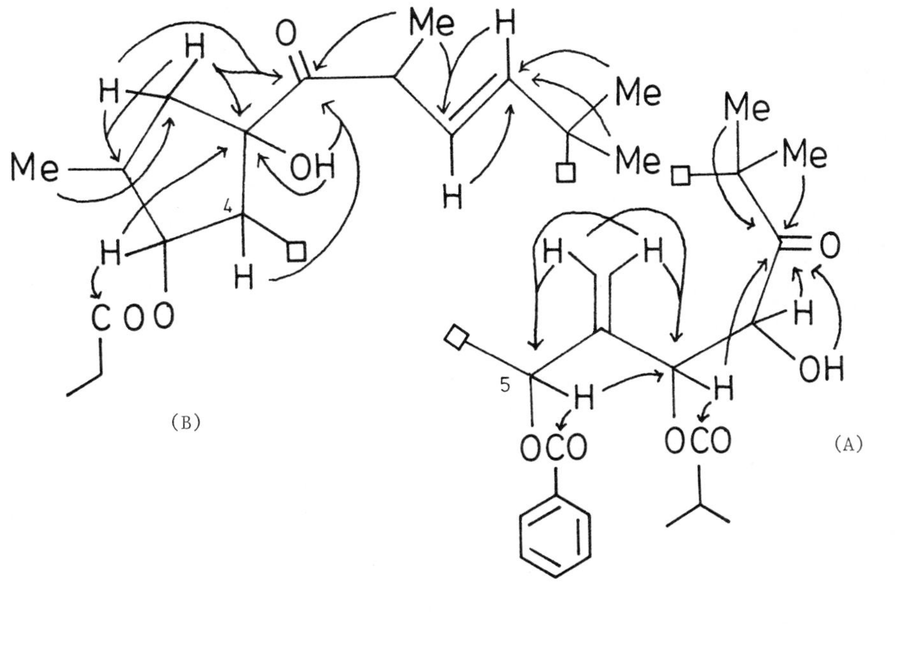

(B)

(A)

The relative stereochemistry of enukokurin as in (8) was determined largely by NOE difference spectroscopy. The experimental NOEs are listed in Table 3. The magnitude of the vicinal couplings along the bottom half of the

TABLE 3.

NOE Difference Results for Enukokurin.

Irradiated	Observed	Irradiated	Observed
Me-16	H-3 (2)	15-OH	H-5 (3)
	H-1 (2)		H-17E (2)
	H-1, H-2 (2)		
		H-13	H-11 (4)
Me-19	H-12 (11)		H-4 (4)
	H-8 (5)		Me-20 (1.4)
	H-11 (2)		H-1, H-2 (2)
	8-OH (0.6)		
		H-8	H-17Z (2)
Me-18	H-11 (6)		H-17E (-0.3)
	H-12 (3)		8-OH (1.7)
			H-7 (1.8)
Me-20	H-12 (3)		H-12 (1.3)
	H-13 (3)		
	H-1 (2)	H-4	H-3 (5)
			H-7 (5)
8-OH	H-8 (1.5)		H-2 (1)

Magnitudes (%) are shown in parentheses.

molecule is similar to kansuinine B (7) (ref. 3) and is consistent with the same stereochemical arrangement. Significant NOEs from the tertiary hydroxyl proton 15-OH to H-5 and H-17E confirm the *trans*-ring junction, the *anti*-relationship of H-4 and H-5 and the assignment of the exomethylene protons discussed above. The large NOE from H-4 to H-7 indicates that the C-7 oxygen function is β. As expected H-4 also has a large NOE to H-3 and a smaller effect to H-2. The β-configuration of H-8 follows from its zero coupling to H-7 and from several NOEs, in particular one from H-17Z. It is clear that the exomethylene must project above the plane of the molecule. A distinction between 3H-18 and 3H-19 is now possible. Irradiation of the methyl signal at δ_H 1.14 resulted in large NOEs at H-8 and H-12. These results identify 3H-19 and also reveal the orientation of the 11,12-double bond. Conversely irradiation of the other t-methyl group (δ_H 1.25, 3H-18) afforded a large NOE at H-11 and a much smaller effect at H-12. The β-configuration of the secondary methyl group follows from its large NOE with H-12. The evidence for the relative configurations of enukokurin and the approximate conformation of the medium ring was completed by irradiation of H-13 which showed *inter alia* NOEs at H-11 and H-4.

Other diterpenoids isolated from E. lateriflora include the 4,20-dideoxy-5-hydroxyphorbol derivative (9) whose ester attachment was determined by 2D long-range δ_C/δ_H correlation. The corresponding 4-deoxyphorbol derivative (10) was isolated as a mixture of esters (benzoate and 2-methylpropanoate) at 12 and 13.

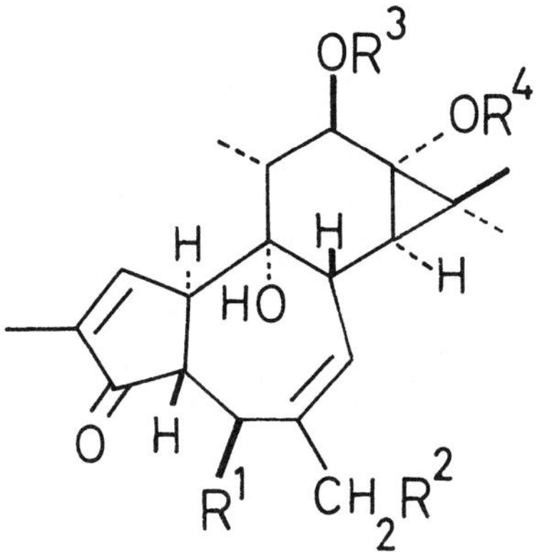

(9) R^1 = OH; R^2 = H; R^3 = benzoyl;
R^4 = 2-methylpropanoyl

(10) R^1 = H; R^2 = OH;
R^3, R^4 = {benzoyl, 2-methylpropanoyl}

I shall now turn to the Meliaceae, a family which has interested me for most of my scientific career. The principal constituents of this family are tetranortriterpenoids (limonoids) (ref. 4). In collaboration with Dr. K.K. Purushothaman and his colleagues at the Captain Srinivasa Murthi Research Institute for Ayurveda in Madras we have recently isolated two new tetranortriterpenoids, trijugins A and B, from Heynea trijuga. Both compounds appeared to be related to methyl angolensate (11) (ref. 4). It is appropriate at this juncture to point out that in methyl angolensate the exomethylene protons 2H-30 are sharp singlets, C-9 correlates with 3H-19, H-12 and 2H-30 (all $^3J_{CH}$) and there is a strong NOE (5%) from {3H-19} to H-9. Trijugin A, $C_{29}H_{34}O_{14}$, has exomethylene protons [δ_H 5.13 (d, J 1.5 Hz) and 5.50 (d, J 1.4 Hz)] which are coupled to an allylic proton H_A [δ_H 3.06 (dt, J 4.2, 1.4 Hz)] which is in turn coupled to a secondary acetate proton [δ_H 5.84 (d, J 4.2 Hz)]. The acetate methyl protons resonate at unusually high field (δ_H 1.75) and must be shielded, presumably by the furan ring. The remaining feature of note is a hemiketal carbon at δ_C 108.4 (s). The other terminus of the hemiketal appears to be attached to C-6 since the AX system arising from H-5 and H-6 [δ_H 3.24 and 4.81 (both d, J 8.3 Hz)] is clearly visible. Rings A and D, the β-substituted furan and the 1,14-ether give the expected spectroscopic properties. At first sight, therefore, trijugin A and the related ketone trijugin B appear to have structures (12) and (13) respectively. However there are some difficulties. If H_A is H-9 its 4.2 Hz coupling with H-12 is a ^4J coupling and the required W geometry requires the acetate to be β, which is not ideal for shielding by the furan. Moreover4.2 Hz is rather large for a ^{4}J coupling of this type. The fact that structures (12) and (13) are untenable was convincingly proved by irradiation of H_A. No NOE was observed with 3H-19. Thus H_A cannot be H-9 in a methyl angolensate structure.

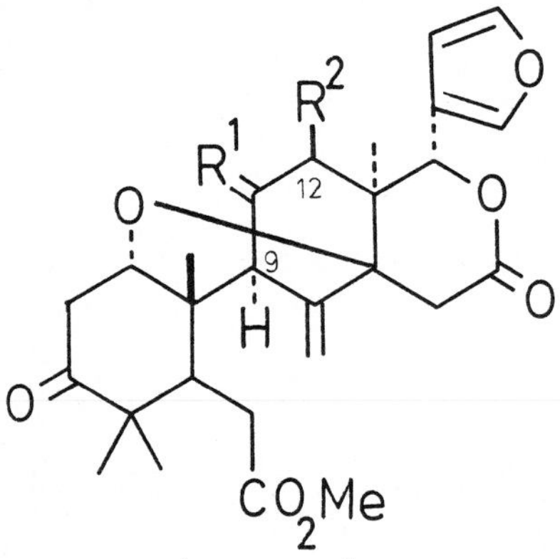

(11) R^1 = H,H; R^2 = H
(13) R^1 = O; R^2 = OH

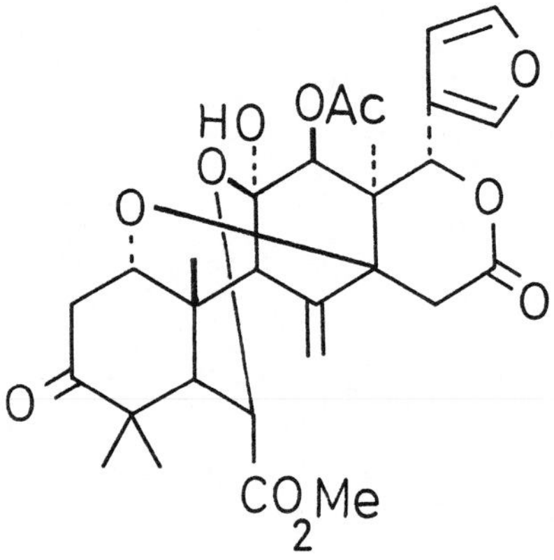

(12)

The definitive structure (14) of trijugin A can be assembled in a logical manner as follows. H-17, identified by its coupling to the furan protons, has a large NOE (8%) to the acetate methine which is therefore H-12. Since this proton is β the acetate must be α, in accord with its shielded nature. H-12 is coupled to H_A i.e. H-11. This coupling must be vicinal since the alternative ^{4}J coupling is inconsistent with the NOEs between H-11 and H-12 (3%) and H-12 and H-17. The allylic coupling of H-11 to the exomethylene protons completes ring C which is therefore cyclopentanoid. Since H-11 is no longer coplanar with the exomethylene protons (cf. methyl angolensate) coupling is observed. Irradiation of 3H-18 results in a 6% NOE at H-11.

The secondary terminus of the hemiketal is at C-6. H-6 and H-5 form an AX spin system and both have large NOEs (6%) with 3H-28. The large NOE (13%) from H-5 to H-6 indicates that their coupling (J 10.5 Hz) arises from a syn rather than an anti relationship and that the configuration at C-6 is R as in other tetranortriterpenoids (ref. 4). The hemiketal carbon must be C-9. The 7% NOE of the hemiketal hydroxyl proton from 3H-19 supports this suggestion and shows that the configuration at C-9 is S. The hemiketal carbon shows 2D long-range δ_C/δ_H correlations with H-6 (^3J via oxygen), H-11 (^2J), H-12 (^3J) and 3H-19 (^3J) which are consistent with its assignment as C-9. These are shown in cross-section (Fig. 2). Prior identification of the methyl groups of trijugin A was achieved using a small J COSY experiment which revealed the ^4J couplings of 3H-28 and 3H-29 and of H-17 and 3H-18.

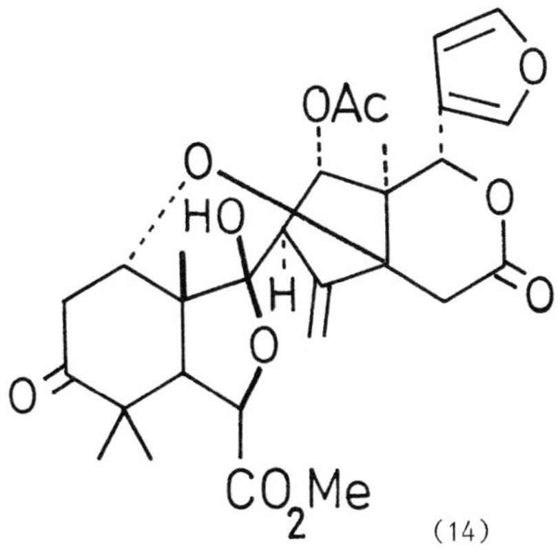

(14)

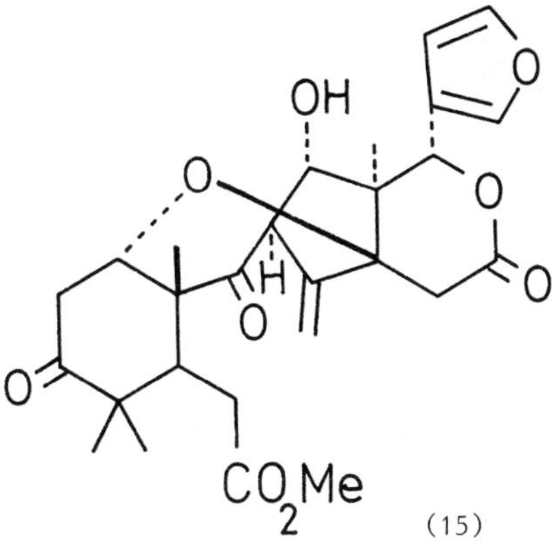

(15)

The structure of trijugin B requires similar modification to (15). Again we observe NOEs from H-17 to H-12 (6%) and 3H-18 to H-11 (4%) and long-range correlations of H-11, H-12 and 3H-19 with the C-9 ketonic carbonyl group.

Trijugins A and B can be regarded as methyl angolensate derivatives but with a modified ring C which may arise by a pinacol rearrangement of an appropriate precursor. Spectroscopically the distinction between the contracted and the normal ring C may be rather subtle and perhaps more careful examination than is usual should be given to compounds of this type. I shall illustrate this point with reference to two tetranortriterpenoids, ekebergolactones A and B, from Ekebergia senegalensis. These compounds were isolated many years ago by the Ibadan group of Bevan, Ekong, and Taylor (ref. 5). We are grateful to Professor D.A.H. Taylor for providing us with material for study.

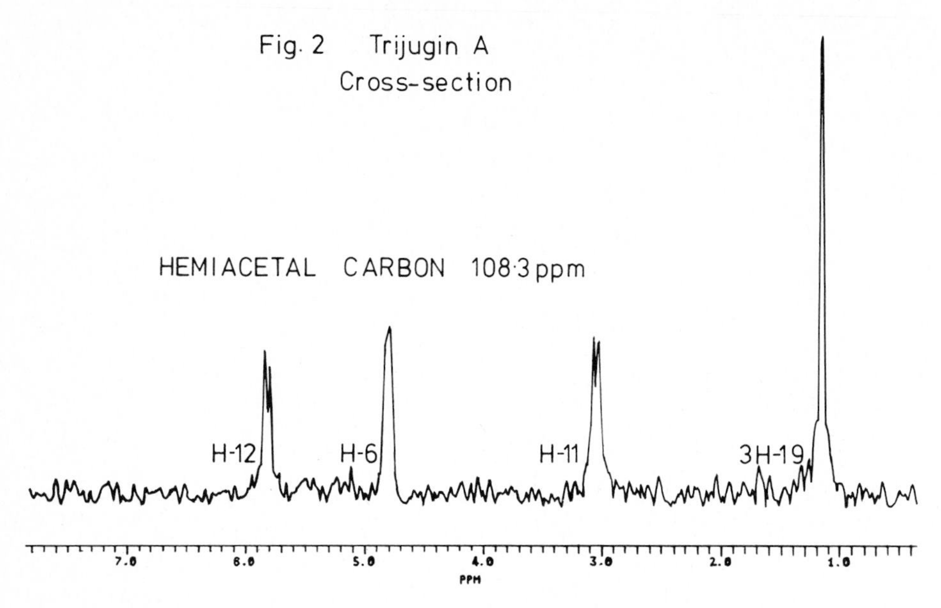

Ekebergolactones A, $C_{40}H_{50}O_{17}$, and B, $C_{37}H_{46}O_{17}$, differ only in the esters attached to the tetranortriterpenoid skeleton. The former has two acetates, a 2-methylpropanoate and a 2-methylbutanoate whereas the latter has three acetates and a 2-methylpropanoate. The presence of four tertiary methyl groups, a carbomethoxyl, an exomethylene concealed as an epoxide and a secondary-tertiary cyclic ether suggested a methyl angolensate structure for the ekebergolactones. The functionality can be satisfactorily accommodated in structure (16). There are several unusual features. This is the first example of an epoxidised exomethylene group in this series. The tertiary hydroxyl group is also uncommon. The stereochemistry of ring A follows from the coupling constants which are all small. The outstanding problems include the position of attachment of the various esters and the configurations at C-8, C-9, C-12 and C-15.

As usual the 2D long-range δ_C/δ_H correlation experiment provided valuable information. Some of the long-range correlations of ekebergolactone B are listed in Table 4. It is immediately clear that the 2-methylpropanoate is attached to C-2. The correlations of the carbomethoxyl carbonyl with H-5 and 2H-6 confirm the acetic acid side-chain. It is also possible to identify unambiguously all the acetate methyls and carbonyls and their position of attachment. The necessity for doing this will become apparent. The only non-protonated carbons which cannot be assigned in this experiment are those associated with the tertiary ether terminus and the tertiary hydroxyl which are too close in chemical shift. The vitally significant correlations involve the ketonic carbonyl and the tertiary epoxide carbon C-8. The ketone has correlations with the tertiary hydroxyl proton, H-12 and, in particular, 3H-19.

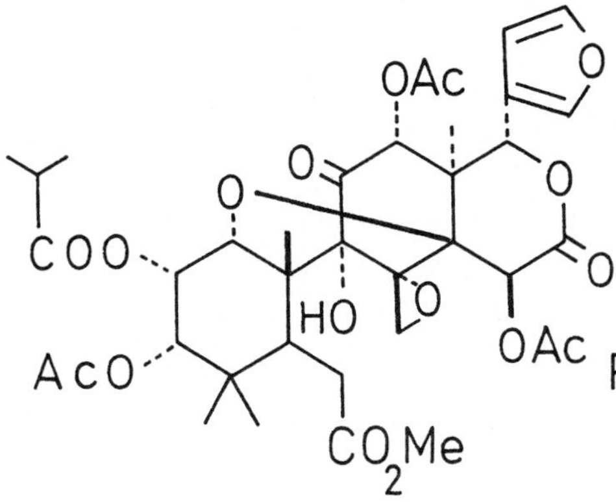

(16)

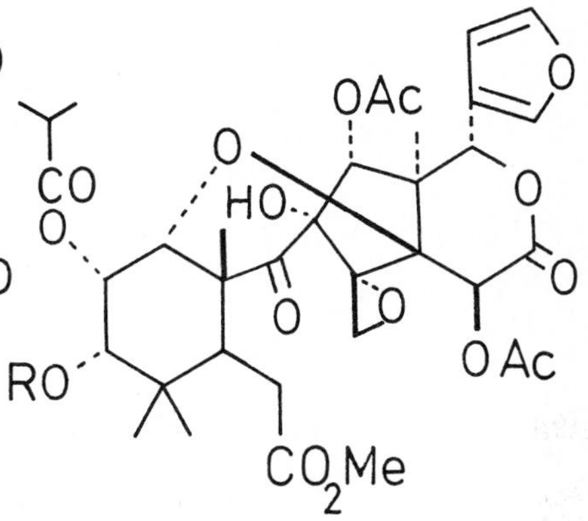

(17) R = Ac

(18) R = 2-methylbutanoyl

TABLE 4.

Some Long-range δ_C/δ_H Correlations of Ekebergolactone B.

δ_C	C	Long-range Correlations
207.0	>C=O	OH, H-12, 3H-19
176.8	C-1'	H-2, H-2', 2 x 3H-3'
173.3	C-7	H-6a, H-6b, H-5, OMe
170.6	3Ac	H-3, middle Ac
169.6	12Ac	H-12, highfield Ac
168.3	15Ac	H-15, lowfield Ac
166.0	C-16	H-15
59.4	C-8	OH, H-12

On the basis of structure (16) this last correlation involves four bonds ($^4J_{CH}$). Our experience with the Heynea compounds and methyl angolensate suggests that it is the $^3J_{CH}$ correlation with 3H-19 which is important. The inevitable conclusion is that ekebergolactone B must have the cyclopentanoid ring C structure (17). The correlation of C-8 with H-12 leads to the same conclusion. Thus although we considered the methyl angolensate type of structure (16) as correct for ekebergolactone B for several years and indeed established the relative stereochemistry on the basis of it we were forced at a late stage to revise our views in favour of the modified structure (17).

Some of the stereochemical points were readily solved by classical NOE experiments. Thus H-15 has a large NOE with 3H-18, indicating that the C-15 acetate is β. Similarly the C-12 acetate is α since H-12 and H-17 have a significant NOE. The third obvious NOE is between H-5 and H-12 and is diagnostic of a methyl angolensate type of structure since formation of the 1,14-ether brings H-5 and H-12 close together.

The 2D NOESY spectrum of ekebergolactone B confirmed the above NOEs and also suggested two interesting results, NOEs between H-17 and the C-3 acetate methyl and between the C-15 acetate methyl and the 2-methylpropanoate, which encouraged us to irradiate the various acetate methyls during NOE difference experiments. This accounts for the necessity to distinguish the acetate methyls in the 2D long-range δ_C/δ_H correlation experiment.

A detailed NOE difference study of ekebergolactone B was carried out. For molecules of intermediate molecular weight NOE difference is a more sensitive and reliable technique than 2D NOESY and provides valuable stereochemical and conformational information. A few of the results are listed in Table 5. Irradiation of H-30a gives positive NOEs to H-15, H-1 and H-30b and a negative NOE to the tertiary hydroxyl proton. The configuration of the

TABLE 5.

Some NOE difference Results for Ekebergolactone B.

{H-30a}	H-15 (1.4), H-1 (5.6), H-30b (19.9), OH (-1.3)
{H-17 }	H-21 (3.9), H-12 (7.4), H-5 (1.7), 3Ac (1.6)
{3Ac}	H-17 (6.0)
{15Ac}	H-2' (5.9)

epoxide as shown is defined by the NOE from H-30a to H-1. The NOE from H-17 to the C-3 acetate methyl, suggested by the NOESY spectrum, was confirmed. In the reverse experiment irradiation of the C-3 acetate methyl gave a substantial NOE to H-17 (6%). The significance of this result is that we now have a simple method for determining the nature of the ester attached to C-3 in ekebergolactone A. Similarly we can easily establish which ester is attached to C-2 in A since irradiation of the C-15 acetate methyl of ekebergolactone B gave a significant NOE at H-2' of the 2-methylpropanoate. Since ekebergolactone A was available only in minor amount it was important to have a ready method for solving the problem of the ester attachments.

In the event these two NOE difference experiments establish the structure of ekebergolactone A as (18). Irradiation of H-17 affected H-2 of the 2-methylbutanoate which is therefore attached to C-3. Irradiation of the C-15 acetate methyl (same chemical shift as in B) gave a NOE at H-2 of the 2-methylpropanoate which is thus at C-2.

Treatment of ekebergolactone B with zinc in acetic acid afforded iso-ekebergolactone B in good yield. A detailed spectroscopic examination of this compound led us to the conclusion that the functionality and relative stereochemistry have not changed but that the compound has undergone an α-ketol rearrangement to the normal methyl angolensate type of structure (16) which was, of course, originally proposed for ekebergolactone B. Because of the presence of the tertiary hydroxyl group the distinction between the five and six membered rings is not easy. Three NOE difference results support the proposed structure - (a) H-30a and H-15 are closer in isoekebergolactone B [3% in (16) and 1% in (17)]. (b) H-30a is closer to H-1 in ekebergolactone (6% in (17) and 1% in (16)]. (c) the magnitude of the NOE between H-5 and H-17 is 2% in ekebergolactone B (17) but - 1% in the iso-compound (16) as a result of the angle change on going from a cyclopentanoid to a cyclohexanoid ring C.

In the final section I wish to turn briefly to the chemistry of liverworts, the Hepaticae family. A fascinating variety of terpenoid and aromatic compounds has been published in recent years (refs. 6-7). I shall restrict my comments to a new sesquiterpenoid alcohol, conocephalenol, which we isolated from _Conocephalum conicum_. Conocephalenol, $C_{15}H_{26}O$, is a tertiary alcohol (δ_C 74.0) and has a tetrasubstituted double bond (δ_C 135.9 and 132.7). It is therefore bicarbocyclic. The most intriguing feature of the ^1H n.m.r. spectrum of conocephalenol is the presence of five methyl groups, four tertiary [δ_H 0.90, 1.00 and 1.26 (6H)] and one secondary [δ_H 1.01 (d, J 7.3 Hz)] [cf. alliacolide (19) (ref. 8)]. It is difficult to establish any connectivity from the ^1H n.m.r. spectrum since, even at 360 MHz, there is considerable overlap of resonances and second order character. Since we were fortunate to have a reasonable amount of conocephalenol we were able to solve the problem using Freeman's 2D INADEQUATE experiment (ref. 9). The spectrum (Fig. 3) revealed the carbon skeleton (20) for conocephalenol. This unusual carbon skeleton has been reported for brasilenol (21) (ref. 10), a marine natural product, which has recently been synthesised (ref. 11). Work continues to establish the relative and absolute configurations of conocephalenol.

Acknowledgements: It has been a pleasure to work with Dr. C.O. Fakunle (Ife, Nigeria), Dr. K.K. Purushothaman (Madras, India) and Dr. L.J. Harrison, Dr. J. Singh and Dr. D.S. Rycroft (Glasgow). I am deeply grateful to them for their collaboration and friendship.

(19)

(20)

(21)

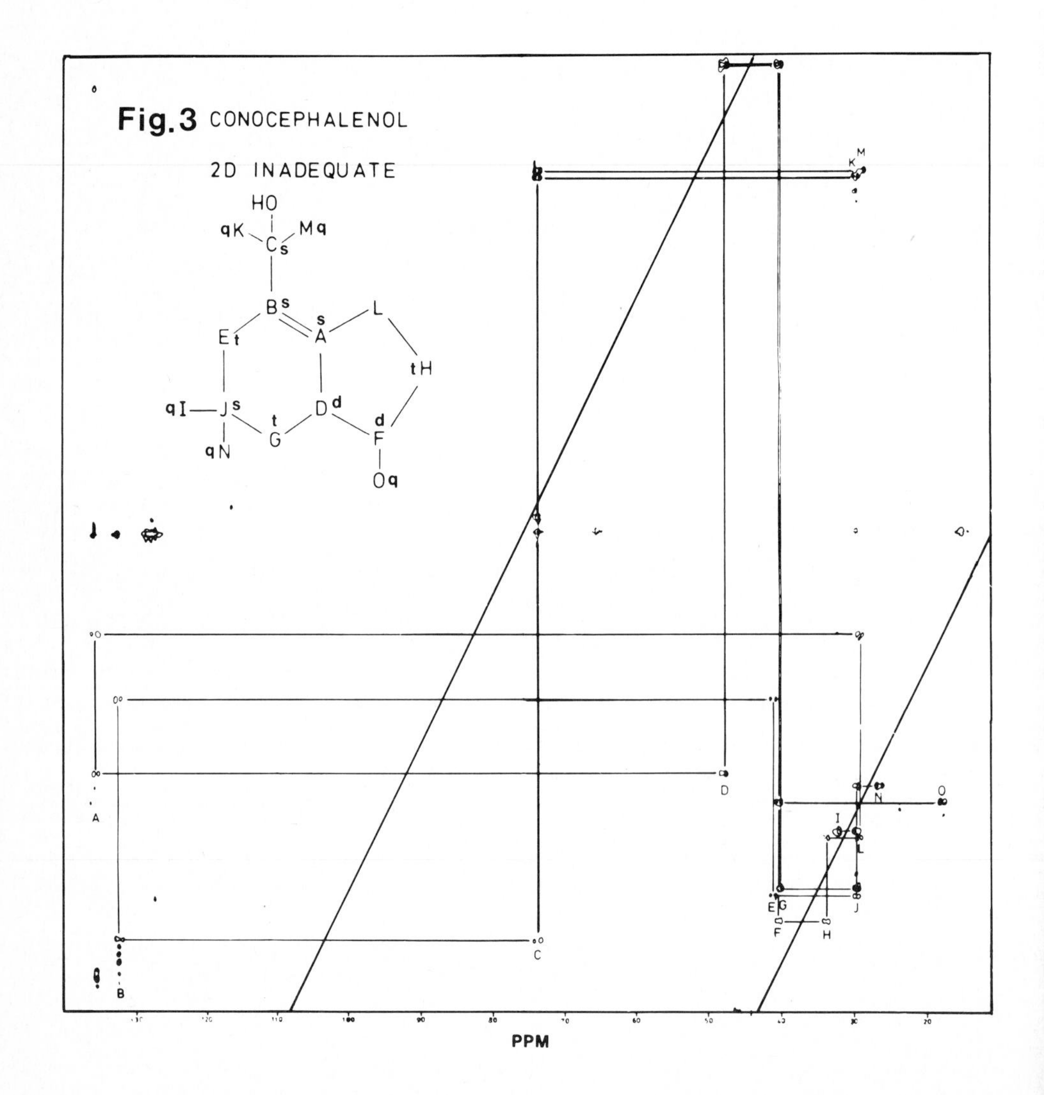

REFERENCES

1. F.J. Evans and S.E. Taylor, Prog. Chem. Org. Nat. Prod., 1983, 44, 1.
2. J.D. Connolly, C.O. Fakunle and D.S. Rycroft, J. Chem. Research (S), 1984, 368.
3. D. Uemura, C. Katayama, E. Uno, K. Sasaki, Y. Hirata, Y.-P. Chen and H.-Y. Hsu, Tetrahedron Letters, 1975, 1703.
4. D.A.H. Taylor, Prog. Chem. Org. Nat. Prod., 1984, 45, 1.
5. C.W.L. Bevan, D.E.U. Ekong and D.A.H. Taylor, Nature, 1965, 206, 1323.
6. Y. Asakawa, Prog. Chem. Org. Nat. Prod., 1982, 42, 1.
7. S. Huneck in New Manual of Bryology, R.M. Shuster (Ed.), Hattori Botanical Laboratory, Nichinan, Japan, 1983, Vol. 1, p.1.
8. A.G. Avent, J.R. Hanson and B.L. Yeoh, J. Chem. Research (S), 1986, 422.
9. T.H. Mareci and R. Freeman, J. Magn. Reson., 1982, 48, 158.
10. M.O. Stallard, W. Fenical and J.S. Kittredge, Tetrahedron, 1978, 34, 2077.
11. A.E. Greene, F. Coelho, E.J. Barriero and P.R.R. Costa, J. Org. Chem., 1986, 51, 4250.

BIOLOGICALLY ACTIVE SUBSTANCES FOUND IN HEPATICAE

Yoshinori Asakawa

1. INTRODUCTION

The bryophytes are taxonomically placed between algae and pteridophytes, and there are about 24000 species in the world. They are further divided into three classes, Musci (mosses), Hepaticae (liverworts) and Anthocerotae (hornworts). The Hepaticae contain oil bodies which are easily extracted with organic solvents whilst the other two classes contain no oil bodies. Up to the last 15 years, the phytochemistry of the bryophytes has been neglected. Generally, the bryophytes have been considered to be a small plant group almost useless for human diet. However, many bryophytes have been used as medicinal plants in China, Europe and North America. A large thalloid liverwort *Marchantia polymorpha* has been used as diuretic, antihepatic and antidotal drugs (ref. 1, 2). A concentrated decoction of the dried liverwort *Conocephalum conicum* has been used against disease of gallstones (ref. 1). A beautiful moss *Rodobryum giganteum* is useful as antipyretic, diuretic and for treatment of the disease of neurasthenia, psychosis and blood vessel of heart (ref. 3, 4). An epiphytic liverwort *Frullania tamarisci* subsp. *obscura* has been used as antiseptic for eyes (ref. 2). *Bryum argenteum* possesses antipyretic and antirhintic activities (ref. 2). Generally, the bryophytes are not damaged by fungi, insects and earthworms. Some bryophytes contain intensely pungent, bitter or saccharine-like substances (ref. 5). Many liverworts emit characteristic fragrant odor (ref. 6). Furthermore, some liverworts show allergenic contact dermatitis, plant growth regulatory, antimicrobial and antifeedant activities (ref. 7).

Although the bryophytes contain chemically and pharmacologically interesting substances, the chemical study has not been carried out fully until recent decade. Among the chemical constituents, we have been essentially interested in the bioactive substances present in the bryophytes. Up to the present time, we have studied about 600 species of the Hepaticae from the view point of phytochemistry, pharmacology and their potential application as a source of cosmetics and medicinal or agricultural drugs.

Most of the liverworts mainly contain terpenoids and lipophilic aromatic compounds which constitute the oil bodies (ref. 7-9). It has been clarified that the biological activities of the liverworts are due to these substances.

The purpose of this paper is to describe the isolation and chemical structures of biologically active substances obtained from the Hepaticae which have been collected in Europe, Asia, Oceania and North America (ref. 10-11).

2. Biological activity

2.1. Fragrant odor

Some liverworts emit volatile terpenoids or simple aromatic compounds which are responsible for intense turpentine, mushroomy, sweet-woody, sweet-mossy, fungal-or carrot-like odor. In order to apply these substances to cosmetics, we investigated the chemical structures of the flavor of some liverworts. *Conocephalum conicum* emits sweet mushroomy odor. (+)-Bornyl acetate (1) and (-)-sabinene (2) were isolated from the ether extract as the major components, together with 1-octen-3-ol and its acetate (ref. 11, 13). The fragrance of this liverwort is due to the mixture of these simple compounds. A very small thalloid liverwort *Targionia hypophylla* emits intense fragrance when it is crushed. Careful chromatography of the methanol extract on silica gel-Lobar column gave *cis*- and *trans*-pinocarveyl acetates (3,4) which comprise the characteristic fragrance of *T. hypophylla* (ref. 14). The fresh *Trichocolea tomentella* emits slightly intense sulfur-like odor. Surprisingly, this liverwort accumulates sulfur. Two sesquiterpenes, tamariscol (5) and bicyclohumulenone (6) were isolated from *Frullania tamarisci* subsp. *tamarisci* and *Plagiochila acanthophylla* subsp. *japonica*, respectively (ref. 15, 16). Both of them emit mysterious sweet-mossy odor and they have been highly estimated in some European, American and Japanese perfumary companies as excellent sources of perfume.

2.2. Hot and bitter taste

Some genera of the Hepaticae produce intensely pungent or bitter substances which show interesting biological activity described later. Polygodial (7) isolated from *Porella vernicosa* complex: *P. arboris-vitae, P. fauriei, P. gracillima, P. obtusata* subsp. *macroloba, P. roerii* and *P. vernicosa* show intense pungency (ref. 5,7). The ether extracts of these species contain *ca.* 10-30% of polygodial. Four sacculatane-type diterpenoids, sacculatal (8), 19-hydroxysacculatal (9), 18-hydroxysacculatal (10) and isosacculatal (11) whose structures have an additional isoprene unit at C-15 of polygodial (7), were isolated from *Trichocoleopsis sacculata* (ref. 7). The three diterpene dials (8-10) also show persistent hot taste. The same pungent compound (8) and the two related sacculatanes (11, 12) were also isolated from *Pellia endiviifolia* (ref. 7). Isosacculatal (11) dramatically lacks hot taste. *Chiloscyphus polyanthos* elaborates intensely pungent substances. Sephadex LH-20 chromatography of the ether extract resulted in the isolation of two hot taste eudesmane-type sesquiterpene lactones, *ent*-diplophyllolide (13) and *ent*-7α-hydroxydiplophyllolide (14) as the major components (ref. 7). From the crude extract of *Diplophyllum albicans*, diplophyllin (15) was isolated as the major compo-

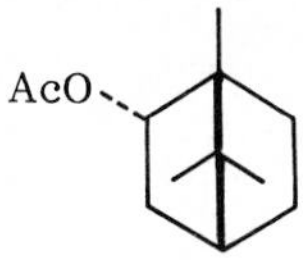

AcO OAc OAc H

HO H

(1) 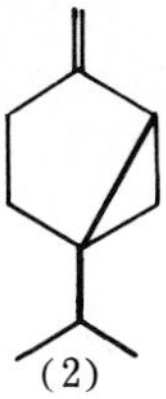 (2) 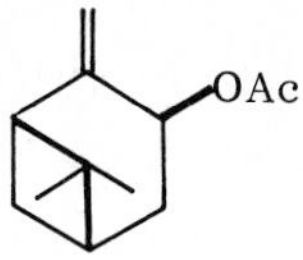 (3) 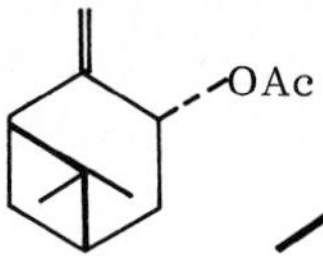 (4) 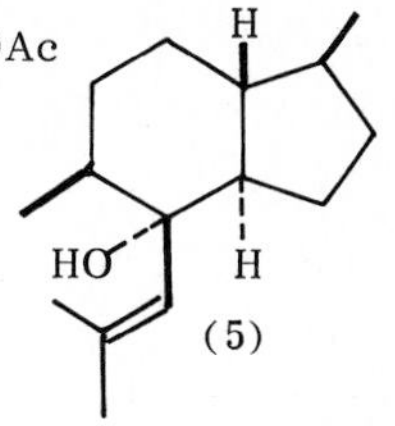 (5)

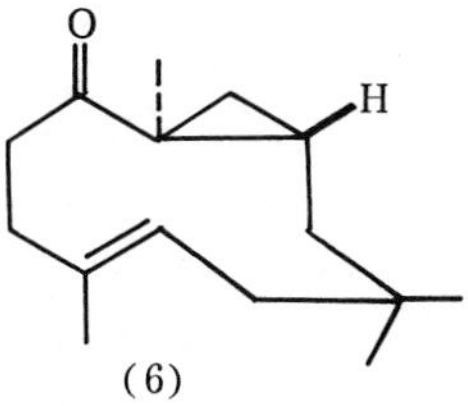

O CHO CHO CHO

H CHO CHO CHO

(6) 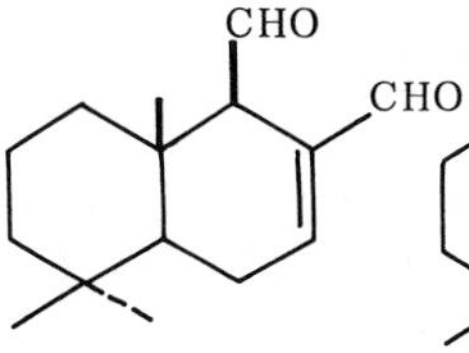 (7) 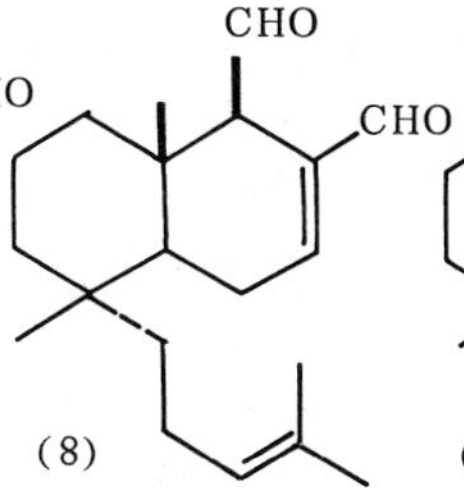 (8) 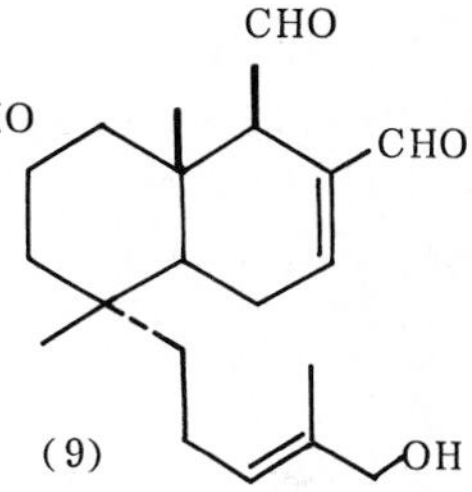 (9) OH

CHO CHO O

CHO CHO O

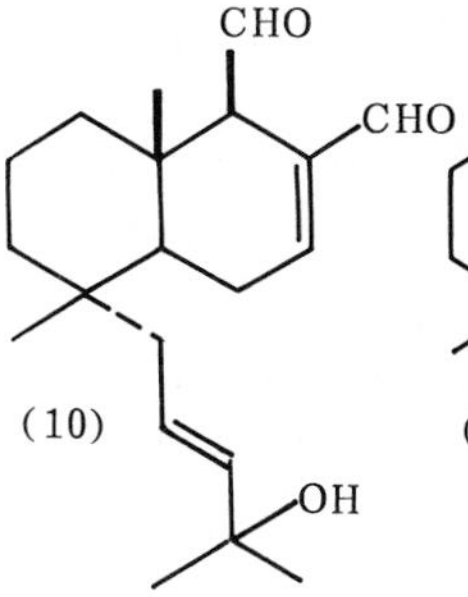

(10) 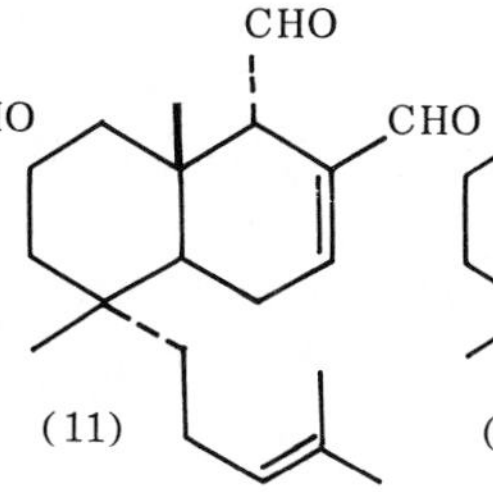 (11) 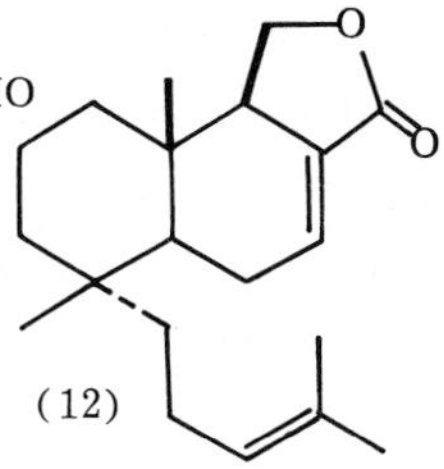 (12)

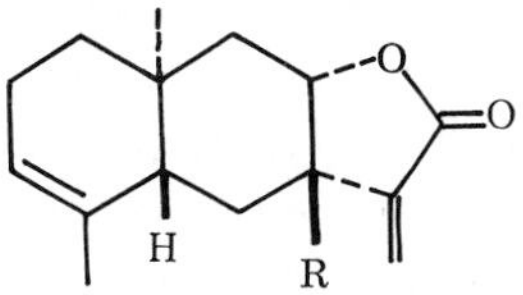

OH

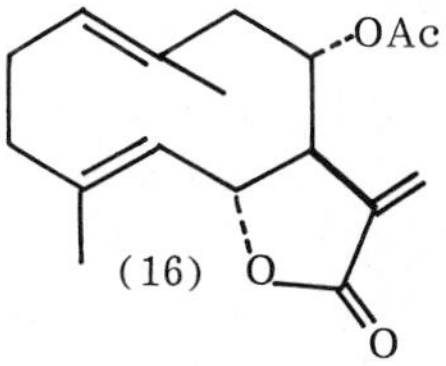

O O OAc

O O

H R (15) (16) O

O

(13) R=H

(14) R=OH

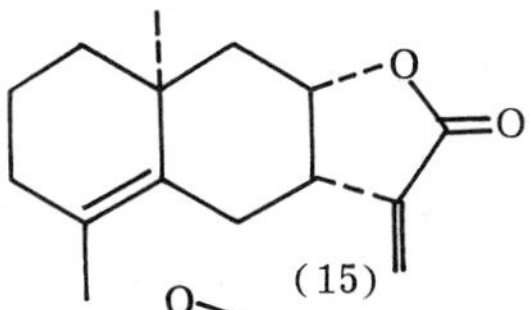

O

AcO H O OHC CHO

O H OAc

R_2

H H

R_1O

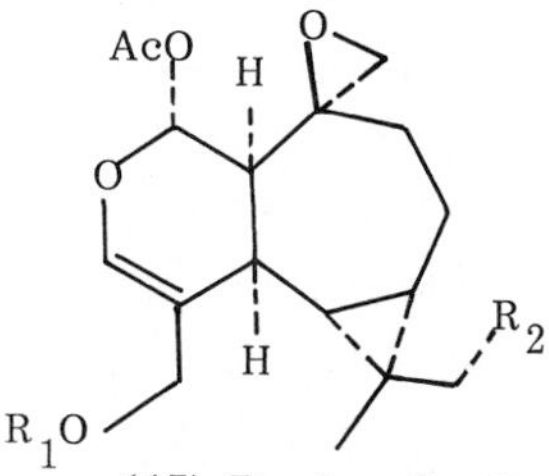

(17) R_1=Ac, R_2=H

(18) R_1=R_2=H

(19) R_1=Ac, R_2=OAc

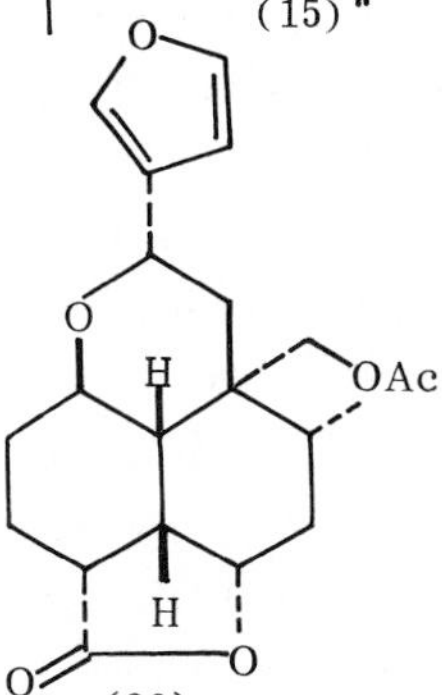

O O

(20) 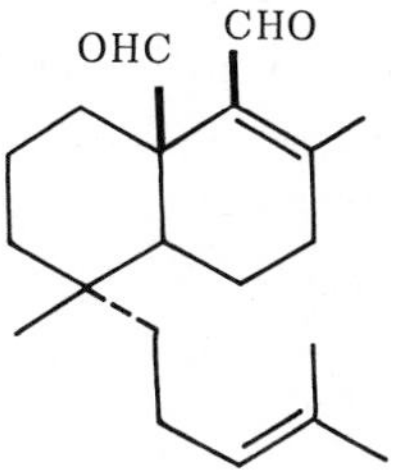 (21)

nent, together with the pungent diplophyllolide (13) (ref. 7). The double bond isomer (15) of 13 is almost tasteless. *Wiesnerella denudata* biosynthesizes a hot taste substance. Pungency is due to tulipinolide (16) which has been isolated from the ether extract as the major component (ref. 7). This is the first record of sesquiterpene lactones possessing powerful hot taste, although more than 500 sesquiterpene lactones have been found in plant kingdom so far.

There are about 3000 species of *Plagiochila* species most of which contain surprisingly intense pungent substances. The two pungent sesquiterpene hemiacetals, plagiochiline A (17) and plagiochiline I (18) were isolated from the ether extract of *P. yokogurensis*, together with their related tasteless compounds (ref. 7). Compound (17) is widely distributed in European, South American and Japanese *Plagiochila* species. This is the first report of the isolation of 2,3-secoaromadendrane-type sesquiterpenoids from natural sources. Plagiochiline B (19) isolated from *Plagiochila hattoriana* shows bitter taste (ref. 7). More intensely bitter substances are contained in some Lophoziaceae species. Gymnocolin (20), a *cis*-clerodane-type diterpene lactone which shows persistent bitter taste and induces vomiting has been isolated from *Gymnocolea inflata* (ref. 7, 11). *Porella perrottetiana* contains bitter principles. From the ether extract, perrottetianal (21) was isolated as the bitter taste substance (ref. 7). *Jungermannia infusca* is rich source of kaurene-type diterpenoids. The methanol extract of the fresh material shows amazingly intense bitterness. This taste is due to a new kaurane glucoside, infuscaside (22) (ref. 17).

2.3. Allergenic contact dermatitis

In Europe, Canada and North America, it has been known that some occupational allergies are associated with handling wood on which some epiphytic liverworts grow. The hapten of this allergy is the mixture of eudesmane- and eremophilane-type sesquiterpene lactones found in *Frullania tamarisci* subsp. *tamarisci*, *F. dilatata*, *F. nisquallensis* and the other *Frullania* species. The patch test of the isolated sesquiterpene lactones, (+)-frullanolide (23), (-)-frullanolide (24), (+)-oxyfrullanolide (25), *cis*-ß-cyclocostunolide (26) and (+)-eremofrullanolide (27) indicated that patients sensitive to *Frullania* are all sensitive to at least one of the sesquiterpene lactones (23-27) (ref. 7). Potentially allergenic eudesmanolides, germacranolide, guaianolides and drimanolides have been isolated not only from leafy liverworts, but also from thalloid liverworts (ref. 7). The pungent polygodial (7) caused allergy on the skin of Guinea pig (ref. 18). A large beautiful New Zealand liverwort *Schistochila appendiculata* causes allergenic contact dermatitis. Three long chain alkyl phenols, 3-undecyl phenol (28), 6-undecyl salicylic acid (29) and potassium 6-undecyl salicylate (30) were isolated from the methanol extract. 6-Undecyl catechol (31), 3-tridecyl and 3-pentadecyl phenol (32, 35), 6-tridecyl and 6-pentadecyl salicylic acid (33, 36), and potassium 6-tridecyl and potassium pentadecyl salicylate (34, 37) were detected as the minor components (ref. 19). This is the first example of the isolation of long chain alkyl phenols from the bryophytes although a number of different aromatic compounds have been found in the Hepaticae. The allergenic reaction brought on by *S. appendiculata* might be due to the presence of these long chain phenolic compounds.

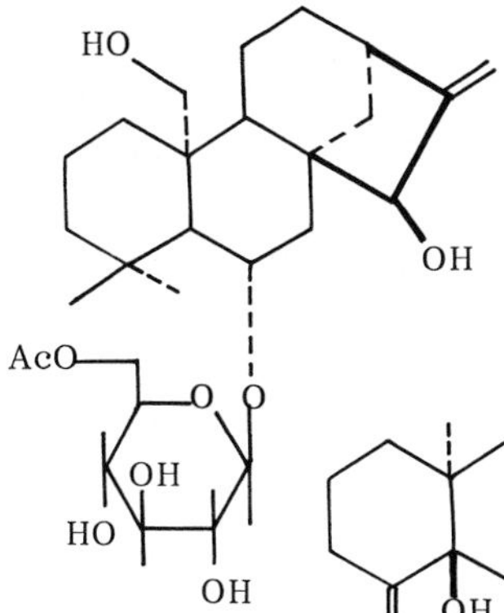

(22) 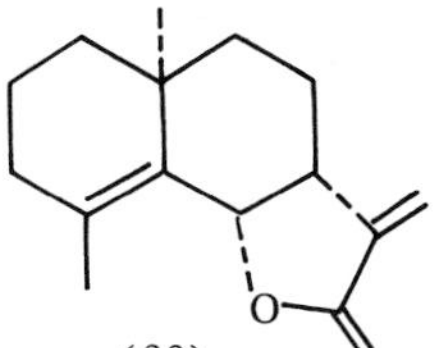 (23) 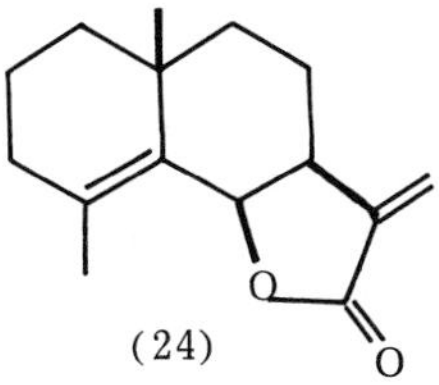 (24) 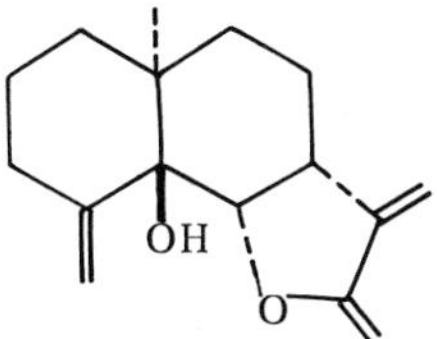 (25) 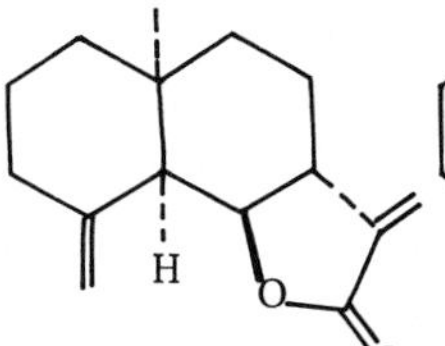 (26) 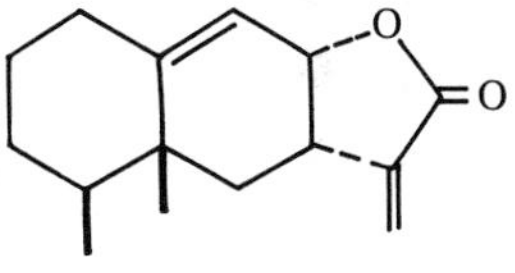 (27)

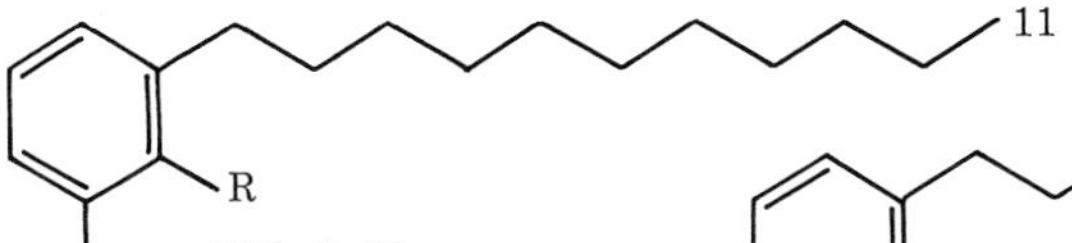

(28) R=H
(29) R=COOH
(30) R=COOK

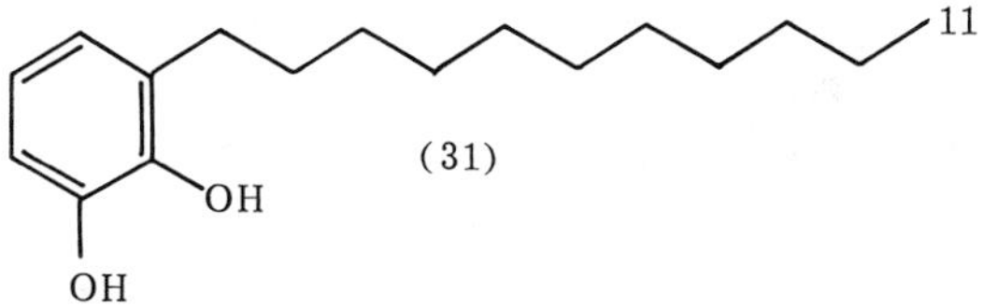

(31)

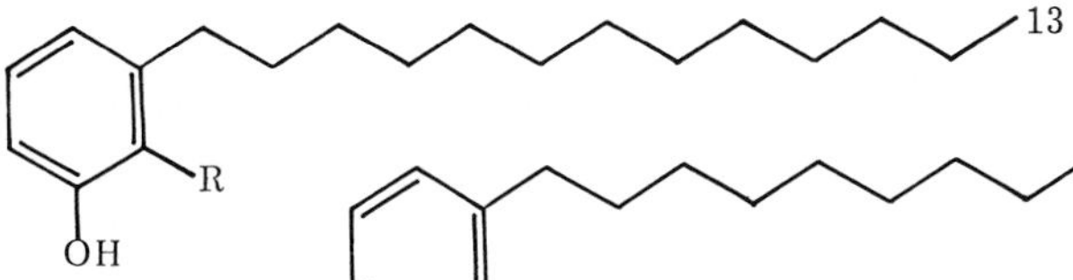

(32) R=H
(33) R=COOH
(34) R=COOK

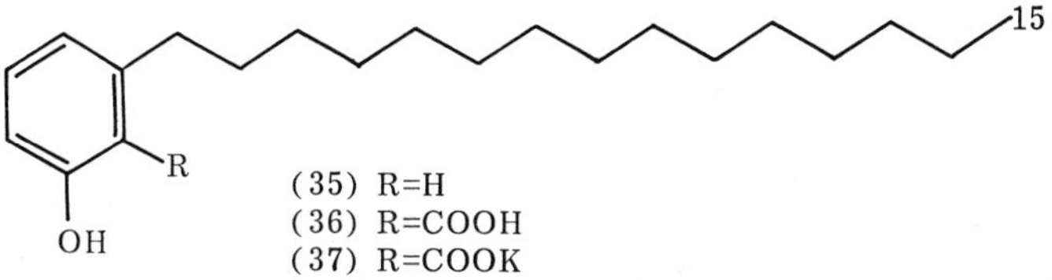

(35) R=H
(36) R=COOH
(37) R=COOK

2.4. Cytotoxicity, 5-lipoxygenase and calmodulin inhibitory activities

The female, male and sterile thalli of Japanese *Marchantia polymorpha*, *M. paleaceae* var. *diptera* and *M. tosana* produce structurally unique cyclic bis(bibenzyl) derivatives named marchantins (38-44, 47-49)(Table 1) of which marchantin A (38)

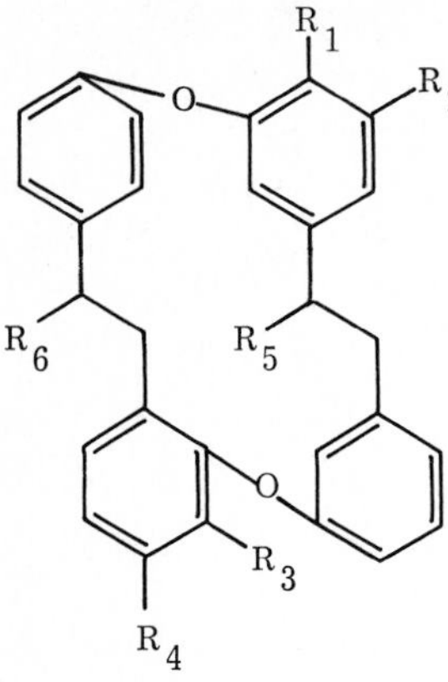

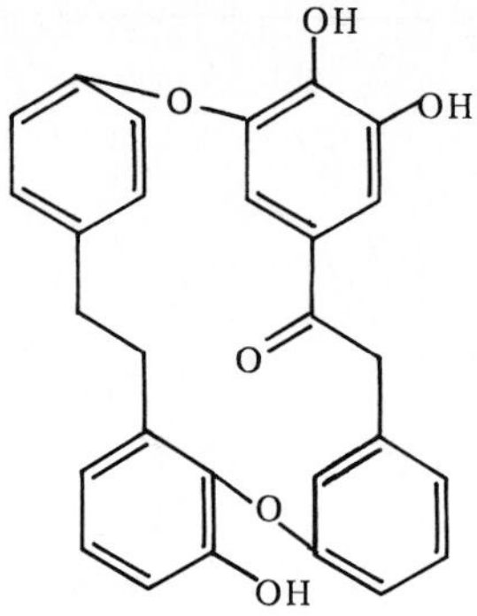

(44) 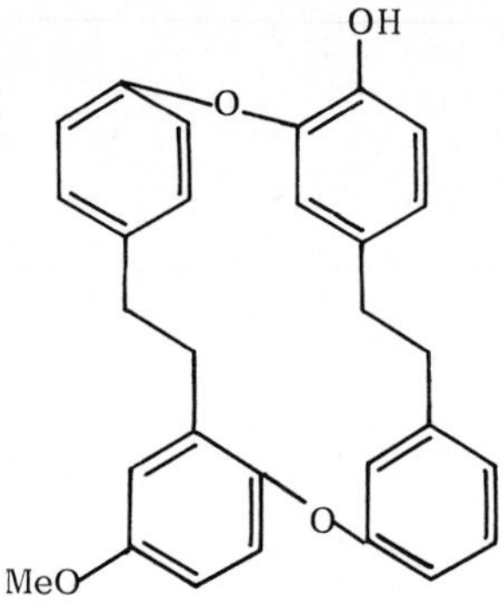 (46)

(38) $R_1=R_2=R_3=OH$, $R_4=R_5=R_6=H$

(39) $R_1=R_2=R_3=R_4=OH$, $R_5=R_6=H$

(40) $R_1=R_3=OH$, $R_2=R_4=R_5=R_6=H$

(41) $R_1=R_2=R_3=R_5=OH$, $R_4=R_6=H$

(42) $R_1=R_2=R_3=OH$, $R_4=R_6=H$, $R_5=OMe$

(43) $R_1=R_2=R_3=R_4=R_5=OH$, $R_6=H$

(45) $R_1=R_3=R_4=OH$, $R_2=R_5=R_6=H$

(47) $R_1=R_2=R_3=OH$, $R_4=R_6=H$, $R_5=OEt$

(48) $R_1=R_2=R_3=R_4=OH$, $R_5=OMe$, $R_6=H$

(49) $R_1=R_2=R_3=R_6=OH$, $R_4=R_5=H$

is the major component (ref. 7, 20-22). For example, about 100 g of marchantin A (38) is obtained from 2 kg of the dried *M. paleaceae* var. *diptera*. The structure of marchantin A was established by chemical degradation of its trimethylether and the extensive 400 MHz ^1H NMR and 100 MHz ^{13}C NMR spectra (2D-COSY, NOE difference, spin-decoupling, ^1H-^{13}C correlation, LSPD etc.) (ref. 21, 22). The stereochemistry of 38 was established by the single crystallographic X-ray analysis of a trimethylether of 38 (ref. 11). The total synthesis of marchantin A (38) has been accomplished in 12 steps by Kodama *et al.* (ref. 23). The structures of the other marchantins were determined by the same NMR technique described above (ref. 21, 22). Marchantin A-C (38-40) exhibit cytotoxicity against KB cell (ED_{50} 8.39-10 μg/ml) (ref. 11, 12, 22). Marchantin A, D (41) and E (42) show 5-lipoxygenase and calmodulin inhibitory activities (Table 2) (ref. 22, 32).

The similar cyclic bis(bibenzyl) derivatives named riccardins (50-55) having a biphenyl linkage (except 51) have been isolated from *Marchantia* (ref. 24), *Riccardia* (ref. 7, 24, 25), *Reboulia* (ref. 26) and *Monoclea* species (Table 1) (ref. 27). The structure of riccardin A (50) was established by the X-ray crystallographic analysis of a triacetate of 50 (ref. 25). Riccardin A and B (51) show cytotoxicity against KB cell (ED_{50} 10 μg/ml) (ref. 11, 12, 25). Kodama *et al.* (ref. 28) synthesized riccardin B (51) by an intramolecular Wittig-type reaction. Iyoda *et al.* (ref. 29) obtained the same compound by nickel-catalyzed intramolecular cycliza-

TABLE 1. Distribution of bis(bibenzyl) derivatives, marchantins, riccardins and perrottetins in the Hepaticae

Compounds	Species *								
	M. po.	*M. pa.*	*M. pd.*	*M. t.*	*R. m.*	*R. p.*	*P. i.*	*R. h.*	*M. f.*
marchantin A (38)	+		+	+					
B (39)	+		+	+					
C (40)	+	+	+	+					
D (41)	+		+						
E (42)	+		+						
F (43)			+						
G (44)	+	+	+						
H (45)							+		
I (46)					+				
J (47)	+								
K (48)	+								
L (49)	+								
riccardin A (50)					+				
B (51)					+				
C (52)	+	+						+	+
D (53)				+					
E (54)									+
F (55)									+
isomarchantin C (56)	+	+							
isoriccardin C (57)	+	+							
perrottetin E (58)						+			+
F (59)						+			
G (60)						+			

* *M. po: Marchantia polymorpha, M. pa: Marchantia palmata, M. pd: Marchantia paleaceae* var. *diptera, M. t: Marchantia tosana, R. m: Riccardia multifida, R. p: Radula perrottetii, P. i: Plagiochasma intermedium, R. h: Reboulia hemisphaerica, M. f: Monoclea forsteri*

tion reaction in high yield. *Marchantia palmata* and *M. polymorpha* collected in India produce isomarchantin C (56) and isoriccardin C (57) whose structures have a different ether or a biphenyl linkage from those of marchantin C (40) and riccardin C (52) (ref. 24). *Radula* species contain linear bis(bibenzyl) ethers. Perrottetins E (58), F (59) and G (60) were isolated from *R. perrottetii* (Table 1) (ref. 30, 31). Perrottetin E (58) has also been found in *Marchantia* and *Monoclea* species (ref. 24, 27). Perrottetins E and F possess cytotoxicity against KB cell (ED_{50} 12 μg/ml) (ref. 11, 30). The former compound has been synthesized in seven steps from 3,4-dihydroxybenzaldehyde (Scheme 1) (ref. 31). In addition to perrottetins (E-G), *Radula complanata, R. perrottetii* and *R. kojana* produce prenyl bibenzyl derivatives (61-79) (Table 3) whose structures have been elucidated by spectroscopic methods and their syntheses (ref. 33). The prenyl bibenzyls (61, 64, 69, 76, 78) possess 5-lipoxygenase and calmodulin inhibitory activities (Table 2) (ref. 32). *Trocholejeunea sandvicensis* elaborates a large amount of cytotoxic pinguisane-type sesquiterpene alcohol, pinguisenol (80) (KB cell, ED_{50}

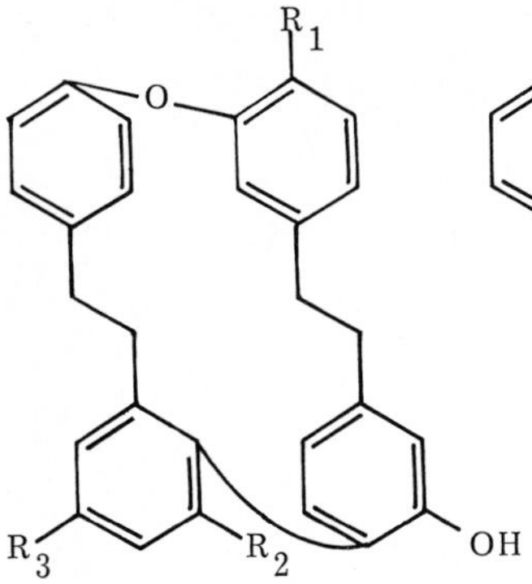

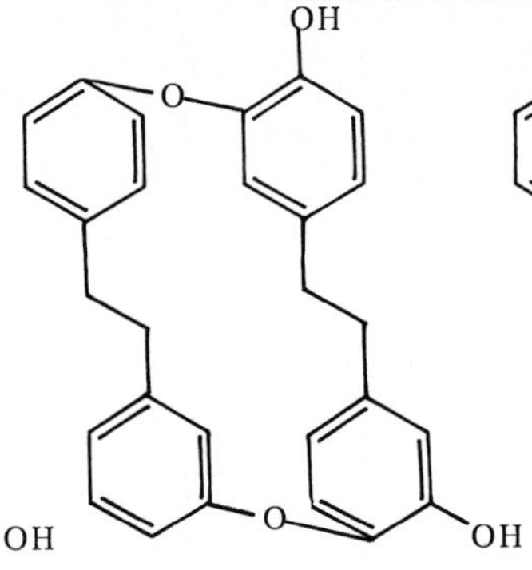

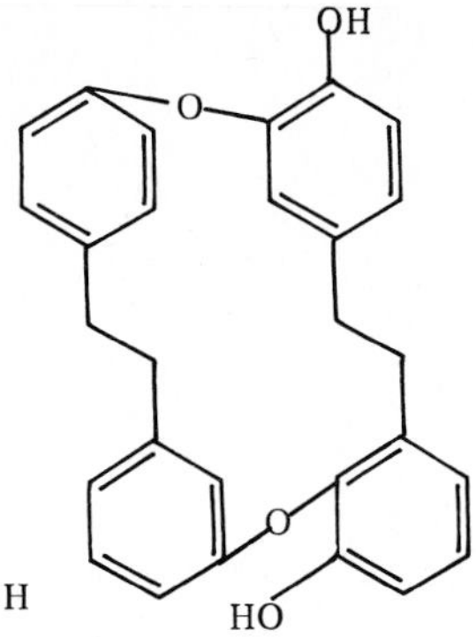

(50) R_1=OH, R_2=H, R_3=OMe

(52) R_1=R_3=OH, R_2=H

(53) R_1=OMe, R_2=H
R_3=OH

(54) R_1=R_2=OH, R_3=H

(55) R_1=OH, R_2=OMe, R_3=H

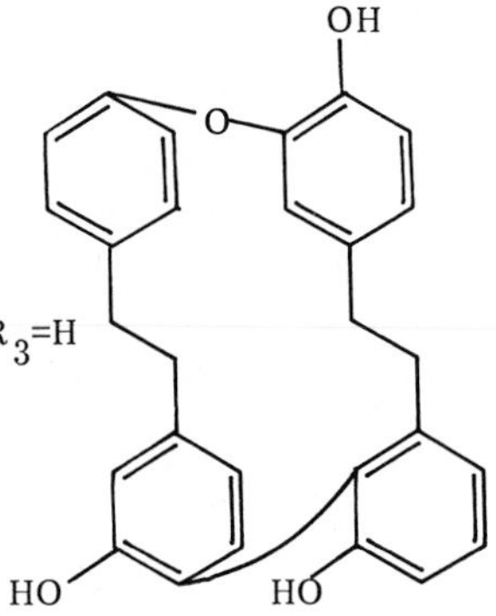

(51)

(56)

(57)

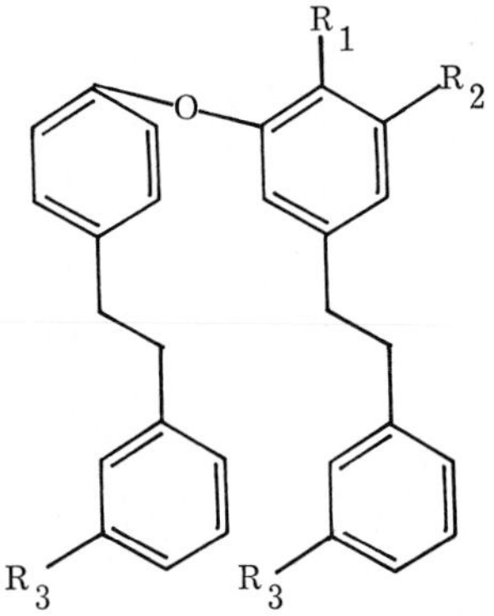

(58) R_1=R_3=OH, R_2=H

(59) R_1=R_2=R_3=OH

(60) R_1=OMe, R_2=R_3=OH

12.0 μg/ml) (ref. 11). Plagiochiline A (17) also shows cytotoxicity (KB cell ED_{50} 2.95 μg/ml) (ref. 11, 12).

TABLE 2. 5-Lipoxygenase and calmodulin inhibitory activities of bisbibenzyls and prenyl bibenzyls

Compounds	5-Lipoxygenase inhibition 10^{-6} mole	Calmodulin inhibition ID_{50} μg/ml
marchantin A (38)	60	1.85
marchantin D (41)	40	6.0
marchantin E (42)	36	7.0
riccardin A (50)	4	20.0
prenyl bibenzyl (61)	76	3.5
prenyl bibenzyl (64)	40	2.0
prenyl bibenzyl (69)	50	4.9
prenyl bibenzyl (76)	11	4.0
prenyl bibenzyl (78)	15	17.0

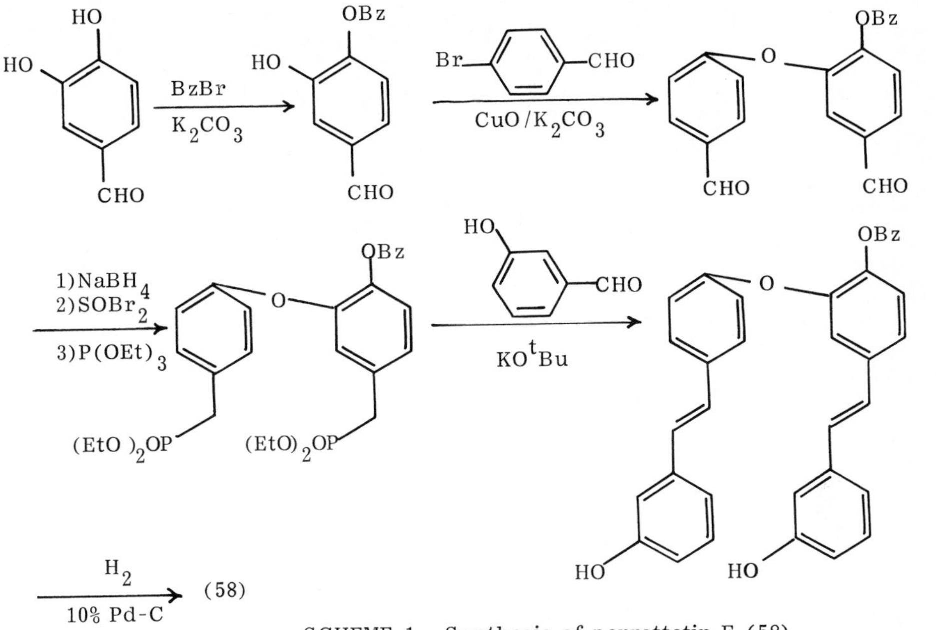

BzBr, K_2CO_3; Br–C$_6$H$_4$–CHO, CuO/K_2CO_3; 1) $NaBH_4$ 2) $SOBr_2$ 3) $P(OEt)_3$; HO–C$_6$H$_4$–CHO, KO^{t}Bu; H_2, 10% Pd-C → (58)

SCHEME 1. Synthesis of perrottetin E (58)

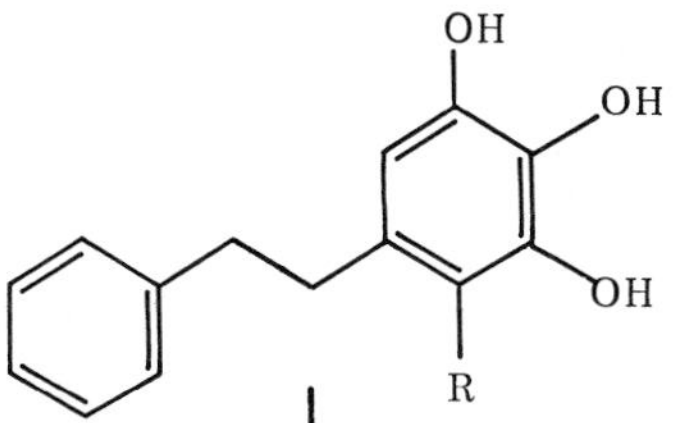

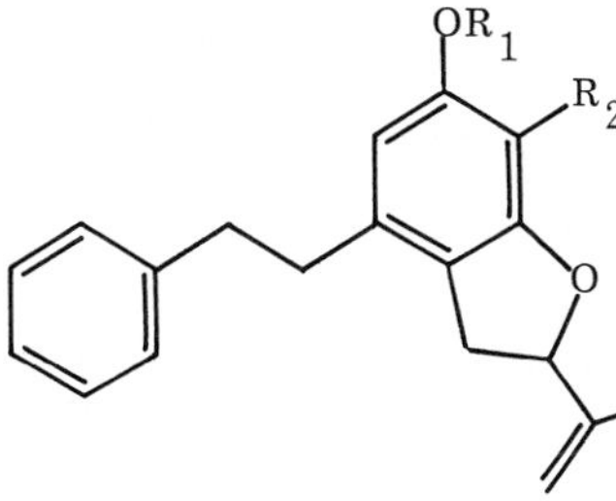

(61) R=

(62) R=

(63) R=

(64) R_1=H, R_2=OH

(65) R_1=R_2=H

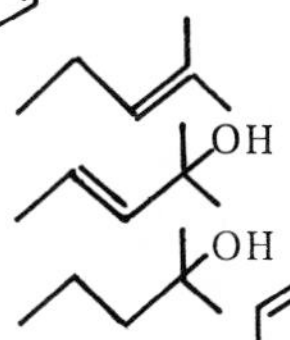

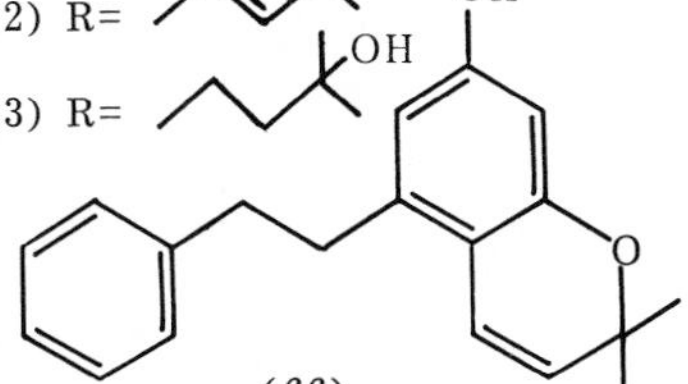

(66)

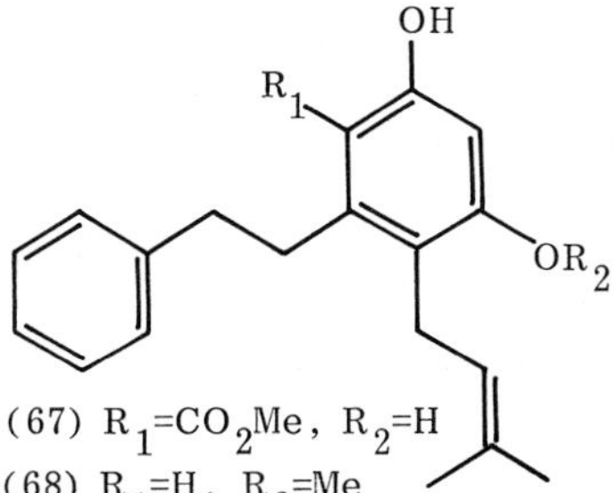

(67) R_1=CO_2Me, R_2=H

(68) R_1=H, R_2=Me

(69) R_1=R_2=H

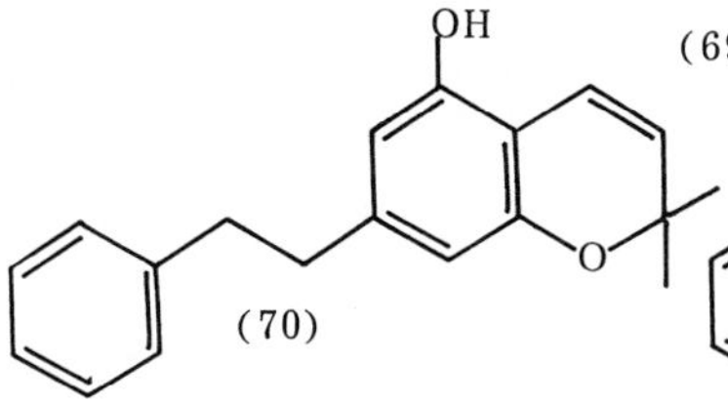

(70)

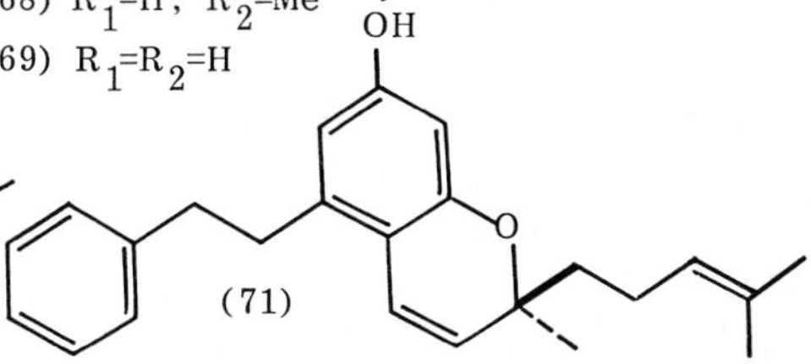

(71)

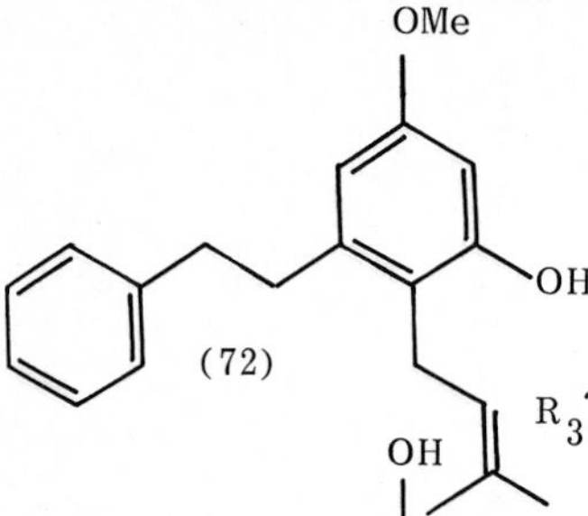

(72)

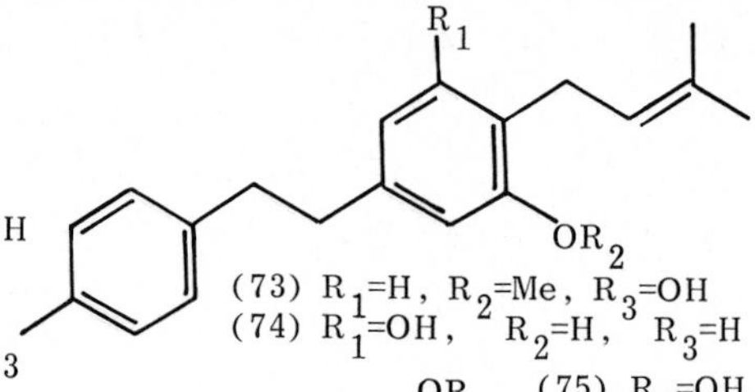

(73) R_1=H, R_2=Me, R_3=OH
(74) R_1=OH, R_2=H, R_3=H
(75) R_1=OH, R_2=Me, R_3=H

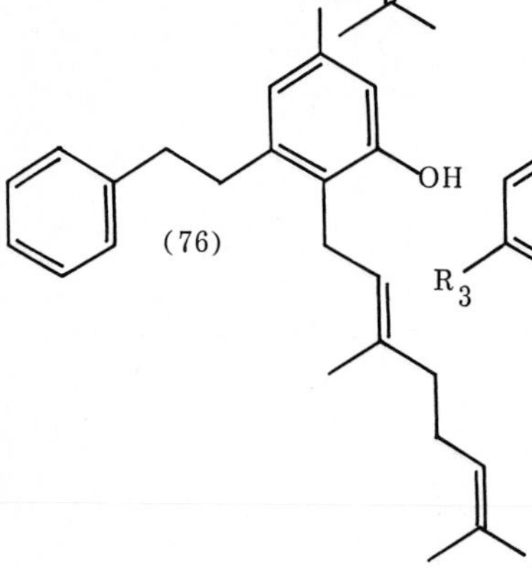

(76)

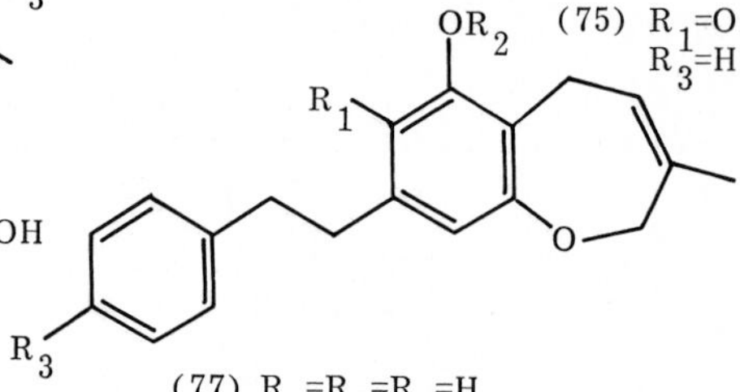

(77) R_1=R_2=R_3=H
(78) R_1=CO_2H, R_2=R_3=H
(79) R_1=CO_2H, R_2=H, R_3=OH

TABLE 3. Distribution of prenyl bibenzyls in *Radula* species

Compounds	*R. perrottetii*	*R. kojana*	*R. complanata*
(61)	+		
(62)	+		
(63)	+		
(64)	+		
(65)	+		
(66)	+	+	
(67)	+		
(68)		+	
(69)	+	+	+
(71)		+	
(72)		+	
(73)		+	
(74)			+
(75)			+
(76)		+	+
(77)			+
(78)			+
(79)			+

2.5. Tumor promoting activity

12-O-Tetradecanoyl-phorbol-13-acetate (TPA) and teleocidin A which are the potent tumor promoters cause irritation of mouse ear, induction of ornithine decarboxylase (ODC) and adhesion of cultured human promyelocytic leukaemia cells (HL-60) (ref. 34). The two pungent sesquiterpenoids, polygodial (7) and plagiochiline A (17) caused the same irritation of mouse ear as those of TPA and teleocidin A (Table 4), however, both of them did not induce ODC. On the other hand, the slight irritant plagiochiline C (81) and sacculatal (8) indicated ODC induction (ref. 11, 32). Thus, these compounds are potential tumor promoters (Table 5). The ODC induction is completely inhibited by 13-*cis*-retinoic acid (Table 5) (ref. 32).

TABLE 4. Irritancy of terpenoids isolated from the liverworts (100 μg/ear of mouse)

Compounds	Sources	Irritation
polygodial (7)	*Porella vernicosa* complex	+++
sacculatal (8)	*Trichocoleopsis sacculata*	+
plagiochiline A (17)	*Plagiochila fruticosa etc.*	++++
plagiochiline C (81)	*Plagiochila fruticosa etc.*	+
teleocidin A	*Streptomyces* species	++++

TABLE 5. ODC activity of irritant compounds and its inhibition by 13-*cis*-retinoic acid

Compounds	ODC activity *	Inhibition by 13-*cis*-retinoic acid
sacculatal (8)	1.7	100%
plagiochiline C (81)	2.5	92
teleocidin A	3.6	91
TPA	7.2	90

* n moles CO_2/30 min/mg protein

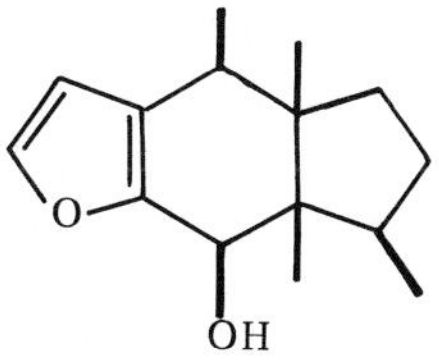

OH

(80)

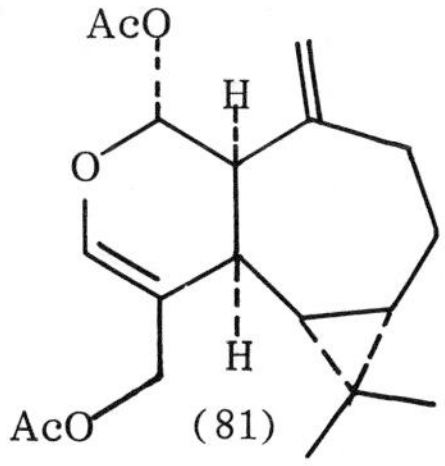

AcO

H

O

H

AcO (81)

2.6. Antimicrobial and antifungal activities

The crude extracts of *Marchantia, Porella, Frullania, Bazzania, Reboulia, Plagiochila* and *Radula* species show antimicrobial and antifungal activities. Marchantin A (38) inhibited the growth of both bacteria and fungi (Table 6) (ref. 12). Polygodial (7), norpinguisone (82) and cinnamolide (83) isolated from pungent or non-pungent *Porella* species showed antimicrobial and antifungal activities. The prenyl bibenzyl derivatives (74-76) isolated from *Radula* species exhibited the growth inhibitory activity against *Staphylococcus aureus* (ref. 11, 35). Lunularic acid (84) which is widespread in the Hepaticae shows dormant, antifungal and thrombo-

TABLE 6. Antimicrobial and antifungal activities of sesquiterpenoids, bis(bibenzyl) and prenyl bibenzyls isolated from the Hepaticae

Compounds	Microorganisms	MCI (μg/ml)
polygodial (7)	*Aspergillus fumigatus*	100
	Aspergillus niger	25
	Candida albicans	100
	Cryptococcus neoformans	100
	Staphylococcus aureus	50-100
	Trichophyton mentagrophytes	50
marchantin A (38)	*Aspergillus fumigatus*	100
	Aspergillus niger	25
	Candida albicans	100
	Cryptococcus neoformans	12.5
	Staphylococcus aureus	3.13
	Trichophyton mentagrophytes	3.13
prenyl bibenzyl (74)	*Staphylococcus aureus*	20
prenyl bibenzyl (75)	*Staphylococcus aureus*	3
prenyl bibenzyl (76)	*Staphylococcus aureus*	20
norpinguison (82)	*Aspergillus niger*	100
cinnamolide (83)	*Microsporum gypseum*	20 (ref. 36)
	Trichophyton mentagrophytes	10 (ref. 36)
	Trichophyton rubrum	20 (ref. 36)

xane synthetase inhibitory activities (ref. 7, 37). The content of 84, however, is very poor in any liverworts. A large amount of 84 was obtained from hydrangenol glucoside (85) which was easily isolated from the higher plant, *Hydrangea macrophylla*. Hydrolysis of 85 with diluted acid gave hydrangenol (86), which was treated with $NaBH_4$-$PdCl_2$ to give lunularic acid (84) in high yield (84%) (Scheme 2) (ref. 38).

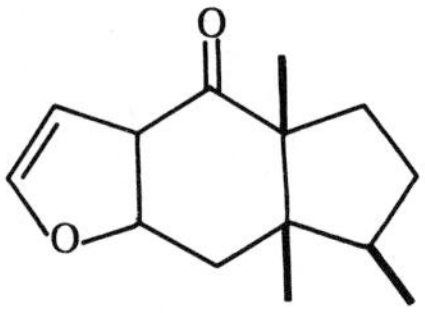

O O O OH COOH OH

(82) 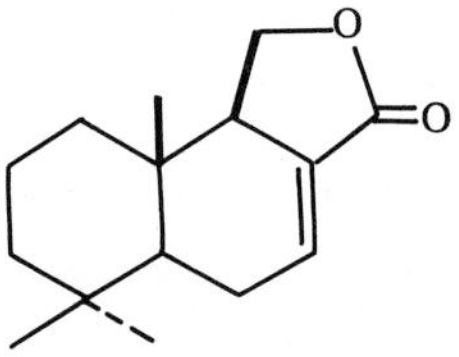 (83) 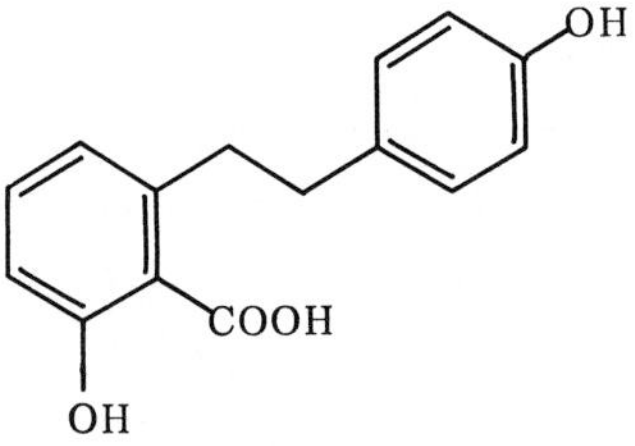 (84)

2.7. Insect antifeedant activity

Generally, the liverworts have not been harmed by the larvae and adults of insects. The pungent and bitter liverworts contain potent antifeedant substances against lavae of African army worm (=*Spodoptera exempta*). Warburganal (= 9α-hydroxypolygodial) isolated from an African pungent tree is the intensely antifeedant substance against African army worm (100 ng/cm^2, leaf disk method)(ref. 39). Plagiochiline A (17), a powerful hot tasty sesquiterpene, showed the strong-

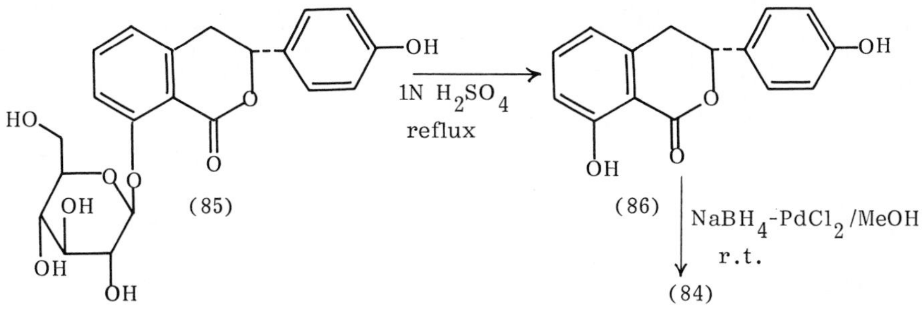

(85) $\xrightarrow[\text{reflux}]{1N\ H_2SO_4}$ (86) $\xrightarrow[\text{r.t.}]{NaBH_4\text{-}PdCl_2/MeOH}$ (84)

SCHEME 2. Preparation of lunularic acid

est antifeedant activity (1-10 ng/cm^2, leaf disk method) against the larvae of the same insect described above (ref. 40). The pungent sacculatal (8), eudesmanolides (13, 14), germacranolide (16), non-pungent sesquiterpene lactones (23, 24, 83) and the bitter diterpene (20) have also antifeedant property against the larvae of *Pieris* species, but they are significantly less active than plagiochiline A (17).

2.8. Plant growth regulatory activity

Most of the crude extracts of the liverworts show inhibitory activity against germination and root elongation of rice in husk, wheat and lettuce. The pungent or non-pungent sesquiterpene lactones, pungnet sesqui- and diterpene aldehydes and 2,3-secoaromadendrane-type sesquiterpenoids described above showed the plant growth inhibitory activity (50-500 ppm) against rice in husk (ref. 7). Polygodial (7) completely inhibited the germination of rice in husk at 100 ppm. At a concentration of less than 25 ppm, it dramatically promoted root elongation of rice in husk (ref. 41).

2.9. Piscicidal activity

The pungent extracts of the liverworts show potent piscicidal activity. Killie-fish (=*Oryzia latipes*) is killed within 2 hr by a 0.4 ppm and within 20 min. by a 7 ppm solution of polygodial (7) or sacculatal (8), and within 4 hr by a 0.4 ppm solution of plagiochiline A (17) or (+)-frullanolide (33), and within 2 hr by a 6.7 ppm solution of diplophyllin (15) (ref. 11, 42). Killie-fish is also killed within 2 hr by a 0.4 ppm solution of non-natural (+)-polygodial (87) which shows the potent hot taste. Polygodial (7) is very toxic against bitterling fish which is killed within 3 min. by a 0.4 ppm solution of 7. On the other hand, isopolygodial (88) and isosacculatal (11) possessing non-pungency, dramatically lack piscicidal activity in even 10000 ppm solution (ref. 12). Thus, it is obvious that occurrence of piscicidal activity of polygodial and sacculatal is significantly related with hot-taste, that is, with the absolute configuration of a formyl group at C-9.

CHO CHO CHO CHO

(87) (88)

Up to the present, more than 150 new compounds have been isolated from the Hepaticae. Most of them are terpenoids and aromatic compounds of which only two nitrogen- and three sulfur-containing compounds have been reported (ref. 7, 43, 44). As described above, some terpenoids or aromatic compounds show interesting biological activity. The species of the Hepaticae chemically studied so far are only 5% of the total Hepaticae. Further chemical and pharmacological studies not only of the Hepaticae,but also of the Musci may provide different types of bioactive compounds and valuable information upon the taxonomy of the bryophytes and evolutional relationship between algae and lower terrestrial green plants (ref. 45).

REFERENCES

1 G. Garnier, L. Bezanger-Beauquesne and G. Debraux, Ressources Medicinales de la Flore Française, Tome I, Vigot Frères Editeurs, Paris, 1961.
2 H. Ding, Zhong guo Yao yun Bao zi Zhi wu, Shanghai Kexue Jishu Chuban she, Shanghai, 1980.
3 H. Ando and A. Matsuo, Applied Bryology, in: W. Schultze-Motel (ed.), Advances in Bryology, Vol. 2, J. Cramer, Vaduz, 1984, pp. 133-224.
4 P. C. Wu, Some Uses of Mosses in China, The Bryol. Times, 13 (1982) 5.
5 M. Mizutani, On the Taste of Some Mosses, Misc. Bryol. et Lichenol., 2 (7) (1961) 100.
6 M. Mizutani, Smell of Some Bryophytes, Misc. Bryol. et Lichenol., 6 (4) (1975) 64.
7 Y. Asakawa, Chemical Constituents of the Hepaticae, in: W. Herz, H. Grisebach and G. W. Kirby (eds.), Progress in the Chemistry of Organic Natural Products, Vol. 42, Springer, Wien, 1982, pp. 1-285.
8 K. R. Markham and L. J. Porter, Chemical Constituents of Bryophytes, in: L. Reinhold, J. B. Harborne and T. Swain (eds.), Progress in Phytochemistry, Vol. 5, Pergamon Press, Oxford, 1978, pp 181-272.
9 S. Huneck, Chemistry and Biochemistry of Bryophytes, in: R. M. Schuster (ed.), New Manual of Bryology, Vol. 1, J. Hattori Bot. Lab., Nichinan, Japan, 1983, pp. 1-116.
10 Y. Asakawa, Biologically Active Substances Obtained from Bryophytes, J. Hattori Bot. Lab., 50 (1981) 123-142.
11 Y. Asakawa, Phytochemistry of Hepaticae: Isolation of Biologically Active Aromatic Compounds and Terpenoids, Rev. Latinoamer. Quim., 14(3) (1984) 109-114.
12 Y. Asakawa, Some Biologically Active Substances Isolated from Hepaticae: Terpenoids and Lipophilic Aromatic Compounds, J. Hattori Bot. Lab., 56 (1984) 215-219.
13 C. Suire, Y. Asakawa, M. Toyota and T. Takemoto, Chirality of Terpenoids Isolated from the Liverwort *Conocephalum conicum*, Phytochemistry, 21 (2) (1982).
14 Y. Asakawa, M. Toyota and A. Cheminat, Terpenoids from the French Liver-

wort *Targionia hypophylla*, Phytochemistry, 25 (11) (1986) 2555-2556.
15 J. D. Connolly, L. J. Harrison and D. S. Rycroft, The Structure of Tamariscol, a New Pacifigorgiane Sesquiterpenoid Alcohol from the Liverwort *Frullania tamarisci*, Tetrahedron Lett., 25 (13) (1984) 1401-1402.
16 A. Matsuo, H. Nozaki, M. Nakayama, Y. Kushi, S. Hayashi, T. Komori and N. Kamijo, (+)-Bicyclohumulenone, a Novel Sesquiterpene Ketone of the Humulane Group from *Plagiochila acanthophylla* subsp. *japonica* (Liverwort): X-Ray Crystal and Molecular Structure of the *p*-Bromobenzoate Derivative, J. Chem. Soc. Chem. Commun., (1979) 174-175.
17 Y. Asakawa, M. Toyota and F. Nagashima, " in preparation".
18 J. -L. Stampf, C. Benezra and Y. Asakawa, Stereospecificity of Allergenic Contact Dermatitis (ACD) to Enantiomers. Part III. Experimentally Induced ACD to a Natural SesquiterpeneDialdehyde, Polygodial in Guinea Pigs, Arch. Dermatol. Res., 274 (1982) 277-281.
19 Y. Asakawa, T. Masuya, M. Tori and E. O. Campbell, Long Chain Alkyl Phenols from the Liverwort*Schistochila appendiculata*, Phytochemistry, 27 (1987) " in press".
20 Y. Asakawa, M. Toyota, R. Matsuda, K. Takikawa and T. Takemoto, Distribution of Novel Cyclic Bisbibenzyls in *Marchantia* and*Riccardia* species, Phytochemistry, 22(6) (1983) 1413-1415.
21 M. Tori, M. Toyota, L. J. Harrison, K. Takikawa and Y. Asakawa, Total Assignment of ^{1}H and^{13}C NMR Spectra of Marchantins Isolated from Liverworts and Its Application to Structure Determination of Two New Macrocyclic Bis(bibenzyls) from *Plagiochasma intermedium* and *Riccardia multifida*, Tetrahedron Lett., 26 (39) (1985) 4735-4738.
22 M. Tori, T. Masuya, K. Takikawa, M. Toyota and Y. Asakawa, The Structure and NMR of Macrocyclic Bis(bibenzyls) Isolated from Liverworts, in: S. Nozoe (ed.), 28th Symposium on the Chemistry of Natural Products, Sendai, Japan, October 7-10, 1986, pp. 9-16.
23 M. Kodama, Y. Shiobara, K. Matsumura and H. Sumitomo, Total Synthesis of Marchantin A, a Cyclic Bis(bibenzyl) Isolated from Liverworts, Tetrahedron Lett., 26 (7) (1985) 877-880.
24 Y. Asakawa, M. Tori, K. Takikawa, H. G. Krishnamurty and S. K. Kar, Novel Cyclic Bis(bibenzyls) and Related Compounds from the Indian Liverworts *Marchantia polymorpha* and *Marchantia palmata*, Phytochemistry, 27 (1987) " in press".
25 Y. Asakawa, M. Toyota, Z. Taira, T. Takemoto and M. Kido, Riccardin A and B, Two Novel Cyclic Bis(bibenzyls) Possessing Cytotoxicity from the liverwort *Riccardia multifida* (L.) S. Gray, J. Org. Chem., 48 (13) (1983) 2164-2167.
26 Y. Asakawa and R. Matsuda, Riccardin C, a Novel Cyclic Bibenzyl Derivative from *Reboulia hemisphaerica*, Phytochemistry, 21 (8) (1982) 2143-2144.
27 Y. Asakawa, M. Toyota and F. Nagashima, " in preparation".
28 Y. Shiobara, H. Sumitomo, M. Tsukamoto, C. Harada and M. Kodama, Synthesis and Structure Confirmation of Riccardin B, a Macrocyclic Bis(bibenzyl) from the Liverwort, *Riccardia multifida*, Chemistry Lett., (10) (1985) 1587-1588.
29 M. Iyoda, M. Sakaitani, H. Otsuka and M. Oda, Synthesis of Riccardin B by Nickel-Catalyzed Intramolecular Cyclization, Tetrahedron Lett., 26 (39) (1985) 4777-4780.
30 Y. Asakawa, M. Toyota, Z. Taira and T. Takemoto, Biologically Active Cyclic Bisbibenzyls and Terpenoids Isolated from Liverworts, in: N. Otake (ed.), 25th Symposium on the Chemistry of Natural Products, Tokyo, Japan, October 19-22, 1982, pp. 337-344.
31 M. Toyota, M. Tori, K. Takikawa, Y. Shiobara, M. Kodama and Y. Asakawa, Perrottetins E, F, and G from *Radula perrottetii* (Liverwort)-Isolation, Structure Determination, and Synthesis of Perrottetin E, Tetrahedron Lett., 26 (49) (1985) 6097-6100.
32 Y. Asakawa, M. Toyota, M. Tori, H. Fujiki, M. Suganuma and T. Sugimura, Possible Tumor Promoters, and 5-Lipoxygenase and Calmodulin Inhibitors Isolated from Some Liverworts, International Symposium on Organic Chemistry of Medicinal Natural Products (IUPAC), Shanghai, China, November 10-14,

1985, Abstracts B-021.

33 T. Hashimoto and Y. Asakawa, " in preparation".

34 H. Fujiki and T. Sugimura, New Potent Tumour Promoters: Teleocidin, Lyngbyatoxin A and Aplysiatoxin, Cancer Surveys, 2 (4) (1983) 539-556.

35 Y. Asakawa, K. Takikawa, M. Toyota and T. Takemoto, Novel Bibenzyl Derivatives and *ent*-Cuparene-type Sesquiterpenoids from *Radula* species, Phytochemistry, 21 (10) (1982) 2481-2490.

36 L. A. Canonica, A. Corbella, P. Gariboldi, G. Jommi, J. Krepinsky, G. Ferrari and C. Casagrande, Sesquiterpenoids of *Cinnamomasma fragrans* Baillon, Structure of Cinnamolide, Cinnamosmolide and Cinnamodial, Tetrahedron, 25 (1969) 3895-3902.

37 Y. Goda and U. Sankawa, Thromboxane Synthetase Inhibitors from *Allium bakeri* Regel, 105th Annual Meeting of Pharmaceutical Society of Japan, Kanazawa, Japan, April 3-5, 1985, Abstracts pp. 468.

38 Y. Asakawa and T. Hashimoto, " in preparation".

39 K. Nakanishi and I. Kubo, Studies on Warburganal, Muzigadial and Related Compounds, Israel J. Chem., 16 (1977) 28-31.

40 Y. Asakawa, M. Toyota, T. Takemoto, I. Kubo and K. Nakanishi, Insect Antifeedant Secoaromadendrane-type Sesquiterpenes from *Plagiochila* Species, Phytochemistry, 19 (10) (1980) 2147-2154.

41 Y. Asakawa, S. Huneck, M. Toyota, T. Takemoto and C. Suire, Mono- and Sesquiterpenes from *Porella arboris-vitae*, J. Hattori Bot. Lab., 46 (1979) 163-167.

42 Y. Asakawa, L. J. Harrison and M. Toyota, Occurrence of a Potent Piscicidal Diterpenedial in the Liverwort *Riccardia lobata* var. *yakushimensis*, Phytochemistry, 24 (2) (1985) 261-262.

43 Y. Asakawa, M. Toyota and L. J. Harrison, Isotachin A and Isotachin B, Two Sulphur-containing Acrylates from the Liverwort *Isotachis japonica*, Phytochemistry, 24 (7) (1985) 1505-1508.

44 Y. Asakawa, K. Takikawa, M. Tori and E. O. Campbell, Isotachin C and Balantiolide, Two Aromatic Compounds from the New Zealand Liverwort *Balantiopsis rosea*, Phytochemistry, 25 (11) (1986) 2543-2546.

45 Y. Asakawa, Chemical Relationship between Algae, Bryophytes and Pteridophytes, J. Bryol., 14 (1986) 59-70.

POLYSACCHARIDES AND GLYCOCOMPLEXES OF PROTOZOA

P.A.J. GORIN and L.R. TRAVASSOS

1. INTRODUCTION

The presence of polysaccharides and glycocomplexes in protozoa has only been properly recognized during the past 10 years. This was probably due to difficulties in tneir culture and in isolation of components in sufficient quantities for analysis. Yields obtained on extraction are very low when compared with those of other organisms, since protozoa do not have a cell wall. Their carbohydrate-containing membrane represents only a small proportion of the total cell and the intracellular carbohydrate content is, in most cases, also low. Now, the chemical structures of the sugar moieties of surface components are now being recognized as being antigenically and immunogenically important and analyses are facilitated by improved procedures and accompanying refinements in culture methodology. Of great interest are pathogenic organisms that cause Chagas' disease, sleeping sickness, and leishmaniases, diseases that afflict tens of millions of people, mainly in tropical and sub-tropical regions. Only the last named has effective remedies.

The first protozoan polysaccharides to be identified were the reserve glucans amylose, amylopectin, and glycogen in the period 1950-1960 (1-10). These included those of large ciliates, isolated from the contents of the colon and caecum of the horse and from sheep's rumen. Now, it seems that similar glucans are present in virtually all protozoa. In 2 of the above investigations, heterocomplexes were also prepared. One was obtained from Trichomonas foetus and contained fucose, rhamnose, xylose, galactose, and glucosamine components (8) and the other was from Oochromonas malhamensis, which had units of glucose, galactose, and mannose (10). These gave a clue to the complex structures that were later encountered.

2. PROTOZOAN GLUCANS

As glucans were the first polysaccharides isolated from protozoa, they are described in a separate group. After a preliminary communication in 1949, Lwoff et al. (1) reported the growth of Polytomella coeca in a medium containing sodium acetate and/or ethanol as carbon sources, sugars being ineffective, and component starch granules were isolated by grinding followed by differential centrifugation. The starch was characterized by Bourne et al. (2), using chemical and enzymatic methods, as a mixture of amylose (19%) and amylopectin (81%), a composition close to that of some plant starches.

Large ciliates were prepared, free from bacteria and plant material, from the contents of sheep's rumen and the caecum and colon of the horse. Oxford (3) isolated, from sheep's rumen, a mixture containing 2 species of Isotricha and 1 species of Dasytricha. Extraction was carried out with the sodium salt of a secondary alkyl sulfate (Teopol XL) and the liberated granules obtained via differential sedimentation followed by centrifugation. Granules were formed readily in the presence of glucose, fructose, or sucrose. Forsyth and Hirst (4) characterized the starch as amylopectin with 1 nonreducing end-group for every 22 glucose residues. A structurally similar starch was present in the horse colon and caecum (5). Their contents were filtered through muslin, the filtrate left for a few hours, and sedimented cells extracted with chloral hydrate to give amylopectin in 49% yield.

Chilomonas paramecium, when grown on an acetate-containing medium, gave rise to a starch with 45% of amylose with 55% of amylopectin having an average chain-length of 22 units, also reminiscent of some plant starches (6).

Glycogen was first found in a protozoon by Manners and Ryley in 1952 (7). Tetrahymena pyriformis contained 16% of glycogen, which was isolated by extraction with 30% aqueous KOH at 100° with subsequent precipitations with ethanol and acetic acid. Its mol. wt. was 10^7 and the average chain-length 13 glucose residues, similar in structure to mammalian glycogens. Other glycogens have been found in Trichomonas foetus, a parasite of the reproductive tract of cattle, and in Trichomonas gallinae. T. foetus contains 5-10% of the reserve glucan, having a mol. wt. of 3×10^6 and a chain length of 15 units. The glycogen of T. gallinae has a similar mol. wt., but has a more highly branched stucture with a chain length of 9 units (9).

A β-D-glucopyranan with predominant (1→3)-linkages and 17% of branching was found in Oochromonas malhamensis (10).

3. POLYSACCHARIDES OF LOWER TRYPANOSOMATIDS

True polysaccharides, other than the above glucans, are found in *Herpetomonas* and *Crithidia* spp., which are considered to be lower trypanosomatids as their habitat is in invertebrates and since they do not have a cycle involving vertebrates (11). The chemical structures of polysaccharides of some of the species have been determined and are of interest since they may be used as markers in taxonomy. In this respect, however, some caution must be used as cell composition depends on the stage of development of the cell, which in turn depends on the composition of the culture medium.

3.1 Polysaccharides of *Crithidia fasciculata*

C. fasciculata occurs in the hindgut of the mosquito. In 1972, Cosgrove and Hanson (12) obtained a cell-surface polysaccharide containing arabinose and galactose by aqueous alkaline extraction. Brooker (13) showed that a periodate susceptible (Thiery's reagent) material or materials stained at the cell surface. This could correspond to the arabinogalactan and/or a mannan, the latter being isolated by Gottlieb et al. (14) in 1972, following phenol-water extraction.

In 1978, Gottlieb (15) investigated the 2 polysaccharides in more detail. The strain "Anopheles" 11745, similarly extracted, gave rise to a mixture of polysaccharides and RNA, which was fractionated on a column of DEAE-cellulose. Mannan was eluted with water and was found not to produce antibodies in rabbits or to agglutinate with ConA, as do β-D-mannans. Its mol. wt. was ∿14000, as determined by gel chromatography, or 13000 by ferricyanide end-group determination. RNA was then eluted from the column with M LiCl, which was followed by 1% Triton X-100. This gave an arabinogalactan containing fraction with 60% purity. Protein (3%) was also present and this proved to be linked to the carbohydrate by virtue of a double electrophoretic band of 200 kDa, which was positive for carbohydrate and protein (16). On hydrolysis, the arabinogalactan fraction gave the monosaccharide components in a 1:1 molar ratio, had a mol. wt. of 2×10^5 but its detailed chemical structure was not determined. Whether it is the main cell-surface antigen was not clear, but immunocytochemical and cell agglutination investigations showed that it is present at the surface of the main part of the cell, although it is less abundant in the flagellar pocket region. Detailed chemical investigations on the 2 polysaccharides were carried out by Gorin et al. (17) in

1979. Growth of the same strain on a sucrose-based medium, followed by extraction of cells harvested in the early exponential phase, furnished ethanol insoluble material with mannose (88%), galactose (8%), and arabinose (4%) as monosaccharide components. The ^{13}C-n.m.r. spectrum of the preparation (Fig. 1) was that of (1→2)-linked β-D-mannopyranan (repeating structure I), isolated by Previato et al. (18) in a parallel study on Crithidia deanei (see Section 3.2). Such a structure shows why the mannan of Gottlieb (15) did not agglutinate with ConA, which is specific for nonreducing ends and 2-O-substituted units of α-linked D-mannopyranose.

-β-D-Manp-(1→2)- I

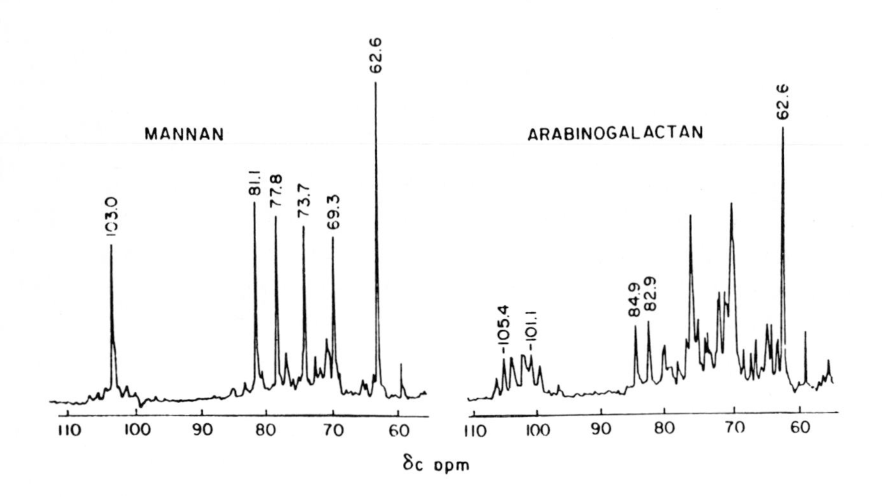

Fig. 1. ^{13}C-n.m.r. spectra of C. fasciculata mannan and arabinogalactan in D_2O at 70°: chemical shifts are based on that of Me_4Si = 0.

Cells harvested at a late stage of growth gave rise to, on alkaline extraction, a preparation with a greatly different monosaccharide composition of D-galactose (56%), D-arabinose (39%), and mannose (5%). The proportion of these components was reversed to such an extent that the mannan signals did not appear in the ^{13}C-n.m.r. spectrum (Fig. 1). Structures similar to this, having D-arabinose, were unknown, the closest polysaccharide being a D-arabino-D-galactan from Mycobacteria, in which it has the furanosyl form (19). A main chain of (1→3)-linked β-D-galactopyranosyl units was suggested by formation, on partial acid hydrolysis, of oligosaccharides having this structure with up to 4 units in length. Methylation analysis of the arabinogalactan showed nonreducing end-groups (39%) and 2-*O*- (or 4-*O*) substituted units (9%) of arabinopyranose with 3-*O* (11%) and 2,3-di-*O*-substituted galactopyranosyl residues (29%). The correspondence of the percentage of nonreducing end-groups to that of di-*O*-substituted units shows that it is a true polysaccharide. Its specific rotation of −25° is close to that of 0° of methyl β-*D*-galactopyranoside showing that the arabinopyranosyl units have a small rotational contribution and thus have the α-configuration. In accord with the data, the structural components are a (1→3)-linked β-*D*-galactopyranosyl main chain, which is unsubstituted (II; ~11%), and substituted at O-2 with α-*D*-arabinopyranosyl (III; ~29%) and 2-*O*- (or 4-*O*) α-*D*-arabinopyranosyl-α-*D*-arabinopyranosyl units (~9%). Minor components were present in the arabinogalactan preparation corresponding to nonreducing end-groups (3%) and 2-*O*-substituted units of mannopyranose (6%) and although traces of β-mannan may be present, the presence of α-mannan was indicated by the liberation of mannose with α-*D*-mannosidase. Nonreducing end-groups of galactofuranose were also detected by methylation analysis and a very low-field C-1 signal in the ^{13}C-n.m.r. spectrum at δ 106.6 showed a β-configuration (20). This is a common component of protozoan polysaccharides and glycocomplexes.

```
                                        α-D-Arap
                                           1
                                           ↓
                                           2
-β-D-Galp-(1→3)-   II            -β-D-Galp-(1→3)-   III
```

3.2 The (1→2)-linked β-*D*-mannopyranan of *Crithidia deanei*

Crithidia deanei is found in the gut of the insect *Zelus leucogrammus* and was found, by Previato et al., in 1979 (18), to contain a linear (1→2)-linked β-*D*-mannopyranan (repeating structure I), which was a previously unknown structure. Extraction of cells grown on a proline-based medium with hot aqueous KOH provided, in 6% yield, the mannan whose ^{13}C-n.m.r. spectrum was comparable to that of Fig. 1. It contained 6 signals, indicating a structure with 1 linkage type only, at 103.0, 81.1, 77.8, 73.7, 69.3, and 62.6. Methylation analysis showed 2-*O*-substituted units of mannopyranose and the specific rotation of −50° was much lower than +88° of the α-mannan of baker's yeast (21), indicating that they had the β-configuration. Cells of *C. deanei*, grown on a sucrose-based medium, provided a polysaccharide containing units of glucose, mannose, and fucose, the latter being misidentified as rhamnose, in a 1:2.8:1 molar ratio.

3.3 Polysaccharides of *Crithidia* spp. and the taxonomic significance of β-mannan components

Polysaccharides appear to have a role as markers in the taxonomy of kinetoplastid flagellates in addition to the properties of DNA components (22) and the distribution of surface proteins (23). A prominent candidate as a marker is the (1→2)-linked β-*D*-mannopyranan of *C. deanei* and *C. fasciculata* and which is also found in *Crithidia harmosa* and *Crithidia luciliae* (24). Baker (25) has suggested that *Crithidia* and *Herpetomonas* are part of a common evolutionary line in the phylogenetic tree of Trypanosomatidae and in accord with this *Herpetomonas samuelpessoai* produces the mannan (26), whereas related *Leptomonas samueli* does not (27). Wallace (28) has proposed that *Crithidia*, *Herpetomonas*, and *Leishmania* are in a common evolutionary line different from that of *Blastocrithidia*, and *Rhynchoidomonas*. However, while the mannan is found in *Leishmania amazonensis amazonensis* (29), it has not been reported in *Leishmania tarentolae* (30, 31). Since the polysaccharide composition depends on the stage of development of the cell, the absence of the mannan is less significant than its presence, an observation that can be applied to *Trypanosoma mega* (32, 33) and *Trypanosoma cruzi* (34). Polysaccharide composition may also be dependent on the sub-species, as occurs with those of *Herpetomonas muscarum* (11).

C. fasciculata, *C. harmosa*, and *C. luciliae* are closely related as they produce arabinogalactans, whereas *C. deanei* and a *Crithidia* sp., isolated in Newfoundland by Dr. Seymour Hutner, do not (24). These arabinogalactans contain nonreducing end-groups of arabinopyranose and 2-*O* and 2,3-di-*O*-substituted galactopyranosyl units, as shown by methylation analysis. Minor structural differences occur, as 2-*O*- (or 4-*O*) substituted arabinopyranosyl residues are not present in the arabinogalactans of proline-grown cells of *C. harmosa* and *C. luciliae* and of the latter grown on sucrose. On the other hand, sucrose-grown cells of *C. harmosa* contain arabinogalactan with this conponent, which is in agreement with its ^{13}C-n.m.r. spectrum, which is identical to that of *C. fasciculata* (Fig. 1). In the case of the mannan, it is present in

C. luciliae grown on sucrose or proline, whereas it occurs only in proline-grown *C. fasciculata*.

The structural differences of the polysaccharide of the *Crithidia* sp. was evidenced on methylation analysis, which showed nonreducing end-groups of xylopyranose (14%), fucopyranose (17%), and mannopyranose (4%) with 2-*O* (19%) and 2,3-di-*O*-substituted mannopyranosyl (11%), 4-*O*-substituted fucopyranosyl (9%), and 2-*O*- (or 4-*O*; 7%) and 2,4-di-*O*-substituted xylopyranosyl residues (8%).

3.4 Polysaccharides of *Herpetomonas* spp., principally *H. samuelpessoai*

Ten valid species of *Herpetomonas* are known and of these only the polysaccharides of *H. samuelpessoai* and 2 sub-species of *H. muscarum* have been investigated.

The polysaccharide components of cells and flagella of *H. samuelpessoai*, an inhabitant of the gut of *Zelus leucogrammus*, have been analyzed in detail by chemical and cytochemical methods. In 1979, Mendonça-Previato et al. (26) showed that alkaline extraction of proline-grown cells liberated mostly (1→2)-linked β-*D*-mannopyranan, whereas glucuronoxylan was obtained from cells grown on sucrose. Also, under certain conditions, a galactose-containing polysaccharide was formed. In proline-grown cells, oligosaccharide components were indicated by the difference in the xylose to mannose ratio in whole cells (31:69) and in polysaccharide isolated via alkaline extraction, followed by ethanol precipitation (83:17). Nonreducing oligosaccharides were obtained on hot aqueous $NaBH_4$-NaOH extraction and detected, on a

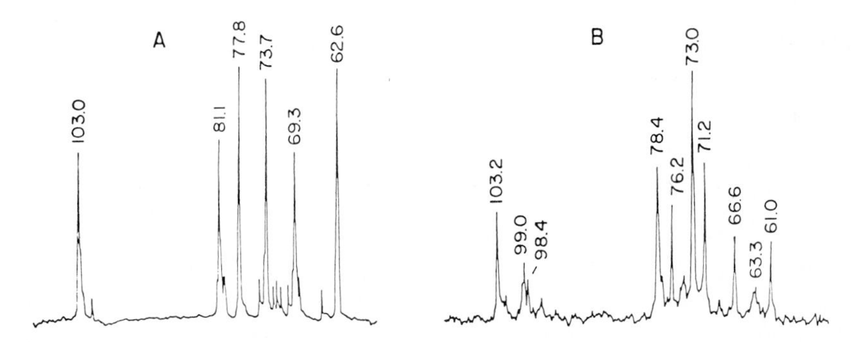

Fig. 2. ^{13}C-n.m.r. spectra of mannan (A) and glucuronoxylan (B) components of *H. samuelpessoai* in D_2O at 70°: numerical values are in p.p.m., based on Me_4Si = 0.

paper chromatogram, as a series of spots starting with mannitol, followed by mannobiitol and continuing with successively higher molecular weight homologues. The ethanol precipitate of this material gave a ^{13}C-n.m.r. spectrum (Fig. 2A) corresponding to the β-mannan, with an additional signal at δ 101.0 arising from terminal units. The homologous series of oligosaccharides was present in the free form and not as glycoconjugates, since reducing sugars were liberated from cells on autoclaving.

Glucuronoxylan was obtained, following alkaline extraction of sucrose-grown cells, and purified by fractional precipitation from water using ethanol. Methylation analysis and examination of oligosaccharides, obtained by partial acid hydrolysis (IV, V, and VI), indicated some of the structures present, although that of the main-chain was not determined. However, it is clearly different from plant glucuronoxylans, which have (1→4)-linked β-D-xylopyranosyl main-chains. Its ^{13}C-n.m.r. spectrum (Fig. 2B) contained a C-1 signal at δ 103.2, arising from β-xylopyranosyl residues and 2 others at δ 98.4 and 99.0, which correspond to the α-anomers of xylopyranosyl and glucopyranosyl(uronic acid) units.

D-GlcAp-(1→2)-α-D-Xylp-(1→2)-D-Xyl IV

D-GlcAp-(1→2)-β-D-Xylp-(1→4)-D-Xyl V

-(1→4)-α-D-Xyl VI

The flagella of H. samuelpessoai contain mannose, xylose, glucuronic acid, and glucose residues. The mannose units have the α-configuration as mannose was liberated on treatment with α-D-mannosidase. Amylose was present, as evidenced by the typical blue color formed with iodine solution and by the liberation of glucose with glucoamylase. In contrast with these traces of glucose, another study reported a glucose content of 22% (35).

With the aid of the detailed investigation, the cytochemical results of de Souza et al. (36), published in 1976, could be further interpreted. Carbohydrate was demonstrated at the cell surface with Thiery's reagent and as a ConA-positive reaction, the latter being due to α-D-mannopyranosyl units and not β-mannan, glucuronoxylan, or amylose. Intense staining was observed in the pellicular membrane of the flagellar pocket but was less intense in the membrane of the cell.

McGhee and Cosgrove (11) showed that *Herpetomonas muscarum muscarum* agglutinates with lectins specific for α-*D*-mannopyranosyl, *N*-acetylglucosaminyl, but not fucopyranosyl units. A structural variation was observed in the sub-species *H. muscarum ingenoplastis*, whose polysaccharide contains fucopyranosyl residues. In both organisms, the agglutination factors are distributed over the cell and flagellar surfaces.

3.5 Polysaccharides of *Leptomonas samueli* and *Leptomonas collosoma*

Partial chemical structure of carbohydrates of promastigotes of *Leptomonas samueli* have been determined by Palatnik et al. (27) and compared with those of *H. samuelpessoai*. A number of similarities were noted, including the presence of xylose, mannose, and glucuronic acid, whose proportions varied with the age of culture. However, the resemblance was not complete as the β-mannan was not detected. Also, the specific rotation of a fraction (IF) obtained by phenol extraction, followed by chloroform-methanol-water treatment, which removed soluble impurity, was +34°, as compared with +69° and differences were present in the ^{13}C-n.m.r. spectra. The most pronounced similarities were the presence of 2-*O*- (or 4-*O*) substituted xylopranosyl units having the β-configuration, according to a C-1 signal at δ 104.0, and the electrophoretic behaviour of IF's and the corresponding soluble fractions. Their monosaccharide compositions were also identical.

The resemblance of some structures of IF and those of hemicellulose, which both occur in the intestine of the host insect, *Zelus leucogrammus*, has been suggested as an example of molecular mimicry, supporting the concept of parasite fitness.

The membrane of *Leptomonas collosum* contains only a small proportion of xylose, when compared with glucose, mannose, and galactose components (37). Thus, of the protozoa mentioned above, only the polysaccharides of *H. samuelpessoai*, *L. samueli*, and the *Crithidia* sp. are rich in xylose.

4. POLYSACCHARIDES AND GLYCOCOMPLEXES OF HIGHER TRYPANOSOMATIDS

Trypanosoma and *Leishmania* spp. are higher trypanosomatids, residing, during their life cycle in vertebrate, as well as invertebrate hosts (11). They have been extensively investigated, as they are responsible for human diseases, the most serious being

Chagas' disease, sleeping sickness, and kala-azar. More recently, the carbohydrate components of these organisms have been examined because of their antigenic and immunogenic properties. As a group, they contain comparitively few polysaccharides, but glycocomplexes and phosphorodiester containing polysaccharides are frequently encountered. Nowadays, advanced chemical techniques are available for analysis of such materials, as are agglutination reactions involving lectins and immunological cross-reactions. The most attention has been paid to _T. cruzi_ and _Leishmania_ spp., contrasting with the relatively few detailed chemical investigations that have appeared on African trypanosomes, such as _T. brucei_.

4.1 Glycocomplexes of _Trypanosoma cruzi_

The main monosaccharide components of _T. cruzi_ are galactose and mannose. The first investigation was carried out by von Brand et al. (38) in 1959, who vigorously extracted blood forms with hot concentrated aqueous alkali and obtained a precipitate by treatment with excess ethanol. The product gave, on electrophoresis in borate, a single band that was carbohydrate- and phosphorus-positive. However, their report that galactose was the sole hydrolysis product seems erroneous. In 1969, Gonçalves and Yamaha (39) obtained a product, following chloral hydrate extraction, which was antigenic and contained units of galactose, mannose, glucose, glucosamine, and xylose.

Five years elapsed before the beginning of a series of in-depth studies on _T. cruzi_ glycocomplexes by the group of Walter Colli, in São Paulo. Firstly, agglutination studies on epimastigotes showed that ConA reacts at the cell surface, indicating the location of carbohydrate containing substances (40). These were isolated by phenol-water extraction, followed by addition of the cooled water layer to excess ethanol. The resulting precipitate, on polyacrylamide gel electrophoresis, gave 4 distinct bands, designated A to D, A to C proving to be glycoproteins by their susceptibility to the action of pronase. Band D arose from a substance rich in carbohydrate and lipid (41) and was tentatively called lipopeptidophosphoglycan, or LPPG, a name which stuck (42-44). It could be freed from glycoproteins, on a larger scale, by their removal with a mixture of chloroform-methanol-water (10:10:3) and was found to be the major antigen of the cell. Preliminary studies on its chemical structure showed it to contain 60% of carbohydrate with mannose, galactose, and glucose in a 35:22:1 molar ratio with 0.8% of glucosamine, 9.5% of protein, 2% of phosphorus, and 2 ninhydrin-positive components suggested to be sphingosine bases (42, 43). Ferguson et al. (45) later showed its apparent mol. wt. to be 10^4. Components of bands A to C contained galactopyranosyl units since successive galactose oxidase and potassium borotritiide treatment resulted in incorporation of label, whereas LPPG contained galactofuranosyl residues, since periodate oxidation gave rise to formaldehyde. As structural confirmation, reduction of oxidized material with potassium borotritiide, followed by acid hydrolysis, provided labeled arabinose (46).

In the light of these experiments, it is significant that both epi- and trypomastigote forms react with ConA at the cell surface, consistent with the presence of terminal and/or 2-O-substituted α-D-mannopyranosyl units with the possibility of α-D-glucopyranosyl nonreducing end-groups (41, 47). ConA induced agglutination of epimastigotes was inhibited by glycoproteins of bands A to C, indicating that these are present in the membrane (48). The cell surface location was confirmed by Gottlieb (49), since epi- and trypomastigotes absorbed rabbit anti-complex serum, the antigen being visualized by fluoroscein and ferritin conjugated antibodies. The cell surface distribution was even.

The minor components of bands A to C seem to correspond to those of Ferguson et al. (45) at GP-24, GP-31, and GP-37, according to their approximate mol. sizes in kDa. They collectively contain galactose, mannose, glucosamine, and a low proportion of glucose.

A variety of other components of different mol. sizes, as determined by electrophoresis, have been isolated from T. cruzi. In 1983, Mendonça-Previato et al. (50) extracted epimastigotes of the Y strain by successive freezing and thawing to give a mixture that was chromatographed on a column of P-10. Of the 4 fractions obtained, designated I to IV, fraction I gave one protein- and carbohydrate-positive band with mol. size 25 kDa and was apparently different from the above GP-24. According to methylation analysis, it contained nonreducing end-groups of mannopyranose (38%) and galactopyranose (11%) with 32% of 2-O-substituted mannopyranosyl units. Since the 49% of the former was not balanced by O-methylated fragments arising from di- or mono-O-substituted units, the carbohydrate must have a low molecular weight or glucosamine or phosphate mono- or diester components, which did not lend themselves to the analysis technique. At the same time Scharfstein et al. (51) demonstrated the antigenis activity of fraction I by precipitation with human antiserum. In 1979, Scott and Hudson (52) isolated GP-90 and showed that it is capable of inducing protective immunity. A trace glycoprotein, GP-72, was isolated by antibody affinity chromatography, taking advantage of its recognition by an anticlonal antibody specific for epimastigotes (53). Its carbohydrate content was 52% and units of galactose, mannose, glucosamine, fucose, xylose, and ribose were present.

The detailed chemical structure of LPPG was studied by de Lederkremer et al. (54) in 1985. Methylation data on LPPG and a partly hydrolyzed sample, in which labile galactofuranosyl units had been removed, were obtained. These indicated that LPPG contains principally nonreducing end-groups of galactofuranose (23.5%) with 2-O (25%), 3-O (22%), 6-O (17%), and 2,3-di-O-substituted units of mannopyranose (3%). Comparison with the nonreducing end-groups (39%) and 2-O (36.5%) and 6-O-substituted units of mannopyranose (20%) of the partly hydrolyzed material indicated that (1→3)-linkages were present in LPPG between galactofuranosyl and mannopyranosyl units. As the partial hydrolyzate was not eluted immediately from a Bio-Gel P-10 column, it cannot have the approximate trisaccharide structure indicated by the methylation data and must contain other components. The discrepancy was resolved by Previato et al. (55), who found it to contain glucosamine and 2-aminoethyl phosphonic acid (2-AEP) as a

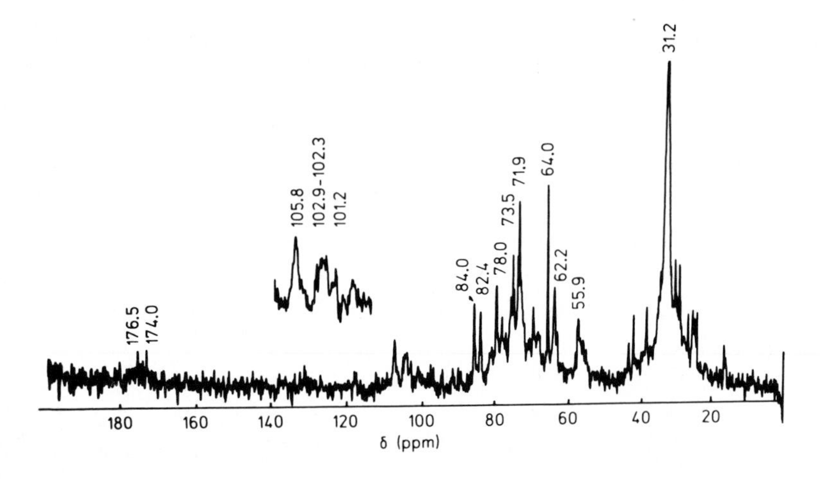

Fig. 3. ^{13}C-n.m.r. spectrum of LPPG in D_2O.

phosphate ester. In the ^{13}C-n.m.r. spectrum of LPPG (Fig. 3), there is a low-field C-1 signal at δ 105.8, corresponding to β-D-galactofuranosyl linkages, and this was resolved into 2 signals using a higher-field spectrometer (55). This type of structure was observed, in 1981 (34), in an oligosaccharide, obtained directly from epimastigotes by extraction with hot aqueous alkali, and whose spectrum contained a signal at δ 106.6 (Fig. 4). With the aid of standards of β-D-galactofuranose linked to various positions of methyl α-D-mannopyranoside, the exact shift served to show that the linkage was (1→3). Another feature of the LPPG spectrum is the presence of a signal at δ 55.9, which can arise from $\underline{C}H_2NH_2$ and/or $\underline{C}HNH_2$ groups, and which was shown to arise principally from the former by a carbon-proton coupled experiment (55).

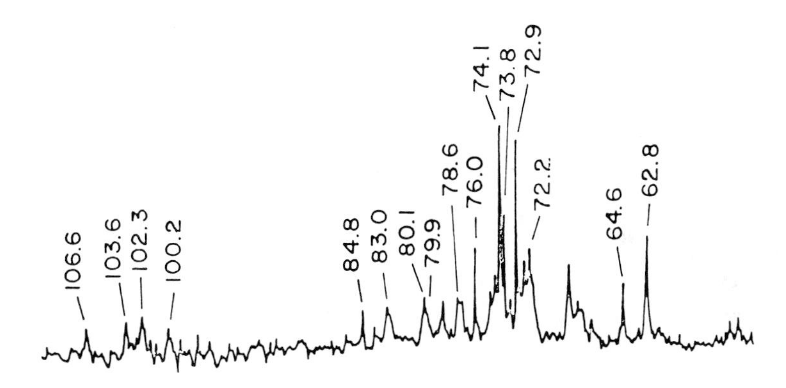

Fig. 4. ^{13}C-n.m.r. spectrum of material obtained from epimastigotes of *T. cruzi* by the action of hot aqueous alkali. The solvent was D_2O at 70° and shifts are expressed as in p.p.m., based on that of Me_4Si = 0.

In early studies, 2-AEP and glucosamine were detected in very small quantities, but it is now clear that they are significant components of LPPG. Originally, it was observed that there was a close correspondence of phosphorus content and the amount of inorganic phosphate liberated on hydrolysis (43). This agreed with the report of Ferguson et al. (56) that 2-AEP was liberated, by acid hydrolysis of cells, in trace amounts. However, Mendonça-Previato et al. (50) treated LPPG with hot aqueous $NaBH_4$-NaOH liberating an oligosaccharide of approximately 24 units and which contained nonreducing end-groups of galactopyranose (16%), mannopyranose (8%), and galactopyranose (4%) with 2-O (19%), 3-O

(47%), and 2,3-di-O-substituted mannopyranosyl residues (16%). Its ^{13}-C-n.m.r. spectrum (Fig. 5) showed it to consist of carbohydrate with assignable signals at δ 106.5 and 106.8 (C-1's of β-D-Gal*f*-(1→3), linked to α-D-Man*p*) with unusual high field signals at δ 23.4 and 28.7 (coupled) and 37.1, corresponding to 2-AEP.

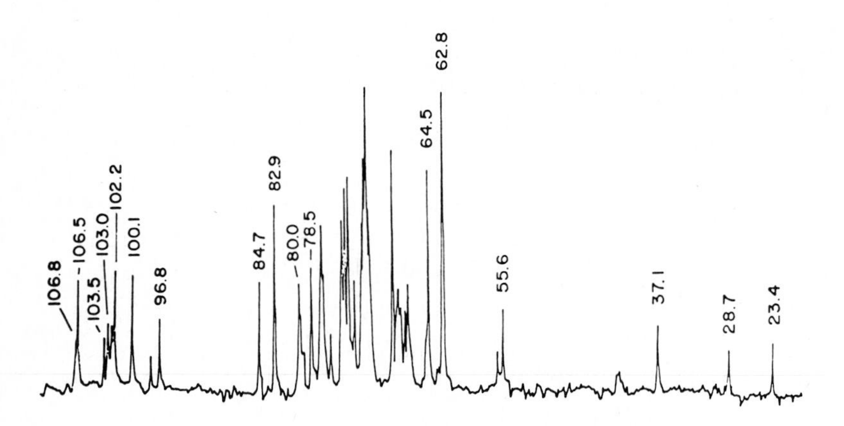

Fig. 5. ^{13}C-n.m.r. spectrum of oligosaccharide in D_2O at 70°, formed from LPPG by the action of hot aqueous $Na^+BH_4^-$-NaOH: numerical values are δ in p.p.m., based on that of Me_4Si = 0.

$$R{-}O{-}\overset{\overset{\displaystyle O}{\|}}{\underset{\underset{\displaystyle OH}{|}}{P}}{-}CH_2CH_2NH_2 \qquad \text{VII}$$

A better concept of the quantity of this material, which should be present in the form of a phosphate ester (VII), is gained from the high field signals of the ^{13}C-n.m.r. spectrum (Fig. 6A) of material, obtained in 15% yield on mild acid hydrolysis of LPPG (55).

Thus, the hydrolytic conditions used by Ferguson et al. (56) were too weak to completely liberate 2-AEP. Also, the ^{31}P-n.m.r. spectrum, as found by Lederkremer et al. (57) contains small signals of 2-AEP (δ-25.00) and phosphate monoesters (δ -4.66, -4.10, and -3.22), with a very large one of phophorodiester (δ 0.87). 2-AEP occurs, in the free and combined form, in many organisms, including the protozoon Tetrahymena thermophila (58), the sea anemone Anthopleura elegantissima as a phosphate ester of glycerol (59), and the snail Megalobulimus paranaguensis whose albumen gland contains a β-D-galactopyranan lightly esterified at O-6 (60).

The glucosamine content of LPPG, originally estimated at 0.8% (42), is much higher, being 9% of the total sugar liberated, including inositol, under the conditions of a conventional amino acid analysis (45). Also, in Fig. 5 2-amino-2-deoxy-α-D-glucopyranosyl signals are present at δ 96.8 and 97.8 (C-1) and at δ 55.6 and 56.6 (C-2). Corresponding signals can be seen in Fig. 6A at δ 96.8 and 55.5 and their size indicates a level of ~6% in LPPG. In neither spectrum was detected a signal of N-acetyl.

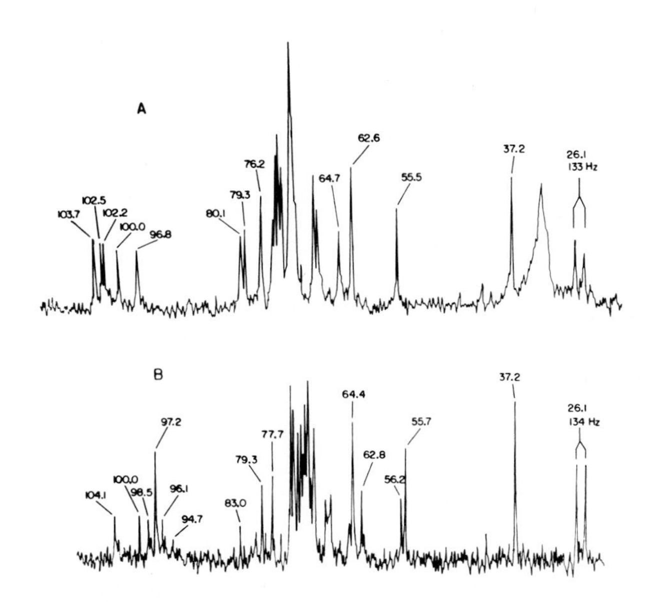

Fig. 6. ^{13}C-n.m.r. spectrum of material obtained from LPPG by mild acid hydrolysis (A) and that obtained by its further treatment under strong hydrolytic conditions (B): the solvent was D_2O at 70° and numerical values are δ in p.p.m., based on Me_4Si.

Evidence is accumulating to show that glucosamine and 2-AEP are closely associated in the LPPG molecule (55). Fig. 6A contains a signal at δ 64.7 that cannot arise from C-6 of β-D-galactofuranosyl units, which are absent, or phosphate mono- or diester which were not detected in the partial hydrolyzate by ^{31}P-n.m.r. spectroscopy. The remaining possibility is that it arises from C-6 of a hexopyranosyl unit attached to 2-AEP as a phosphate ester. The partial hydrolyzate was treated with 1.6 M

TFA at 100°, which is strong enough to hydrolyze mannopyranosyl units, but not those of resistant 2-amino-2-deoxy-α-D-glucopyranose (61) or esters of 2-AEP. Chromatography of the product on a column of Bio-gel P-2 gave a fraction (25% yield) whose elution characteristics were those of a hexose-containing trisaccharide, which gave glucosamine and mannose on hydrolysis with 6 M TFA at 100°. Its ^{13}C-n.m.r. spectrum (Fig. 6B) contained a prominent C-1 signal at δ 97.2, those of C-2 at δ 55.7 and 56.2, and a C-6 signal at δ 64.4, arising from glucosaminyl units. The signal at δ 62.8 has a shift typical of C-6 of mannopyranosyl and not glucosaminyl units (62). Particularly evident is that the major signals belong to the units of glucosamine and to 2-AEP, suggesting that it could be located at O-6 as an ester.

Analysis of an oligosaccharide, obtained from epimastigotes of T. cruzi by treatment with hot aqueous alkali, furnished a preview of some of the structures of LPPG, from which it was partly derived (34). It also proved useful in antigenic studies on the protozoon. It was isolated via precipitation with ethanol and contained galactose, mannose, glucose, and inositol in a molar ratio of 22:50:4:24. Methylation analysis gave rise to fragments corresponding to nonreducing end-groups of mannopyranose (∿ 26%) and galactofuranose (9%) with 2-O-substituted mannopyranosyl (17%) and 4-O-substituted glucopyranosyl units (27%). The presence of glycogen is suggested, as 4,6-di-O-substituted glucopyranosyl units are present and since the fragment arising from its nonreducing end-groups would be hidden by the acetate of 2,3,4,6-tetra-O-methylmannitol. Glycogen was detected in a greater than expected proportion since lower molecular weight O-methylated fragments were lost in the Haworth dialysis procedure. The 2-O-substituted α-D-mannopyranosyl units are present as short chains of up to 3 units, according to the mixture of oligosaccharides formed via partial acetolysis. As mentioned above, the galactofuranosyl units were shown to have the β-configuration by ^{13}C-n.m.r. spectroscopy, which also demonstrated the (1⟶3)-linkage with adjacent mannopyranosyl residues.

In 1983, Mendonça-Previato et al. (50) studied the effect of inhibitors on the agglutination of rabbit serum raised against LPPG with the homologous antigen. The level was 100% with material obtained from LPPG by reductive alkaline treatment, 85% for methyl 3-O-β-D-galactopyranosyl-α-D-mannopyranoside, and 20% for methyl β-D-galactofuranoside. In 1986, Schnaidman et al. (63) showed that the disaccharide structure is an antigenic determinant in humans. Chagasic serum plus the oligosaccharide, formed on alkaline hydrolysis of cells, has its agglutination level 100% inhibited by the (1→3)-linked methyl glycoside, as compared with 75% inhibition with the (1→6)-isomer. A comparable (1→3)-linked structure occurs in the glucogalactomannan of Dactylium dendroides (64) and its agglutination was only 33% inhibited with the (1→3)-linked methyl glycoside as compared with 81% for the (1→6)-isomer. The polysaccharide was ineffective in challenge experiments, using lab. animals, against infection with T. cruzi.

Milani (65) recently tested synthetic agglutination inhibitors having structure VIII and incorporating sugar units

MONO-(or DI)-SACCHARIDE-HYDROXYAZELAIC ACID-POLYLYSINE VIII

β-D-Galp, β-D-Galf, β-D-Galf-(1→3)-α-D-Manp, and α-D-Manp-(1→2)-α-D-Manp (66), which are present in T. cruzi glycocomplexes. All inhibitors were effective against the agglutination of epimastigotes with various human Chagasic sera. In the case of agglutination with trypomastigotes, the (1→2)-linked mannobiose derivative was an effective inhibitor, those with galactosyl units being less active. The (1→3)-linked complex was inactive, indicating the absence of this epitope on the surface, although it may be masked by sialic acid residues, which are present in glycocomplexes of T. cruzi (67-69). Apparently, sialic acid is not synthesized directly by the protozoon, but is transferred from substrates such as fetuin or sialyl lactose present in the serum (70, 71). It is possible that a mannobiosyl epitope is a component of epi- and trypomastigotes since they react with ConA (47). It could also be present in the GP-75 complex, which is a common component of epimastigotes and blood and culture forms of trypomastigotes (72).

Parodi and Cazzulo (73) isolated bound oligosaccharides from T. cruzi, following incubation with ^{14}C-labeled glucose. They contained $GlcMan_{7-9}GlcNAc$, having structures related to $Man_7GlcNAc$-β-(1→4)-GlcNAc, isolated from Crithidia fasciculata (74). However, as the former structure differs from those found in T. cruzi, the biosynthetic pathway leading to glycocomplexes should be further investigated.

4.2 Polysaccharides and Glycocomplexes of other *Trypanosoma* spp.

Glycocomplexes of other Trypanosomatidae have not been structurally examined in sufficient detail to make many interspecies comparisons. However, non-acetylated glucosaminyl units were found in the variant surface glycoprotein of *T. brucei* (75, 76) and polysaccharides of *T. mega* (32) and *T. conorhini* (77). The presence of 2-AEP seems to be unique to *T. cruzi*, although certain multiplets in the ^{1}H-n.m.r. spectrum of the *T. brucei* glycoprotein (75) are not inconsistent with this structure. This parasitic protozoon is undoubtedly the most important of this group, but only preliminary studies have been carried out on the chemical structure of its carbohydrate components. To date, it is known to contain a glycoprotein of 65 kDa, having galactose, mannose, glucose, and glucosamine in a molar ratio of 11:11:4:5, according to chemical staining and lectin reactions (78). The glycosidic linkages of the glucosaminyl units being highly resistant to acid hydrolysis (76).

T. mega has the african toad, *Bufo regularis*, as its vertebrate host. Growth on a medium containing brain-heart infusion gave cells which were extracted with hot aqueous alkali and a product isolated following ethanol precipitation. It contained units of L-fucose, xylose, D-mannose, D-galactose, and *myo*-inositol. The proportion of xylose and mannose was higher in cells at the exponential phase of growth, whereas fucose and galactose predominated in older cells, galactose representing 59% of the preparation. It contained a true polysaccharide with nonreducing end-groups of glucopyranose (10%) and fucopyranose (14%) with 2-*O* (10%), 3-*O* (12%), and 2,3-di-*O*-substituted galactopyranosyl units (31%). According its specific rotation of -2°, the predominant galactopyranosyl units have the β-configuration (32). In another investigation (33), *T. mega* was, after removal of lipids, extracted with phenol-water and liberated material purified by gel filtration chromatography. A fraction was obtained containing carbohydrate (36%), protein (13%), phosphate (10%), and hexosamine (4%) and whose electrophoresis profile contained a sharp band at M_r ∿55000 (carbohydrate and protein positive) and a diffuse one at ∿22000 (carbohydrate and lipid positive). Units of galactose, mannose, glucose, and fucose were present in a 8:2:1:3 molar ratio and methylation analysis indicated nonreducing end-groups of fucopyranose (25%) and mannopyranose (10%) with 2,3-di-*O*-substituted galactopyranosyl units (10%). As in other cases, 4-O-substituted glucopyranosyl units were identified at a higher level than expected, the appropriate O-methylated fragment representing 32% of the mixture. A fraction soluble in chloroform-methanol-water was obtained from the above product and had a somewhat similar chemical structure, except that the content of nonreducing end-groups of fucopyranose was very low.

A carbohydrate-rich fraction was obtained from epimastigote forms of T. conorhini, consisting of neutral carbohydrate (67%), protein (15%), phosphorus (5%), and hexosamine. Mannose, galactose, and xylose units were present in a molar ratio of 1:1.8:1.8 and, according to chemical analyses and lectin agglutination experiments, the fraction contained nonreducing end-groups of β-D-galactopyranose (24%), α-D-mannopyranose (15%), and β-D-xylopyranose (10%) with 3-O-substituted galactopyranosyl units (14%). Also present in significant proportions were 3-O-substituted galactopyranosyl and 2,6-di-O-substituted mannopyranosyl units. ^{13}C-n.m.r. spectroscopy gave a signal at δ 64.7, consistent with O-6 substituted by phosphate mono- or diester. A C-1 signal at δ 96.6 indicates that 2-amino-2-deoxy-α-glucopyranosyl units were present, their great resistance to acid hydrolysis also being observed. T. conorhini had antigenic determinants in common with those of T. cruzi, as the above fraction gave a precipitation line with rabbit serum raised against whole cells of the latter (77).

4.3 Polysaccharides and Glycocomplexes of Leishmania spp.

Manson-Bahr and Southgate (79) noted the existence of a considerable degree of cross immunity between L. adleri, which infects lizards, and L. donovani, the causative agent of kala-azar, and some degree of cross immunity between L. adleri and L. braziliensis, which causes American muco-cutaneous leishmaniasis, and between L. enrietti (guinea pig parasite), and L. tarentolae (lizard parasite). The results were in agreement with serological studies of Adler (80). These observations are likely due to structural similarities of carbohydrate moieties at the cell surface and, so far, a number of investigations have been carried out on their chemical structures.

Wallace, in 1966 (28), found that the aqueous alkaline extract of L. donovani contained a polymer having units of galactose (34%), mannose (50%), and glucose (16%). Dwyer (81) et al. (82, 83) reported reactions at the cell surface with the lectins, ConA and PHA.

Semprevivo and MacCleod (84) isolated material called glycopeptidophosphosphingolipid from a culture of L. donovani on a defined medium. Turco and collaborators isolated and characterized the lipophosphoglycan (LPG), present on the cell surface of this protozoon. The glycocomplex was extracted from promastigote forms by the action of H_2O-EtOH-Et_2O-pyridine-NH_4OH (15:15:5:1:0.017) and comprised more than one half of polymeric carbohydrate. It was very labile to acid hydrolysis, forming a phosphorylated galactosyl mannose. A free disaccharide was obtained by the action of phosphatase: acid hydrolysis gave rise to mannose and galactose (85) and methylation analysis indicated a 4-O-galactopyranosyl-mannose structure. Examination of a tetrasaccharide, also liberated on acid hydrolysis, showed the location of the phosphate ester and the cofigurations of the glycosidic linkages. Conventional ^{1}H-n.m.r. spectroscopy indicated β-galactopyranosyl and α-mannopyranosyl structures and two dimensional interproton correlated spectroscopy (COSY) of the di- and tetrasaccharide was consistent with phosphorylation at O-6 of the galactopyranosyl units. These data are consistent with repeating structure IX for the polysaccharide. Approximately 16 of such disaccharide phosphate units are present in the overall structure, which is anchored to the membrane by a derivative of phosphatidylinositol (86).

-β-Galp-(1→4)-α-Manp-(1→P—6)- IX

Another component of L. donovani, also grown on a BHI-based medium, was investigated by Banerji et al. (87). Extraction of cells with phenol-water, followed by fractionation on a column of Sephadex G-100 gave an electrophoretically homogeneous product. However, on Sephadex G-150 chromatography, a major and 2 minor fractions were obtained. A complex structure was proposed for the main component, with a main chain of 4-O-substituted α-D-glucopyranosyl units substituted at O-3 with α-D-Manp-(1→3)-Galp, Araf-(1→2)-Galp, and Araf-(1→4)-Manp side chains in a 1:1:2 ratio. Such a structure would account for the reaction of ConA at the cell surface.

According to Handman and Goding (88), L. major, the cause of Old World cutaneous leishmaniasis, has bound and excreted molecules that can be biosynthetically labeled with glucose, galactose, phosphate, and sulfate, but not with amino acids. As the bound form can assimilate labeled palmitate, a glycolipid structure was suggested. Galactose, mannose, and phosphate were liberated on acid hydrolysis, suggesting a possible resemblance to LPG. However, a difference does exist, as the glycolipid is susceptible to galactose oxidase and binds to ricin, consistent with unsubstituted nonreducing end-groups of β-D-galactopyranose.

The carbohydrate components of such glycolipids are possibly dominant antigens in man (89). It is of great interest that Mitchell and Handman (90) found that the L. major glycolipid can immunize mice against infection, contrasting with the free carbohydrate, which is a suppressogenic, disease-promoting antigen.

L. adleri, appeared to be responsible for natural human immunity, in Kenya, against kala-azar (91). In an investigation of its carbohydrate components, Palatnik et al. (92) treated promastigotes with phenol-water and the residue from the aqueous layer was then extracted with chloroform-methanol-water. The soluble complex contained neutral carbohydrate (61%), glucosamine (11%), phosphate (17%), and protein (25%). (The insoluble fraction also contained carbohydrate and 14% of phosphate). The main structures were nonreducing end-groups (30%) with 2-O (26%) and 6-O-substituted units of mannopyranose (25%). The cell residue, remaining after phenol-water extraction, was treated with hot aqueous alkali, giving rise to a polysaccharide with units of galactose, mannose, glucose, and xylose, but little phosphorus. Its main structures, as determined by methylation analysis, were nonreducing end-groups (14%) and 3-O-substituted units of galactopyranose (25%) and 4-O-substituted glucopyranosyl residues (17%), although the latter percentage is too high to be accurate. Another fraction was obtained by aqueous extraction of cells and contained carbohydrate (26%), protein (48%), and phosphate (2%). The principal carbohydrate components were nonreducing end-groups (51%), with 2-O (16%) and 2,6-di-O-substituted units of mannopyranose (16%).

The carbohydrate-containing substances of L. tarentolae have been examined in detail. The first investigation was by Barreto-Bergter et al. (30). Cells were grown on modified Eagle's medium, which contained fetal calf serum (medium A), and on another which had fetal calf serum and brain-heart infusion (medium B). Those of medium A were extracted with hot aqueous alkali furnishing polymeric material, which contained galactose (14%), mannose (68%), glucose (6%), and inositol (12%). The main component structures, according to methylation analysis, were nonreducing end-groups (13%) and 4-O-substituted units of glucopyranose (31%), with 20% of 2-O-substituted residues of mannopyranose, much less than would be expected from the mannose content of the preparation. In contrast, cells grown on medium B gave rise to a product containing 8% of nonreducing end-groups and 13% of 5-O-substituted arabinofuranosyl units (see ref. 87). Xavier et al. (31) disrupted cells of L. tarentolae and liberated material was fractionated on a ConA-Sepharose column, giving a bound fraction containing protein (74%) and carbohydrate (26%). The latter gave mannose (18%), glucose (22%), and galactose (60%), on acid hydrolysis. Also, a phenol-water extract of cells was treated with hot aqueous alkaline borohydride furnishing a fraction having galactose (82%), mannose (12%), and glucose (6%). It was a predominantly linear β-D-galactopyranan with (1→3)-linkages. the only other structure of note being 2-O-substituted mannopyranosyl units (11%).

L. tropica major, which causes Oriental cutaneous leishmaniasis, formed a predominant exocellular glycoconjugate of 20-67 kDa, according to electrophoresis. It contained galactopyranosyl units, phosphorus, and sulfate, and was not present in L. mexicana, the causative agent of diffuse cutaneous leishmaniasis, or L. donovani. However, similar, but antigenically different glycocomplexes are were produced (93).

L. mexicana amazonensis forms a β-D-mannopyranan, which is (1→2)-linked, leading to the taxonomic inferences previously mentioned (Section 3.3), and is the only Leishmania species in which it has been so far found. Aqueous alkaline extraction of cells of L. mexicana mexicana gave polymeric material containing galactose (29%), mannose (46%), glucose (20%), and fucose (5%; 94).

Saline extraction of L. enrietti gave an antigen which migrated as a single electrophoretic band and was composed of D-mannose, D-galactose, D-glucose, and apparently N-acetyl-galactosamine. Agglutination occurred with ConA (95).

5. CONCLUSIONS

The role of carbohydrate moieties, in protozoa, as antigenic and immunogenic determinants has been established. It only remains to be seen if knowledge of these structures can be utilized for diagnosis of diseases and in the production of vaccines. In chemical analyses, refinements are needed in the methylation technique, which is difficult to apply to carbohydrates containing phosphate ester groups and those with glucosaminyl units. There seems little doubt, however, that n.m.r. spectroscopy will be extensively used. The ^{13}C technique can be applied to the detection of amino sugars and in the characterization of sugar units containing esters of 2-AEP, and phosphate mono- and diesters (55, 60, 96, 97). ^{31}P-n.m.r. spectroscopy can be used to distinguish between 2-AEP, organic phosphate mono- and diester, and inorganic phosphate (57, 98).

REFERENCES

1 A. Lwoff, M. Ionesco, and A. Gutman, Biochim. Biophys. Acta, 4 (1950), 270-275.

2 E.J. Bourne, M. Stacey, and J.A. Wilkinson, J. Chem. Soc., (1950) 2694-2698.

3 A.E. Oxford, J. Gen. Microbiol., 5 (1951) 88-90.

4 G. Forsyth and E.L. Hirst, J. Chem. Soc., (1953) 2132-2135.

5 G. Forsyth, E.L. Hirst and A.E. Oxford, J.Chem. Soc., (1953) 2030-2033.

6 A.R. Archibald, E.L. Hirst, D.J. Manners, and J.F. Ryley, J. Chem. Soc., (1960) 556-560.

7 D.J. Manners and J.F. Ryley, Biochem. J., 52 (1952) 480-482.

8 J.G. Feinberg and W.J.J. Morgan, Brit. J. Exp. Pathol., 34 (1953) 104-118.

9 D.J. Manners and J.F. Ryley, Biochem. J., 59 (1955) 369-372.

10 A.R. Archibald, D.J. Manners, and J.F. Ryley, Chem. and Ind. (London), (1958) 1516-1517.

11 R.B. McGhee and W.B. Cosgrove, Microbiol. Reviews, 44 (1980) 140-178.

12 W.B. Cosgrove and W.L. Hanson, Amer. Zool., 2 (1972) 401.

13 B.E. Brooker, Parasitology, 72 (1976) 259-267.

14 M. Gottlieb, P. Lanzetta, and J. Berech Jr., Exptl. Parasitol., 32 (1972) 206-210.

15 M. Gottlieb., Biochim. Biophys. Acta, 541 (1978) 444-458.

16 M.L.S. Gütter, master's thesis, Escola Paulista de Medicina, São Paulo, 1988 pp. 1-98.

17 P.A.J. Gorin, J.O. Previato, L. Mendonça-Previato, and L.R. Travassos, J. Protozool., 26 (1979) 473-478.

18 J. O. Previato, L. Mendonça-Previato, and P. A. J. Gorin, Carbohydr. Res., 70 (1979) 172-174.

19 A. Misaki, M. Seto, and I. Azuma, J. Biochem., 76 (1974) 15-27.

20 P.A.J. Gorin and M. Mazurek, Carbohydr. Res., 48 (1976) 171-186.

21 W.N. Haworth, R.L. Heath, and S. Peat, J. Chem. Soc., (1941) 833-842.

22 B.A. Newton, in: W.H.R. Lumsden and D.A. Evans (Eds), Biology of the Kinetoplastida, Academic Press, London and New York, vol.1, 1976, pp. 405-434.

23 E.P. Camargo, C.L. Barbieri, and J.V. Jankevicius, J. Immun. Methods, 52 (1982) 245-253.

24 J.O. Previato, L. Mendonça-Previato, R.Z. Lewanczuk, L.R. Travassos, and P.A.J. Gorin, Exptl. Parasitol., 53 (1982) 170-178.

25 J.R. Baker, in: A.E.R. Taylor (Ed), Evolution of Parasites: Symposia of the British Society of Parasitology, Blackwell, Oxford, 1965, pp. 1-27.

26 L. Mendonça-Previato, P.A.J. Gorin, and J.O. Previato, Biochemistry, 18 (1979) 149-154.

27 C.B. Palatnik, J.O. Previato, P.A.J. Gorin, and L. Mendonça-Previato, Comp. Biochem. Physiol., 86 (1987) 593-599.

28 F.G. Wallace, Exptl. Parasitol., 18 (1966) 124-193.

29 J.O. Previato, M.T. Xavier, P.A.J. Gorin, and L. Mendonça-Previato, J. Parasitol., 70 (1984) 449-450.

30 E. Barreto-Bergter, J. O. Previato, and P. A. J. Gorin, Carbohydr. Res., 97 (1981) 156-160.

31 M.T. Xavier, J.O. Previato, P.A.J. Gorin, and L. Mendonça-Previato, Comp. Biochem. Physiol., 88 (1987) 101-104.

32 E. Barreto Bergter and P.A.J. Gorin, Comp. Biochem. Physiol., 77 (1984) 413-417.

33 R.L. dos Santos, P.A.J. Gorin, and E. Barreto-Bergter, J. Protozool., 34 (1987) 298-302.

34 P. A. J. Gorin, E. M. Barreto-Bergter, and F. S. da Cruz, Carbohydr. Res., 88 (1981) 177-188.

35 M.J.G. Esteves, C.S. Alviano, J. Angluster, and W. de Souza, Eur. J. Cell Biol., 20 (1979) 113-115.

36 W. de Souza, M.M. Bunn, and J. Angluster, J. Protozool., 23 (1976) 329-333.

37 R.C. Hunt and D.J. Ellar, Biochim. Biophys. Acta, 339 (1974)

173-189.

38 T. von Brand, P. McMahon, E.J. Tobie, M.J. Thompson, and E. Mossetig, Exptl. Parasitol., 8 (1959) 171-181.

39 J.M. Gonçalves and T. Yamaha, Amer. J. Trop. Med. Hyg., 72 (1969) 39-44.

40 M.J.M. Alves and W. Colli, J. Protozool., 21 (1974) 575-578.

41 M.J.M. Alves and W. Colli, FEBS Letters 52 ((1975) 188-190.

42 R.M. de Lederkremer, M.J.M. Alves, G.C. Fonseca, and W. Colli, Biochim. Biophys. Acta, 444 (1976) 85-96.

43 R.M. de Lederkremer, C.T. Tanaka, M.J.M. Alves, and W. Colli, Eur. J. Biochem., 74 (1977) 263-267.

44 R.M. de Lederkremer, O.L. Casal, C.T. Tanaka, and W. Colli, Biochem. Biophys. Res. Commun., 85 (1978) 1268-1274.

45 M.A.J. Ferguson, D. Snary, and A.K. Allen, Biochim. Biophys. Acta, 842 (1985) 39-44.

46 R.M. de Lederkremer, O.L. Casal, M.J.M. Alves, and W. Colli, FEBS Letters, 116 (1980) 25-29.

47 E. Chiari, W. de Souza, A.J. Romanha, C.A. Chiari, and Z. Brener, Acta Tropica, 35 (1978) 113-121.

48 M.J.M. Alves, J.F. da Silveira, C.W.R. de Paiva, C.T. Tanaka, and W. Colli, FEBS Letters, 99 (1979) 81-85.

49 M. Gottlieb, Exptl. Parasitol., 45 (1978) 200-207.

50 L. Mendonça-Previato, P. A. J. Gorin, A. F. Braga, J. Scharfstein, and J.O. Previato, Biochemistry, 22 (1983) 4981-4987.

51 J. Scharfstein, M.M. Rodrigues, C.A. Alves, W. de Souza, J.O. Previato, and L. Mendonça-Previato, J. Immunol., 131 (1983) 972-976.

52 D. Snary and L. Hudson, FEBS Letters, 100 (1979) 166-170.

53 D. Snary, M.A.J. Ferguson, M.T. Scott, and A.K. Allen, Molec. Biochem. Parasitol., 3 (1981) 343-356.

54 R.M. de Lederkremer, O.L. Casal, A. Couto, and W. Colli, Eur. J. Biochem., 151 (1985) 539-542.

55 J.O. Previato, L. Mendonça-Previato, M.T. Xavier, M. Mazurek, and P.A.J. Gorin, to be published.

56 M.A.J. Ferguson, A.K. Allen, and D. Snary, Biochem. J., 207 (1982) 171-174.

57 R.M. de Lederkremer, O.L. Casal, M.J.M. Alves, and W. Colli, Biochem. International, 10 (1985) 89-96.

58 J.S. Smith and N.M. Lepak, Arch. Biochem. Biophys., 213 (1982) 565-572.

59 J.S. Kittredge, E. Roberts, and D.G. Simonsen, Biochemistry, 1 (1962) 624-628.

60 J.D. Fontana, J.H. Duarte, C.B.H. Gallo, M. Iacomini, and P.A.J. Gorin, Carbohydr. Res., 143 (1985) 175-183.

61 R. J. Ferrier and P. M. Collins, in: idem (Eds) Monosaccharide Chemistry, Penguin, New York, 1972. pp. 1-318.

62 M. Bock and C. Pedersen, Adv. Carbohydr. Chem. Biochem., 41 (1988) 27-66.

63 B.B. Schnaidman, N. Yoshida, P.A.J. Gorin, and L.R.Travassos, J. Protozool., 33 (1986) 186-191.

64 C. Kemmelmeier, and G.T. Zancan, Exptl. Mycol., 5 (1981) 339-348.

65 S.L. Milani, master's thesis, Escola Paulista de Medicina, São Paulo, 1987, pp. 1-212.

66 D.S. Tsui and P.A.J. Gorin, Carbohydr. Res., 156 (1986) 1-8.

67 M.E.A. Pereira, M.A. Loures, F. Villalta, and A.F.B. Andrade, J. Exptl. Med., 152 (1980) 1375-1382.

68 T. Souto-Padron, T.U. de Carvalho, E. Chiari, and W. de Souza, Acta Trop., 41 (1984) 215-225.

69 A.N. Confalonieri, N.F. Martin, B. Zingales, and W. Colli, Biochem. International, 7 (1983) 215-222.

70 J.O. Previato, A.F.B. Andrade, M.C.V. Pessolani, and L. Mendonça-Previato, Mol. Biochem. Parasitol., 16 (1985) 85-95.

71 M.M. Piras, D. Henriquez, and R. Piras, Mol. Biochem. Parasitol., 22 (1987) 135-143.

72 N. Nogueira, S. Chaplan, J.D. Tydings, J. Unkeless, and Z. Cohn, J. Exp. Med., 153 (1981) 629-639.

73 A.J. Parodi and J.J. Cazzulo, J. Biol. Chem., 257 (1982) 7641-7648.

74 A.J. Parodi, L.A.Q. Allue, and J.J. Cazzulo, Proc. Natl. Acad. Sci., U.S.A., 78 (1981) 6201-6205.

75 M.A.J. Ferguson, M.G. Low, and G.A.M. Cross, J. Biol. Chem., 260 (1985) 14547-14555.

76 A.-M. Strang, J.M. Williams, M.A.J. Ferguson, A.A. Holder, and A.K. Allen, Biochem. J., 234 (1986) 481-484.

77 M.C.V. Pessolani, L. Mendonça-Previato, A.F.B. Andrade, P.A.J. Gorin, and J.O. Previato, Mol. Biochem. Parasitol., 26 (1987) 193-202.

78 J.G. Johnson and G.A.M. Cross, J. Protozool., 24 (1977) 584-591.

79 P.E.C. Manson-Bahr and B.A Southgate, J. Trop. Med. Hyg., 67 (1964) 79-84.

80 S. Adler, Adv. Parasitol., 2 (1964) 35-96.

81 D.M. Dwyer, Science, 184 (1974) 471-478.

82 D.M. Dwyer and M. Gottlieb, Mol. Biochem. Parasitol., 10 (1984) 139-150.

83 D. Zilberstein and D.M. Dwyer, Mol. Biochem. Parasitol., 13 (1984) 327-336.

84 L.H. Semprevivo and M.E. MacCleod, Biochem. Biophys. Res. Commun., 103 (1981) 1179-1185.

85 S.J. Turco, M.A. Wilkerson, and D.R. Clawson, J. Biol. Chem., 259 (1984) 3883-3889.

86 S.J. Turco, S.R. Hull, P.A. Orlandi Jr., S.D. Shepherd, S.W. Homans, R.A. Dwek, and T.W. Rademacher, Biochemistry, 26 (1987) 6233-6288.

87 N. Banerji, A.K. Das, P.C. Majumder, S.B. Bhattacharya, and A.K. Sen, Carbohydr. Res., 159 (1987) 328-335.

88 E. Handman and J.W. Goding, EMBO J., 4 (1985) 329-336.

89 M. Landner, S. Frankenburg, G. M. Slutzky, and C. L. Greenblatt, Parasite Immunol., 5 (1983) 278-286.

90 G. Mitchell and E. Handman, Parasite Immunol., 8 (1986) 255-263.

91 B.A. Southgate and P.E.C. Manson-Bahr, J. Trop. Med. Hyg., 70 (1967) 29-33.

92 C.B. Palatnik, J.O. Previato, P.A.J. Gorin, and L. Mendonça-Previato, Molec. Biochem. Parasitol., 14 (1985) 41-54.

93 E. Handman, C.L. Greeblatt, and J.W. Goding, EMBO J., 3 (1984) 2301-2306.

94 J.E. Fiorini, H.P. Azevedo, J. Angluster, W. de Souza, I. Roitman, and C.S. Alviano, in: L.R. Travassos (Ed), VIIIth Annual Meeting on Basic Research on Chagas' Disease, Caxambú, Brazil, 9-10 November, CNPq, 1981, p. 84.

95 G. I. Pardoe, R. Hahn, H. Jaquet, and W. Wojnarowski, Experientia, 31 (1975) 723.

96 P.A.J. Gorin, Can. J. Chem., 51 (1973) 2105-2109.

97 P.A.J. Gorin and M. Mazurek, Can. J. Chem., 52 (1974) 3070-3076.

98. A.J.R. Costello, T. Glonek, M.E. Slodki, and F.R. Seymour, Carbohydr. Res., 42 (1975) 23-37.

CHEMICAL STRUCTURE, REACTION MECHANISMS AND APPLICATIONS OF GLUCOAMYLASES

J. H. PAZUR

1 INTRODUCTION

Enzymes are unique and highly specific proteins with a wide spectrum of catalytic activities for a multitude of biological reactions. Enzymes are natural products of all living cells and are essential for the maintenance of the life process. Early studies were concerned with the identification, the purification, the chemical structure, and the biological functions of enzymes of all types. More recently, enzymes with industrial applications have received attention and much research has been conducted in this area, particularly in the emerging field of biotechnology (ref. 1). Examples of such enzymes are glucoamylase, thermostable α-amylase, glucose isomerase, pectinase, glucose oxidase, ligninase, cyanide hydratase, and a variety of bacterial proteases and lipases (ref. 2). A most important biotechnological advance in the application area has been the development of the enzymatic process for the conversion of starch to glucose with glucoamylase (ref. 3) and the further conversion of the glucose to fructose with glucose isomerase (ref. 4) to yield high fructose sweeteners. The high fructose sweeteners are being used in increasing amounts as ingredients in foods, confectionaries and soft drinks, and already have captured a large share of the sweetener market. The glucose obtained from starch with glucoamylase is used in the manufacture of many other products of commerce, as for example, citric acid, lactic acid, glutamic acid, ethyl alcohol, sorbitol and crystalline glucose.

Glucoamylase, a (1,4) (1,6)-α-D-glucan glucohydrolase (EC. 3.2.1.3) (ref. 5), is elaborated by many strains of fungi and yeasts. Most of the organisms produce multi-molecular forms or isozymes of glucoamylase and one isozyme generally predominates in the culture filtrates (ref. 6). The recommendation of the Enzyme Commission for identifying isozymes by numbering the isozymes in accordance with the electrophoretic mobilities (ref. 5) will be followed. Thus the isozyme with the fastest mobility will be designated as isozyme I, the second fastest will be isozyme II and so forth. Some strains of the organisms elaborate several isozymic forms of glucoamylase (refs. 6, 7). The chemical properties such as molecular weights and isoelectric points of isozymes of a set may be quite different but the biological properties such as substrate specificity and mechanism of action are very similar. The number, the types and the yield of glucoamylase isozymes produced by a specific fungal strain will be determined by the genetic constitution of the organism and the culture conditions used for growing the cells.

Glucoamylase preparations derived from the *Aspergillus* and the *Rhizopus* groups of fungi are available commercially and are used for starch conversions to glucose (refs. 8, 9). The enzymes from these strains have been extensively studied from the standpoint of production, purification, structure, reaction mechanisms, and industrial applications. The purification of a glucoamylase to homogeneity (ref. 6) was achieved only after the advent of ion exchange chromatography for the purification of proteins (ref. 10). Studies on structure, reaction mechanisms, electrophoretic properties and other aspects of the glucoamylase molecule followed. Oligosaccharides of defined structure also became available around this time and the mechanism of enzyme action could be investigated by use of these substrates (ref. 11).

An important discovery in the early studies on the purification of glucoamylase was the observation that many glucoamylase-producing fungi also synthesized a glucosyltransferase (ref. 12), an enzyme which had been identified in fungal filtrates earlier (ref. 13). During starch conversion with the enzyme mixture the glucosyltransferase converts some glucose units of starch to unfermentable oligosaccharides containing α-D-(1,6) linkages. The production of the latter oligosaccharides is undesirable since the formation of these compounds reduces the yield of glucose obtainable from starch. Genetic mutation techniques (ref. 14) and adsorption of the transferase on resins (refs. 15, 16) have been used successfully to remove the transferase from the glucoamylase preparations.

Recently, the techniques of recombinant DNA technology have been applied to produce recombinant cells containing the glucoamylase gene and capable of synthesizing glucoamylase but not the transferase (refs. 17-19). In the recombinant DNA studies, immunological methods have been used to detect glucoamylase synthesis. Antibodies specific for glucoamylase have now been isolated in pure form by affinity chromatography methods from antisera of animals immunized with the enzyme (ref. 20). These antibodies are available for genetic engineering studies which should proceed at an accelerated pace. All of the topics enumerated in the preceeding are within the scope of this chapter and will be discussed to varying degrees. Much of the discussion will center on isozyme I of the glucoamylase from *Aspergillus niger* as this glucoamylase has been studied in greatest detail. Data obtained with glucoamylases from other organisms including *Aspergillus oryzae*, *Aspergillus awamori*, *Rhizopus delemar*, and *Rhizopus niveus* will also be considered.

2 HISTORICAL

Since early times, starch-hydrolyzing enzymes from animal tissues, from cereal grains and from microorganisms have been used to convert starch to reducing sugars, especially in baking and fermentation applications. It is not always advantageous to use whole cells or tissue extracts in the above applications since such preparations may contain many types of enzymes some of which catalyze undesirable side reactions. Pure enzymes are superior for most uses because of ease of manipulation, high specificity and predictability in function of enzymes. However, the large scale application of enzymes developed slowly in modern technology and has occurred only within the last few decades. Among the enzymes that are utilized are the bacterial proteases, fungal amylases, microbial lipases, and recently microbial isomerases. As mentioned earlier, of the amylases, glucoamylase has become very important for use in starch conversions. Glucoamylase is capable of converting starch directly and quantitatively to glucose and results in high yields of glucose from starch.

The earliest report on the use of a glucoamylase type enzyme for starch conversions was in a patent issued in 1940 (ref. 21). The identity of the enzyme was not established in the early study but in view of the products formed from starch and the rate of formation of these products, the enzyme was most likely glucoamylase. In the studies described in the patent, a variety of the fungal strains of *Aspergillus oryzae*, *Aspergillus flavus*, *Aspergillus niger*, *Aspergillus wentii*, *Monilia sitophila*, *Rhizopus nigricans*, and *Rhizopus tritici* were used. Subsequently additional strains of fungi and several yeasts have been found to produce glucoamylase. These are *Aspergillus awamori*, *Aspergillus phoenicis*, *Rhizopus delemar*, *Coniophora cerebella*, *Mucor rouxianus*, *Humicola lanuginosa*, *Rhizopus niveus*, *Terula thermophila*, *Mucor miehe*, *Sporotrichum thermophile*, *Endomyces bispora*, *Endomyces fibuligosa*, and *Endomyces* IFO 0111.

The early methods for purifying glucoamylase utilized the conventional techniques of protein purification including ammonium sulfate fractionation, organic solvent precipitation (ref. 22) as well as preparative paper chromatography (ref. 23) and paper electrophoresis (ref. 24). The substrates utilized for studying enzyme mechanisms were starch and the two components of starch, amylose and amylopectin. Criteria for purity of the enzymes employed in the early studies were not always presented and consequently much of the data on the mechanism of enzyme action, the chemical structure and related aspects of the enzyme molecule were obtained with preparations of an unknown degree of purity. Often the results of such studies did not give the information needed to develop technological applications but did show that the glucoamylase possessed many desirable properties for starch conversion.

Thus non-purified glucoamylases could not be used to establish the rate of the hydrolysis of different glucosidic linkages in starch, the nature of the intermediates and the occurrence of a single-chain or a multi-chain mechanism of enzyme action. For the latter, pure enzymes were needed as well as pure substrates of defined structures and known arrangement of α-D-(1,6) and α-D-(1,4) glucosidic bonds. Results of experiments on the purification of glucoamylase and on the substrate specificity of the pure enzyme are presented in the next section.

3 METHODS

3.1 Purification of glucoamylase

In order to conduct experiments on the mechanisms of action of glucoamylase and on the determination of the structure of the enzyme it was necessary to prepare a glucoamylase in pure state and to establish molecular homogeneity of the enzyme by suitable methods. With the advent of ion-exchange chromatography for separating proteins (ref. 10) it became possible to effect a purification of glucoamylase utilizing this chromatography technique (ref. 6). The glucoamylase of *Aspergillus niger* was used in the purification experiments. A lyophilized sample of the glucoamylase preparation was dissolved in citrate phosphate buffer of pH 8 to yield a 2% solution and then this solution was introduced on a DEAE-cellulose column which had been equilibrated with the citrate-phosphate buffer. The column was washed with buffer of differing pH by a stepwise change in the buffer from pH 8 to pH 6 to pH 4. Fractions of 10 ml were collected from the columns and each fraction was assayed for the production of reducing sugars from starch. The results are plotted in Fig. 1. It can be seen that there are four groups of fractions

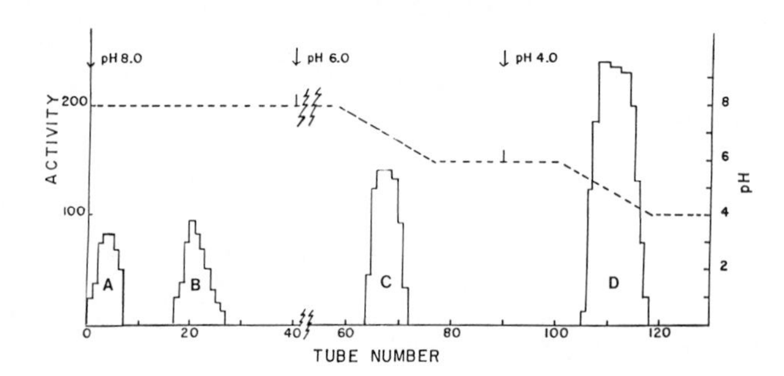

Fig. 1. Elution pattern of the starch and maltose hydrolyzing enzymes of *A. niger* on chromatography on a DEAE-cellulose column: (A) glucosyltransferase, (B) α-amylase, (C) glucoamylase, isozyme II, and (D) glucoamylase, isozyme I (ref. 25).

with starch hydrolyzing activity and thus four amylolytic enzymes. The enzyme in fractions of group A appeared in the void volume. The enzyme in fractions of group B was slightly retarded but eluted with the buffer at pH 8; the enzyme in fractions of group C eluted at pH 6, and the enzyme in fractions of group D eluted at pH 4.5. By determining the nature of the reducing sugars produced by the enzymes from starch by paper chromatographic methods it was concluded that group A fractions contained a glucosyltransferase, group B fractions contained an α-amylase, groups C and D fractions contained glucoamylase. The enzymes in fractions C and D possessed the same enzymatic activity but differed in chromatographic behavior and in electrophoretic mobility. These forms of glucoamylase are isozymes and this observation represents an early detection of isozyme forms of an enzyme long before the extensive studies on lactic acid dehydrogenase isozymes. The isozymes have been numbered I and II in accord with the electrophoretic mobilities (ref. 5).

The results of DEAE-cellulose chromatography experiments (Fig. 1) and the enzyme assays of the fractions from the column on maltose and starch substrates showed that glucosyltransferase was a different protein from glucoamylase. The synthesis of the two proteins is very likely under different genetic control. The transferase is an undesirable enzyme in starch conversions and in view of biosynthesis control mechanisms it should be possible to eliminate the transferase from glucoamylase preparations by genetic mutation or by recombinant DNA technology. Advances in these areas are discussed in a later section. The results in Fig. 1 together with enzyme assays showed that the original glucoamylase preparation contained an α-amylase as well as the glucosyltransferase and the glucoamylase isozymes.

3.2 Purity of glucoamylase

The criteria of purity for the glucoamylase isolated by the DEAE cellulose chromatography method include homogeneity on ultracentrifugation, homogeneity on gel electrophoresis and constancy of enzyme activity in the fractions from the ion-exchange column. The ultracentrifuge patterns were obtained in a model E Beckman ultracentrifuge by a standard method and are reproduced in Fig. 2 (ref. 26). The patterns show a uniformity in molecular size of the purified glucoamylase for all the time periods of analysis. Impurities were not detected in the preparation by this centrifugation method.

Electrophoretic gel patterns of the purified glucoamylase from *Aspergillus niger* and from *Rhizopus niveus* were prepared (ref. 27). Duplicate gels of each enzyme preparation were obtained and one set (A and C) was stained for protein components with Coommasie Blue G-250 (ref. 28) and the other set (B and D) was stained for carbohydrate components with periodate and Schiffs

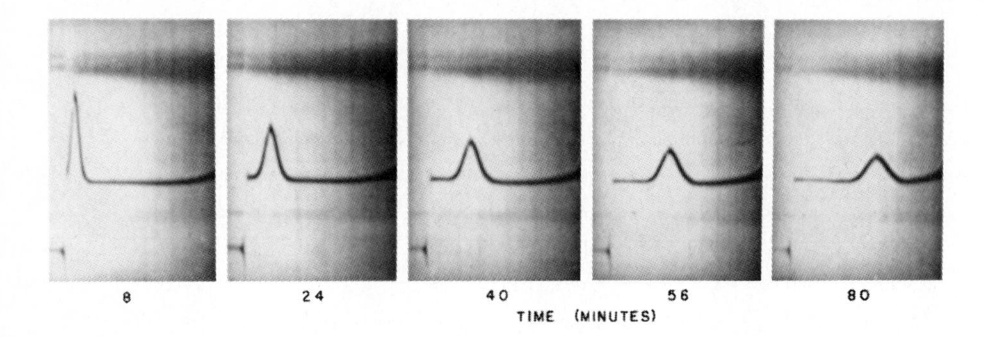

Fig. 2. Sedimentation velocity photographs obtained in the Model E, analytical ultracentrifuge with purified glucoamylase (ref. 26).

reagents (ref. 29). These results are shown in Fig. 3. It is noted in the figure that a single protein and a single carbohydrate band was obtained in each glucoamylase preparation as revealed by specific spray reagents. The migration rates of the protein and the carbohydrate components in each set of gels were identical. However a large difference in migration rates for the *A. niger* and the *R. niveus* glucoamylases does exist. Quite clearly both preparations are electrophoretically homogeneous and also both enzymes are glycoproteins as shown by the results with the protein and the carbohydrate stains.

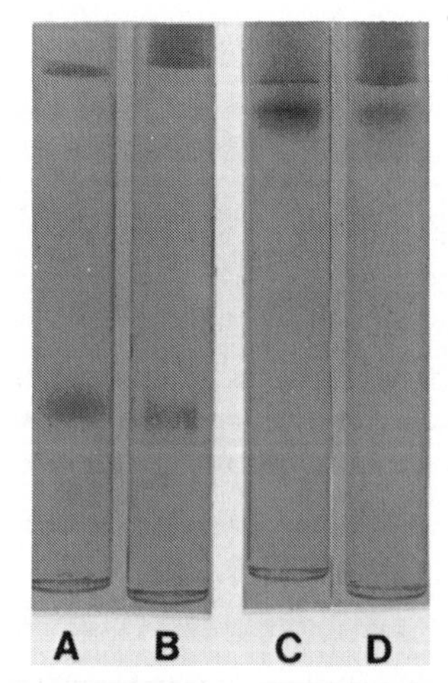

Fig. 3. Electrophoretic patterns in polyacrylamide gels of glucoamylase from *A. niger* (A and B) and *R. niveus* (C and D) stained with protein reagents (A and C) and with carbohydrate reagents (B and D).

The distributions of enzyme activity and protein in the fractions from the DEAE-cellulose column were determined and the ratios of the two values were calculated. The enzyme activities were obtained from the assays on the starch substrate and the protein concentrations were measured by the UV absorbance at 280 nm. The ratios of enzymatic activities to the protein concentrations were constant in all the fractions from the DEAE-cellulose column containing enzyme activity and thus substantiating homogeneity in the glucoamylase preparation.

3.3 Glucoamylase action on starch and glucosyl oligosaccharides

In composition starch is a mixture of two polymers, amylose and amylopectin, both of which are composed of glucopyranose units. The glucosidic bonds in amylose and the majority of the bonds of amylopectin are α-D-(1,4). A small percentage of the bonds of amylopectin are α-D-(1,6) resulting in a branched structure for this polymer. It has also been reported that some amylopectins may contain a few α-D-(1,3) glucosidic bonds (ref. 30) and the susceptibility of these bonds as well as the α-D-(1,6) bonds to hydrolysis with glucoamylase was determined. It was observed in early studies with glucoamylase that the enzyme converted starch directly to glucose. In order to check the extent of conversion of starch to glucose by glucoamylase, the purified enzyme was tested for ability to hydrolyze starch, amylose and amylopectin. After incubation of the substrate with the pure glucoamylase for 18 hrs, the enzymic digests were analyzed for glucose by a reducing sugar method. From the yield of glucose and the concentration of the polysaccharide used initially it was calculated that 97% of starch, 94% of amylose and 98% of amylopectin were converted to glucose. Thus starch and the two component fractions were all converted quantitatively to glucose. Reports have been published that some glucoamylases do effect an incomplete hydrolysis of starch. Data are recorded in the literature stating that only 80% of starch was hydrolyzed by glucoamylases from *A. niger* and from *Endomyces* yeasts (refs. 24, 31). A substantiation of incomplete hydrolysis of starch by the glucoamylases (ref. 32) has been published but other reports in the literature (refs. 9, 33-35) support the original finding of a complete hydrolysis of starch to glucose by glucoamylases.

Since the amylopectin component of starch contains α-D-(1,6) linkages and perhaps some α-D-(1,3) linkages (ref. 30), the complete conversion of amylopectin to glucose requires that such linkages be hydrolyzed by the glucoamylase. Two oligosaccharides were used to verify that the α-D-(1,6) linkages are indeed hydrolyzed and one oligosaccharide was used to test the action of the enzyme on α-D-(1,3) bonds. The latter results are presented in a later table. In the former experiments isomaltose (α-D-glucopyranosyl-

(1,6)-D-glucose) (ref. 36) and panose (α-D-glucopyranosyl-(1,6)-α-D-glucopyranosyl-(1,4)-D-glucose) (ref. 37) were used. Digests of these oligosaccharides with the pure glucoamylase were prepared and analyzed by paper chromatography for hydrolytic products as a function of time (ref. 11). Photographs of the chromatograms stained for carbohydrates with the silver nitrate reagent are reproduced in Fig. 4.

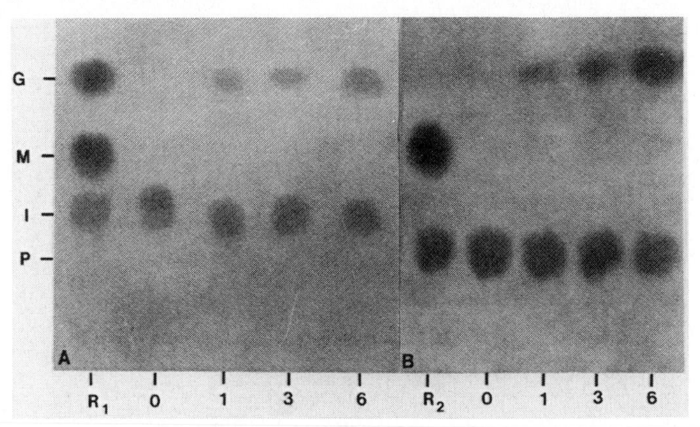

Fig. 4. Identification of hydrolytic products released from isomaltose (A) and panose (B) by glucoamylase; G = glucose, M = maltose, I = isomaltose, P = panose, R_1 and R_2, reference compounds, numbers = hours of enzyme digestion (ref. 11).

It will be noted in Fig. 4 that the isomaltose is indeed hydrolyzed to glucose since the concentration of the glucose increased with time of incubation while the concentration of isomaltose decreased. The trisaccharide, panose, is also hydrolyzed to glucose but without the accumulation of the intermediate, maltose, in the digest. The latter is most likely due to the differences in rates of hydrolysis of the α-D-(1,4) linkage in maltose and the α-D-(1,6) linkage in isomaltose as recorded later. Quite clearly, these results establish that the α-D-(1,6) glucosidic bond is hydrolyzed by the glucoamylase in oligosaccharides and presumably in amylopectin and starch.

An oligosaccharide mapping technique was used to obtain additional evidence for the mechanism of glucoamylase action on starch oligosaccharides (ref. 38). The results are shown in Fig. 5. For these experiments malto-oligosaccharides of varying number of glucose units joined by α-D-(1,4) bonds and labeled with ^{14}C at position 1 were prepared. The malto-oligosaccharides were separated on paper chromatograms in one direction in a solvent system of n-butyl alcohol, pyridine and water (6:4:3 by volume). A solution of pure enzyme was then sprayed on the oligosaccharides on the paper and the enzyme was allowed to act

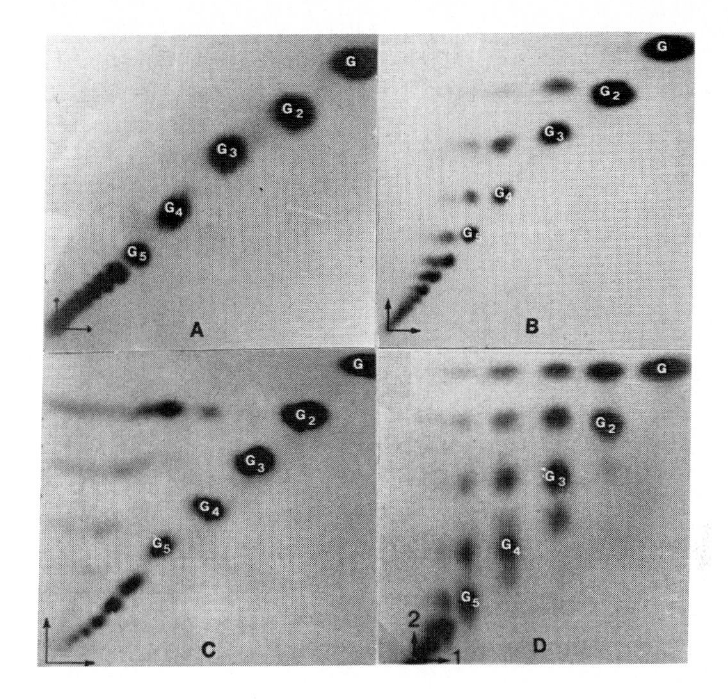

Fig. 5. Oligosaccharide maps; reference malto-oligosaccharides-1-^{14}C (A), action of glucoamylase (B), action of α-amylase (C) and action of glucosyltransferase (D) on malto-oligosaccharides-1-^{14}C (ref. 38).

on the oligosaccharides for 15 minutes. The chromatogram was then developed in the second direction in the same solvent system. Radioautograms were prepared from the finished chromatograms. Frame A in the figure is a photograph of a radioautogram of the reference labeled oligosaccharides with no enzyme treatment and showing the mobility and the purity of each saccharide. The chromatogram in Frame B has been sprayed with glucoamylase, that in Frame C with α-amylase and that in Frame D with a glucosyltransferase. Comparisons of the radioautograms in Frame A and Frame B show that glucoamylase hydrolyzed each malto-oligosaccharide by removal of a single glucose unit from the non-reducing end of the oligosaccharide and the other product was an oligosaccharide with one less glucose unit. Since the glucose was not radioactive, this glucose did not show up on the radioautogram. However the oligosaccharide was radioactive and appeared as a dark spot on the radioautogram (Fig. 5, frame B). The new oligosaccharide possessed an Rf value equal to that of an oligosaccharide with one less glucose unit than the substrate. The oligosaccharide products of α-amylase action are shown in Frame C and these are quite different from those in frame B. The α-amylase converts the radioactive oligosaccharides largely to maltose. The radioautogram in Frame D shows the products of action of glucosyltransferase on the labeled oligosaccharides. The products include hydrolytic and transfer products produced by transfer of glucose units of 1,4-linked oligosaccharides to the 6 position of an acceptor molecule. The transfer products moved at a slower rate than the substrate oligosaccharides on the paper chromatogram.

The results with the α-D-(1,3) disaccharide of glucose and other glucosides showing the hydrolysis rates of the various bonds by glucoamylase are shown in Table 1. These analyses were performed essentially by methods that have been published (ref. 26). The products were identified by paper chromatographic analysis. Since some of the substrates were found to be hydrolyzed at extremely slow rates, the amount of enzyme used in these experiments was increased. The substrates, their structure, the units of enzyme employed and the rates of hydrolysis of the substrates relative to the rate of hydrolysis of starch which has been assigned a value of 100 are recorded in Table 1.

TABLE 1

Relative Rates of Hydrolysis of α-D-glucosides by Purified Glucoamylase

α-D-Glucoside	Structure	Units of enzyme	Relative rate
Starch	α-D-(1,4), α-D-(1,6)-D-Glucan	1	100
Maltoheptaose	$[\alpha\text{-}(1,4)\text{-D-Glucopyranosyl}]_6$-D-glucose	1	100
Maltose	α-D-Glucopyranosyl-(1,4)-D-glucose	10	15
Maltobionic acid	α-D-Glucopyranosyl-(1,4)-D-gluconic acid	10	5
Nigerose	α-D-Glucopyranosyl-(1,3)-D-glucose	50	1
Isomaltose	α-D-Glucopyranosyl-(1,6)-D-glucose	100	0.5
Arabinosyl glucoside	α-D-Glucopyranosyl-(1,3)-D-arabinose	100	0.3
Maltulose	α-D-Glucopyranosyl-(1,4)-D-fructose	100	0.2
Sucrose	α-D-Glucopyranosyl-β-D-fructofuranoside	100	0.2
Glycerol glucoside	α-D-Glucopyranosyl-(1,1)-D-glycerol	100	0.1
Turanose	3-α-D-Glucopyranosyl-D-fructose	150	0.09
Phenyl glucoside	Phenyl α-D-glucoside	100	0.05
Methyl glucoside	Methyl α-D-glucoside	200	0.01
Trehalose	α-D-Glucopyranosyl-α-D-glucopyranoside	200	0.01

The data in Table 1 show that the α-D-(1,3) disaccharide of glucose (nigerose) was hydrolyzed by glucoamylase at a relatively fast rate. Such linkages in starch could therefore be hydrolyzed by the enzyme. The data also show that glucoamylase hydrolyzes most types of α-glucosidic linkages in a variety of glucosides but at different rates. The linkages in starch and maltoheptaose are hydrolyzed at highest rates while maltose, maltulose and nigerose were hydrolyzed at intermediate rates. Isomaltose was hydrolyzed at a slow rate but this disaccharide was indeed hydrolyzed showing that the α-D-(1,6) linkage was susceptable to hydrolysis. The other glucosides in the table were hydrolyzed very slowly with methyl α-glucoside and trehalose being hydrolyzed at the slowest rate.

In order to verify that the hydrolysis of the two major types of linkages in starch, the α-(1,4) and the α-(1,6), occurred at the same site of the enzyme, experiments on pH optimum, temperature optimum and competitive inhibition were conducted. The experiments were performed with maltose and isomaltose as substrates. Fig. 6 shows the pH and temperature effects on the hydrolysis of the two substrates (ref. 11). Quite clearly the pH optimum and the temperature stability were the same for both substrates and accordingly both substrates are most likely hydrolyzed at the same active site of the enzyme. The inhibition results for the enzymatic hydrolysis of maltose in the presence of isomaltose are presented in Table 2.

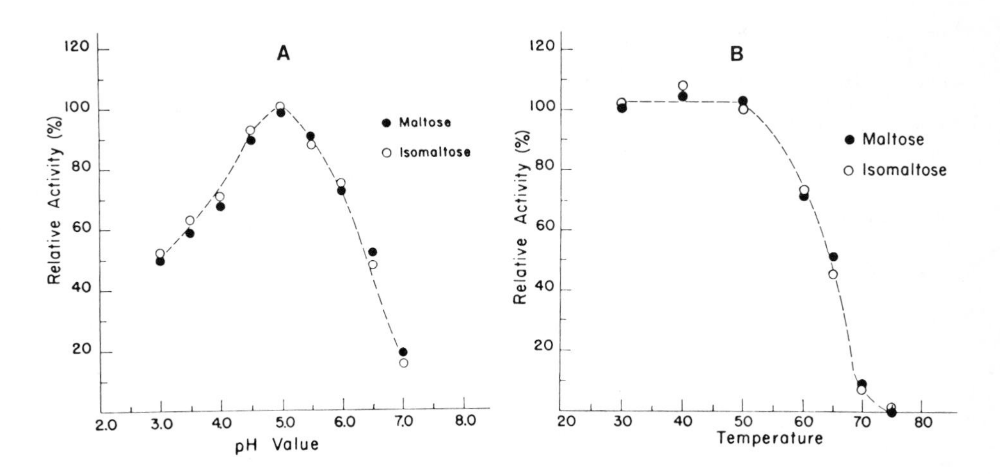

Fig. 6. The relative rate of hydrolysis of maltose and isomaltose at various pH values (A) and at different temperatures (B) (ref. 11).

TABLE 2

Hydrolysis of Maltose-1-^{14}C in the Presence of Isomaltose by Glucoamylase

Time hrs.	Residual maltose concentration		
	0 Isomaltose	0.025 M Isomaltose	0.10 M Isomaltose
0	0.150	0.150	0.150
1	0.129	0.132	0.135
2	0.114	0.119	0.125
4	0.080	0.086	0.095
8	0.034	0.039	0.055

In the inhibition experiments the rate of hydrolysis of radioactive maltose in the presence of non-radioactive isomaltose was determined by measuring the radioactivity of the glucose that was liberated by the enzyme (ref. 11). The results show that with an increase in isomaltose concentration the extent of hydrolysis of the maltose decreased. At highest concentrations of isomaltose the inhibition of maltose hydrolysis was over 20% in the 8 hr period. These data are interpreted to show that maltose and isomaltose are hydrolyzed at the same active site of the enzyme molecule and as a result the isomaltose functions as a competitive inhibitor for the hydrolysis of maltose.

In the development of the glucoamylase starch conversion process, it was found that glucosyltransferase (ref. 36) which contaminates glucoamylase preparations was responsible for low yields of glucose from starch. This transferase catalyzes transfer reactions to form non-fermentable oligosaccharides and also catalyzes reversion reactions to form isomaltose. A number of investigators have reported that glucoamylase may also effect reversion reactions (refs. 8, 39, 40). The formation of transfer and reversion products by the transferase and by glucoamylase can be measured by use of radioactive substrates (ref. 41). Data from such experiments are presented in Table 3.

The formation of transfer products was measured in the maltose digests and the formation of reversion products in the glucose digests. From the data in Table 3 it has been calculated that 35% of the maltose was converted to transfer products by the transferase and less than 2% by the glucoamylase. It was also calculated that the transferase catalyzed reversion reactions which were five times greater than those catalyzed by the glucoamylase. It was therefore concluded that glucoamylase does not catalyze transfer reactions but may catalyze reversion reactions at a very slow rate.

TABLE 3

Radioactivities (CPM) of Products in Digests of Maltose-1-^{14}C and of Glucose-1-^{14}C with Glucosyltransferase and with Glucoamylase (ref. 41)

Time (h)	With Maltose-1-^{14}C				With Glucose-1-^{14}C	
	Glucose	Maltose	Isomaltose	Panose	Glucose	Isomaltose
			Glucosyltransferase			
0	750	29,000	120	90	44,400	35
1	4,900	22,600	920	2,500	44,300	59
3	8,800	16,100	1,700	4,700	43,900	104
6	12,800	11,100	2,500	6,300	43,700	312
12	14,600	6,500	3,500	6,100	43,400	635
24	17,700	3,000	4,300	5,900	42,700	1,280
48	-	-	-	-	41,900	2,020
			Glucoamylase			
0	130	30,200	60	30	38,700	95
1	2,500	27,300	190	50	38,800	119
3	5,300	25,100	210	90	38,600	109
6	9,100	22,400	240	110	38,500	140
12	14,800	15,200	320	240	38,400	152
24	28,100	2,100	380	210	38,200	320
48	-	-	-	-	38,100	402

3.4 Chemical structure of glucoamylase

3.4.1 Amino acid composition

Results of ion-exchange chromatography of a glucoamylase preparation from *Aspergillus niger* (Fig. 2) showed that the organism produces two major isozymes (I and II) of glucoamylase. The paper electrophoretic mobilities of the two isozymes are shown in Fig. 7 (ref. 42). Most of the structural data reported in this chapter have been obtained with isozyme I of the glucoamylase of *A. niger* since this isozyme is generally produced in higher yields.

The isozyme I of glucoamylase of *A. niger* which was purified by DEAE-cellulose chromatography was shown to be homogeneous by ultracentrifugation, gel electrophoresis and ratio of enzyme activity to protein concentration. This sample was used for amino acid analysis utilizing standard methods of sample preparation and analysis in a Beckman amino acid analyzer. The enzyme was first hydrolyzed in 6 N HCl for 18 hrs and the solvent and acid were evaporated under nitrogen. The residue was dissolved in citrate buffer of pH 2.2 and aliquots of this solution were analyzed for neutral amino acids and for basic amino acids on two different columns. From the analytical values, the weight of the sample used for analysis, the molecular weight of 94,000, and the carbohydrate content (14%) of the enzyme, the amino acid composition was calculated. These values and the values of amino acid analysis of a glucoamylase subjected to alkaline borohydride β-elimination reaction are

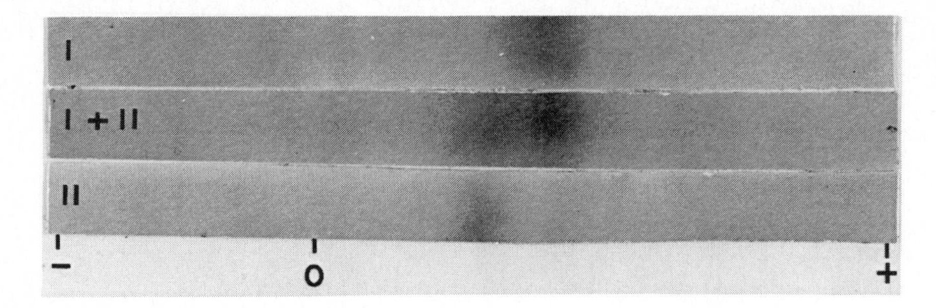

Fig. 7. A photograph of paper electrophoretic strips of the isozymes (I and II) of glucoamylase from *A. niger* (ref. 42).

recorded in Table 4. Of note in the native enzyme sample is the low content of methionine, cystine and histidine and the high content of threonine and serine. The amino acid sequence of isozyme 1 of glucoamylase has been determined (ref. 43) and these data are discussed in the section on biosynthesis.

TABLE 4

Amino Acid Residues of Native Glucoamylase (ref. 42) and of the Modified Glucoamylase and After Reduction with Alkaline Sodium Borohydride (ref. 44)

Residue	Native	Reduced	Difference
Aspartic	83	86	+3
Threonine	104	76	−28
Serine	119	102	−17
Glutamic	56	57	+1
Proline	29	31	+2
Glycine	59	72	+13
Alanine	78	101	+23
Cystine/2	8	6	−2
Valine	46	46	0
Methionine	3	3	0
Isoleucine	27	25	−2
Leucine	53	51	−2
Tyrosine	30	29	−1
Phenylalanine	27	26	−1
Lysine	16	18	+2
Histidine	6	6	0
Arginine	23	22	−1
Tryptophan	30	29	
α-Amino butyric	0	10	+10

3.4.2 Glycoprotein structure

Of significance is the observation that the acid hydrolysate of glucoamylase turned brown in color typical of a glycoprotein. Evidently the glucoamylase is a glycoprotein and contains carbohydrate residues. Several lines of evidence have established that the carbohydrate residues are indeed structural units of the enzyme molecule. First the carbohydrates were identified by paper chromatography to be mannose, glucose and galactose after acid hydrolysis of the glucoamylase (ref. 45). The monosaccharides were not present in the solution prior to hydrolysis. Second the carbohydrate component, the protein moiety and the enzymatic activity of the native glucoamylase migrated at identical rates on density gradient centrifugation and on paper electrophoresis indicating a chemical linkage between the carbohydrate and the protein moieties. These results are presented in Fig. 8 (ref. 42). Results of gel electrophoresis have been presented in Fig. 3. Additional evidence for a glycoprotein structure for glucoamylase are presented in later sections.

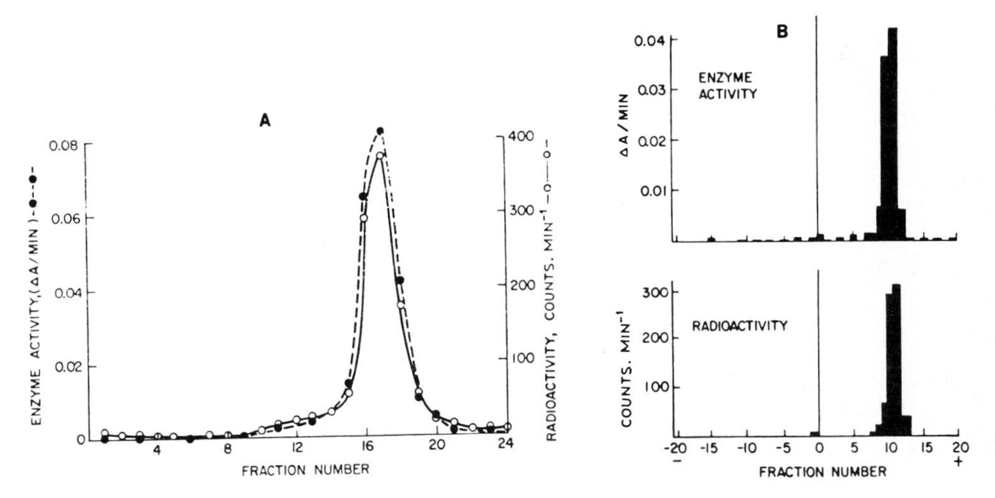

Fig. 8. Distribution of enzyme activity, protein and radioactivity on density gradient centrifugation A and on paper electrophoresis (B) in a sample of radioactive glucoamylase I (ref. 42).

3.4.3 Alkaline borohydride β-elimination

In order to obtain information on the type of linkages of carbohydrate to protein alkaline borohydride β-elimination experiments were performed. In these experiments samples of glucoamylase were subjected to β-elimination reactions in 0.1 N NaOH and a 2% solution of sodium borohydride. At the end of the reaction the borohydride was destroyed by acidification and the reaction mixture was divided into two fractions. One fraction was dialyzed to remove the alkali and the carbohydrates released from the enzyme in the reaction sequence. This modified glucoamylase was then recovered by lyophilization and subjected to acid hydrolysis and amino acid analysis. The other fraction was used for reduction and identification of carbohydrates by gas liquid chromatography (GLC). In the amino acid calculation a correction factor for the loss of carbohydrate as determined colorimetrically was applied for obtaining the amino acid composition of the modified enzyme. These values and the values for the native glucoamylase are recorded in Table 4.

Also recorded in the table are the number of residues that were lost or gained on reductive β-elimination reactions with glucoamylase. It will be noted that values for threonine and serine were decreased significantly in the reduced enzyme while glycine and alanine increased and α-amino butyric acid appeared. The remaining amino acids were the same within experimental error. The total loss in serine and threonine was 45 residues while the gain in glycine, alanine and α-amino butyric acid was 46. From these data it was concluded that the glucoamylase contains 45 or 46 carbohydrate chains attached O-glycosidically to threonine and serine residues of the polypeptide chain of the enzyme. As stated earlier the carbohydrate residues were mannose, glucose and galactose.

3.4.4 <u>Structure and distribution of carbohydrate chains</u>

The quantitative values for the carbohydrates constituents of glucoamylase were obtained by GLC analysis of the alditol acetates prepared from the hydrolysate of glucoamylase (ref. 27). The total carbohydrate content of the native glucoamylase I was 14% consisting of 11.1% mannose, 2.6% glucose and 0.3% galactose. The identity of the carbohydrates was established following hydrolysis, reduction and acetylation and analysis by GLC. The patterns for unhydrolyzed, hydrolyzed glucoamylase and a glucoamylase sample which had been subjected to alkaline β-elimination reactions are shown in Fig. 9. Pattern A in the figure shows that free carbohydrates were not present in the glucoamylase sample. Pattern B shows that mannose, galactose and glucose were present after acid hydrolysis of the enzyme. These derivatives have been identified by comparison of retention times with values for standard alditol acetates. Pattern C shows the products in the alkaline reductive β-elimination reaction establishing that free mannose was liberated from the enzyme in the β-elimination reaction. Therefore some of the carbohydrate chains linked to the polypeptide of the enzyme are single mannose residues. Methylation analysis, the results of which are recorded in Fig. 10, show that the remaining carbohydrate chains are di-, tri-, and tetra-saccharides.

The number and types of monosaccharide and oligosaccharide chains were determined from the methylation data presented in Fig. 10 on the native glucoamylase. The methylation was performed by the Hakomori method (ref. 46) and the analysis by GLC and mass spectrometry by the Lindberg method (ref. 47). The large peak in Fig. 10 is due to a mixture of tetramethyl mannose and tetramethyl glucose which migrate at identical rates in the GLC method

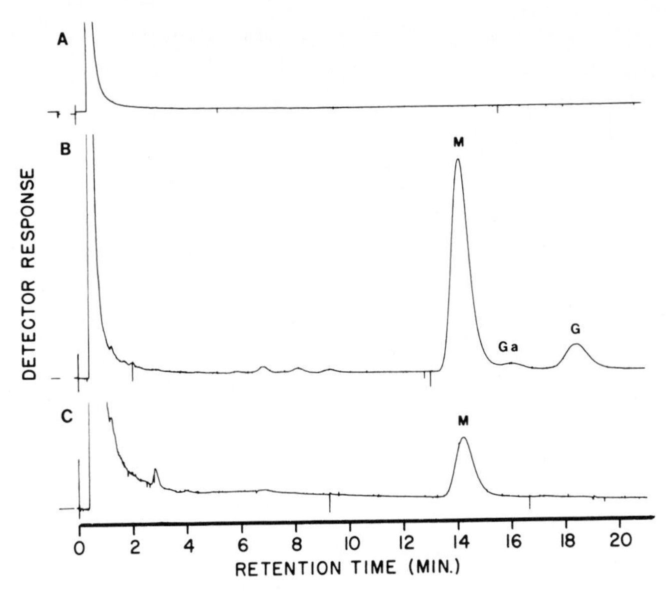

Fig. 9. The alditol acetates of monosaccharide constituents of glucoamylase and β-eliminated glucoamylase. A = control of enzyme, B = acid hydrolysate of glucoamylase, C = β-eliminated monosaccharide, M = mannitol hexaacetate, Ga = galactitol hexaacetate, and G = glucitol hexaacetate (ref. 27).

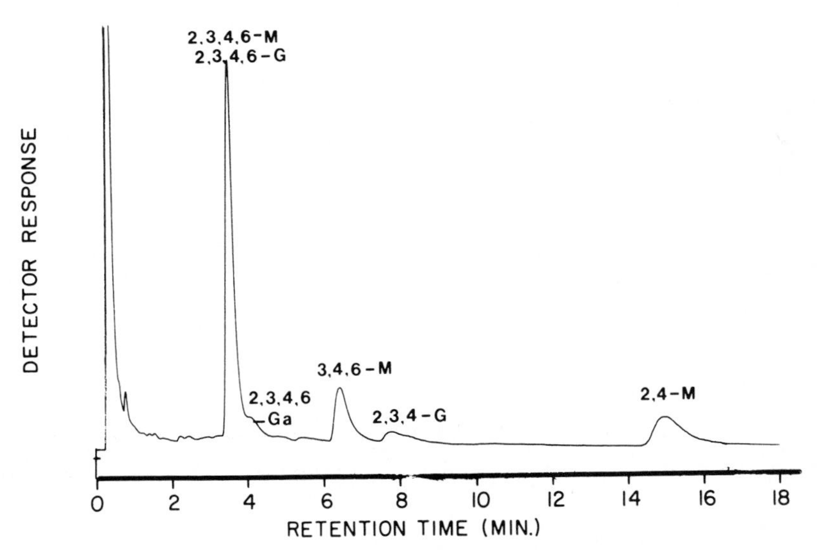

Fig. 10. GLC pattern of the methylated alditol acetates obtained from fully methylated glucoamylase: 2,3,4,6-Man = 1,5-di-O-acetyl-2,3,4,6-tetra-O-methylmannitol; 2,3,4,6-Glc = 1,5-di-O-acetyl-2,3,4,6-tetra-O-methylglucitol; 2,3,4,6-Gal = 1,5-di-O-acetyl-2,3,4,6-tetra-O-methylgalactitol; 3,4,6-Man = 1,2,5-tri-O-acetyl-3,4,6-tri-O-methylmannitol; 2,3,4-Glc = 1,5,6-tri-O-acetyl-2,3,4-tri-O-methylglucitol; and 2,4-Man = 1,3,5,6-tetra-O-acetyl-2,4-di-O-methylmannitol (ref. 27).

employed. Consequently the quantitative value for glucose has been obtained by a difference method. Also the mannose and the glucose are terminal units of the carbohydrate chains of the enzyme. A small amount of tetramethyl galactose was obtained in the methylation analysis indicating that a few chains are terminated by galactose units. The remainder of the methylated derivatives arise from internal residues and these derivatives are 3,4,6 trimethyl mannose, 2,3,4 trimethyl glucose and 2,4 dimethyl mannose.

The quantitative data on the yield of methyl derivatives can be obtained by integration of the areas under the peaks in the pattern in Fig. 10. The values given in moles of derivative per mole of enzyme are recorded in Table 5. From the nature of the derivatives and the quantitative values, the

TABLE 5

Partially Methylated Alditol Acetates from Methylated Glucoamylase

Separated by GLC on OV-225 at 190° and the retention times are relative to 1,5-di-O-acetyl-2,3,4,6-tetra-O-methylglucitol.	Retention time	Moles
1,5-Di-O-acetyl-2,3,4,6-tetra-O-methylmannitol	0.99	49
1,5-Di-O-acetyl-2,3,4,6-tetra-O-methylglucitol	1.00	6
1,5-Di-O-acetyl-2,3,4,6-tetra-O-methylgalactitol	1.16	2
1,2,5-Tri-O-acetyl-3,4,6-tri-O-methylmannitol	1.82	11
1,5,6-Tri-O-acetyl-2,3,4-tri-O-methylglucitol	2.19	5
1,3,5,6-Tetra-O-acetyl-2,4-di-O-methylmannitol	4.22	13

oligosaccharide structures for the chains of glucoamylase have been proposed. These structures are shown in Fig. 11 (ref. 27). Structures for the oligosaccharide chain attached to glucoamylase have been proposed more recently by other investigators also on the basis of the methylation analysis data (ref. 48). Minor differences have been reported for the trisaccharide and tetrasaccharide structures between the two sets of data.

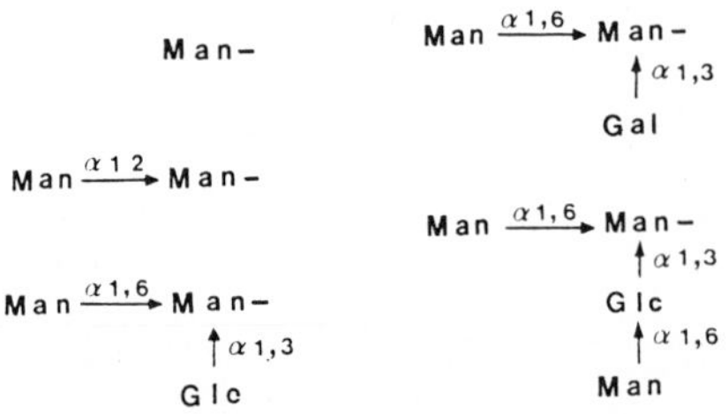

Man-

Man $\xrightarrow{\alpha 1,2}$ Man-

Man $\xrightarrow{\alpha 1,6}$ Man- ↑α1,3 Glc

Man $\xrightarrow{\alpha 1,6}$ Man- ↑α1,3 Gal

Man $\xrightarrow{\alpha 1,6}$ Man- ↑α1,3 Glc ↑α1,6 Man

Fig. 11. Diagram of the possible structures for the carbohydrate chains of glucoamylase with mono-, di-, tri-, and tetra-saccharide structures.

The carbohydrate chains of glucoamylase range in degree of polymerization from one to four monosaccharide units and the glycosidic linkages are (1,2), (1,3) or (1,6) or combinations thereof. On the basis of the methylation data, the structures shown in Fig. 11 seem to be the most logical for the carbohydrate chains of glucoamylase. Further, from the methylated derivatives in Table 5 it has been calculated that the glucoamylase molecule contains 21 monosaccharide units of mannose residues, 11 disaccharides of mannose, 8 trisaccharides of mannose, glucose or galactose, 5 tetrasaccharides of mannose and glucose. The configuration of the glycosidic linkage was determined by susceptability to α-mannosidase hydrolysis and by NMR measurements on the native enzyme. The mannose residues are joined by the α linkage as shown by the results on the chromatogram reproduced in Fig. 12. The NMR data also show that the linkage of glucose units is α while the linkage of galactose units is probably α (ref. 49).

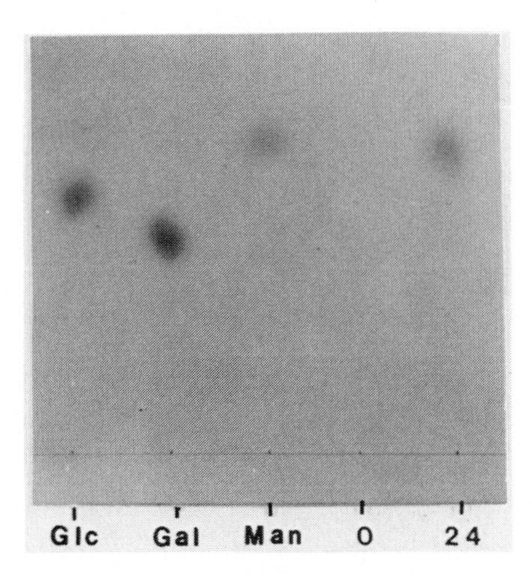

Fig. 12. Photograph of a paper chromatogram of reference compounds and products liberated by α-mannosidase from glucoamylase; abbreviations are standard carbohydrate abbreviations and the 0 lane is an analysis for carbohydrate at 0 time and the 24 lane after digestion for 24 hrs.

The carbohydrate chains are linked O-glycosidically to serine and threonine residues of the polypeptide chain of the enzyme since on β-elimination the number of serine and threonine residues decreased and the number of alanine and glycine residues increased while α-amino butyric acid also appeared in the hydrolysate of the β-eliminated glucoamylase. About half of the mannose residues appeared in the β-elimination mixture as the hexaacetate. The remainder of the carbohydrates were released as oligosaccharides but these were not isolated. The number and types of oligosaccharides liberated from the enzyme have been recorded in an earlier section. The preceeding results have been used to construct a diagrammatic model for the glucoamylase molecule as shown in Fig. 13. It should be noted in the figure that the carbohydrate residues and chains are distributed randomly along all of the polypeptide chains of the enzyme, the evidence for which is presented in the immunological section. This evidence does not support a cluster arrangement suggested by others (ref. 50). Also, the carbohydrate chains of amylases from other organisms were located in several enzyme fragments of those enzymes and the distribution conforms to a random arrangement of the carbohydrate chains along the protein chains (refs. 51, 52).

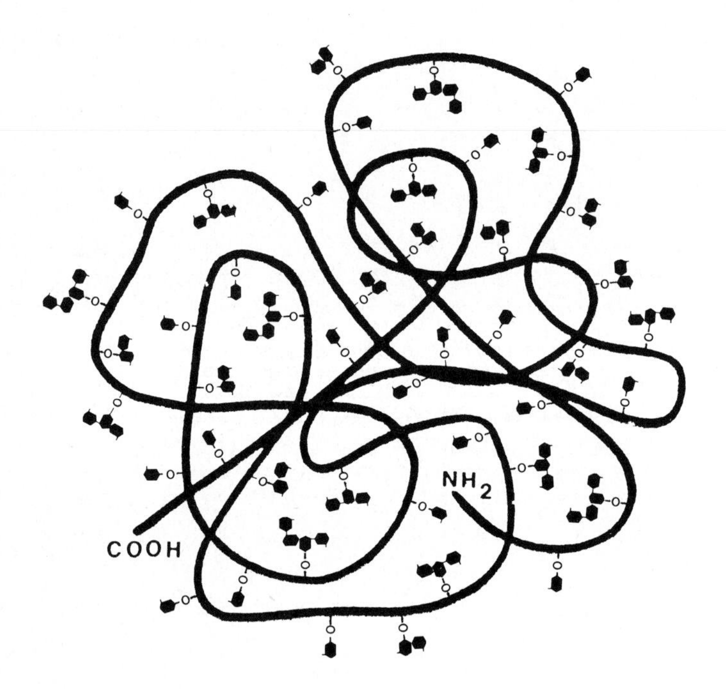

Fig. 13. Diagrammatic representation of the structure of glucoamylase; solid lines = polypeptide chain, hexagons = monosaccharide residues.

3.4.5 Carbohydrate content of glucoamylases

Glucoamylases from fungi and yeasts contain carbohydrate residues and are glycoproteins. Many of these enzymes have been analyzed for carbohydrate content and most have been found to contain mannose, glucose and galactose as the carbohydrate components. A few of the glucoamylases contain glucosamine, the latter most likely in the N-acetyl form as well as mannose. The nature and the quantitative value for the carbohydrates in a few glucoamylases are recorded in Table 6. It should be noted that the enzyme from Aspergillus oryzae contains the highest amount (30%) of carbohydrate and that the enzymes from the Rhizopus group contain glucosamine. In general mannose is the most abundant carbohydrate and is also the linkage carbohydrate in many glucoamylases forming O-glycosidic bonds to serine and threonine of the protein. Whether the N-acetyl glucosamine functions as the linkage carbohydrate to asparagine of the polypeptide chain is not yet known. If so, these glucoamylases probably possess N-glycosidic linkages as well as O-glycosidic linkages. For the glucoamylase from Aspergillus niger, the arrangement of the carbohydrate chains on the polypeptide chain has been determined (Figs. 11 and 13) but the arrangement of the carbohydrates in other glucoamylases is not yet known. It should be mentioned that the list in Table 6 is not intended to be all inclusive and some important omissions may have been made.

TABLE 6

Carbohydrate Content of Glucoamylases of Different Fungi

Enzyme source	Isozyme Type	Carbohydrate Content					Ref
		Total	Man	Glc	Gal	GlcNAC	
Aspergillus niger	I	10.2	8.0	2.0	0.2		(45)
	I	14.1	11.2	2.6	0.3		(42)
	I	18.0					(53)
	II	18.2	15.0	3.0	0.2		(45)
	II	23.5	19.3	3.7	0.6		(42)
	II	22.0					(53)
Aspergillus oryzae	I	30.0					(54)
Aspergillus phoenicis	I	12.0	8.0	2.8	1.2		(55)
Aspergillus awamori	I		7.6				(56)
Paecilomyces varioti			13.0				(57)
Rhizopus delemar		13.0	10.0			3.0	(58)
Rhizopus niveus	I	11.0	9.2			1.8	(59)
	II	9.1	6.9			2.2	(59)
	III	13.5	11.4			2.1	(59)

3.4.6 Molecular weight and subunit structure

With regard to the molecular size of glucoamylase, there have been conflicting reports on the molecular weights and subunit structure of the glucoamylases. Data from ultracentrifugation studies have been obtained with the Model E ultracentrifuge (ref. 26) and by density gradient centrifugation (ref. 60). The photographs of the pattern from ultracentrifugation in the

Model E centrifuge have been shown earlier (Fig. 2). These data were used to calculate a molecular weight of glucoamylase to be 97,000±5,000 (ref. 26). Results of density gradient ultracentrifugation of fetuin, glucoamylase and glucose oxidase, all of which are glycoproteins containing 22%, 14% and 16% carbohydrates, respectively, are shown in Fig. 14 (ref. 61).

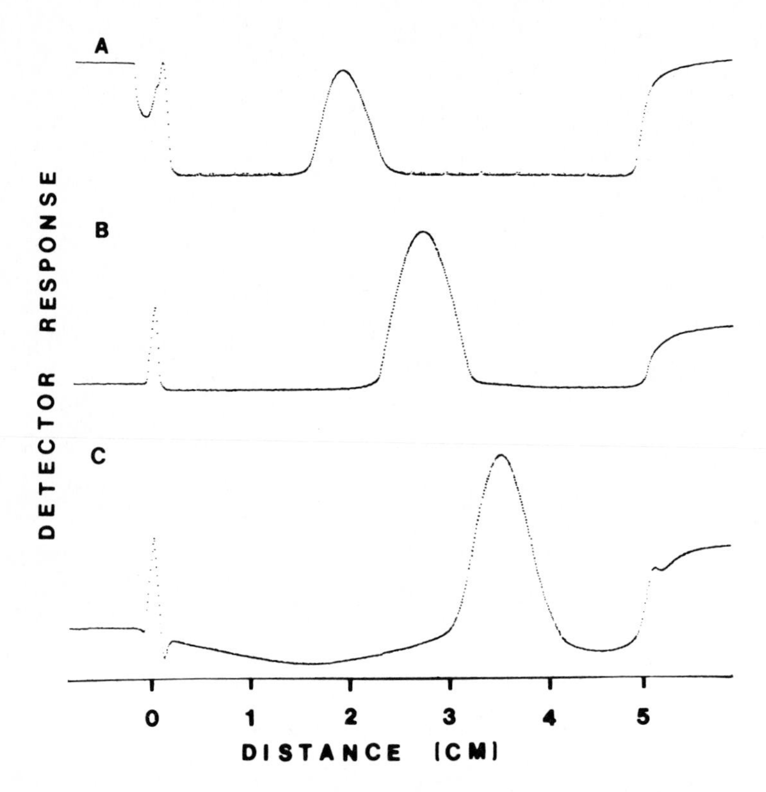

Fig. 14. Density gradient centrifugation patterns for fetuin (A), glucoamylase (B) and glucose oxidase (C) (ref. 61).

The centrifugation data from density gradient measurements were used to calculate the molecular weight of glucoamylase to be 94,000 (refs. 61, 62) which is in good agreement with the value obtained by the Model E ultracentrifuge. Literature values for the molecular weight value of glucoamylase from A. niger reported by other investigators from measurements by SDS electrophoresis (ref. 63), gel filtration (ref. 64, 65) and analytical ultracentrifugation (ref. 66) are not in agreement with the above values and not in agreement with each other; 61,000 (ref. 17) 52,000 (ref. 53), 82,000 (ref. 43), 63,000 (ref. 67), 75,000 (ref. 68) and 69,000 (ref. 54). Values for glucoamylases from other organisms also vary considerably; 59,000 (ref. 69), 74,000 (ref. 57), and 90,000 (ref. 52). Values from ultracentrifugation data are consistently higher and are in better agreement than the other values. Accordingly, the value of 94,000 obtained by the density gradient centrifugation method as the molecular weight of glucoamylase has been used in calculations requiring a M.W. value.

A second area of conflicting reports on glucoamylase structure is in the area of subunit structure. Some investigators claim that glucoamylase is made up of subunits (ref. 67) while others claim the opposite (refs. 42, 53). Results of dissociation experiments with glucose oxidase and glucoamylase are shown in Fig. 15 (ref. 25).

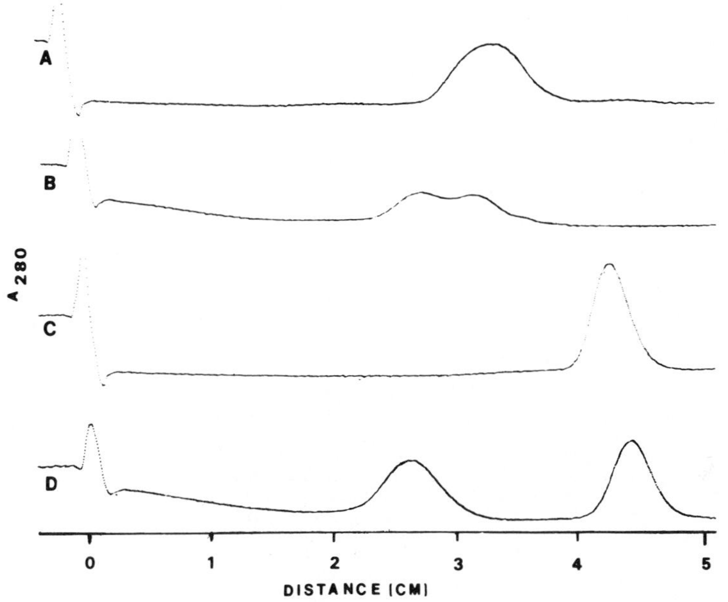

Fig. 15. Density gradient centrifugation patterns for glucoamylase (A), glucoamylase treated with mercaptoethanol and SDS (B), native glucose oxidase (C) and glucose oxidase treated with mercaptoethanol and SDS (D) (ref. 25).

In Fig. 15, the presence of subunits for glucose oxidase (Pattern D) is readily apparent from the density gradient centrifugation patterns. However subunits of glucoamylase are not apparent (Patterns A and B). A is a centrifugation pattern for native glucoamylase while pattern B in the pattern for a glucoamylase sample treated with mercaptoethanol and SDS. The glucoamylase molecule has undergone some unfolding as noted by the appearance of a second UV peak in pattern B near the peak for the native enzyme. Molecular weight calculations showed that the component is not a subunit but rather unfolded glucoamylase since the sedimentation rates differ by less than 15%.

In the unfolded state, the enzyme molecule would be rod shaped and it would sediment at a slower rate than the folded molecule of the glucoamylase. On the basis of such results, it has been concluded that glucoamylase is not made up of subunits.

3.4.7 Reaction with cyanogen bromide

Fragmentation of glucoamylase by cynogen bromide has been used to obtain more detailed information on the structure of glucoamylase. As recorded earlier in Table 2, there are three methionine residues per molelcule of glucoamylase. Fragmentation of the enzyme has been conducted by a standard cynogen bromide method and the results have been reported (ref. 61). The fragments for the enzyme were separated on electrophoresis on duplicate gels and one gel was stained to reveal protein and the other to reveal carbohydrate components. The photograph of the gels is reproduced in Fig. 16 and it can be seen that six fragments were produced on cyanogen bromide cleavage. This result indicates that a single cleavage at each methionine residue of the enzyme had occurred. The cleavage fragments were separable on polyacrylamide gels (Fig. 16) to yield six fragments from the enzyme. The gels stained for protein and for carbohydrate show that each of the six fragments contained protein and carbohydrate components. Such a result would be obtained if the

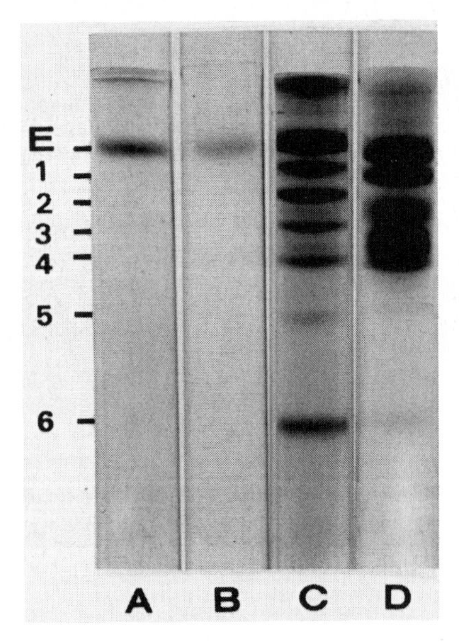

Fig. 16. SDS gel electrophoresis patterns for native glucoamylase (gels A and B) and CNBr treated glucoamylase (gels C and D) stained for proteins (gels A and C) and for carbohydrate (gels B and D), E = enzyme and 1 to 6 CNBr fragments (ref. 61).

distribution of the carbohydrate chains was random along the polypetide chain of the enzyme. This type of structure is shown in the diagram in Fig. 13.

3.5 Antigenicity and anti-glucoamylase antibodies

Antibodies are extremely sensitive and highly specific for detecting minute amounts of corresponding antigens. These substances have been very valuable for identifying many types of biological molecules and pure antibodies should prove to be a highly selective method for identifying glucoamylase synthesized in recombinant DNA cells. Antibodies directed at glucoamylase were first produced in 1963 (ref. 45) and used for determining similarities of the isozymes of glucoamylase from *A. niger*. These antibodies were produced in rabbits which had been immunized intramuscularly with solutions of pure glucoamylase. The serum was tested for reactivity with the *A. niger* glucoamylase as well as glucoamylases from several other fungal strains by agar diffusion tests. The results of the agar diffusion tests are shown in Plate A of Figure 17. It should be noted in the figure that the glucoamylase from all *Aspergillus* strains reacted with the antibodies but the glucoamylase from *Rhizopus* did not. Also the different isozymes of glucoamylase from *Aspergillus* strains reacted with the antibodies. These results indicate structural similarities of glucoamylases from the same group of organisms.

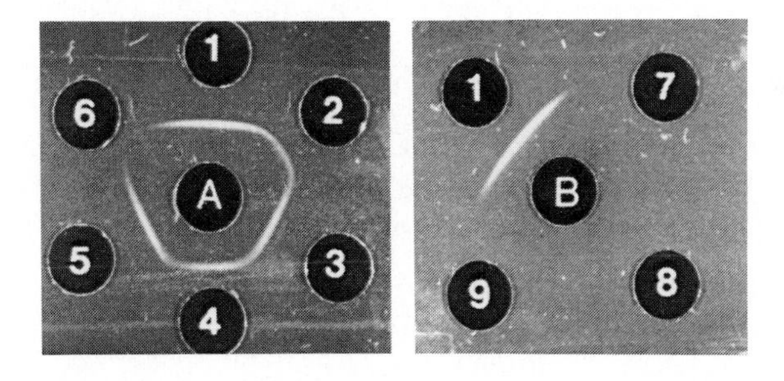

Fig. 17. Agar diffusion patterns of anti-glucoamylase serum (A) and anti-glucoamylase antibodies (B) with glucoamylase I of *A. niger* (well 1), glucoamylase II of *A. niger* (well 2), glucoamylase I of *A. awamori* (well 3), glucoamylase II of *A. awamori* (well 4), glucoamylase of *A. oryzae* (well 5) and glucoamylase of *R. niveus* (well 6). Plate B shows patterns of the pure antibodies with glucoamylase I of *A. niger* (well 1) and periodate oxidized glucoamylase from *A. niger* (well 7) and with glucoamylase from *R. niveus* (well 8) and buffer solution (well 9).

The antibodies directed against glucoamylase of *A. niger* have been isolated in pure form by affinity chromatography on a glucoamylase-Sepharose 4B column and elution with ammonium thiocyanate (refs. 20, 25). The elution patterns for the anti-glucoamylase antibodies from a glucoamylase-Sepharose column and from a mannosyl-Sepharose column are shown in Fig. 18. From the glucoamylase-Sepharose column the thiocyanate solution effected the elution of a UV absorbing component but no UV absorbing material was eluted from the mannosyl-Sepharose column by the mannose. The UV absorbing fractions eluting with the thiocyanate from the former column were collected separately, combined and mixed with an equal volume of saturated ammonium sulfate. The precipitate which formed on refrigeration overnight was separated by centrifugation and dissolved in a small volume of phosphate buffer of pH 7.0. The agar diffusion test on this material (Well 1, plate B of Fig. 17) showed that the sample reacted in the precipitin test with glucoamylase in a comparable manner as the original serum. To obtain information on the immunodeterminant groups of glucoamylase the enzyme was subjected to oxidation in 0.04M sodium periodate for 1 to 2 hrs. The oxidized antigen was recovered by dialysis and lyophilization. Solutions of the oxidized antigen were no longer reactive with the antibodies (Fig. 17, well 7). Wells 8 and 9 show no reactivity with pure Rhizopus glucoamylase or with the buffer solution. Since periodate oxidizes carbohydrate residues in the glucoamylase and does not degrade the polypeptide chain under mild oxidation conditions it was concluded that the antibodies are specific for carbohydrate moieties of the enzyme. Therefore the carbohydrate chains are the hapten groups of glucoamylase. Cross

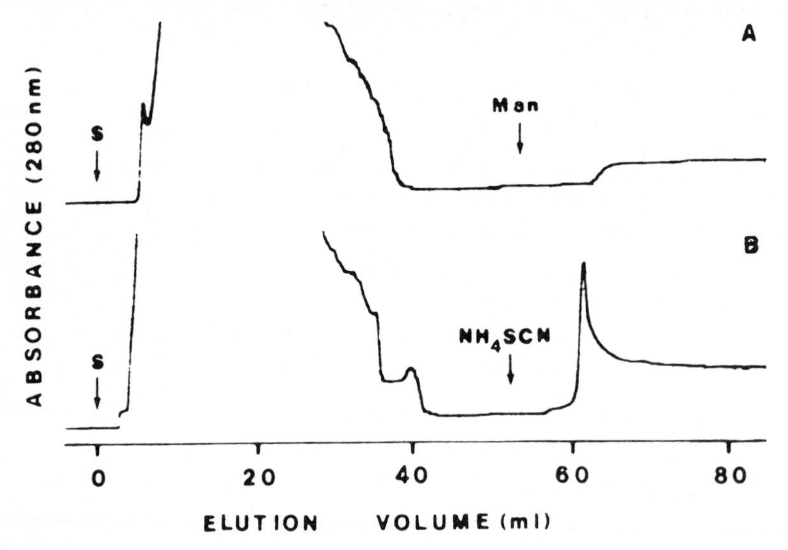

Fig. 18. Elution patterns of anti-glucoamylase antibodies from mannosyl-Sepharose (A) and glucoamylase-Sepharose (B), elution in frame (A) was with mannose and in frame (B) was with ammonium thiocyanate (ref. 20).

reactivities of the antibodies with glucoamylases from strains of the same group of organisms and with isozymes do occur but not with glucoamylases from unrelated organisms. These results are shown by the results in Fig. 17. Cross reactivities of anti-glucoamylase immune serum and other glucoamylases have been observed by other investigators (ref. 70).

Since the anti-glucoamylase antibodies are specific for the carbohydrate units of glucoamylase, the antibodies were used in agar diffusion tests with the CNBr fragments from glucoamylase in order to detect those fragments which contained carbohydrate chains. The results are presented in Fig. 19. It can be seen that the six CNBr fragments which were isolated from a reaction of CNBr with glucoamylase all reacted with the antibodies. Thus all the fragments contain carbohydrate chains. This result is consistent with a random distribution of the carbohydrate chains along the polypeptide chain of the glucoamylase.

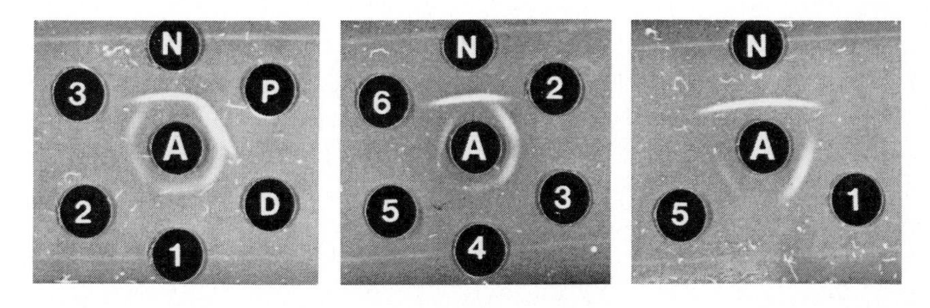

Fig. 19. Agar diffusion patterns of anti-glucoamylase antibodies with glucoamylase and the CNBr fragments from glucoamylase; N = native glucoamylase, P = partially denatured glucoamylase, D = denatured glucoamylase and numbers 1 to 6 = the CNBr fragments (ref. 61).

The purified antibodies were checked by isoelectrofocusing on polyacrylamide gels in a pH gradient of pH 5 to 8 to assess the protein homogeneity in the preparation. Duplicate samples of the antibodies were applied separately on different gels. One finished gel was stained with a protein stain and the other gel was embedded in fluid agar. When the agar had solidified a trough was cut about 2 cm from the gel and a solution of glucoamylase was placed in the trough. Diffusion of antibodies and antigen was allowed to proceed for 24 to 48 hrs at room temperature. Such a technique was developed to investigate anti-lactose isoantibodies as described in another study (ref. 71). The protein species which were separated by isoelectric focusing and which react with the antigen can be readily detected by this method. Fig. 20 shows a photograph of the gels containing glucoamylase and subjected to isoelectric focusing followed by agar diffusion.

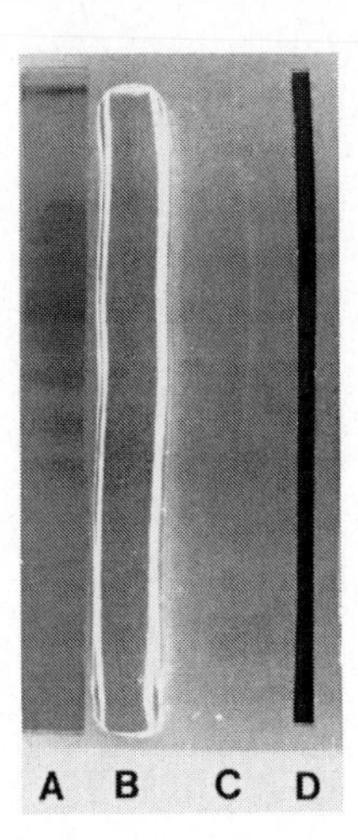

Fig. 20. Isoelectrofocusing and agar diffusion of anti-glucoamylase antibodies, gel A was stained for protein, gel B was embedded in agar, area C is the area of precipitin band formation, and area D is a trough containing the antigen.

In this figure it will be noted that the purified antibodies consisted of nine different proteins as revealed by gel A which had been subjected to isoelectric focusing and stained with a protein dye. These proteins possessed different isoelectric points and migrated to different points in the gel. Gel B contained antibodies and was not stained with the dye but embedded in agar. On diffusion for 24 hrs the components in the gel and the antigen in the trough D yielded a long precipitant band (Area C). The precipitin band was formed opposite each of the nine proteins and thus each protein possessed antibody activity. A set of such proteins has been termed isoantibodies by analogy to the isozyme terminology. All proteins of a set are induced by the same immunodeterminant group of the antigen and all react with this determinant in the precipitin reaction to form the antigen-antibody complex. The biological significance of isoantibodies of this type is not yet known.

4 BIOSYNTHESIS

4.1 Production of glucoamylase

Glucoamylase is produced by a fermentation process in submerged culture by a procedure similar to that originally described (ref. 72). The enzyme can also be produced by growing appropriate cultures on a semi-solid mold media (ref. 73) and extracting the enzyme from the media. In the submerged culture method a fermentor containing a sterile culture medium is aseptically inoculated with an actively growing liquid culture of a strain of the desired organism. The fermentation medium for growing the organisms in submerged culture contains a mineral mixture, ground corn (14 to 20%) and calcium carbonate (ref. 74). The inoculated medium is aerated, agitated and maintained at the desired temperature for optimum microbial growth. Incubation continues until the maximum yield of glucoamylase is attained as measured by a suitable enzyme assay. At this point the microorganisms and other insoluble materials are removed by filtration or centrifugation. The filtrate is concentrated by lyophilization and the glucoamylase is precipitated from this solution with an appropriate organic solvent. Culture filtrates may contain substantial amounts of glucosyltransferase which can be removed by adsorption on ion exchange resins (refs. 14, 15).

A process utilizing semi-solid medium has also been used. The medium of choice consists of moistened, cooked and sterilized wheat or rice bran, and a mineral mixture (ref. 75). This medium is inoculated aseptically with a spore culture of the glucoamylase producing strain of organism. The inoculated medium is incubated in shallow trays in a well ventilated room under carefully controlled conditions of temperature and humidity. After completion of growth the moist bran is first dried and then ground to a powder. The dried bran is extracted with water and the water evaporated to concentrate the enzyme. The glucoamylase is precipitated with organic solvents from the concentrated solution.

4.2 Biosynthesis from radioactive carbohydrates

Several types of experiments using radioactive tracers have been performed with the view of elucidating biosynthetic pathways for glucoamylase in the intact organism (ref. 42, 76). Actively growing cultures of *A. niger* which produce high levels of glucoamylase have been used to ensure adequate production of enzyme. *Aspergillus niger* strain NRRL 330 was employed in these experiments. The medium used for growing the fungi contains a glucose or mannose carbohydrate source (3%), nutrient broth (0.8%), sodium nitrate (0.3%), potassium phosphate (0.1%), magnesium sulfate (0.05%), potassium chloride (0.05%) and a small amount of a trace element mixture in a volume of 500 ml (ref. 42). The vegetative form of the organism was allowed to grow in the submerged culture for periods up to 85 hrs. The mycelium was removed by filtration through cheese cloth and washed with a small amount of water. The extracts were combined and stirred with four volumes of ethyl alcohol to precipitate the glucoamylase. The precipitated glucoamylase was collected by centrifugation, dissolved in citrate-phosphate buffer of pH 8.0 and used for additional purification and evaluation studies.

Radioactive carbohydrate precursors have been used to investigate the addition of the carbohydrates to the protein portion of the enzyme (ref. 42). In these experiments *Aspergillus niger* was grown on a medium containing 0.5 millicuries of D-glucose-1-^{14}C and the constituents listed above. The glucoamylase was isolated from the mycelium of the organism. This glucoamylase sample was purified further by chromatography on DEAE cellulose as outlined in section 3.1. The two isozymes of glucoamylase were obtained by this procedure and the yield of isozyme I was 50 mg while the yield of isozyme II was 20 mg. Since isozyme I was obtained in a greater yield, this glucoamylase has been used to isolate the carbohydrate for degradation studies. The total radioactivity of the glucoamylase obtained from the glucose-1-^{14}C was 1.1 x 10^5 CPM and the specific activity was 2,270 CPM per mg. Radioactive glucoamylase was also isolated from *A. niger* cells grown on mannose-1-^{14}C and used for degradation studies.

The carbohydrates of the labeled glucoamylase were liberated by acid hydrolysis of the enzyme in 0.5 N hydrochloric acid for 5 hr at 100°C. The compounds were isolated by a preparative chromatographic method in a solvent system of 6 parts n-butyl alcohol, 4 parts pyridine and 3 parts water (ref. 36). Radioautograms were prepared on no-screen x-ray film and these showed that radioactive compounds were present at Rf values of 0.65, 0.59 and 0.54. These values are identical to those obtained with reference mannose, glucose and galactose, respectively. The glucose and galactose were further identified by the glucose oxidase and the galactose oxidase clinistix test strips. The areas of the chromatograms containing the carbohydrates were eluted separately with a small volume of water. The activity of each monosaccharide was determined and the positions of the ^{14}C labelling were located. The location of the label was by a microbiological method utilizing *Leuconostoc mesenteroides* (ref. 77). The fermentation products produced from these hexoses by the organism were carbon dioxide from the C-1, ethanol from C-2 and C-3 and lactic acid from carbons 4, 5 and 6. The fermentation products were isolated by standard methods and aliquots of the solutions were used for radioactivity measurements in a liquid scintillation counter. Reference solutions of glucose-1-^{14}C, mannose-1-^{14}C and mannose-2-^{14}C were also degraded. The results are presented in Table 7.

TABLE 7

Distribution of ^{14}C in Labeled Monosaccharides Recovered from Glucoamylase-^{14}C Grown on Mannose-1-^{14}C or Glucose-1-^{14}C and Determined with *Leuconostoc mesenteroides*

Compound	Radioactivity CPM	C-1	C-2+C-3	C-4+C-5 +C-6
Glucose-1-^{14}C (Reference)	16,500	14,000	25	40
Mannose-1-^{14}C "	22,400	18,900	80	30
Mannose-2-^{14}C "	17,000	230	11,000	770
Mannose-^{14}C (Glucoamylase)	19,900	16,800	435	950
Glucose-^{14}C "	3,760	3,270	-	-
Galactose-^{14}C	1,120	-	-	-

On the basis of the results in Table 6 it can be seen that the mannose and glucose isolated from the glucoamylase obtained from cells grown on radioactive carbohydrates were labeled predominately at carbon 1. It is likely that galactose was also labeled at carbon 1 but the labeling pattern of this compound was not determined. The labelling patterns of the hexoses show that the incorporation of intact carbohydrate units into glucoamylase had occurred and such incorporation was most likely by the nucleotide diphosphate hexose pathway (ref. 76). Guanosine diphosphate mannose, uridine diphosphate glucose and uridine diphosphate galactose would be the logical precursors of the carbohydrates of the glycoprotein. The transfer of sugar moieties from these nucleotides to the glycoprotein would be catalyzed by appropriate transferases of the organism. The enzymes needed for the synthesis of the nucleotide diphosphate hexoses have been identified in extracts of the mycelium of *A. niger* (ref. 76). These experiments were performed with enzyme extracts from freshly grown cells of the organism and with the appropriate nucleotide triphosphate and hexose-1-phosphate as substrates and cosubstrates. The products synthesized in such reaction mixtures were separated on paper chromatograms and identified by the characteristic Rf values of the compounds to be guanosine diphosphate mannose, uridine diphosphate glucose and uridine diphosphate galactose.

In another type of experiment on biosynthetic pathways with a radioactive precursor (ref. 76), guanosine diphosphate mannose-1-^{14}C was tested as a donor of the mannose moieties of glucoamylase. In this experiment a modified glucoamylase was used as the acceptor molecule and the modification was achieved by the removal of some of the mannose residues from the native enzyme by treatment with an α-mannosidase (Fig. 12). The modified glucoamylase was recovered and used as the acceptor by incubating this preparation with radioactive guanosine diphosphate mannose and an enzyme extract from *A. niger*. After incubation for 6 hrs at room temperature, the reaction was stopped by heat and the products were fractionated into high and low molecular weight components by chromatography on Sephadex G-25. Assays for glucoamylase activity showed that the glucoamylase was present in fractions 14-16. These fractions also gave positive tests for protein and for radioactivity. The unreacted guanosine diphosphate-mannose-1-^{14}C was present in fractions 29-33. The radioactivity found in the glucoamylase in fractions 14-16 ranged from 1100 to 1600 CPM in several experiments. Thus a transfer of the mannosyl units from guanosine diphosphate mannose to partially modified glucoamylase had occurred. It is likely that glucose and galactose units would also be transferred from uridine diphosphate glucose or galactose to the glycoprotein by specific transferases in the cellular extracts. The regulatory mechanisms which specify the number and types of carbohydrate residues that will be transferred, the sequence of the amino acids in the peptide at the acceptor site and the chemical nature of the linkages of carbohydrate to protein need yet to be determined.

4.3 Biosynthesis by recombinant DNA methods

Recently the biosynthesis of glucoamylase by recombinant DNA methods has been reported by several groups of investigators (refs. 17-19, 78-82). The first group utilized the glucoamylase genes from *Aspergillus niger* (ref. 17). Prior to these studies it was necessary to determine the complete amino acid sequence of glucoamylase in order that comparisons between the native and the recombinant glucoamylase could be made. The native enzyme and various cleavage fragments were used for determining the complete amino acid sequences of two isozymes of native glucoamylase (refs. 43, 78) and the nucleotide sequence of the gene was used for determining the sequence of recombinant glucoamylase (ref. 17). The synthesis of glucoamylase by recombinant methods requires the participation of several types of nucleic acids, enzymes, cofactors, and the array of amino acids which constitute the glucoamylase molecule. Isozyme I and II of the glucoamylase of *A. niger* have been synthesized in recombinant cells from an *E. coli* PBR 327 plasmid annealed with *A. niger* DNA. On the basis of the nature of the gene products and the type of plasmid used it was proposed that the regulation of the biosynthesis of the two isozymes of glucoamylase resides in a single gene. In the single glucoamylase gene, several intervening nucleotide sequences are present as well as the sequence for the glucoamylase. The gene has been located by appropriate mapping techniques and it was found that the intervening sequences carry the initiation and stop codons. Segments of the nucleotide chain can undergo splicing events and produce a second gene unit which is used for the synthesisis of isozyme II of glucoamylase (ref. 17).

The nucleotide fragment directing the synthesis of isozyme I of glucoamylase results in the formation of a primary translation product of 640 amino acid residues which is a pro-protein unit of this isozyme. Since it was determined by amino acid sequencing that the glucoamylase contains only 616 amino acids a polypeptide of 24 residues must be removed from the pro-protein to yield the active glucoamylase molecule. The amino acid sequence of the resulting glucoamylase was deduced from nucleotide sequence and it was found that this sequence and that determined by chemical methods on the native enzyme are similar. Isozyme II of glucoamylase was also synthesized as a pro-protein which was processed by removal of a peptide fragment to yield the active enzyme (ref. 78).

The second group of investigators utilized Aspergillus awamori as the organism in recombinant DNA studies (ref. 18, 79). This organism when grown on a hexose carbohydrate source produces high levels of glucoamylase but little glucoamylase when grown on a pentose or triose carbohydrate source. Isozyme I of this glucoamylase was used for the recombinant studies with this organism. In the initial experiments the total cellular nucleic acids were isolated from A. awamori mycelium by appropriate methods (ref. 18). The mRNA from the mycelium was translated in vitro utilizing the rabbit reticulocyte lysate system for identifying the types of proteins synthesized. For molecular cloning experiments, a genomic library was prepared from A. awamori DNA by digesting the nucleic acids to completion with several types of endonucleases. The library was screened using a glucoamylase cDNA probe. EcoRI and Sal I fragments were also used for identifying the 5-prime end of the glucoamylase gene. E. coli PBR 322 plasmid was used to produce a new plasmid with A. awamori DNA fragments. The new plasmid was used to produce recombinant cells which synthesized glucoamylase.

In an attempt to elucidate the molecular basis for the carbon source dependence of A. awamori for glucoamylase production, the nature of mRNA unique for glucoamylase synthesis on different carbon sources was examined. Total cellular mRNA was isolated from A. awamori cells and was used for synthesising proteins with the rabbit reticulocyte lysate system. The mRNA specific for glucoamylase was identified by immunoprecipitation of the protein products with serum containing anti-glucoamylase antibodies. These experiments showed that a translatable mRNA was present in the RNA preparation from cells grown on a starch carbon source but not from cells grown on a pentose or triose carbon source. An EcoRI digest of genomic DNA was then prepared and was molecularly cloned to lambda charon 4A and used to identify genome sequences flanking the glucoamylase gene. These results and the results of other investigators (ref. 17) have been used to construct the composite restriction endonuclease map shown in Fig. 21.

Other groups of investigators used the glucoamylase genes from several strains of Rhizopus and yeasts to produce recombinant cells which synthesize glucoamylase (refs. 19, 80-82). In the yeast studies, several Saccharomyces cerevisiae transformants were constructed with the glucoamylase gene. Some of these transformants produced high levels of glucoamylase which was secreted into the culture media, others did not (ref. 19). All of the fungal and yeast transformants which produced glucoamylase yielded enzyme molecules with similar properties to those of glucoamylases isolated from the original organisms. All of these glucoamylases need to be characterized structurally and enzymologically more fully.

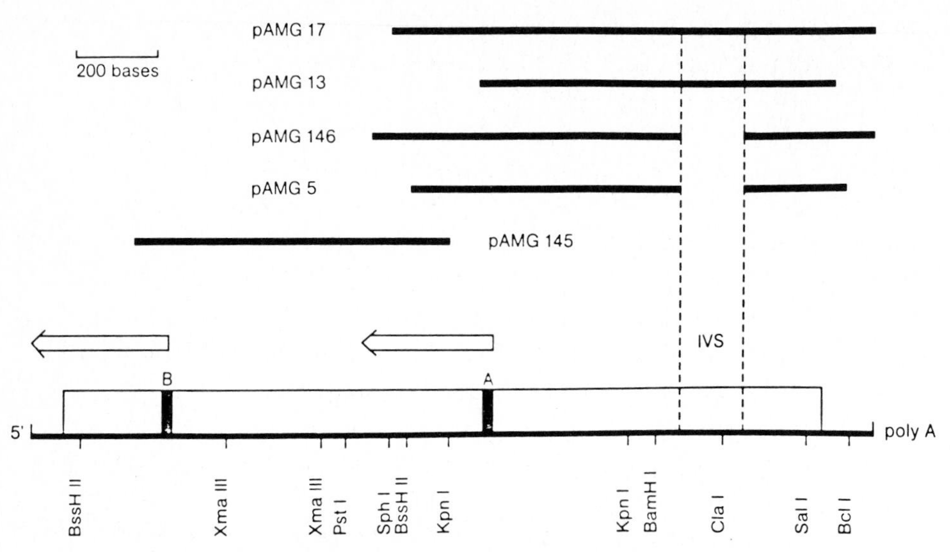

Fig. 21. The restriction endonuclease map of the Aspergillus genome surrounding the glucoamylase gene (ref. 17).

Some of the results of the genetic engineering experiments which have been reported need further verification and documentation before technological uses can follow. To illustrate, recombinant glucoamylases have not been isolated in sufficient quantities for chemical characterization of the enzyme. Results of suitable enzyme assays with these glucoamylases on conventional substrates have not been reported. Little data have been provided on the yield of glucoamylase from the transformants and this type of information is needed for the development of technological uses. The glycoprotein nature of glucoamylase synthesized by recombinant DNA methods has not been established. Neither the identification of the carbohydrate units attached to the recombinant glucoamylase nor the manner of attachment of the carbohydrate units to proteins has been elucidated. The immunoprecipitation results with the gene products and with anti-glucoamylase serum which have been reported are difficult to reconcile with the fact that such antibodies are specific for the carbohydrate chains of glucoamylase.

There is a discrepancy in the literature as to the exact molecular weight of the native and recombinant glucoamylases. Accordingly, a determination of exact amino acid composition of the enzyme cannot be made until the molecular weight value is known. Also the amino acid sequence cannot be determined until the amino acid composition and the molecular weight have been precisely determined. The genetic engineering experiments described in this section may be of theoretical value, however, at the present time results of these experiments are of little value for producing glucoamylase for industrial uses.

5 APPLICATIONS

The emphasis in this section of the chapter is on the applications of glucoamylase for the conversion of starch to glucose and the subsequent conversion of glucose to high fructose sweeteners, amino acids, organic acids, alcohols, and crystalline glucose. The high fructose sweeteners, very important products of glucoamylase application, are used as an ingredient in foods, soft drinks, alcoholic beverages, candies, ice cream, and many low calorie products. The organic acids and crystalline glucose are used in foods but also in pharmaceutical and medicinal products. Glucoamylase is used as a reagent in analytical methods for the determination of glycogen, starch and serum amylase and for the treatment of glycogen storage diseases.

Carbohydrate sweeteners were originally made from starch by acid hydrolysis but are presently manufactured by the enzymatic process utilizing glucoamylase and glucose isomerase. Starch is a renewable agricultural resource and consequently the supply of starch is virtually inexhaustable. Much research has been done on many aspects of starch chemistry such as the structure, the genetic manipulation of the plant to produce starches of varying ratios of amylose to amylopectin and the mechanism of enzymatic synthesis and degradation (ref. 83).

The industrial manufacture of syrups by the acid or enzyme method results in a product containing glucose, malto-oligosaccharides and branched oligosaccharides which are resistant to hydrolysis. Syrups produced with glucoamylase contain primarily glucose and some malto-oligosaccharides but little of the branched oligosaccharides. Data on the carbohydrate types and concentrations in a number of syrups are recorded in Table 8.

In Table 8 the column labeled DE refers to dextrose equivalent of the syrup and the mono-, di-, tri-, etc. refer to the monosaccharide, the disaccharide, the trisaccharide, etc. The DE terminology relates to the reducing power of a syrup in comparison to dextrose (glucose). Thus a DE of 40 means that 100 parts of syrup possess the same reducing power as 40 parts of pure glucose.

TABLE 8

Carbohydrate Composition (%) of Some Corn Syrups (ref. 84)

Conversion	D.E.	Mono-	Di-	Tri-	Tetra-	Penta-	Hexa-	Hepta-	Higher
Very Low	10	2.3	2.8	2.9	3.0	3.0	2.2	2.1	81.7
Low	20	5.5	5.9	5.8	5.8	5.5	4.3	3.9	63.5
Low	30	10.4	9.3	8.6	8.2	7.2	6.0	5.2	45.1
Low	40	16.9	13.2	11.2	9.7	8.3	6.7	5.7	28.3
Intermediate	50	25.8	16.6	12.9	10.0	7.9	5.9	5.0	15.9
Intermediate	60	36.2	19.5	13.2	8.7	6.3	4.4	3.2	8.5
High	70	45.1	21.4	12.7	7.5	5.1	3.6	2.2	4.5
Very High	95	92.0	4.0	3.0	1.0	-	-	-	-

The syrup which is designated as very high conversion was produced by an enzymatic method employing glucoamylase. The low conversion syrups are produced by hydrolysis of starch in dilute acid and for short periods. It is quite apparent from the data in the table that there is a wide variation in glucose content and in the amount of various oligosaccharide in the corn syrups prepared under different hydrolytic conditions.

An important advance in the industrial use of starch and glucose was the development of a method for obtaining crystalline glucose from hydrolysates of starch (ref. 85). Crystalline glucose has a number of properties which are desirable in applications in the pharmaceutical, medicinal and the food industries. In the original method developed for the crystallization of glucose acid hydrolysates of starch were used. From such hydrolysates glucose crystallized with difficulty because the products of side reactions during acid hydrolysis of starch interferred with crystallization. These products were formed by isomerization to other aldehydes, dehydration to anhydro compounds and reversion to oligosaccharides. Consequently the maximum yield of glucose obtained from starch by this method was about 85 to 90%.

The enzymatic process is now used exclusively for hydrolyzing starch (refs. 3, 84, 86) and has replaced the acid hydrolysis method. The biological process possesses a number of advantages over the acid catalyzed hydrolysis of starch. First the enzyme catalyzed reactions can be carried out at much lower temperatures than acid-catalyzed hydrolysis which require high temperature. Second, enzyme catalyzed reactions can be effected at pH values near neutrality which, of course, is not possible in acid hydrolysis. Third, enzyme catalyzed reactions are highly specific and hydrolyze specific bonds of the starch molecule and accordingly cleavage of selected bonds can be effected with enzymes. Fourth, the enzymatic process converts starch to glucose in higher yield than the acid catalyzed process.

Considerable research has been done on the preparation of an immobilized glucoamylase for use in starch conversions with the objective of obtaining superior enzyme preparations. Methods have been developed for immobilizing glucoamylase on a variety of inert supports (refs. 87-89). Immobilized glucoamylase if prepared in a stable form and by economical means would have a number of advantages for use in starch conversions. The insoluble supports that have been used for preparing immobilized enzymes are resins (refs. 87, 88) and activated ground glass or porous silica (ref. 89). However the advantages of using immobilized glucoamylase may be outweighed by losses of the enzyme from the supports (ref. 89) and by incomplete hydrolysis of starch by immobilized enzyme preparations (ref. 84). Immobilized glucose isomerase is used for the isomerization of glucose to fructose (ref. 4) since immobilized glucose isomerase is easier to prepare and use. Attempts to improve immobilized glucoamylase materials are still in progress and preparations suitable for use in starch conversions should be forthcoming in the future.

The glucose which is made from starch with glucoamylase is used in the manufacture of many products such as high fructose sweeteners, glutamic acid, industrial alcohol, sorbitol, citric acid, and others. Many of these products are used in foods, confectionaries, beverages, and pharmaceuticals. The fructose sweeteners contain at least 45% fructose and as high as 90% fructose. Since fructose is nearly twice as sweet as sucrose the same sweetness can be obtained by use of only half as much fructose. Therefore, high fructose sweeteners can be used in less quantity than sucrose in foods and accordingly yield food products which are lower in calories.

Glutamic acid is a dicarboxylic amino acid of which the derivative, glutamine, has been found to be an excellent flavor enhancer. The glutamic acid required for the synthesis of glutamine is made by a fermentation process in which the glucose derived from starch by glucoamylases is converted to glutamic acid by fermentation with a specific microorganism. This organism is *Corynebacterium* (ref. 1) which possesses the enzyme systems for converting glucose via glycolysis to pyruvic acid and the pyruvic acid via the TCA cycle to α-ketoglutaric acid. The strain of organism has been selected to produce low levels of α-ketoglutaric dehydrogenase and high levels of glutamic dehydrogenase. Therefore, the normal conversion of the α-ketoglutaric acid to succinic acid is much reduced and a conversion of the keto acid to glutamic acid is accordingly increased. The glutamic acid is released from the cell by the addition or removal of components from the media. Removal of biotin or the addition of saturated fatty acids, detergents or penicillum promotes the secretion of glutamic acid from the cell. Lysine, methionine and aspartic acid can also be synthesized by the fermentation process with other organisms.

Glucoamylase has been used in the production of ethyl alcohol to be used as a fuel additive. The glucoamylase converts starch from cereal grains to glucose and the latter is fermented to ethyl alcohol (ref. 73). Ethyl alcohol is perhaps the most important organic chemical of commerce. Much of the alcohol used industrially is produced synthetically from petroleum refining by-products. However the biological fermentation of glucose derived from starch or cellulose is an alternate source of this important compound.

During the production of beverages from the starch in cereal grains, unfermentable oligosaccharides with α-D-(1,6) and α-D-(1,4) linkages are produced due to incomplete hydrolysis. These oligosaccharides are not hydrolyzable by the normal fermentation enzymes and remain unchanged in the beverage. Glucoamylase is capable of hydrolyzing the oligosaccharides. Thus the addition of glucoamylase to the fermentation mixture will result in the hydrolysis of the resistant saccharides followed by a fermentation of the hydrolytic products. A lower carbohydrate content results in the beverage and the product will have a lower calorie value.

Citric acid, a tricarboxylic acid, is used in soft drinks and food products to adjust and maintain the proper acidity in the product. This acid can be made by fermentation of glucose with the appropriate strains of *Aspergillus niger*. Under proper conditions, the organism converts glucose to citric acid by reactions of the glycolytic and the TCA cycles. Citric acid is secreted in the filtrate and is isolated from this filtrate. Other organic acids can be made by a fermentation process and include lactic acid and acetic acid.

Glucoamylase has been used in analytical methods developed for determining glycogen and starch (ref. 90, 91) and for measuring serum amylase in the circulatory system of humans with certain diseases (ref. 92). In the analytical method the polysaccharides are first converted quantitatively to glucose by the glucoamylase and the glucose is then measured colorimetrically with a phenol or another carbohydrate specific reagent. The precision of the method for the determination of the polysaccharides is high and the sensitivity is equally high. A method based on use of glucoamylase has been developed for assay of serum amylase in patients with abnormal serum amylase levels (ref. 92). Glucoamylase has been tested for the treatment of individuals with glycogen storage diseases (ref. 93, 94). In these diseases the glycogen accumulates in the tissues and causes pathological conditions. Glucoamylase has been injected into tissues of such patients to promote the hydrolysis of the glycogen to glucose. The glucose is removed from the tissue by the circulatory system of the patient and is subsequently metabolized. The experiments have been only partially successful and additional studies are needed to develop methods of utilizing glucoamylase for the treatment of glycogen storage diseases.

6 CONCLUDING REMARKS

Glucoamylase is a remarkable enzyme in structure, in substrate specificity, in action mechanism and in applications. The enzyme converts starch quantitatively to glucose and is used industrially in starch conversions for the manufacture of syrups, sweeteners, organic acids and industrial alcohols. Glucoamylase is a natural product produced by many strains of fungi, particularly of the *Aspergillus* and *Rhizopus* groups and a few strains of yeasts of the *Endomyces* group. The enzyme is generally synthesized in isozymic forms but the precise function of isozymes is not yet known. In structure glucoamylase is a glycoprotein consisting of a single polypeptide chain of about 800 amino acid residues with serine and threonine being especially high and 90 carbohydrate residues being attached to some of these amino acid residues. The carbohydrate chains represent from 15 to 20% of the enzyme molecule and are composed principally of mannose, glucose and galactose and occasionally of N-acetyl glucosamine. The most abundant isozyme of the glucoamylase from A. niger contains 45 carbohydrate chains ranging in molecular size from single to tetrasaccharide units and attached to the polypeptide chain. A high proportion of the chains are single residues and all of the chains are linked O-glycosidically to serine and threonine of the enzyme. The carbohydrate chains are randomly distributed along the polypeptide chain of the enyzme and not clustered in a short segment of the chain.

Antibodies directed at glucoamylase are induced in rabbits immunized with a vaccine of the enzyme. These antibodies have been isolated in pure form by affinity chromatography on a glucoamylase-Sepharose column and elution with ammonium thiocyanate. Several lines of evidence have shown that the antibodies are anti-carbohydrate antibodies directed at the carbohydrate chains of glucoamylase. The antibodies, especially the pure antibodies, should be very useful for following the biosynthesis of glucoamylase in recombinant DNA cells.

The complete amino acid sequence of glucoamylase from A. niger has been reported. However, in view of variations in values for amino acid composition and in molecular weight recorded in the literature for this enzyme, additional studies are needed before the exact amino acid sequence of the glucoamylase is elucidated. The biosynthesis of glucoamylase in the whole cells requires the participation of appropriate types of nucleic acids, enzymes, cofactors and the amino acids which constitute the glucoamylase. Reports on the cloning of bacterial and yeast cells with glucoamylase genes to obtain recombinant cells that synthesize glucoamylase have appeared. Such reports need to be documented further with additional verification studies. Information on the synthesis of the carbohydrate portion of the enzyme has been obtained by use of radioactive substrates. These results show that nucleotide diphosphate hexoses participate in the pathway of biosynthesis of the carbohydrate moieties. Much remains to be learned about the biosynthesis of the carbohydrate portion of the enzyme as well as biosynthesis of a functional glucoamylase by recombinant DNA methods.

Glucoamylase hydrolyzes the glucosidic linkages in starch, glucosyl oligosaccharides and glucosides with the liberation of glucose from these substrates. The enzyme is capable of hydrolyzing all types of α-glucosidic linkages in these substrates, including α-D-(1,4), α-D-(1,3), α-D-(1,2), and α-D-(1,6). However differences in the rates of hydrolysis of these linkages do exist. The action mechanism of glucoamylase on oligosaccharides and starch has been shown to be a multi-chain mechanism in which single glucose units are removed from the non-reducing ends of the chains and all the chains of the substrate are shortened simultaneously. Since some linkages are hydrolyzed more slowly than others these linkages will persist longer in the enzymatic digests. Starch does contain a considerable number of α-D-(1,6) linkages which are hydrolyzed more slowly than the α-D-(1,4) linkages. Therefore the hydrolysis of starch occurs in stages, first a rapid rate for terminally linked α-D-(1,4) linkages, then a slow rate of hydrolysis when α-D-(1,6) linkages are encountered and then again a rapid rate when new α-D-(1,4) linkages become exposed. This multi-stage hydrolysis continues until all of the starch molecule is hydrolyzed to glucose.

Many of the applications of glucoamylase are related to the ability of the enzyme to hydrolyze starch quantitatively to glucose. The manufacture of glucose sweeteners and crystalline glucose are examples of two early applications. More recently, high fructose sweeteners containing 90% or more of fructose have been introduced and used to produce low calorie foods. Other products made from glucose derived from starch by glucoamylase action are glutamic acid, lysine, lactic acid, citric acid, sorbitol, and alcohols. Selective fermentation processes have been developed for converting glucose to these products. In the future, additional advances in glucoamylase research should be forthcoming in the areas of structure, stability, immobilization and synthesis by recombinant DNA methods. Such information will contribute greatly to knowledge of this important enzyme and ultimately result in new applications and ingeneous new products.

7 REFERENCES

1 D.E. Eveleigh, The microbial production of industrial chemicals, in: D. Flanagan, B.P. Hayes and D. Morrison (Eds.), Industrial Microbiology and the Advent of Genetic Engineering, Scientific American, U.S.A., (1981) 71-79.

2 P.L. Layman, Industrial enzymes battling to remain specialties, Chem. Eng. News, Sept. 15 (1986) 11-14.

3 L.A. Underkofler, Development of a commercial enzyme process: glucoamylase, Adv. Chem. Series, 95 (1969) 343-358.

4 R.L. Antrim, W. Colilla and B.J. Schnyder, Glucose isomerase production of high fructose-syrups, Appl. Biochem. Bioeng., 2 (1979) 97-155.

5 Enzyme Nomenclature Commission, Elsevier, Amsterdam, The Netherlands, 1973, pp. 23-25, and Academic Press Inc., Orlando, FL, U.S.A., 1984,. pp. 308.

6 J.H. Pazur and T. Ando, The action of an amyloglucosidase of Aspergillus niger on starch and malto-oligosaccharides, J. Biol. Chem., 234 (1959) 1966-1970.

7 A. Tanaka, Y. Fukuchi, M. Ohnishi, K. Hiromi, S. Aibara and Y. Morita, Fractionation of isozymes and the determination of the subsite structure

of glucoamylase from Rhizopus niveus, Agric. Biol. Chem., 47 (1983) 573-580.

8 L.A. Underkofler, L.J. Denault and E.F. Hou, Enzymes in the starch industry, Die Starke, 17 (1965) 179-184.

9 E.R. Kooi and F.C. Armbruster, Production and use of dextrose, in: R.L. Whistler and J. N. BeMiller (Eds.), Starch Chemistry and Technology, Academic Press Inc., New York, U.S.A., 2 (1970) 553-568.

10 H.A. Sober, F.J. Gutter, M.M. Wyckoff and E.A. Peterson, Chromatography of proteins. II. Fractionation of serum protein on anion-exchange cellulose, J. Amer. Chem. Soc., 78 (1956) 756-763.

11 J.H. Pazur and T. Ando, The hydrolysis of glucosyl oligosaccharides with α-D-(1,4) and α-D-(1,6) bonds by fungal amyloglucosidase, J. Biol. Chem., 235 (1960) 297-302.

12 J.H. Pazur and T. Ando, The isolation and the mode of action of a fungal transglucosylase, Arch. Biochem. Biophys., 93 (1961) 43-49.

13 J.H. Pazur and D. French, The transglucosidase of Aspergillus oryzae, J. Amer. Chem. Soc., 73 (1951) 3536.

14 F.C. Armbruster, U.S. Patent, 3,012, 944 (1961): Chem. Abstr., 56 (1962) 5224-5225.

15 K. Watanabe and T. Fukinbara, Studies on saccharogenic amylase produced by Aspergillus awamori, (2) Elimination of maltose-oligosaccharide-transglucosidase from culture filtrate, J. Ferment. Tech., 40 (1964) 332-337.

16 M.Z. Sternberg, The separation of proteins with heteropolyacids, Biotech. Bioeng., 12 (1970) 1-17.

17 E. Boel, I. Hjort, B. Svensson, F. Norris, K.E. Norris and N.P. Fiil, Glucoamylases G1 and G2 from Aspergillus niger are synthesized from two different but closely related mRNAs, The EMBO Journal, 3 (1984) 1097-1102.

18 J.H. Nunberg, J.H. Meade, G. Cole, F.C. Lawyer, P. McCabe, V. Schweickart, R. Tal, V.P. Wittman, J.E. Flatgaard and M.A. Innis, Molecular cloning and characterization of the glucoamylase gene of Aspergillus awamori, Mol. Cell. Biol., 4 (1984) 2306-2315.

19 T. Ashikari, N. Nakamura, Y. Tanaka, N. Kiuchi, Y. Shibano, T. Tanaka, T. Amachi and H. Yoshizumi, Rhizopus raw-starch-degrading glucoamylase: its cloning and expression in yeast, Agric. Biol. Chem., 50 (1986) 957-964.

20 J.H. Pazur, K.R. Forry, Y. Tominaga and E. M. Ball, Anti-glycosyl antibodies: antibodies directed against the carbohydrate moieties of a glycoprotein, Biochem. Biophys. Res. Commun., 100 (1981) 420-426.

21 J.K. Dale and D.P. Langlois, U.S. Patent 2,201,609 (1940) Syrup and method of making the same, United States Patent Office 1-5, Chem. Abstr., 34 (1940) 6474.

22 L.L. Phillips and M.L. Caldwell, A study of the purification and properties of a glucose-forming amylase from Rhizopus delemar, gluc amylase, J. Am. Chem. Soc., 73 (1951) 3559-3563.

23 S.A. Barker and J.G. Fleetwood, 977, Studies on Aspergillus niger. Part VIII. The purification of glucamylase, J. Chem. Soc., (1957) 4857-4864.

24 Y. Tsujisaka, J. Fukumoto and T. Yamamoto, Specificity of crystalline saccharogenic amylase of moulds, Nature, 181 (1958) 770-771.

25 J.H. Pazur, Y. Tominaga and S. Kelly, The relationship of structure of glucoamylase and glucose oxidase to antigenicity, J. Protein Chem., 3 (1984) 49-62.

26 J.H. Pazur and K. Kleppe, The hydrolysis of α-D-glucosides by amyloglucosidase from Aspergillus niger, J. Biol. Chem., 237 (1962) 1002-1006.

27 J.H. Pazur, Y. Tominaga, L.S. Forsberg and D. L. Simpson, Glycoenzymes: An unusual type of glycoprotein structure for a glucoamylase, Carbohydr. Res., 84 (1980) 103-114.

28 B.J. Davis, Disc electrophoresis -II Method and application to human serum proteins, Annal. N.Y. Academy Sciences, 121 (1964) 404-427.

29 G. Fairbanks, T.L. Steck and D.F.H. Wallach, Electrophoretic analysis of the major polypeptides of the human erythrocyte membrane, Biochemistry, 10 (1971) 2606-2617.

30 M.L. Wolfrom and A. Thompson, Degradation of amylopectin to nigerose, J. Amer. Chem. Soc., 77 (1955) 6403.
31 T. Fukui and Z. Nikuni, Preparation and properties of crystalline glucoamylase from Endomyces species IFO 0111, Agr. Biol. Chem., 33 (1969) 884-891.
32 J.J. Marshall and W.J. Whelan, Incomplete conversion of glycogen and starch by crystalline amyloglucosidase and its importance in the determination of amylaceous polymers, FEBS Letters, 9 (1970) 85-88.
33 R.W. Kerr, F.C. Cleveland and W.J. Katzbeck, The action of amylo-glucosidase on amylose and amylopectin, J. Amer. Chem. Soc., 73 (1951) 3916-3921.
34 K.L. Smiley, M.C. Cadmus, D.E. Hensley and A. A. Lagoda, High-potency amyloglucosidase-producing mold of the Aspergillus niger group, Appl. Micro., 12 (1964) 455.
35 N.J. King, The glucoamylase of Coniophora cerebella, Biochem. J., 105 (1967) 577-583.
36 J.H. Pazur and D. French, The action of transglucosidase of Aspergillus oryzae on maltose, J. Biol. Chem., 196 (1952) 265-272.
37 S.C. Pan, L.W. Nicholson and P. Kolachov, Enzymic synthesis of oligosaccharides-A transglycosidation, Arch. Biochem. Biophys., 42 (1953) 406-420.
38 J.H. Pazur and S. Okada, A novel method for the action patterns and the differentiation of α-(1,4)-glucan hydrolases, J. Biol. Chem., 241 (1966) 4146-4151.
39 M. Nakamura and S. Kuroiwa, Back polymerization by saccharifying amylase, Internatl. Chem. Eng. Japan, 4 (1964) 530-534.
40 K. Watanabe and T. Fukimbara, Studies on saccharogenic amylase produced by Aspergillus awamori (V) The action of acid-stable saccharogenic amylase on starch and glucosyl substrates, J. Fermentation Tech., 44 (1966) 25-33.
41 J.H. Pazur, A. Cepure, S. Okada and L.S. Forsberg, Comparison of the action of glucoamylase and glucosyltransferase on D-glucose, maltose and malto-oligosaccharides, Carbohydr. Res., 58 (1977) 193-202.
42 J.H. Pazur, H. R. Knull and A. Cepure, Glycoenzymes: structure and properties of the two forms of glucoamylase from Aspergillus niger, Carbohydr. Res., 20 (1971) 83-96.
43 B. Svensson, K. Larsen, I. Svendsen and E. Boel, The complete amino acid sequence of the glycoprotein, glucoamylase G1, from Aspergillus niger, Carlsberg. Res. Commun., 48 (1983) 529-544.
44 J.H. Pazur, H.R. Knull and D.L. Simpson, Glycoenzymes: a note on the role for the carbohydrate moieties, Biochem. Biophys. Res. Commun., 40 (1970) 110-116.
45 J.H. Pazur, K. Kleppe and E.M. Ball, The glycoprotein nature of some fungal carbohydrases, Arch. Biochem. Biophys., 103 (1963) 515-518.
46 S.-I. Hakomori, A rapid permethylation of glycolipid and polysaccharide catalyzed by methylsulfinyl carbanion in dimethyl sulfoxide, J. Biochem. (Tokyo), 55 (1964) 205-208.
47 H. Bjorndal, C. G. Hellerquist, B. Lindberg and S. Svensson, Gas-liquid chromatography and mass spectrometry in methylation analysis of polysaccharides, Angew Chem. Internat. Ed., 9 (1970) 610-619.
48 A. Gunnarsson, B. Svensson, B. Nilsson and S. Svensson, Structural studies on the O-glycosidically linked carbohydrate chains of glucoamylase G1 from Aspergillus niger, Eur. J. Biochem., 145 (1984) 463-467.
49 K. Dill and A. Allerhand, Studies of the carbohydrate residues of glycoproteins by natural abundance carbon 13 nuclear magnetic resonance spectroscopy, J. Biol. Chem., 254 (1979) 4524-4531.
50 B. Svensson, K. Larsen and I. Svendsen, Amino acid sequence of tryptic fragments of glucoamylase G1 from Aspergillus niger, Carlsberg Res. Commun., 48 (1983) 517-527.
51 J.F. McKelvy and Y.C. Lee, Microheterogeneity of the carbohydrate group of A. oryzae α-amylase, Arch. Biochem. Biophys., 132 (1969) 99-110.
52 S. Hayashida and E. Yoshino, Formation of active derivatives of glucoamylase I during the digestion with fungal acid protease and α-mannosidase, Agric. Biol. Chem., 42 (1978) 927-933.

53 B. Svensson, T.G. Pedersen, I. Svendsen, T. Sakai and M. Ottesen, Characterization of two forms of glucoamylase from Aspergillus niger, Carlsberg Res. Commun., 47 (1982) 55-69.
54 M. Ohga, K. Shimizu and Y. Morita, Studies on amylases of Aspergillus oryzae cultured on rice. Part II. Some properties of glucoamylases, Agr. Biol. Chem., 30 (1966) 967-972.
55 D.R. Lineback and W.E. Baumann, Properties of a glucoamylase from Aspergillus phoenicis, Carbohydr. Res., 14 (1970) 341-353.
56 S. Hayashida, T. Nomura, E. Yoshino and M. Hongo, The formation and properties of subtilisin-modified glucoamylase, Agr. Biol. Chem., 40 (1976) 141-146.
57 Y. Takeda, H. Matsui, M. Tanida, S. Takao and S. Chiba, Purification and substrate specificity of glucoamylase of Paecilomyces varioti AHU 9417, Agr. Biol. Chem., 49 (1985) 1633-1641.
58 J.H. Pazur and S. Okada, Properties of the glucoamylase from Rhizopus delemar, Carbohydr. Res., 4 (1967) 371-379.
59 T. Takahashi, Y. Tsuchida and M. Irie, Purification and some properties of three forms of glucoamylase from a Rhizopus species, J. Biochem., 84 (1978) 1183-1194.
60 J.H. Pazur, K. Kleppe and J. S. Anderson, The application of density-gradient centrifugation for the isolation of enzymes, Biochim. Biophys. Acta, 65 (1962) 369-372.
61 J.H. Pazur, B. Liu, S. Pyke and C. R. Baumrucker, The distribution of carbohydrate side chains along the polypeptide chain of glucoamylase, J. Protein Chem., 6 (1987) 517-527.
62 R.G. Martin and B.N. Ames, A method for determining the sedimentation behavior of enzymes: Application to protein mixtures, J. Biol. Chem., 236 (1961) 1372-1379.
63 K. Weber and M. Osborn, The reliability of molecular weight determinations by dodecyl sulfate-polyacrylamide gel electrophoresis, J. Biol. Chem., 244 (1969) 4406-4412.
64 S. Hjerten and R. Mosbach, "Molecular-sieve" chromatography of proteins on columns of cross-linked polyacrylamide, Anal. Biochem., 3 (1962) 109-118.
65 Bio-Rad, Chromatography Electrophoresis Immunochemistry HPLC bulletin, (1985) pp. 32-39, Biorad, Richmond, CA.
66 H.K. Schachman, Ultracentrifugation, diffusion and viscometry, Methods Enzymology, 4 (1957) 32-103.
67 I.M. Freedberg, Y. Levin, C.M. Kay, W.D. McCubbin and E. Katchalski-Katzir, Purification and characterization of Aspergillus niger exo-1,4-glucosidase, Biochim. Biophys. Acta, 391 (1975) 361-381.
68 D.R. Lineback, L.A. Aira and R.L. Horner, Structural characterization of the two forms of glucoamylase from Aspergillus niger, Cereal Chem., 49 (1972) 283-297.
69 A. Tsuboi, Y. Yamasaki and Y. Suzuki, Two forms of glucoamylase from Mucor rouxianus. I. Purification and crystallization, Agr. Biol. Chem., 38 (1974) 543-550.
70 P. Manjunath and M.R. Raghavendra Rao, Comparative immunochemical studies on fungal glucoamylases, Indian J. Biochem. Biophys., 17 (1980) 388-390.
71 J.H. Pazur, M.E. Tay, B.A. Pazur and F.J. Miskiel, Observations on a set of isomeric anti-lactose antibodies directed against a group D streptococcal polysaccharide, J. Prot. Chem., 6 (1987) 387-399.
72 E.H. Le Mense, J. Corman, J.M. Van Lanen and A.F. Langlykke, Production of mold amylases in submerged culture, J. Bacteriol., 54 (1947) 149-159.
73 L.A. Underkofler, G.M. Severson and K.J. Goering, Saccharification of grain mashes for alcoholic fermentation, Ind. Eng. Chem., 38 (1946) 980-985.
74 H.M. Tsuchiya, J. Corman and H.J. Koepsell, Production of mold amylases in submerged cultures, II, Factors affeting the production of alpha amylase and maltase by certain Aspergilli, Cereal Chem., 27 (1950) 322-330.
L.A. Underkofler, G.M. Severson, K.J. Goering and L.M. Christensen, Commercial production and use of mold bran, Cereal Chem., 24 (1947) 1-22.

76 J.H. Pazur, D.L. Simpson and H.R. Knull, Biosynthesis of glucohydrolase 1, a glycoenzyme from Aspergillus niger, Biochem. Biophys. Res. Commun., 36 (1969) 394-400.
77 M. Gibbs, P.K. Kindel and M. Busse, Determination of isotope distribution in hexoses, Methods Carb. Chem., 2 (1963) 496-509.
78 B. Svensson, K. Larsen and A. Gunnarsson, Characterization of a glucoamylase G2 from Aspergillus niger, Eur. J. Biochem., 154 (1986) 497-502.
79 M.A. Innis, M.J. Holland, P. C. McCabe, G.E. Cole, V.P. Wittman, R. Tal, K.W.K. Watt, D.H. Gelfand, J.P. Holland and J.H. Meade, Expression, glycosylation, and secretion of an Aspergillus glucoamylase by Saccharomyces cerevisiae, Science, 228 (1985) 21-26.
80 I. Yamashita, K. Suzuki and S. Fukui, Nucleotide sequence of the extracellular glucoamylase gene STA1 in the yeast Saccharomyces diastaticus, J. Bacteriol., 161 (1985) 567-573.
81 I. Yamashita, K. Suzuki and S. Fukui, Proteolytic processing of glucoamylase in the yeast Saccharomyces diastaticus, Agric. Biol. Chem., 50 (1986) 475-482.
82 Y. Tanaka, T. Ashikari, N. Nakamura, N. Kiuchi, Y. Shibano, T. Amachi and H. Yoshizumi, Glucoamylase produced by Rhizopus and by a recombinant yeast containing the Rhizopus glucoamylase gene, Agric. Biol. Chem., 50 (1986) 1737-1742.
83 R.L. Whistler, J.N. Be Miller and E.F. Paschall, Starch Chemistry and Technology, 2nd Edition, Academic Press Inc., Orlando, FL (1984).
84 R.V. MacAllister, Nutritive sweeteners made from starch, Advances in Carbohydrate Chemistry and Biochemistry, 36 (1979) 15-56.
85 W.B. Newkirk, Dextrose from starch, U.S. Patent 1,508,569 (1924) Chem. Abstr., 18 (1924) 3736.
86 D.P. Langlois, Application of enzymes to corn syrup production, Food Tech., 7 (1953) 303-307.
87 M.J. Bachler, G.W. Strandberg and K.L. Smiley, Starch conversion by immobilized glucoamylase, Biotech. Bioeng., 12 (1970) 85-92.
88 K.L. Smiley, Continuous conversion of starch to glucose with immobilized glucoamylase, Biotech. and Bioeng., 13 (1971) 309-317.
89 P.J. Reilly, Starch hydrolysis with soluble and immobilized glucoamylase, Appl. Bioch. and Bioeng., 2 (1979) 185-206.
90 Z. Gunja-Smith, J.J. Marshall and E.E. Smith, Enzymatic determination of the unit chain length of glycogen and related polysaccharides, FEBS Letters, 13 (1971) 309-311.
91 E.Y.C. Lee and W.J. Whelan, Enzymic methods for the microdetermination of glycogen and amylopectin, and their unit-chain lengths, Arch. Biochem. Biophys., 116 (1966) 162-167.
92 J.J. Marshall, A.P. Iodice and W.J. Whelan, A new serum α-amylase assay of high sensitivity, Clinica Chimica Acta, 76 (1977) 277-283, Elsevier, North-Holland Biomedical Press.
93 C. Mercier and W.J. Whelan, Further characterization of glycogen from type-IV glycogen-storage disease, Eur. J. Biochem., 40 (1973) 221-223.
94 F. Huijing, B.L. Waltuck and W.J. Whelan, α-Glucosidase administration: experiences in two patients with glycogen storage disease compared with animal experiments, Enzyme Therapy in Genetic Diseases, 9 (1973) 191-194.

ACKNOWLEDGMENTS

The author acknowledges the many contributions of former and present students, post-doctorates and research associates to the methods described in this chapter, particularly T. Ando, K. Kleppe, A. Cepure, S. Okada, D. Simpson, H. Knull, Y. Tominaga, K. Forry, S. Kelly-Delcourt, N. Li, F. Miskiel, and B. Liu. Special acknowledgments are made to my secretary, Eileen McConnell, and to my editorial assistant, Jean Pazur, for their valuable and expert contributions.

STUDIES IN PLANT TISSUE CULTURE - THE BIOSYNTHESIS OF COMPLEX NATURAL PRODUCTS

J.P. KUTNEY

1. INTRODUCTION

The area of plant cell cultures represents a very diverse field of research, and depending on the focus of the research program, may encompass avenues of interest to botanists, phytochemists, plant cell geneticists, chemists, biochemists, genetic engineers, etc. It is not possible within this review to address all of these aspects but rather to provide a summary of the selected areas which are of more direct interest to the chemical audience. The emphasis of this article will be on the use of plant cell culture methodology for the synthesis and biosynthesis of complex natural products of medicinal interest. To illustrate the methodology and approaches generally employed in such studies, several examples, taken from the author's laboratory, will be taken and discussed in detail.

Excellent books (1-5) and review articles (6-8) summarize the various diverse areas of research conducted within the plant tissue culture area, but it is clear that the large majority of recent studies are focussed on phytochemicals of pharmaceutical interest.

2. METHODOLOGY

The above-mentioned publications provide details of the methodology normally employed in a plant tissue culture program, but it is perhaps appropriate to provide a very brief summary of the overall program as it applies to the production of complex natural products (secondary metabolites).

It is important to recognize, at the outset, that any successful program must address the interdisciplinary nature of the research as shown in Scheme I. The various stages of the "biological" (culture development) component of the program must be carefully monitored by appropriate analytical methods developed by the chemists within the research team.

SCHEME I

OVERALL PROGRAM

CULTURE DEVELOPMENT	NATURAL PRODUCTS CHEMISTRY
1. Primary callus	
2. Cell suspension	Analytical methods:
3. Pre-selecting callus	
4. Selected callus and suspension	1. Chromatography-TLC,HPLC
5. Product yield improvment	2. Radioactive assays
6. Pilot scale	3. Immunoassays

As Scheme I indicates, there are distinct stages of "culture development" and a brief description of the methodology involved is provided. In general, living tissue of the plant (stem, leaves, root sections, flowering portions) is sterilized and placed on agar containing the nutrients essential for cell growth. A typical example of one of the nutrient media (B5) (9) normally employed is shown in Scheme II and the various other nutrients, vitamins and other additives which may be used to achieve proper growth and stimulate the enzymes essential for biosynthesis of the target end products are shown in Schemes III and IV. Normally, a substantial number (perhaps several hundred) of calli are generated via appropriate variations of the constituents shown in Schemes II - IV since it is not known in the initial studies, what combination of nutrients are likely to produce good growth and production of the desired metabolites. Clearly analyses (see later) of the calli for the presence of the desired products must be performed simultaneously with culture development so that the chemical information, thus derived, provides a guide for further biological work.

Once appropriate growth within the callus culture (perhaps doubling in size per week) has been established, transfer into liquid suspension, (stage 2, Scheme I) employing shakers, is performed with appropriate analytical monitoring. Scale-up of these liquid suspensions via large scale Erlenmeyer shake flasks and/or bioreactors (10-50 liters) affords sufficient material (broth and cells) for a complete chemical analysis of the metabolites formed.

In terms of analytical methodology for monitoring development of callus cultures etc., it must be emphasized that the analytical method must be highly

SCHEME II

B-5 Medium

O.L. Gamborg and D.E. Eveleigh, Can. J. Biochem., 46, 417 (1968)

Ingredient	mg/l
$NaH_2PO_4 \cdot H_2O$	150
KNO_3	2500
$(NH_4)_2SO_4$	134
$MgSO_4 \cdot 7H_2O$	250
$CaCl_2 \cdot 2H_2O$	150
Iron	28
KI	0.75
Micronutrients	1.0 ml
Vitamins	10.0 ml
Sucrose	20.0 g
2,4-D	2.0 mg
Final pH	5.5

SCHEME III

Micronutrients and Vitamins added to B-5 and PRL-4 Media

Iron

Sequestrene 330 Fe (Geigy Agric. Chem., Saw Mill River Rd, Ardsley, N.Y.)

Micronutrients

Stock solution: dissolved in 100 ml water:

1 g $MnSO_4 \cdot H_2O$
300 mg H_3BO_3
300 mg $ZnSO_4 \cdot 7H_2O$
25 mg $Na_2MoO_4 \cdot 2H_2O$
25 mg $CuSO_4$
25 mg $CoCl_2 \cdot 6H_2O$

Vitamins

Stock solution; dissolved in 100 ml water:

10 mg nicotinic acid
100 mg thiamine
10 mg pyridoxine
1 g myoinositol

SCHEME IV

Other Additives Normally Employed in Media

Fresh frozen coconut milk
2,4-Dichlorophenoxyacetic acid (2,4-D)
Indole-3-acetic acid
Kinetin (6-furfurylaminopurine)
1-Naphtalene acetic acid
4-Aminobenzoic acid
4-Chlorophenoxyacetic acid
2,4,5-Trichlorophenoxyacetic acid
2-Chlorophenoxyacetic acid
6-Benzylaminopurine

sensitive. In general, concentrations of the target compound produced by the callus is in the range of 10-20 nanograms, at least in the initial studies, so that highly sensitive detection is mandatory. As a result, thin layer chromatography (TLC) is not suitable at this level but high performance liquid chromatography (HPLC) is an excellent method for monitoring culture development. Radioactive assays employing radiolabels and/or radioimmunoassay are too tedious and time-consuming so that HPLC is generally the method of choice. An important advantage of HPLC, apart from its sensitivity, is that it affords information about the number and concentration of components formed within a given series of fermentation studies. Figure 1 illustrates the HPLC monitoring of alkaloid production in a cloned line of *Catharanthus roseus* (coded 953) studied in our laboratory, where the concentration of alkaloids produced varies with age of culture. Obviously this type of analysis is extremely important in determining when to harvest a particular fermentation so

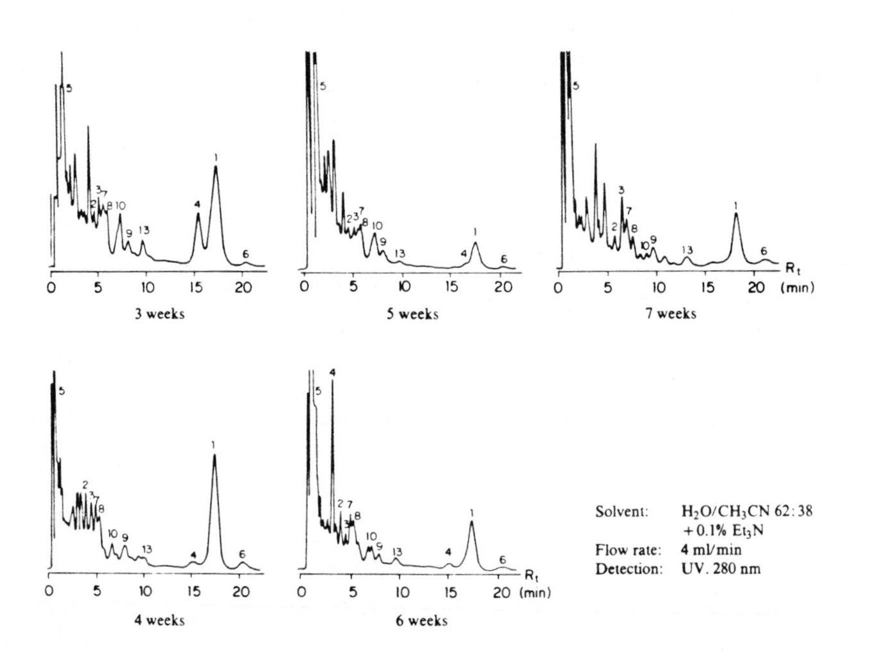

Fig. 1. Reverse phase HPLC of alkaloid mixtures obtained after different growths periods.

as to achieve optimum yields of a desired product. Subsequent isolation and characterization of the alkaloids produced from the 953 line allowed structural assignments (structures 1-14) to the metabolites as shown in the HPLC analyses (Figure 1).

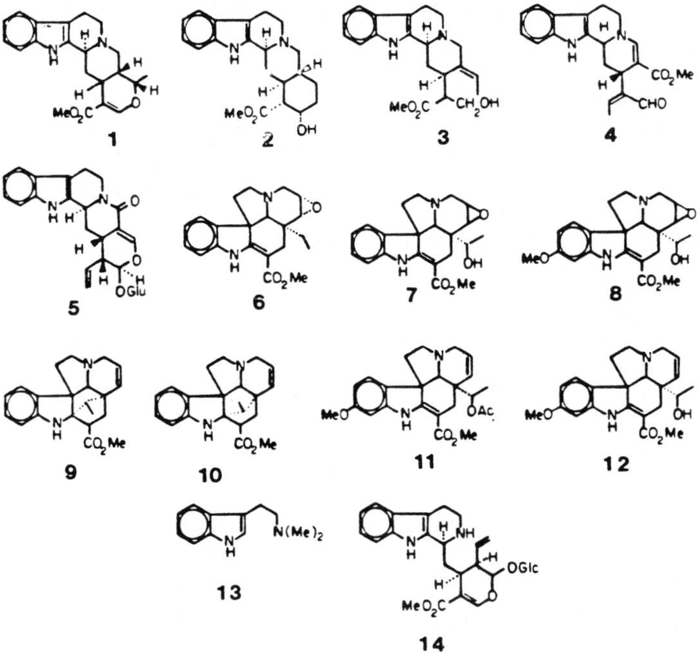

3. SECONDARY METABOLITE PRODUCTION

3.1 Indole Alkaloids

The indole alkaloids, due to their pharmacological properties and generally low concentrations in the plants from which they are derived, have received a great deal of attention in terms of their production via tissue culture methodology. For example, cell suspensions of *Phaseolus vulgaris* produce -carboline (10,11) while cultures of *Peganum harmala* produced harmine 12. Lysergic acid type alkaloids are shown to be present in *Ipomoea* tissue culture (13,14), reserpine is produced by tissue cultures of *Alstonia constricta* (15) and *Rauwolfia serpentina* (14,16). The alkaloids serpentine and ajmalicine have been produced in cell suspension cultures of Catharanthus roseus in various laboratories (2,6,7,8) and studied extensively in terms of production and/or biosynthesis. Vinca minor cell clones issued by protoplast isolation have been shown to produce vincamine, 16-epi-vincamine, vincadifformine and 1,2-dehydroaspidospermidine (17). Good yields of alkaloids with both the canthin-6-one and β-carboline structures have been produced in in cell cultures of Ailanthus altissima (18). As noted earlier, it is not the intention to provide an extensive review of the numerous investigations which have been pursued within a given area by the various research groups since accounts of their researches are already available elsewhere (1-8), but rather to exemplify, in some detail, the types of the results that can be obtained with plant tissue culture methodology and to illustrate to the interested chemist their advantages over living plants as sources of pharmaceuticals. For this purpose, studies from the author's laboratory are presented in more detail.

3.1.1 Studies with Catharanthus roseus

3.1.1.1 Alkaloid Production

The initial direction and approach of our program was stimulated by earlier studies directed toward the synthesis of vinblastine **17**, an important clinical anti-cancer drug. In these studies (19-21), we had shown that the alkaloids catharanthine **15** and vindoline **16** could be coupled to afford 3',4'-anhydrovinblastine **19** and the bisindole products **20** and **21** (Figure 2). A similar approach was also studied in another laboratory (22). A simultaneous study (23-25) in our laboratory and utilizing cell free extracts from C. roseus plants revealed that 3',4'-anhydrovinblastine **19** is also enzymatically formed from catharanthine **15** and vindoline **16** (Figure 3) and that, under appropriate conditions, enzymatic transformation of **19** to the alkaloids vinblastine (**17**, R=CH_3), leurosine **22** and catharine **23** can be achieved (Figure 4).

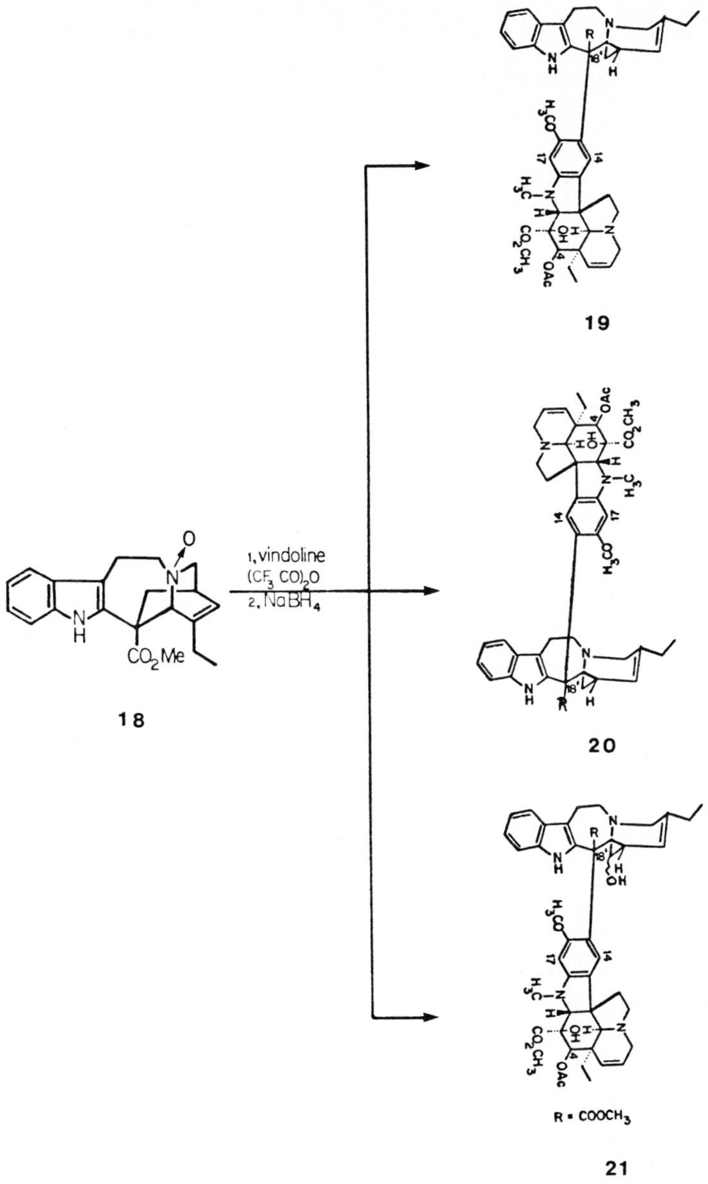

Fig. 2. Summary of results when catharanthine N-oxide **18** is coupled with vindoline **16**.

An independent study by Scott (26) provided results identical to those presented in Figure 3.

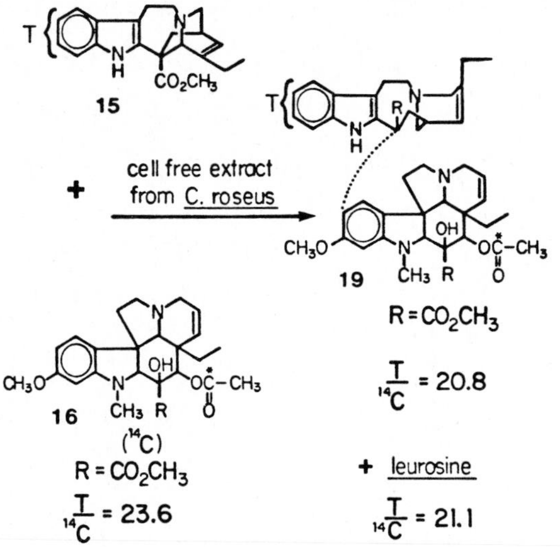

Fig. 3. The biosynthesis of 3',4'-anhydrovinblastine 19 and leurosine 22 from catharanthine 15 and vindoline 16 employing cell free extracts.

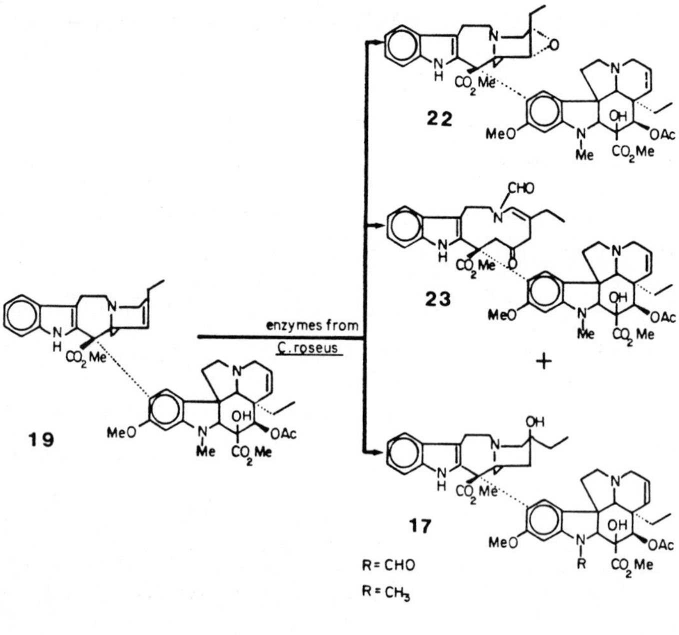

Fig. 4. Enzyme catalyzed conversion of 3',4'-anhydrovinblastine 19 to leurosine 22, catharine 23 and vinblastine (17, R = CH_3) employing cell free extracts.

In summary, the above studies placed particular importance on the two monomeric alkaloids, catharanthine and vindoline, and these compounds became prime targets in our tissue culture program. A number of recent publications (27-41) provide details of the experiments summarized below.

Our initial study (27,28) was undertaken to delineate the variability of serially cultured callus and cell suspension cultures derived from highly uniform explants, i.e. anthers of buds identical in developmental stage. The only variables introduced were the use of 3 periwinkle cultivars and treatment of buds with a mutagen. In a supplementary study the synthesis and accumulation of alkaloids was related to the growth of those periwinkle cultures which were selected for particular alkaloid content.

Callus grown from anthers generally originated at the cut of the filament and in the anther walls, i.e. diploid tissue. When grown to a size of 1-2 g freshweight, about 2 cm in diameter, the callus was cut into small pieces and serially subcultured on fresh agar medium or transferred to liquid medium (Gamborg's B5 medium) (Scheme II) giving rise to a cell suspension. For large scale production, Zenk's alkaloid production medium was employed.

The alkaloid production varied with the cell line and age of the subculture and ranged from 0.1 - 1.5% of cell dry weight. The relative amounts of alkaloids produced was fairly constant under conditions given and appeared cell line specific.

All subcultures of cell lines grown in 7.5 liter Microferm bioreactors followed essentially the pattern shown in Figure 5. After incubation with actively growing cell suspension the mitotic index (MI) dropped to zero within 24 hours and remained there for 2 to 3 days. Thereafter the index rose sharply and reached its maximum (MI 1.8 - 3.0) within 2 days and declined again gradually over the following 10 - 15 days to zero. The cell dry weight over the culture period increased by a factor of 8 to 10 while the variation in pH stayed within half a unit.

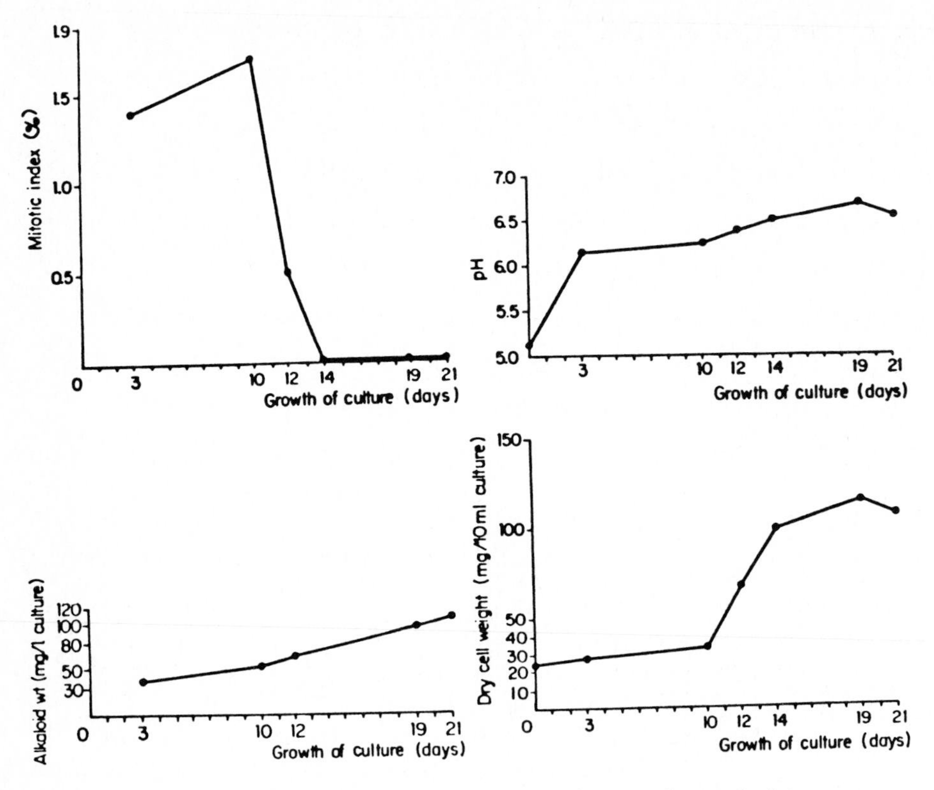

Fig. 5. General growth pattern of C. roseus tissue culture in bioreactor.

During an 8 week culture period alkaloids have been found as soon as 2 weeks after inoculation. Most cell lines showed a maximum accumulation of alkaloids in the 3rd to 5th week of culture. Having established a large number of cell lines capable of alkaloid production, we proceeded to a more detailed study with several of the more promising lines. The results from two such lines coded as "953" and "200GW" are summarized below.

The 953 line (30,33)

Studies with this selected line were performed both in shake flasks and bioreactors employing the 1B5 medium for inoculum growth and then Zenk's alkaloid production medium. On harvesting the culture, the water is removed by freeze drying and the alkaloids are extracted in the conventional manner to provide the data summarized in Table I. The crude alkaloid mixtures were fractionated by intermediate scale reverse phase high performance liquid chromatography (HPLC). Final purification by analytical reverse phase HPLC allowed the isolation of the following alkaloids, characterized by their physical and spectral data and by comparison with authentic materials: ajmalicine 1, yohimbine 2, isositsirikine 3, vallesiachotamine 4, strictosidine lactam **5**, lochnericine **6**, horhammericine **7**, horhammerinine **8**, vindolinine **9**, 19-epivindolinine **10**, 19-acetoxy-11-methoxytabersonine **11**, 19-hydroxy-11-methoxytabersonine **12** and dimethyltryptamine **13**.

TABLE I

Alkaloid yields from batches of 953 line C. roseus cell cultures.

Sample	Culture Method	Weight of freeze dried cells (g)	Weight of basic fraction (g)	% Alkaloid
1	Bioreactor (10 days)	90.5	0.168	0.185
2	Bioreactor (11 days)	110.0	0.178	0.16
3	Bioreactor (22 days)	26.9	0.058	0.21
4	Shake flask (14 days)	40.6	0.065	0.16
5	Shake flask (21 days)	49.66	0.182	0.37

TABLE II

Alkaloid yields from 953 line C. roseus shake flask cultures.

Cultivation Time	Weight of freeze dried cells (g)	Weight of basic fraction (g)	% Alkaloid
3 weeks	65.9	0.15	0.23
4 weeks	51	0.15	0.29
5 weeks	87.6	0.24	0.28
6 weeks	19.8	0.125	0.63
7 weeks	19.7	0.1	0.51

Since general alkaloid formation was not observed during the initial periods of rapid cell growth, it was decided to examine whether the appearance, disappearance or build-up of particular components could be observed over different time periods. The results are given in Tables I and II and Figure 1.

These show that the percentage of alkaloid per gram of cell weight increases with time, with optimum production at 3-4 weeks. Figure 6 supports this observation showing maximum cell dry weight occurring during the same period, coinciding with a zero value of the mitotic index. With respect to the earlier periods of culture growth, Figure 7 demonstrates a more rapid increase in the biosynthesis of ajmalicine **1** and yohimbine **2** (Corynanthe family) than observed for vindolinine **9** (Aspidosperma family). That is, the simple Corynanthe alkaloids ajmalicine **1** and yohimbine **2** reach maximum concentration at a much earlier period in culture growth than the biosynthetically more complex vindolinine **9**. These are presumably derived from a common key intermediate, strictosidine **14**, reflecting the differences in complexity of their biogenesis. Figure 7 also shows that at ca. 25 days, the concentration of these alkaloids begins to equilibrate, coincident with the onset of cell autolysis (Figure 7). Figure 1 shows HPLC traces of the later stages of growth period (3-7 weeks). Each sample contained ajmalicine 1 as the major component (ca. 15%). Further-

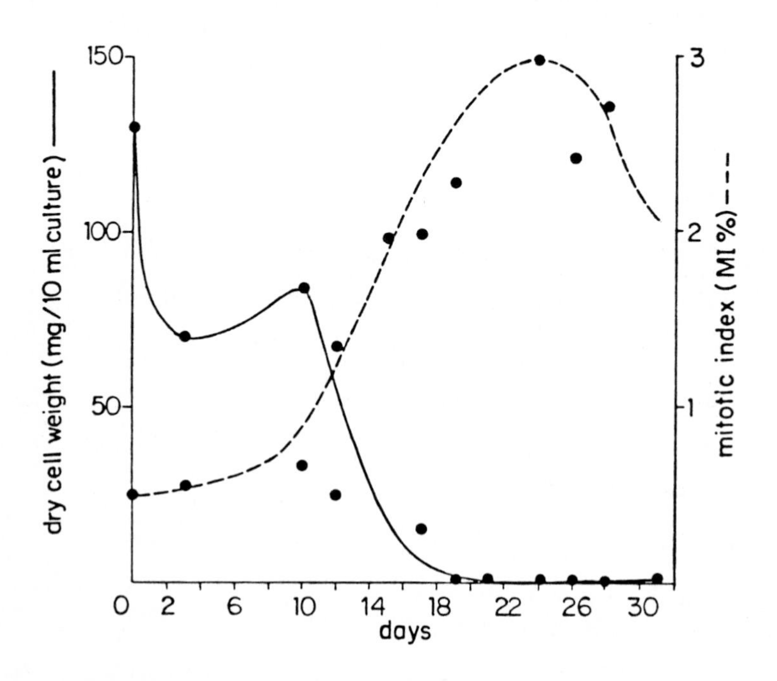

Fig. 6. Dry cell weight (---) and mitotic index ( ).

more, the analytical traces indicate that the other identifiable components of the mixture remained the same throughout this later period with only small changes in their relative concentrations.

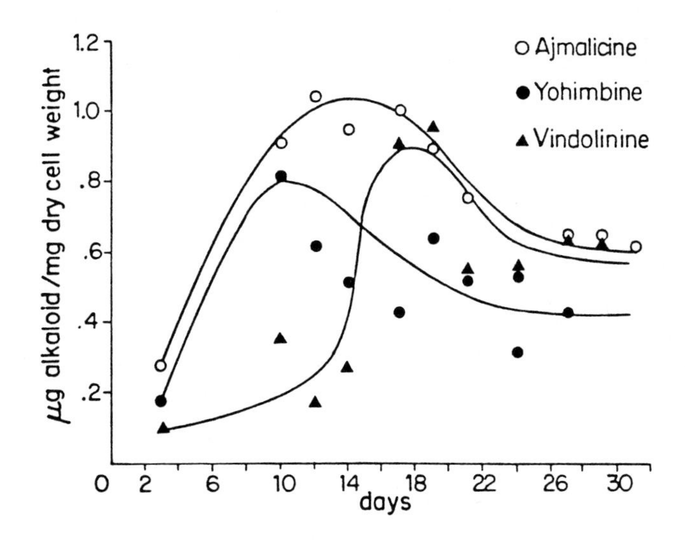

Fig. 7. Content of ajmalicine **1**, yohimbine **2** and vindoline **9** at the earlier periods of culture growth.

The 200GW line (29,34,38)

Another particularly interesting cell line which has been studied is coded as "200GW". The general procedures concerning tissue propagation, HPLC analyses, etc. are very similar to those discussed above. However, this line is uniquely different from the 953 line and produces its own "spectrum" of alkaloids as summarized in Table III. Of particular interest is the alkaloid catharanthine (**15**, 0.005% dry cell wt.) isolated for the first time in our studies. This line originally provided this alkaloid in amounts ca. three times that normally obtainable from *C. roseus* plant material. Recent studies (38) with bioregulators, as discussed below, reveal a marked improvement in yield.

3.1.1.2 Studies with Bioregulators

It is well established that tryptamine **24** and the isoprenoid, secologanin **25** are the biosynthetic building units for a variety of indole alkaloids - the

Corynanthe family, exemplified by structures 1-4, the Strychnos family (not shown), the Aspidosperma (structures 6-12 and 16) and Iboga (structure 15) bases (Scheme V). On this basis, it was attractive to consider the possibility that any compounds known to stimulate isoprenoid biosynthesis would hopefully stimulate the production of secologanin **25**, and, in turn, increase alkaloid levels in the cell.

TABLE III

Alkaloids isolated from the 200GW cell line.

Alkaloid	% Yield from dry cell wt.	% of crude alkaloid mixture
15	0.005	1.35
4	0.015	4.05
epimer of 4	0.026	7.02
1	0.006	1.62
7	0.002	0.54
8	0.005	1.44
9	0.002	0.54
10	0.002	0.54
5	0.224	60.48

% figures refer to _isolated_ yields.

Yokoyama et al (42) have shown that rubber biosynthesis is induced by certain aminoethyl phenylethers, implying that this type of synthetic bioregulators may be general inducers of terpenoid biosynthesis. Scott et al. (43) also demonstrated that some of these compounds are effective in promoting indole alkaloid formation in one of their _C. roseus_ cell lines. Consequently, it was appropriate to evaluate such bioregulators with our previously developed cell lines. For this purpose, the catharanthine producing 200GW cell line was selected and the influence of these bioregulators on ajmalicine **1** and catharanthine **15** production was evaluated.

Five bioregulators, 1,1-dimethylpiperidine bromide **26**, 2-diethylaminoethyl-3,4-dichlorophenylether **27**, 2-diethylaminoethyl-2,4-dichlorophenylether **28**, 2-diethylaminoethyl-β-naphthylether **29** and 2-diethylaminoethyl-3,4-dimethylphenylether **30**, obtained as gifts from Dr. H.

Yokoyama, were selected for the study. The bioregulators were added as the free base (except **26** as the bromide), dissolved in methanol/water and filter-sterilized through a 0.45 μm membrane.

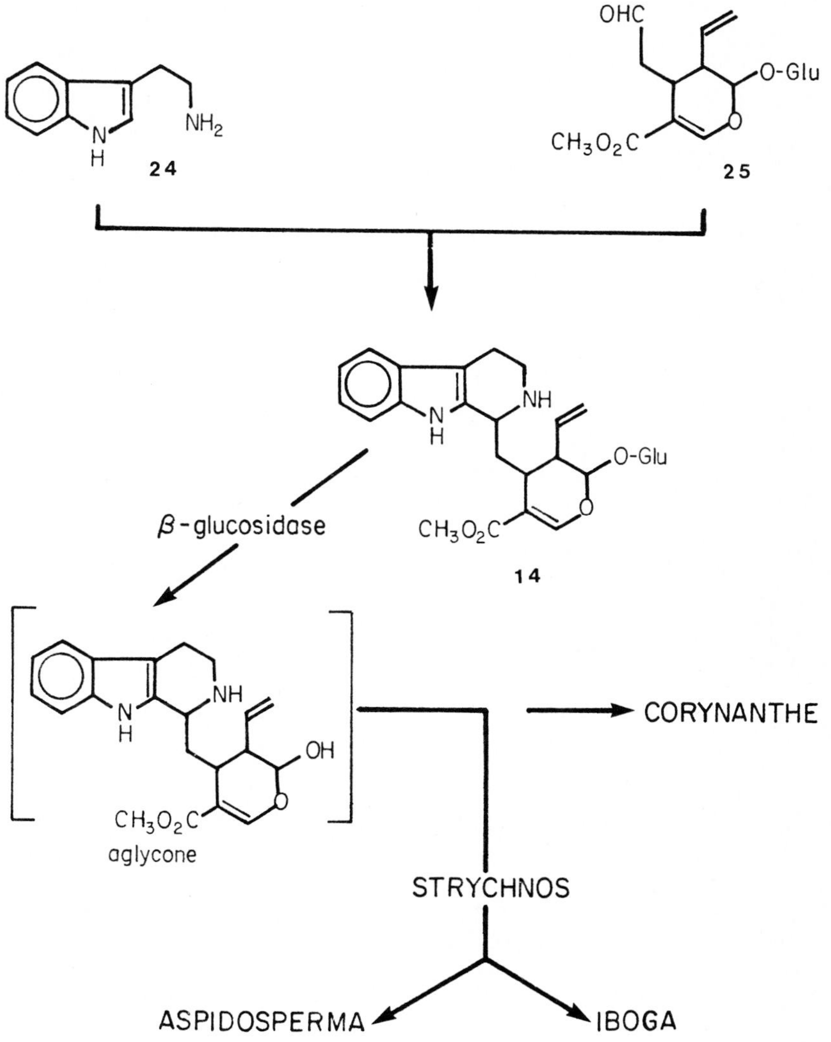
NH2
24
OHC
O-Glu
CH3O2C
25
NH
O-Glu
β-glucosidase
CH3O2C
14
NH
OH
CH3O2C
aglycone
CORYNANTHE
STRYCHNOS
ASPIDOSPERMA
IBOGA

SCHEME V

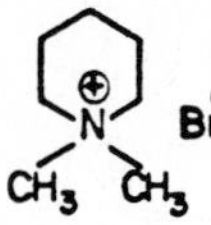

26 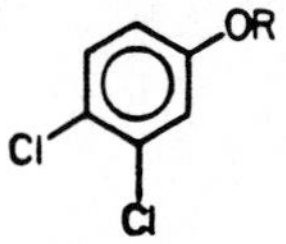 27 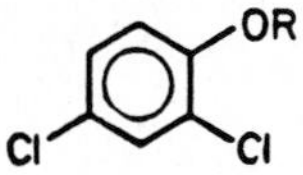 28 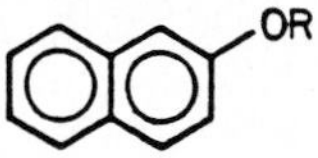 29

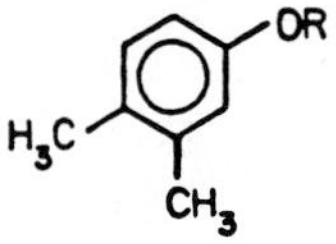

30

R = $CH_2CH_2N(CH_2CH_3)_2$

Shake flask cultures of 200GW cells (500 ml in 1 L flask) were grown in the alkaloid production medium for 7 days when the cultures were near or at stationary phase of growth as monitored by the mitotic activity of the cells. At this stage the average cell dry weight was 25 mg/ml culture. For the first time course experiment all five bioregulators, **26-30**, were examined. These were added separately to the above cultures at a concentration of 2 mg/l and further incubated for four different time periods (12, 18, 24 and 31 days). Parallel control cultures (no bioregulator added) were also set up. Samples were withdrawn at indicated intervals, extracted and analysed for alkaloid content.

A second time course experiment was carried out to investigate the effects of the two most promising bioregulators, **26** and **30**, after shorter incubation time (4, 6, 8 and 10 days) with the cultures. Samples were harvested at indicated intervals. Extraction and analyses of the alkaloid content were then carried out, employing HPLC for evaluation of the two target alkaloids, 1 and 15.

From the first time course experiment, it was shown that the mitotic index of all samples remained at or near zero, while the pH of the culture medium maintained within a fairly constant value (pH = 5.0-5.6). There was no significant fluctuation of biomass in terms of cell dry weight from day 12 to day 31 of incubation in the presence of bioregulator (Figure 8). This indicated that the five bioregulators, when added separately at a concentration of 2 mg/l to cell cultures which had already reached the stationary phase, did not affect further mitotic activity nor inhibit the normal growth of the cultures. Cell biomass continued to increase for 6 to 8 days (Figure 12, from second time course experiment) and then levelled off from day 12 on (Figure 8). Total alkaloid obtained from the cell extracts, expressed as percent of cell dry weight, showed substantial differences (Figure 9) especially after 12 and 18 days of incubation. Of the five bioregulators used, **26**, **29** and **30** were effective in increasing the total alkaloid production from the cell culture as compared with the control sample, while the other bioregulators showed the opposite effect. Yields of the two specific alkaloids, ajmalicine **1** and catharanthine **15**, as monitored by HPLC analyses (Figures 10 and 11) indicated the different effects of the bioregulators on the biosynthesis of these two alkaloids by cultures of the 200GW cell line. The results suggested that bioregulators **26** and **30** affected the best improvement especially for compound **15**.

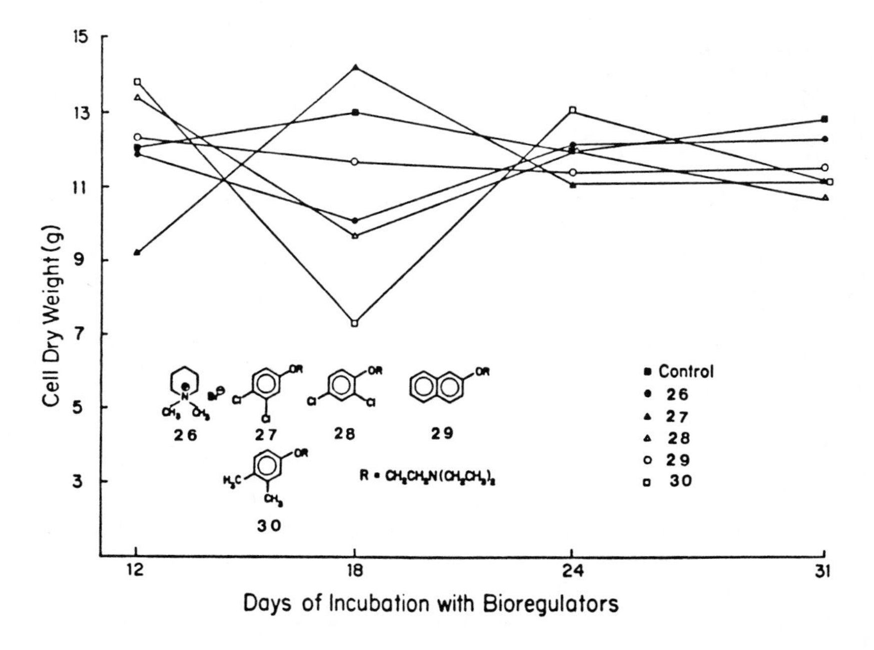

Fig. 8. Effect of bioregulators **26** to **30** on cell dry weight over an incubation period of 12 to 31 days.

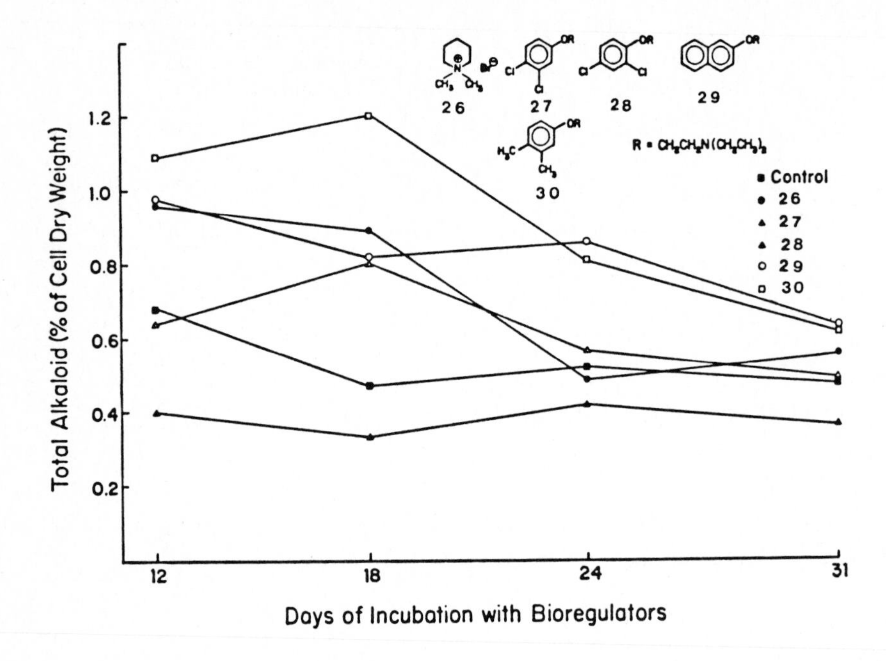

Fig. 9. Effect of bioregulators 26 to 30 on total alkaloid yield over an incubation period of 12 to 31 days.

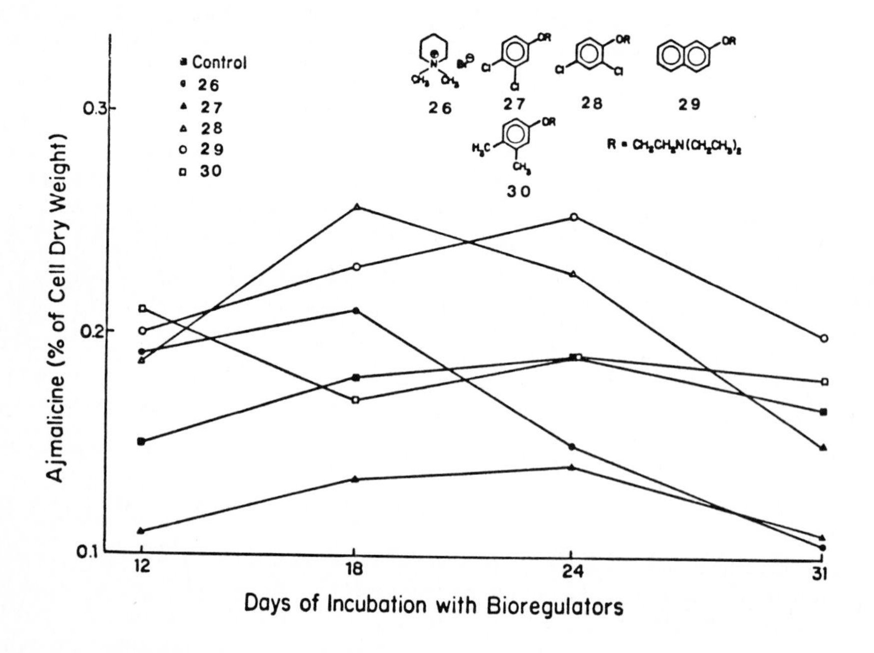

Fig. 10. Effect of bioregulators 26 to 30 on yield of ajmalicine over an incubation period of 12 to 31 days.

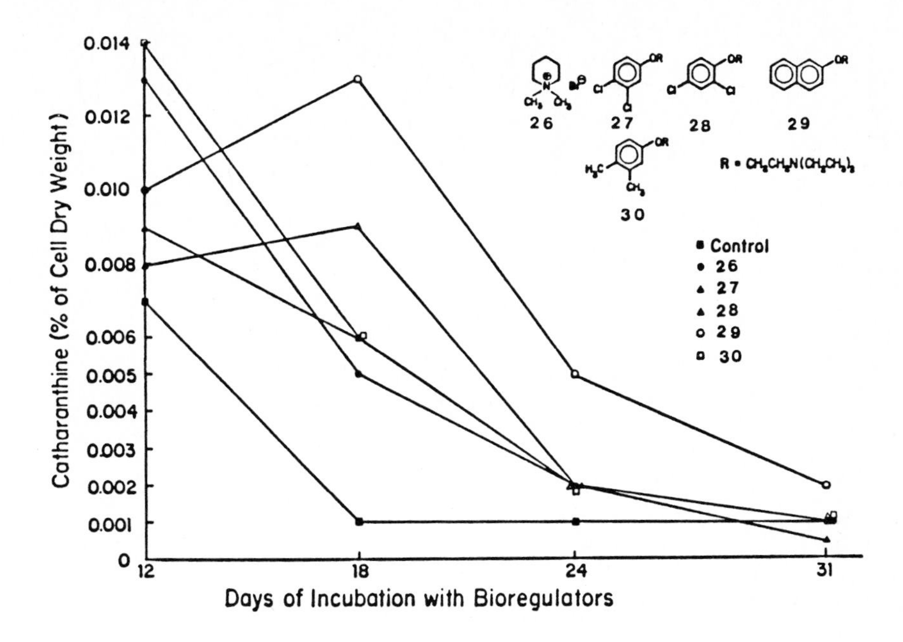

Fig. 11. Effect of bioregulators 26 to 30 on yield of catharanthine over an incubation period of 12 to 31 days.

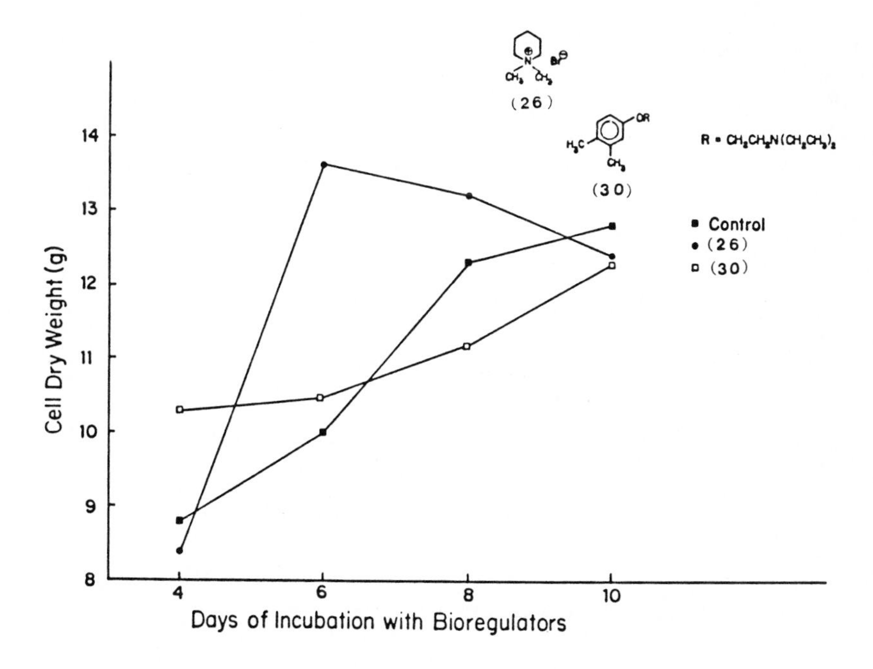

Fig. 12. Effect of bioregulators 26 and 30 on cell dry weight over an incubation period of 4 to 10 days.

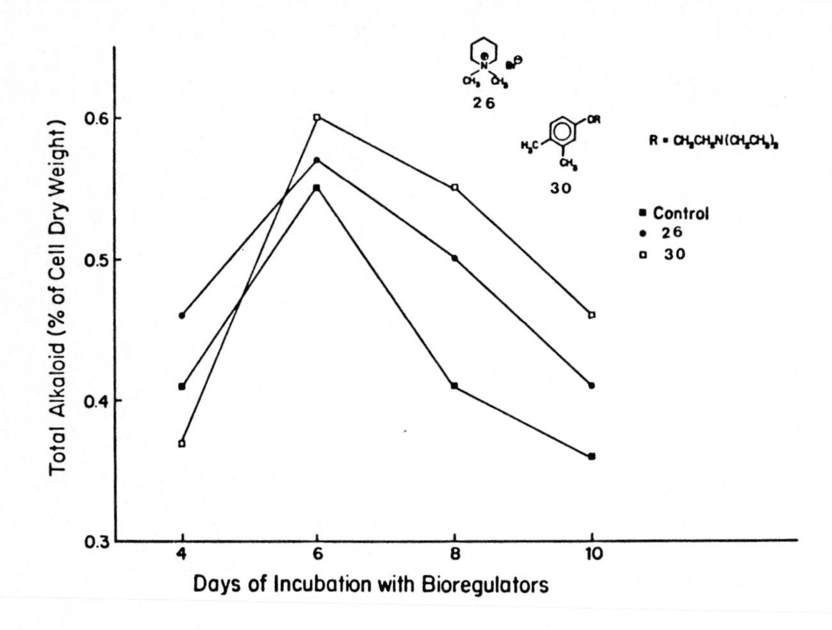

Fig. 13. Effect of bioregulators 26 and 30 on total alkaloid yield over an incubation period of 4 to 10 days.

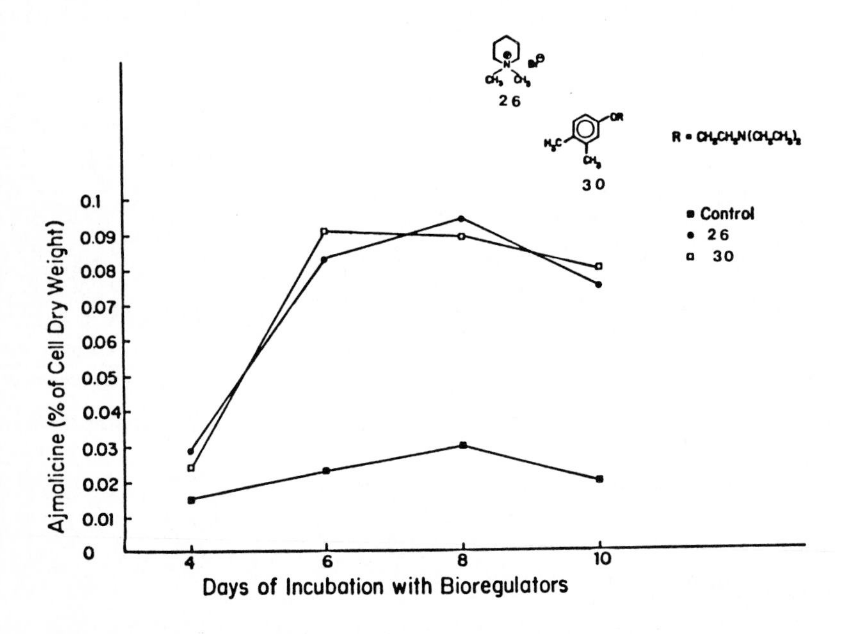

Fig. 14. Effect of bioregulators 26 and 30 on yield of ajmalicine over an incubation period of 4 to 10 days.

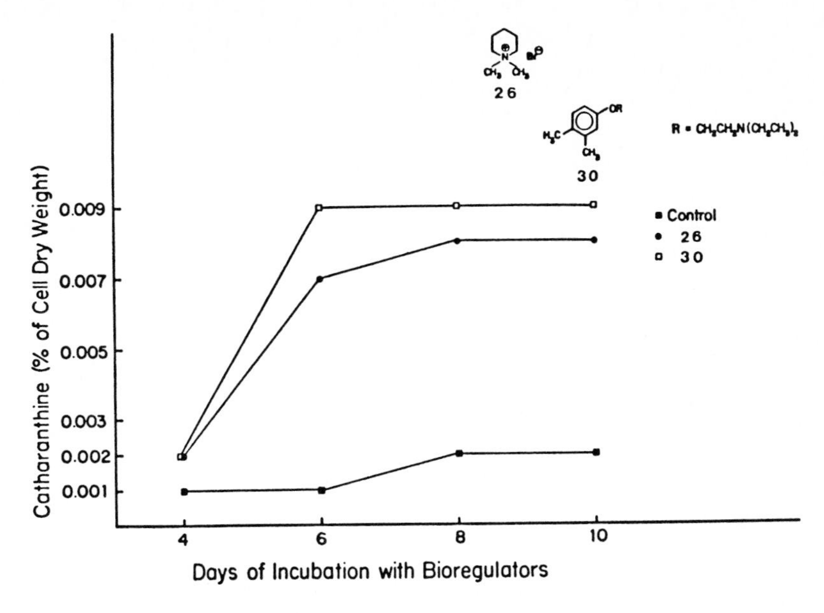

Fig. 15. Effect of bioregulators **26** and **30** on yield of catharanthine over an incubation period of 4 to 10 days.

Results of the second time course experiment carried out to investigate the effects of the two most promising bioregulators, **26** and **30**, after shorter incubation time (4,6, 8 and 10 days) with the cultures are now described. The mitotic index and pH of the cultures again remained quite constant. Increases in biomass were observed after 4 days of incubation and maximum cell dry weight was reached a few days later (Figure 12). The mitotic index indicated that although cell division had ceased at the time of bioregulator addition and did not re-initiate again, dry weight increase continued for 6 to 8 days at the stationary phase (Figure 12). Total alkaloid production expressed as percent of cell dry weight (Figure 13) reached a maximum level in all three cases after 6 days of incubation with or without bioregulator. The actual yield was highest for bioregulator **30**, while **26** also produced more than the control sample. Yields of alkaloids **1** and **15**, as shown in Figures 14 and 15 respectively, indicated both bioregulators provided quite similar effects on the biosynthetic capacity of the 200GW cell line suspension cultures. Comparing with control samples, significant increase in production of **1** and **15** was observed after 4 to 6 days of incubation with the bioregulators. Thus of the five bioregulators, **26** and **30** were found to increase the level of the important indole alkaloid catharanthine. Figure 15 illustrates that catharanthine production is increased nine-fold when **30** is employed as the bioregulator.

The above study illustrates one possible approach directed toward maximizing yields of target products in plant tissue culture. There are other methods, for example, nutrient media alterations, as will be seen later in the case of tripdiolide production, the use of ellicitors, etc., which can be utilized for this purpose.

3.1.1.3. Biotransformation Studies

The above discussion has demonstrated the capabilities of different tissue culture cell lines from C. roseus to produce various types of alkaloids. Another area of potential importance for the purpose of increasing cell yield of desired products, as well as for biosynthetic investigations, concerns the use of selected cell lines for biotransformation of appropriate substrates introduced into the culture medium at various stages of culture growth. Studies involving the transformation of various functional groups within organic compounds by plant tissue culture techniques have been reported (6) but compared to the extensively studied area of microbial transformation, much research is still required with such cultures before a proper understanding of this method can be attained. To this end we have initiated some studies (37) with selected C. roseus cell lines and appropriate substrates available from our earlier investigations.

The substrate 3',4'-anhydrovinblastine **19** available from the synthetic route outlined in Figure 2, was selected for our initial experiments. Several serially cultured cell lines were propagated for the preliminary screening to determine their capability of biotransforming **19** into desirable products. Only four experiments with each line were necessary to ascertain whether productive biotransformation was occurring. For each line, these experiments were: a) two control experiments (24 h and 72 h) involving only cells to determine whether alkaloids isolated were being produced by the cells or were metabolites of the precursor; b) cell line + **19** (3-5 mg), harvested 24 h after addition of substrate, and c) cell line + **19** (3-5 mg), harvested 72 h after addition of substrate. The results of these experiments revealed that several cell lines were metabolizing **19** to other products. For our further studies we chose a cell line coded "916". This line was unique in that it exhibited satisfactory growth characteristics, cell line stability etc. but did <u>not</u> produce any of the alkaloids normally found in the other lines, for example, 953 and 20CGW as discussed above. It was felt that the isolation of the rather complex bisindole products expected in the biotransformation of **19** would be simplified if cell-produced alkaloids were not present in the biotransformation mixture.

In the initial study with the 916 cell line, 3-5 mg of 3',4'-anhydrovinblastine **19** was incubated with the cells in shake flasks for 2, 6, 12, 18, 24, 48 and 72 h. In the sample incubated for 2 h, mainly **19** was found; samples incubated for 6-72 h contained a new, less polar compound. However, the highest concentration of this new product was observed in 24 and 48 h incubation samples, and a 48 h incubation period was selected for a larger scale biotransformation experiment which was performed in a 7.5 liter Microferm bioreactor. The pertinent data of this experiment are summarized in Table IV.

Based on the amount of recovered substrate, the transformation of 3',4'-anhydrovinblastine **19** to leurosine **22** and catharine **23** was 26% and 16% respectively, or, approximately 42% of **19** had been utilized by the cells. It should be noted, however, that no attempts have yet been made to optimize the yields of specific products.

Further studies with two alkaloid producing cell lines (coded CR3 and AC3) and **19** as substrate, allowed a more detailed evaluation of the biotransformation process in terms of metabolic products of the bisindole vinblastine family (41).

TABLE IV

Alkaloids isolated from biotransformation of 3',4'-anhydro-vinblastine (**19**, 300 mg) in C. roseus suspension cell cultures in Microferm bioreactor, cell line 916, 48 h.

Alkaloid isolated	Weight of alkaloid isolated (mg)					% of the substrate added
	Basic super-natant extract	Basic cell material extract	Neutral super-natant extract	Neutral cell material extract	Total	
3',4'-anhydro-vinblastine (19)	-	30.0	16.4	54.0	100.4	33.5
Catharine (23)	28.8	3.1	0.8	-	32.7	10.9
Leurosine (22)	17.0	8.2	17.2	8.6	51.0	17.0

a) Added as the hydrogensulfate salt

TABLE V

Biotransformation of 3',4'-anhydrovinblastine **19** by C. roseus cultures (cell line CR3) of various ages in 1B5 medium.

Age of culture (days)	Wt. of substrate (19) (mg)[b]	Wt. of basic cell extract (mg)	Wt. of basic supernatant extract (mg)	Recovered (19) (%)	Metabolites Identified
5	200	100	92	61.8	(22), (23), + others
10	500	220	220	26.0	(22), (23), + others
13	100	48.1	42.2	7.7	(17), (22), (23), (31), (32), + others
18	100	31.0	54.0	0	(23), (31), (32), + others
25	100	20.8	74.0	0	(23), (31), (32), + others

[a] Incubation time = 15 hr, substrate concentration = 100 mg/l

[b] Added as hydrogensulfate salt.

An important parameter which was first evaluated was the age of the culture in order to assess enzymatic activity in terms of biotransformation of **19**. A summary of such a study with the CR3 line is provided in Table V. It can be seen that the younger culture, for example, 5 days old, has a lower enzymatic activity as seen by a high recovery of **19** after a 15 h incubation period. Leurosine **22** and catharine **23** were the only metabolites which were identified in this latter study. Older cultures (13-25 days) have higher enzymatic activity (no recovery of **19** in 18 and 25 day old cultures) and furthermore possess the capability of producing, in addition to **22** and **23**, the interesting bisindole alkaloids, vinamidine **31**, hydroxyvinamidine **32** and vinblastine **17**. It is interesting to note that the clinical drug vinblastine **17** was only observed in the 13 day old culture study. At this age, the

TABLE VI

Biotransformation of 3',4'-anhydrovinblastine **19** by *C. roseus* cultures cell line AC3) in 1B5 medium.

Biotransformation Medium	Incubation Time (h)	Wt. of substrate (19) (mg)[b]	Wt. of basic cell extract (mg)	Wt. of basic super-natant extract (mg)	Recovered (19) (%)	Metabolites Identified
Whole cells	15	160	182.2	145	Trace <1	(17), 1.1% (22), 8.3% (23), 13.3% (31), 12.0% (32), 1.4%
Spent medium	15	140	-	140	0	(22), 2% (23), 4.7% (31), 3.6% (32), 23%

[a] Age of culture = 18 days

[b] Concentration of substrate = 100 mg/l, added as hydrogensulfate salt.

culture has started to enter into a stationary phase of growth. Scheme VI provides the structures of these metabolic products.

Further study with the other cell line (coded AC3) was performed and, in this instance, it was possible to obtain quantitative data on the metabolic products formed (Table VI). It should be noted that enzymatic activity is

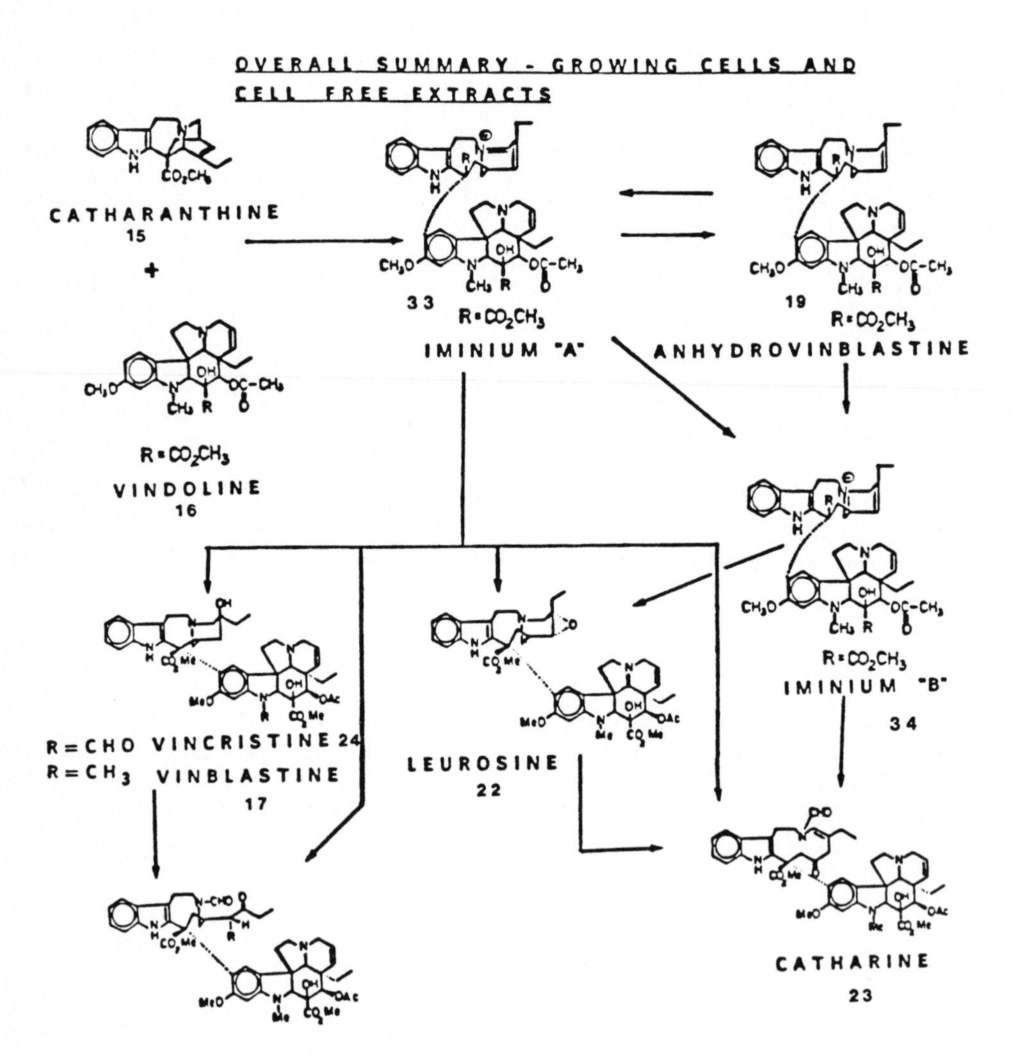

SCHEME VI

found in <u>both</u> the cells and "spent medium", that is, the supernatant liquid obtained after filtration of the cells. It is therefore clear that enzymes presumably produced within the cells <u>or</u> which are membrane bound, can be released into the medium thereby providing the observed enzymatic activity. Although the enzymes within the cells can produce vinblastine **17**, the enzymes in the medium do not convert **19** to this alkaloid.

In summary, the short period of time required for such biotransformations (15 h) is interesting, particularly when compared to plant cell culture production of alkaloids from nutrients present in the growth medium (usually several weeks). The inoculation of suspension cultures with biosynthetically 'advanced' precursors which reduce time periods for the production of target compounds may provide an important avenue for the commercial production of such pharmaceutically important agents. It should be emphasized that optimum conditions have <u>not</u> been established in these experiments so that significantly higher yields are expected from further study.

3.1.1.4. Studies with Cell Free Systems (Enzyme Mixtures)

Plant tissue cultures can provide excellent media for biosynthetic studies either directly with whole cells or with enzyme mixtures available from cell free systems. We have initiated some investigations with such systems in the hope of understanding the biosynthetic pathways involved with the above-mentioned natural products and, in particular, to attempt an evaluation of the enzymes responsible for optimum production of such target compounds.

Brief mention has already been made (Figures 3 and 4) of earlier experiments (23-25) with cell free systems prepared from *C. roseus* leaves but a more detailed discussion is now appropriate in order to relate the results of the most recent investigations as performed in tissue culture.

The purification procedure employed in all of the experiments concerned with *C. roseus* leaves and/or tissue cultures is summarized in Figure 16.

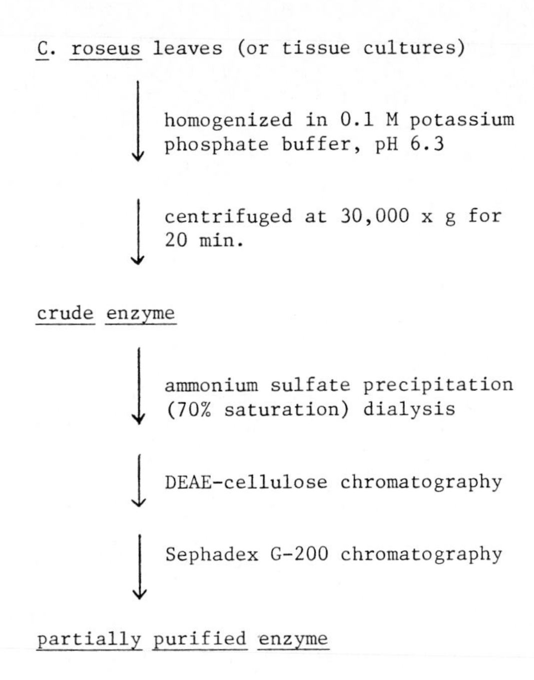

Figure 16. General procedure for preparation of cell free systems from _C. roseus_ leaves and/or tissue cultures.

The crude enzyme thus obtained was utilized in the experiments portrayed in Figures 3 and 4 where important information concerning the late stages of the biosynthetic pathway of the bisindole alkaloids was obtained.

Based on these earlier results, we proceeded to refine the methodology and obtain more information concerned with the enzymes involved in such conversions. Of particular interest to us was the enzyme(s) involved in the coupling of catharanthine **15** and vindoline **16** to 3′,4′-anhydrovinblastine **19** and its subsequent transformation to the other bisindole alkaloids (Figures 3 and 4). Thus we initiated a study directed at the recognition and purification of the relevant enzyme(s) involved in this coupling reaction.

The coupling enzyme activity was determined by monitoring the formation of 3′,4′-anhydrovinblastine **19** and leurosine **22** using radiolabelled tracer techniques with (Ar-^{3}H) catharanthine and vindoline as substrates. We also applied HPLC methodology to analyse the protein contents of the cell free enzyme mixtures. The HPLC system employed two protein columns (Waters Associates I-250 and I-125) which were calibrated with a number of standard proteins. Table VII lists the retention times for the respective protein under the conditions used for the analyses of the cell free extracts.

TABLE VII

HPLC analysis of standard proteins (molecular weight versus retention time).

Protein	Molecular Weight	Log (m.w.)	retention time (min)
Ferritin	450,000	5.65	12.67
Catalase	240,000	5.38	14.07
Aldolase	158,000	5.20	14.60
Albumin (bovin serum)	68,000	4.83	14.66
Albumin (hen egg)	45,000	4.65	15.33
Horseradish peroxidase	40,000	4.60	16.51
DN ase	31,000	4.49	16.91
Chymotrypsinogen A	25,000	4.40	20.74
Cytochrome C	12,500	4.09	28.35

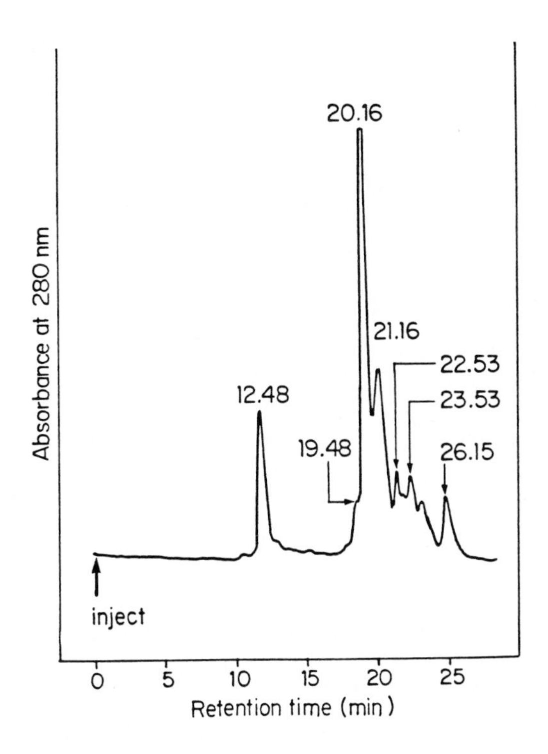

Fig. 17. HPLC analysis of cell free extract prepared from C. roseus leaves.

Figure 17 shows the HPLC profile of the crude enzyme from C. roseus leaves as prepared according to the procedure outlined in Figure 16. It will be noted that a mixture of proteins varying in molecular weights of approximately 15,000 - 450,000 are present. In order to establish a relationship between the molecular size of the enzyme(s) involved in the coupling of catharanthine and vindoline to the bisindole system, we proceeded to further separate the cell free extract (crude enzyme) by precipitation, dialysis and chromatographic techniques to a "partially purified" enzyme stage (Figure 16).

The crude enzyme extract was brought to 70% saturation with ammonium sulfate. The precipitate thus formed was dialysed against phosphate buffer (pH 6.8) and the dialysate was applied on a DEAE-cellulose column equilibrated with potassium phosphate buffer (20 nM, pH 6.8). The elution profile of the DEAE-cellulose chromatography is shown in Figure 18.

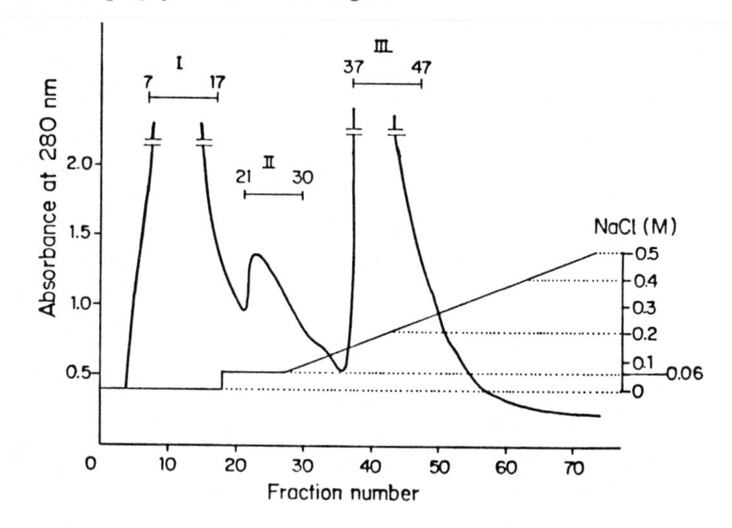

Fig. 18. Elution profile of DEAE cellulose chromatography of cell free extract from C. roseus leaves.

Fractions 21 to 30 were found to possess the coupling enzyme activity and were therefore combined into Fraction II, and concentrated to a small volume by ultrafiltration. This concentrate was then subjected to Sephadex G-200 chromatography which exhibited two peaks as monitored by the uv absorbance at 280 nm (Figure 19). The fractions corresponding to the two peaks II-1

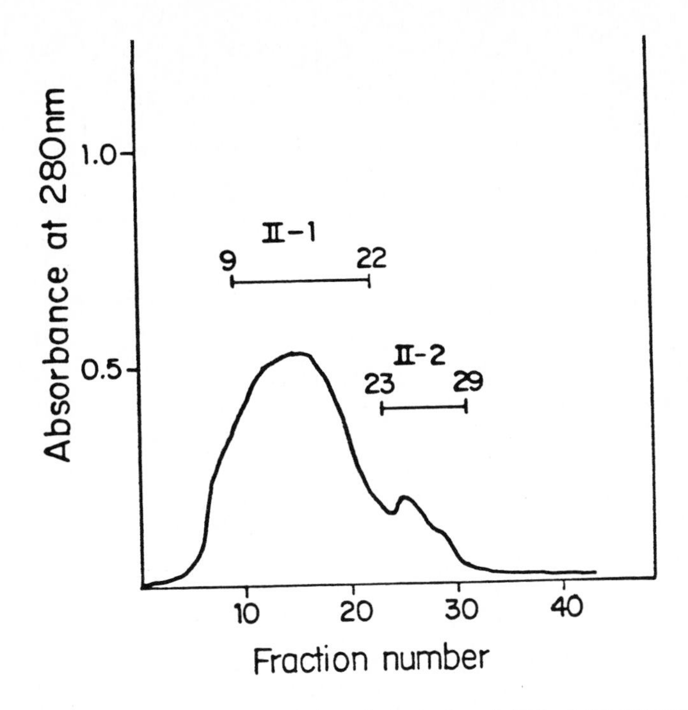

Figure 19. Elution profile of Sephadex G-200 chromatography - Fraction II.

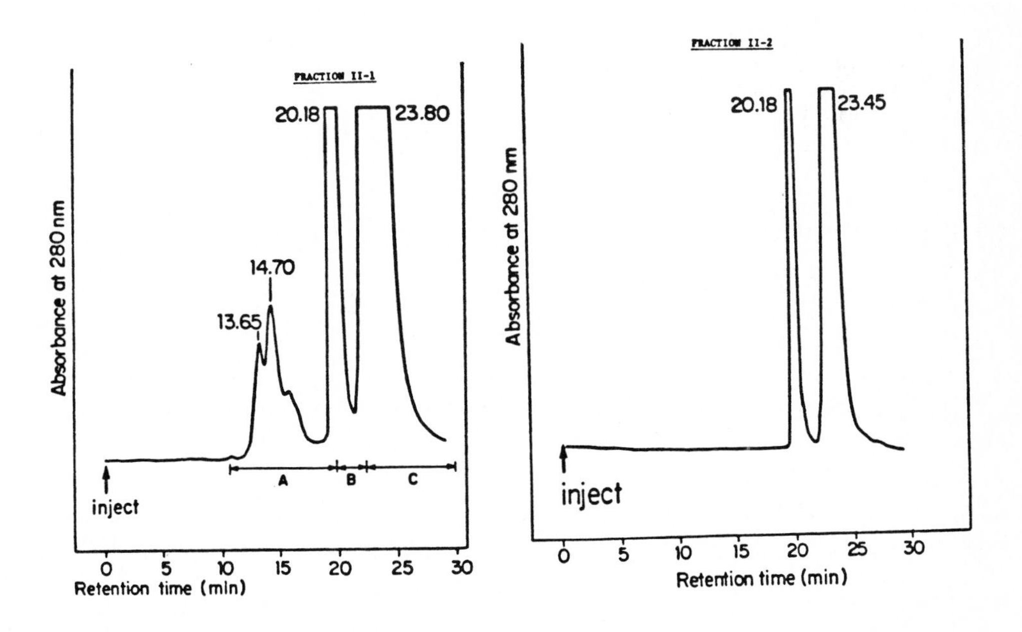

Fig. 20. HPLC profiles of Fractions II-1 and II-2 from Sephadex G-200 chromatography.

(fractions 9 to 22) and II-2 (fractions 23 to 29) were collected and analysed by HPLC (Figure 20) as well as assayed for coupling enzyme activity. Fraction II-1 which possessed the desired coupling enzymes activity was further fractionated by HPLC. Three fractions (A, B and C (See Table VIII)) corresponding to elution peaks of different retention times were collected and the results of their coupling enzyme activity determination are shown in Table VIII.

TABLE VIII

Preparative HPLC - enzyme Fraction II-1 and evaluation of coupling enzyme activity.

Fraction	Retention time (min)	Net activity (dpm)	
		3',4'-anhydrovinblastine	Leurosine
Frs. A	11 - 20	4137	3012
Frs. B	20 - 22.5	57	965
Frs. C	22.5 - 30	0	0

It is clear from these investigations that the enzyme system(s) involved in the biosynthesis of **19** and **22** from the appropriate monomeric alkaloids are present in the short HPLC retention time region (11-20 min., Fraction A in Table VIII). From the calibration standards (Table VII), this indicates proteins of molecular weight greater than 25,000.

Based on the above methodology and results, crude enzyme extracts were prepared from various C. roseus tissue culture lines and HPLC analyses according to the above-mentioned method were performed. Figure 21 summarized the results of four well-developed C. roseus lines from which various alkaloids have been isolated and characterized. The spectrum of alkaloids from the lines coded as "953" and "200GW" was discussed earlier. The "916" line is somewhat unique in that it exhibits normal growth characteristics but does not produce any of the alkaloids normally found in the other lines. A subline coded as "91601" and developed from "916" does produce alkaloids and the HPLC enzyme profiles, particularly in the region of 7-15 minutes retention time, are

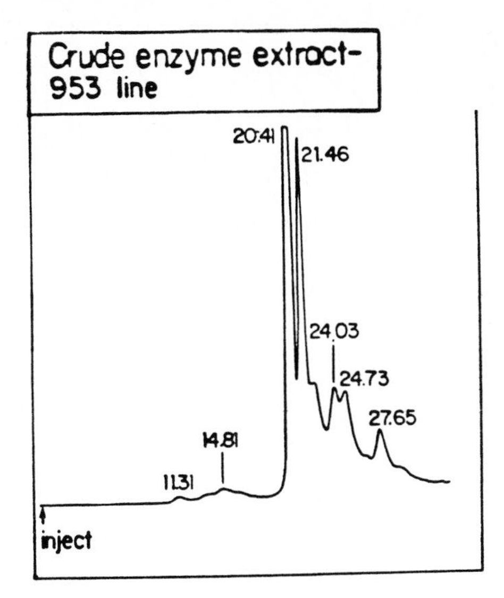

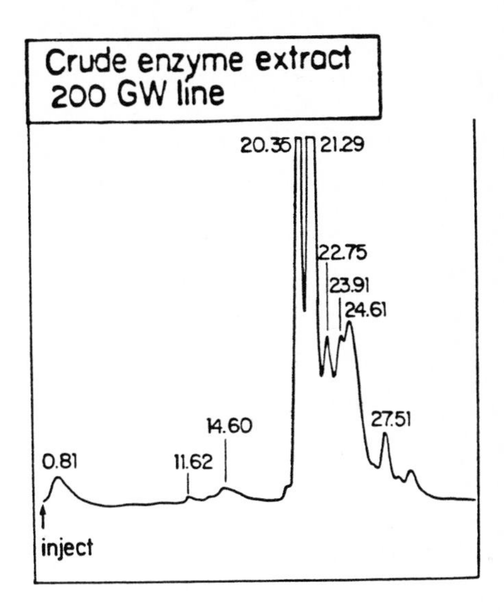

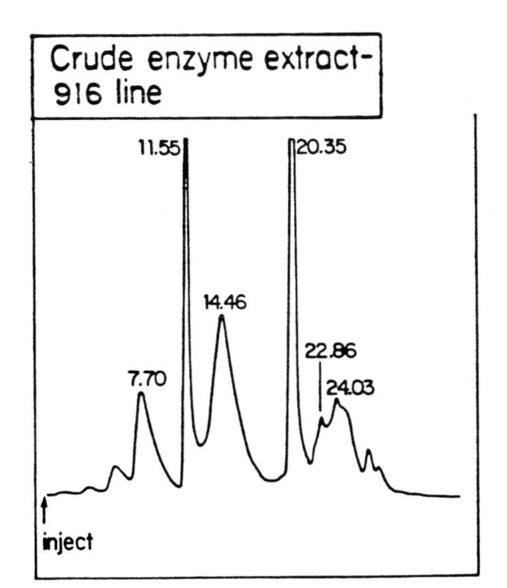

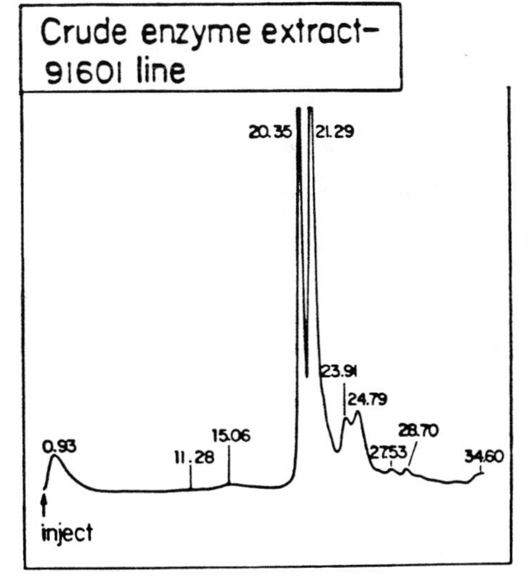

Fig. 21. HPLC profiles of crude enzyme extracts from several C. roseus tissue culture cell lines.

strikingly different. Such data are of considerable value in developing tissue culture lines with optimum production of target compounds and are also useful in biosynthetic investigations.

Extensive and more recent studies employing cell free extracts from our various developed C. roseus cell lines (AC3, for example), and involving **19** as substrate, have been completed (39). For example, the cell free extract, prepared according to the purification procedure outlined in Figure 16 (crude enzyme), from a 9 day old culture grown in a B5 medium and carrying a protein concentration of 0.9 mg/ml of buffer (phosphate, pH 6.3) with FAD as cofactor, was incubated with **19** for a 13 h period. The resulting metabolites which could be identified were leurosine (Scheme VI, **22**, 8%), catharine (**23**, 24%), vinamidine (**31**, 14%) and hydroxyvinamidine (**32**, 2%).

It could be shown in a parallel study with crude enzymes from the AC3 line that several cofactors exhibited a marked influence on the enzymatic transformation of **19**. For example, FAD and FMN increased the rate of conversion of **19** to the iminium intermediate **33** (Scheme VI), the latter being readily monitored by HPLC analysis (reverse phase, C_{18} column). On the other hand, $MnCl_2$ had no effect in this regard.

It should be noted that vinamidine **31** and hydroxyvinamidine **32** represent bisindole alkaloids which are end products of enzymatic oxidation of various earlier formed alkaloids in the culture medium. For example, we have shown that vinblastine **17** is biotransformed to vinamidine **31** under the above conditions. Longer incubation times of **17** with the enzyme mixture provides N-formylvinamidine (**31**, N-CH_3 replaced by N-CHO). Similarly, vincristine **24** affords N-formylvinamidine.

Major emphasis in the most recent studies, and employing such enzyme systems, has been placed on the enzymatic coupling of catharanthine and vindoline (Figure 3 and Scheme VI). Table IX summarizes our studies with enzyme systems derived from varying ages of cultures. It appears that a higher coupling activity is seen in enzyme(s) derived from younger cultures (5-9 days) and furthermore, ultracentrifugation at 150,000 g removes the active enzymes from the crude mixture (compare 18% versus 4.6% coupling in 5 day old culture, for example).

TABLE IX

Coupling of catharanthine **15** and vindoline **16** employing crude enzymes (C.F.E.) derived from varying ages of C. roseus cell line (AC3).

Age of Culture (days)	C.F.E.[b] Protein conc. (mg/ml)	Wt. of total Bisindole alkaloids (mg)	Yield of Coupling (%)	Bisindole alkaloids identified (Relative % of total dimers)			
				(22)	(23)	(31)	(32)
5	crude[c] (0.75)	12.5	18	1.9	27.4	7.2	1.4
5	u-C.F.E.[d] (0.54)	3.2	4.6	7.8	29.7	8.1	2.5
9	crude[c] (0.9)	9.9	14.2	2.1	22.4	7.2	1.4
9	u-C.F.E.[d] (0.65)	1.8	2.6	3.0	11.7	2.2	0.6
12	crude[c] (0.96)	3.9	5.6	1.3	8.7	-	4.9
12	u-C.F.E.[d] (0.82)	2.3	3.3	2.2	10.7	1.3	-

[a] Incubation conditions: catharanthine (40 mg), vindoline (40 mg), FAD (44 mg) $MnCl_2$ (6.8 mg), C.F.E. (120 ml); incubated at 26° for 5 hrs.

[b] C.F.E. prepared in 0.1 M phosphate buffer (pH=6.3)

[c] Crude C.F.E. referred to the supernatant fraction at 20,000g (20 min.)

[d] u -C.F.E. referred to the supernatant fraction at 150,000g (2 hrs.)

Table X compares the coupling activity of crude enzyme versus partially purified enzyme. It is clear that coupling activity is being lost during attempted purification. In this instance, the crude enzyme (Figure 16) was concentrated by ammonium sulfate precipitation followed by desalting on a Sephadex G25 column. The further separation by DEAE cellulose and Sephadex G200 chromatography as shown in Figure 16, was not employed.

Considerable data on cofactors involved in the enzymatic coupling are available from our studies. NADP, NAD, FMN and FAD are important in activating the coupling enzyme(s). With these data in hand, studies with appropriate

TABLE X

Coupling of catharanthine **15** and vindoline **16** by crude and partially purified (P.P.) enzymes (C.F.E.) of <u>C. roseus</u> cell cultures (AC3 line).

C.F.E. (ml)	Wt. of Catharan-thine (mg)	Wt. of Vindoline (mg)	Wt. of FAD (mg)	Wt. of $MnCl_2$ (mg)	Wt. of total dimeric alkaloids (mg)	Yield of coupling (%)	Dimeric alkaloids identified (Relative % of total dimers)			
							(22)	(23)	(31)	(32)
crude (80 ml)	27	27	60	4.6	5.3	10.2	1	12.1	2.9	0.9
P.P. (90 ml)	30	30	67	5	2.9	5.6	4.3	15.3	7.7	2.5

affinity gels (β-NADP agarose, Reactive Red agarose supplied by Sigma and ADP sepharose supplied by Pharmacia) and immobilization of such enzymes were undertaken (40). The specific binding of the proteins onto the gel surface was accomplished by standard techniques (stirring of enzyme mixture in 0.1 M Tris HCl buffer in presence of gel, followed by removal of unbound enzymes by subsequent washing with buffer). In this manner, the results summarized in Table XI were obtained.

Several important features of the results shown in Table XI should be highlighted. The <u>dramatic increase</u> in yield of coupling (70-90% versus 2-18% in Tables IX and X) illustrates the importance of enzyme concentration and/or selective binding to a solid support for high yield biosynthesis of the iminium intermediate 33 (Scheme VI). The yield of the various bisindole products (19, 22, 23 and others) is then dependent on the relative rates of enzymatic

biotransformation of 33.

As will be noted in Table XI, the use of an inert atmosphere (argon versus air) for the coupling reaction, has considerable advantages in terms of overall yields of bisindole products. This is at least in part due to the fact that the substrate catharanthine **15** is unstable in air and converts to other products in competition with coupling. Furthermore, binding of enzymes to β-

TABLE XI

Studies of enzymatic coupling of catharanthine 15 and vindoline 16 by affinity gel bound enzymes.

Exp.	Incubation Conditions			[a] % Yield of Coupling (Total)[b]	% Yield of (19)	% Yield of (22)
	Time (min)	Temp. (^{o}C)	Atmosphere			
NADP-1	90	26	air	28.0	0.7	1.4
NADP-2	25	26	air	9.8	2.3	0.6
NADP-3	90	26	air	44.2	3.9	1.0
NADP-4	40	26	air	37.5	5.3	1.5
NADP-5	80	ice-H_2O	Ar	79.6	16.7	5.6
R.red-1	90	26	air	26.0	1.6	1.7
R.red-2	25	26	air	17.6	1.4	0.2
R.red-3	90	26	air	30.8	3.4	1.5
ADP-1	120	26	air	38.5	4.2	1.7
ADP-2	60	ice-H_2O	air	40.8	10.8	5.2
ADP-3	160	ice-H_2O	Ar	74.4	20.4	5.3
ADP-4	200	ice-H_2O	Ar	90.3	37.7	13.4
ADP-5 (2X)	240	ice-H_2O	Ar	77.7	23.7	8.5

[a] After appropriate incubation, the mixture was reduced with $NaBH_4$ prior to isolation of bisindole coupling products.

[b] Total coupling yield refers to total of bisindole products obtained. Only 3',4'-anhydrovinblastine (19) and leurosine (22) were characterized while other unknown bisindole compounds are not yet elucidated.

NADP agarose and ADP-sepharose offers an advantage over binding to Reactive Red agarose. The implications of these and other factors are not well understood at present and further studies will be required before a better understanding of the overall process is obtained.

In conclusion, the utilization of enzyme systems derived from cell cultures has considerable advantage over that of biotransformation with whole cells. In particular, such factors as greatly reduced incubation times and/or yields of end products must be emphasized.

Finally, in summary it is clear that optimum yields in the enzyme-catalyzed conversion of catharanthine **15** and vindoline **16** via 33, and other subsequent intermediates, to the target clinical drugs vinblastine **17** and vincristine **24** <u>must</u> address the problem of selective enzyme control for the various alternatives which are available, as summarized in Scheme VI. Thus any enzymatic depletion of iminium "A" 33 via the route 33 → 34 → 23 <u>or</u> 33 → 22 **23** will obviously lower yields of **17** and **24** as will the subsequent oxidation of of **17** and **24** to the higher oxidation levels of the vinamidine series, **31** and 32. The understanding of such processes, via enzyme manipulations, made possible through varying growth parameters of the cultures, enzyme concentration and isolation, etc. will hopefully enhance the utilization of tissue cultures for the production of these clinical drugs.

3.2 <u>Diterpenes</u>

Studies on plant tissue cultures for the production of phytochemicals within the diterpene family have been limited to only several areas. This situation is in marked contrast to that involving alkaloids, as discussed above, and steroids, particularly <u>Digitalis</u> glycosides, as discussed in the above mentioned books (1-5) and reviews (6-8).

Miyasaka et al (44) have reported on the production of the diterpene ferruginol by cell suspension cultures of <u>Salvia</u> <u>miltiorrhiza</u>. These authors studied the time-course production of ferruginol and the effects of auxins and light on ferruginol production and on cell growth.

Studies with cell cultures of <u>Thuja</u> <u>occidentalis</u> by Witte et al (45) also report the presence of the diterpenes, dehydroabietane, 2-dehydroferruginol and ferruginol. The latter compounds were recognized by the technique of gas liquid chromatography-mass spectrometry so no information is available on the levels of these diterpenes produced in the culture media.

The cell suspension culture of <u>Cryptomeria</u> <u>japonica</u> studied by Ishikura et al (46) is reported to contain two diterpenes, abietatriene and ferruginol.

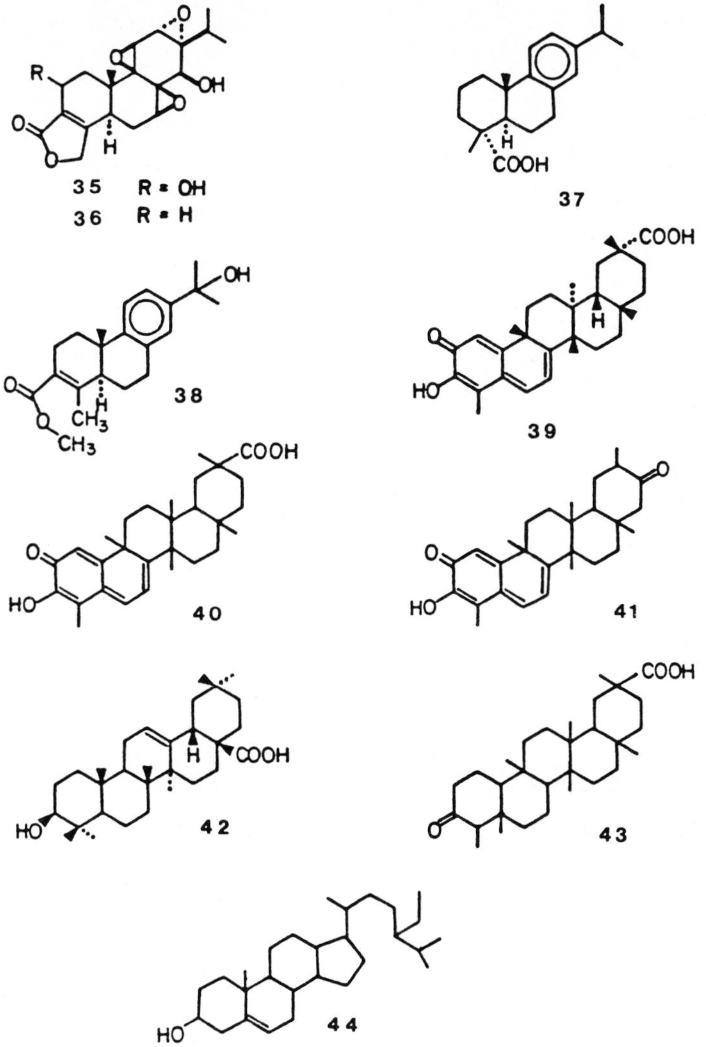

Figure 22. Natural products isolated from Tripterygium wilfordii cultures.

Diterpene production by callus cultures of some plants belonging to the family Cupressaceae has been reported by Ohgaku et al (47). Here again GC and GC-MS analyses were performed to identify such diterpenes as abietatriene, totarol, ferruginol, hinokiol, etc. so levels of production of the specific compounds are not well established.

Dehydroabietane and another unidentified diterpene hydrocarbon has been found in the steam distillates of callus cultures of *Melissa officinalis* L. (47). GC analysis was employed to evaluate the diterpene components which varied in relative proportions depending on the age of the culture. Older cultures tended to reveal a higher content of the dehydroabietane system.

3.2.1 Studies with Tripterygium wilfordii (48-50)

An extensive program in our laboratory and concerned with the plant cell production of the cytotoxic agents tripdiolide **35** and triptolide **36** (Figure 22), natural products isolated from the Chinese plant, *Tripterygium wilfordii*, is discussed in some detail in order to illustrate other approaches which can be employed in order to optimize cell yields of secondary metabolites.

Our research involving the propagation of plant cell cultures of *Tripterygium wilfordii* was stimulated by the research of the late S. M. Kupchan (51) in which it was demonstrated that tripdiolide and triptolide reveal significant activity *in vivo* against the L-1210 and P-388 leukemias in the mouse and *in vitro* against cells derived from human carcinoma of the nasopharynx (KB). Since the extracts of *T. wilfordii* are also used in Chinese herbal medicine, there have been recent extensive investigations on the chemistry (52) and pharmacology (53-56) of such extracts.

3.2.1.1 Secondary Metabolite Production

Employing stem and leaf explants from *T. wilfordii* plants and utilizing culture methodology similar to that already mentioned above in the *C. roseus* area, we were able to establish callus and stable cell suspension cultures which produce the target compounds tripdiolide **35** and triptolide **36** (48-51). Analysis for these compounds within the cultures was performed via a rapid thin layer chromatographic assay employing fluorimetric detection (57). HPLC analysis could not be directly employed since the target compounds do not possess a UV chromophore in the readily accessible region of the spectrum. In addition, a KB bioassay (see Table XII) was performed to locate cytotoxic compounds.

TABLE XII

Effect of various medium supplements on the growth and tripdiolide production by T. wilfordii (TRP4a) cell suspension cultures.

PRL-4 medium supplemented with[a]				Time of Growth					
				5 weeks			6 weeks		
Co	K	D	NA	Growth assessment	TLC assay[b]	KB assay[c]	Growth assessment	TLC assay[b]	KB assay[c]
0	0.5	2		+	?	1.7	+	?	7.5
0	1	2		+	-	3.0	+	-	43
0	1.5	2		+	-	0.5	+	-	9
0	0.5		0.15	+	-	1.3	+	-	1.3
0	0.5		0.5	++	+	1.01	++	-	<1
0	0.5		1	++	-	1.1	+++	?	21.5
0	0.5		2	++	?	4.5	++ (a few roots)	+	9
0	0.5		2.5	++	-	1.25	++++	-	28.5
0	1		0.15	+	-	<1	+	-	27
0	1		0.5	++	-	1.45	++	?	12.5
0	1		1	++	-	1.45	+++ (a few roots)	-	26.5
0	1		2	++	-	1.8	++	-	26
0	1		2.5	++	++	<1	+++ (many roots)	++	23.5
10	0	2		++	-	43.5	++	-	29
10	0.5	2		+++	+	28	++	-	-
10	1	2		+++	?	2.6	+++	-	22

/continued ...

TABLE XII. (continued)

PRL-4 medium supplemented with[a]				Time of Growth					
				5 weeks			6 weeks		
Co	K	D	NA	Growth assessment	TLC assay[b]	KB assay[c]	Growth assessment	TLC assay[b]	KB assay[c]
10	1.5	2		+++	?	10.05	++	?	14.5
30	0	2		++++	+?	49	+++	+?	1.8
30	0.5	2		++++	-	14	++++	-	6.6
30	1	2		++++	-	10.5	+++	-	6.2
30	1.5	2		+++	-	42	+++	-	30.5
60	0	2		++++	-	12	+++	-	20.5
60	0.5	2		++++	-	12	++++	-	12
60	1	2		++++	+?	12	++++	-	25
60	1.5	2		++++	-	34	+++	-	37

[a] Units for the concentration of supplements are: coconut milk (Co) - ml/l of broth; kinetin (K), 2,4-dichlorophenoxyacetic acid (D) and 1-napthaleneacetic acid (NA) - mg/l of broth.

[b] TLC assays were qualitative.

[c] KB assays are expressed as ED_{50} values in μg/ml. These assays were performed at Arthur D. Little, Cambridge, Massachusetts. The assays were done using KB cells (human epidermoid carcinoma of the nasopharynx-type 9 KB-5 as developed under a program sponsored by NCI, NIH, Bethesda, Maryland. ED_{50} represents the calculated effective dose which inhibits growth of 50% of control growth.

The cell line, designated TRP4a, was grown initially in shake flasks and subsequently in a 60 liter Chemapac bioreactor from which isolation and chemical characterization of the secondary metabolites produced could be completed. Figure 22 illustrates the structures of these compounds. It is noted that, in addition to **35** and **36**, two other diterpenes, the known compound, dehydroabietic acid **37**, and a novel unsaturated carboxylic acid ester **38**, of possible biogenetic interest with respect to **35** and **36**, were isolated from the cultures. Furthermore, the cultures produced the interesting triterpene quinone methides, celastrol **39**, an isomer **40**, a structure **41** closely related to tingenone, the triterpene carboxylic acids oleanolic acid **42** and polpunonic acid **43**, and finally β-sitosterol **44** as a steroidal component.

3.2.1.2 Parameter Studies to Optimize Tripdiolide (Td) Production

In the earlier discussion concerning alkaloid production in *C. roseus* cultures, a study concerning the influence of bioregulators on alkaloid yield and yields of specific alkaloids, catharanthine **15** and ajmalicine **1**, was presented. Within the tripdiolide area, we have made an extensive investigation of another approach, that is, the influence of various other culture parameters (changes of nutrient media, age of inoculum, etc.) so as to ascertain their effect on yields of **35**. The following discussion summarizes the data obtained.

3.2.1.2.1 Initial Medium Studies on the Production of Tripdiolide (Td)

The influence of hormones and hormone-like compounds on the production of Td **35** by TRP4a cell suspension cultures was examined. The PRL-4 medium of Eveleigh and Gamborg (9) (without casein hydrolysate) was used as the basic medium. This was supplemented with various levels of coconut milk (Co), kinetin (K), 2,4-dichlorophenoxyacetic acid (D) and 1-naphthaleneacetic acid (NA) as listed in Table XII. Cultures were harvested and analysed after 5 and 6 weeks of incubation. These times were selected since preliminary screening experiments carried out using PRD_2Co_{100} (PRL-4 medium supplemented with 2,4-dichlorophenoxyacetic acid, 2 mg/l of broth, coconut milk, 100 mg/l of broth) indicated that high cytotoxicity activities were generally associated with samples extracted from 4 to 6 week old cultures. At this point of the project only qualitative TLC and cytotoxicity (KB) analyses were available to monitor the level of tripdiolide **35**. These results together with the growth assessment of the cultures are shown in Table XII. Cultures grown in medium with $Co_0K_{1.0}NA_{2.5}$ gave the strongest indication of Td on TLC, but copious root formation in the culture made it less desirable for future use in large scale fermentations. Instead, the medium supplemented with kinetin (0.5 mg/l) and naphthaleneacetic acid (0.5 mg/l) ($PRNA_{0.5}K_{0.5}$) was selected as it produced a positive TLC analysis for Td plus good KB assay results.

3.2.1.2.2. Effect of Inoculum Size

Parallel time course (42 days) experiments using TRP4a in $PRNA_{0.5}K_{0.5}$ broth were set up with three different inoculum sizes, 10% (standard), 50% and 100%. Duplicate samples were harvested and analysed at 7 day intervals starting at day 21 of incubation. Positive TLC analyses for Td were shown by all the broth extracts. However, a significant difference in Td concentration could not be determined. Likewise, KB assays did not discriminate among the cultures since all samples gave ED_{50} values of less than one. The results, overall, showed that inoculum sizes larger than 10% did not prevent production of Td. However, there was no apparent indication of higher yield. For practical purposes, subsequent experiments employed the standard inoculum size of 10%.

3.2.1.2.3. Course Study in $PRNA_{0.5}K_{0.5}$ Medium

At this point, we developed a rapid TLC assay of Td using fluorimetric detection (57) which was accurate for Td concentrations of 0.2 μg to 3.6 μg. Therefore, a detailed time course study, including Td measurement, using cell suspension cultures was carried out in $PRNA_{0.5}K_{0.5}$ medium. Triplicate samples (500 ml) were harvested at weekly intervals and analysed individually. Other culture parameters monitored at the same time included cell dry weight (cells were weighed after filtration through Miracloth (Calbiochem), and lyophilization), pH and refractive index. These results are shown in Figure 23.

Significant Td formation occurred after 14 days. A maximal concentration of 2.3 mg/l of culture broth was attained around day 35. Generally, Td was found to be present in both the cell extracts as well as the broth. However, because of interference by the co-occurrence of several quinonemethide

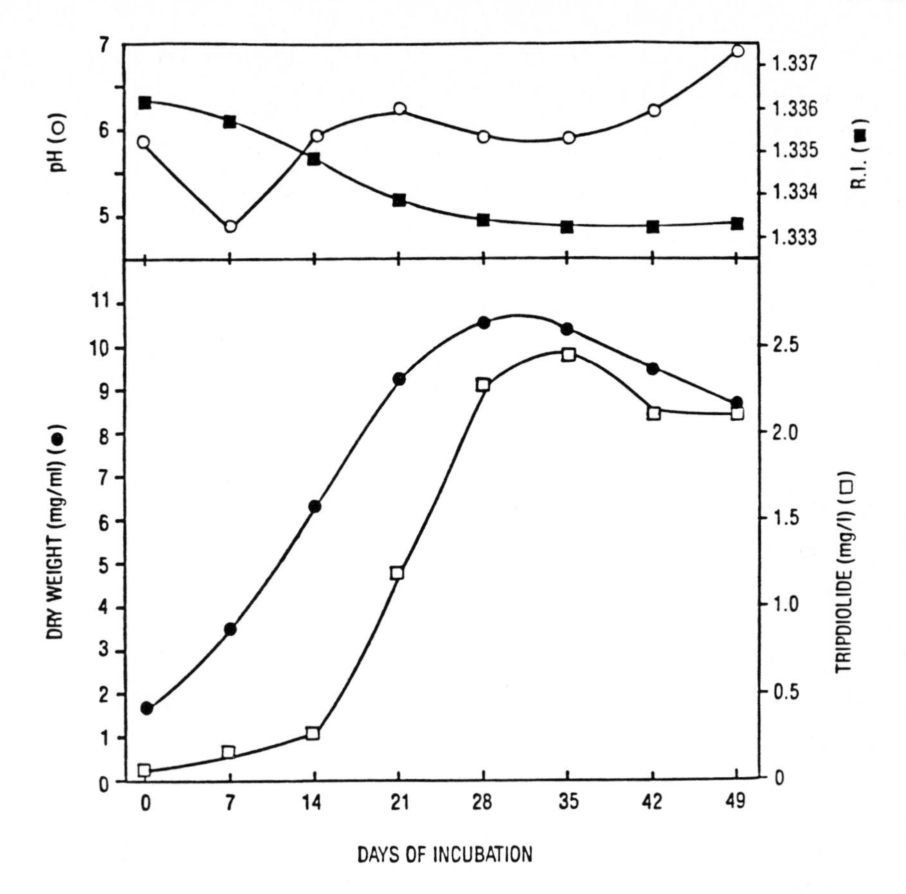

Fig. 23 TRP4a culture growth and Td production in $PRNA_{0.5}K_{0.5}$ broth using 21 day old inoculum originally grown in PRD_2Co_{100} broth.

compounds **39-41** in the cell extracts, only broth extracts were analysed by the fluorimetric method for the concentration of tripdiolide. Biomass, in terms of cell dry weight, increased from 1.7 mg/ml at day 0 to the peak of 10.5 mg/ml at day 28. After this time the cells appeared to enter a stationary phase.

3.2.1.2.4. Effect of Younger Inoculum

In all previous experiments inocula were from 18-22 day old stock cultures (PRD_2Co_{100}) that had reached early stationary phase as assessed by biomass measurement. One idea for shortening the time for Td production was to use younger inoculum for the production phase. To test this approach, a time course experiment was performed using the $PRNA_{0.5}K_{0.5}$ medium and 11 day old inocula. Cell suspension cultures (3 x 500 ml) were harvested at weekly intervals and analysed for Td. Cell dry weight, pH and refractive index of each sample were also recorded. Results are shown in Figure 24.

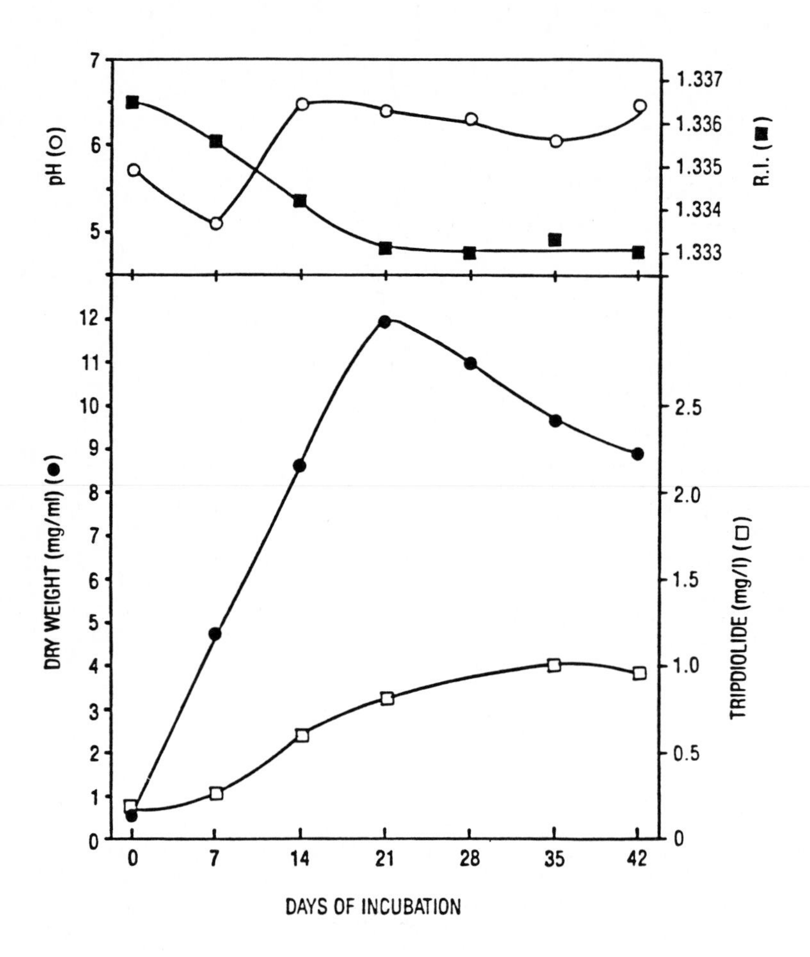

Fig. 24. TRP4a culture growth and Td production in $PRNA_{0.5}K_{0.5}$ broth using 11 day old inoculum originally grown in PRD_2Co_{100} broth.

Growth was rapid after a short lag period so that a maximal dry weight of 11.9 mg/ml was reached by day 21. Use of inocula still in growth phase and containing residual sucrose likely accounts for these differences in growth rate and cell yields. The peak concentration of Td again occurred after about 35 days of incubation. However, this level (1 mg/l) was less than half that obtained from cultures using older inocula. Therefore, subsequent experiments employed inocula about 3 weeks old.

3.2.1.2.5. Influence of Medium Composition on Tripdiolide Production

Although we had established that formation of Td by TRP4a cells in $PRNA_{0.5}K_{0.5}$ medium can be achieved with a peak level of over 2 mg/l, it was of interest to examine different production media in the continuing effort to improve the yield of the desired compounds. The effects of two other basal media (those of Murashige and Skoog (58), (MS) and Hildebrandt and Schenk (59), (SH)), on growth and Td production of TRP4a cells were compared with $PRNA_{0.5}K_{0.5}$ medium in a series of parallel experiments. The media were prepared as $MSNA_{0.5}K_{0.5}$ (2% sucrose) and $SHNA_{0.5}$. Suspension cultures (250 ml each) were harvested at appropriate times over a 45 day incubation period and analysed for Td concentration and cell dry weight. Results are shown in Figure 25.

A more detailed time course experiment in $MSNA_{0.5}K_{0.5}$ (2% sucrose) medium was carried out and afforded a tripdiolide concentration of greater than 3.0 mg/l after 37 days of incubation. Good biomass production, in terms of cell dry weight, was also obtained. These results strongly indicated that $MSNA_{0.5}K_{0.5}$ (2% sucrose) is a more effective Td production medium than similarly supplemented PRL-4 or SH media for cultivation of TRP4a cell suspension cultures in shake flasks.

Comparison of the three basal media (PRL-4, MS, SH) reveal several major differences in their compositions. Some possible key components are: (a) concentration of available nitrogen in the forms of NH_4^+ ion or NO_3^- ion. Both are present in much higher concentrations in the MS medium. (b) Concentration of calcium chloride ($CaCl_2.2H_2O$) is also higher in MS (440 mg/l). (c) Concentration of thiamine is much lower in MS (0.1 mg/l) than in the other two media (10 mg/l and 5 mg/l). (d) Glycine (2 mg/l) is only present in MS. In addition, there are other differences in the micronutrients (e.g. Mn^{2+}, Zn^{2+}, Cu^{2+}, and Co^{2+}).

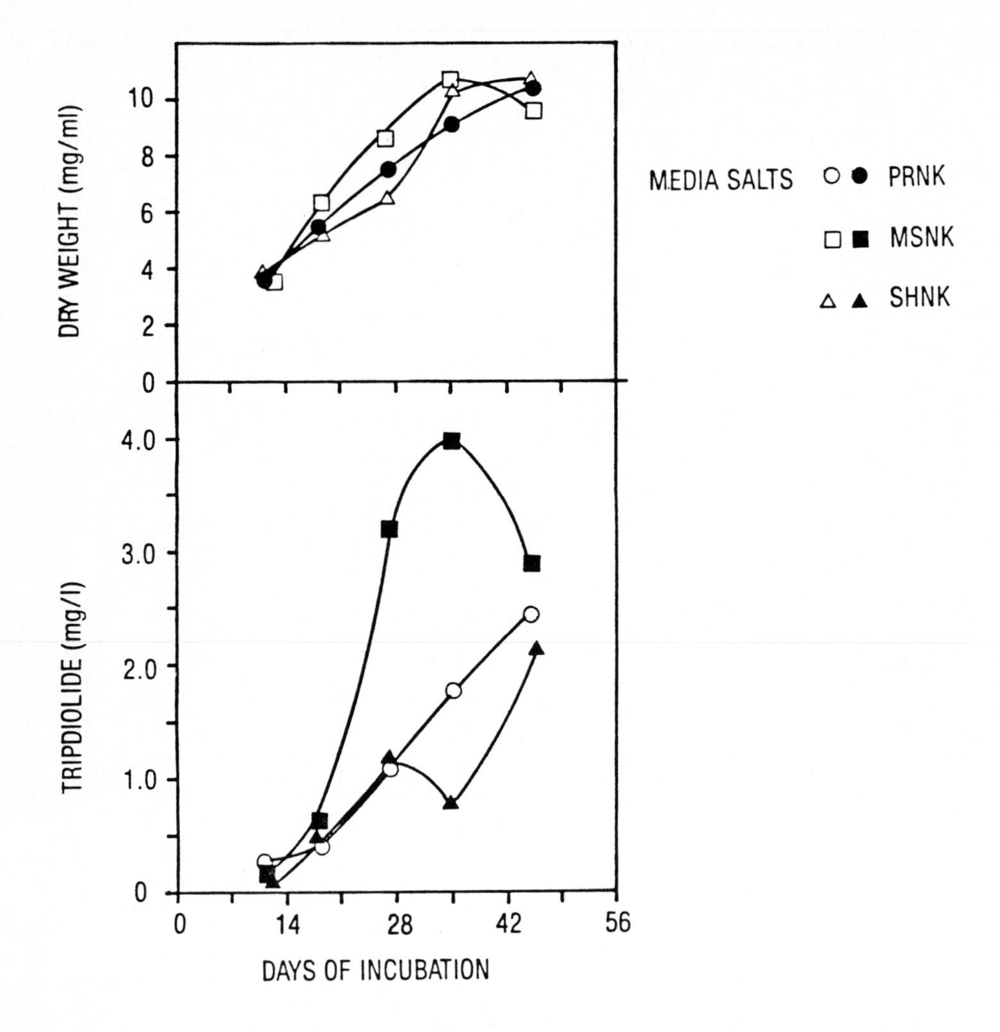

Figure 25. Comparison of Three Different Basal Media For Growth and Td Production by TRP4a.

The effect of different levels of some of these components was studied using $MSNA_{0.5}K_{0.5}$ medium as the basal. Results of some of these preliminary experiments with TRP4a cultures are described below.

Effect of Ammonium Nitrate Concentration

Three different levels of ammonium nitrate (850 mg/l, 1650 mg/l (standard), and 2450 mg/l) were evaluated. Cultures (2 x 250 ml) were harvested at appropriate times after 28 days of incubation. Dry biomass yield and Td concentration were monitored (Figure 26A). Td production was highest in medium with 1650 mg/l of ammonium nitrate. Biomass yield was greatest when ammonium nitrate at 850 mg/l was used.

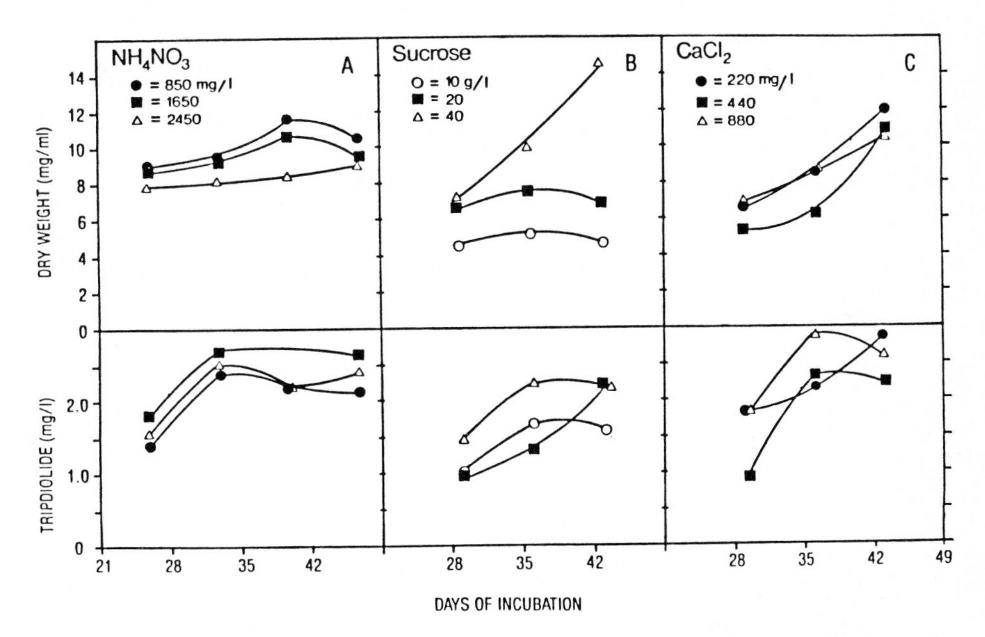

Fig. 26. Effects on Td and dry biomass production of different medium concentrations of NH_4NO_3 (Panel A), Sucrose (Panel B), and $CaCl_2$ (Panel C).

time period. Cultures grown in 4% sucrose showed the best biomass yield while the 1% sucrose samples afforded the lowest yield.

Effect of Calcium Chloride Concentration

Three different calcium chloride concentrations were used (220 mg/l, 440 mg/l (standard) and 880 mg/l). Cultures (2 x 125 ml) were harvested and analysed at appropriate times after 28 days of incubation. Biomass yield was very similar for all three calcium chloride concentrations (Figure 26C) while Td production was maximal with 880 mg/l of calcium chloride at day 43 (2.6 mg/l).

In conclusion, it is clear that variations in nutrient media supplements, age of inoculum, etc., can have considerable effect on culture growth and product yield. Although Td production, presently at a level of 4.0 mg/l, is clearly not yet fully optimized, this yield is about 36 times greater than that reported for the plant by Kupchan (51).

Effect of Sucrose Concentration

Three sucrose concentrations were tested, namely, 1% (10 g/l), 2% (20 g/l) (standard) and 4% (40 g/l). TRP4a cultures (2 x 125 ml) were procured and analysed at appropriate time intervals after 28 days of incubation. Results, as shown in Figure 26B, indicated Td production reached the same level (2.3 mg/l) at day 43 in both the extracts from cultures with 2% and 4% sucrose, whereas 1% sucrose afforded a lower Td concentration (1.8 mg/l) after the same time period. Cultures grown in 4% sucrose showed the best biomass yield while the 1% sucrose samples afforded the lowest yield.

Effect of Calcium Chloride Concentration

Three different calcium chloride concentrations were used (220 mg/l, 440 mg/l (standard) and 880 mg/l). Cultures (2 x 125 ml) were harvested and analysed at appropriate times after 28 days of incubation. Biomass yield was very similar for all three calcium chloride concentrations (Figure 26C) while Td production was maximal with 880 mg/l of calcium chloride at day 43 (2.6 mg/l).

In conclusion, it is clear that variations in nutrient media supplements, age of inoculum, etc., can have considerable effect on culture growth and product yield. Although Td production, presently at a level of 4.0 mg/l, is clearly not yet fully optimized, this yield is about 36 times greater than that reported for the plant by Kupchan (51).

Misawa (8) and his colleagues have also reported studies on plant tissue culture of _T. wilfordii_. The level of Td in the culture was very low (95 μg/l). A report by Dujack (60) also provides a description of their results with _T. wilfordii_ but no characterization of secondary metabolites produced is reported.

4. SUMMARY

The area of plant tissue culture is rapidly expanding as scientists from various disciplines recognize the potential for such methodology within the framework of their respective interests. Academic scientists realize their importance in biosynthetic studies of complex natural products where control of culture growth, the various media supplements, pH and culture age, can provide a consistent and reproducible medium for precursor and biotransformation studies. Apart from the much higher incorporation of precursors when compared to that normally achieved in living plant systems, the advantage of using culture media for the _controlled_ production of relevant enzymes is obvious. The isolation of enzymes produced in such culture media is much easier than from the living plant and the production of larger amounts of important enzymes, via large scale fermentor experiments, clearly affords an opportunity for detailed studies relating to enzyme characterization and/or enzyme immobilization - an area of increasing importance in industrial applications.

Industrial chemists must consider the use of such culture methods in selected areas of medicinal agents, for example, and even at this early stage of plant tissue culture developments, one can cite examples of commercial production via this technique. It is perhaps appropriate to remind the skeptics that it took many years of intensive research and effort before the area of microbial fermentation, particularly when combined with chemistry, became of tremendous importance for the presently known industrial purposes. Our knowledge of the use of plant cell cultures, particularly in large scale production, is at an infancy stage when compared to microbial technology which began its developments in the 1950's. With advances in cloning, genetic engineering techniques, etc., it is predictable that the plant cell area will see considerable growth in academic, government and industrial laboratories.

Acknowledgments.

Since a large portion of the present review represents research results from the author's laboratory, it is appropriate to acknowledge the many enthusiastic and dedicated group of researchers which made this presentation possible. A substantial portion of the earlier studies in the C. roseus area involved a collaborative program between the author's laboratory and the National Research Council of Canada, Plant Biotechnology Institute of Saskatoon. The senior collaborators in Saskatoon are Drs. W.G.W. Kurz and F. Constabel, and they, along with their associates, K. B. Chatson, H. Evans, P. Gaudet-LaPrairie, S. Rambold and J. Rushkowsky, are responsible for development of some of the cell culture lines (953, 200GW) discussed above and for the large scale propagation of these lines in order to produce the natural products involved. Development of other cell lines discussed occurred in the author's laboratory, as well as development of all the analytical methods (HPLC) and isolation and chemical characterization of all the alkaloids discussed. All studies concerning crude enzyme preparations (cell free extracts), and partially purified enzyme systems were performed at UBC, after enzyme isolation and HPLC techniques for protein monitoring and recognition of bisindole metabolites had been established. The extensive chemical synthetic experiments, as required in the biotransformation studies, were also performed at UBC. The research workers involved are: B. Aweryn, J. Balsevich, B. Botta, C. Boulet, C. Buschi, L.S.L. Choi, M. Gumulka, W. Gustowski, G.M. Hewitt, T. Honda, P. Kolodziecjczyk, G.C. Lee, N.G. Lewis, M. McHugh, T. Matsui, J. Nakano, T. Nikaido, J. Onodera, I. Perez, P. Salisbury, T. Sato, S.K. Sleigh, K.L. Stuart, R. Suen, H. Tsukamoto, and B.R. Worth. Financial aid was provided by a grant to the author from the Natural Sciences and Engineering Research Council of Canada, an NRC Research Contract under the Fermentation Technology Program, (00-310-SX-8-3011) and, most recently, through a collaborative program with Allelix Inc., Mississauga, Ontario, under the NRC/PILP program.

Studies with T. wilfordii represent a collaborative program between the author's laboratory and that of Prof. P. M. Townsley, Department of Food Science at this University. The development of the cell culture methods was performed jointly at Food Science by P.M. Townsley, W. T. Chalmers, D. J. Donnelly, K. Nilsson and F. Webster; G.G. Jacoli, Canada Department of Agriculture, Vancouver; and by P. J. Salisbury and G. Hewitt at the Chemistry Department. Analytical methods, isolation and characterization of the metabolites and development of synthetic methodology were performed at the Chemistry Department by M. H. Beale, L.S.L. Choi, E. Chojecka-Koryn, R. Duffin, M. Horiike, H. Jacobs, N. Kawamura, T. Kurihara, R.D. Sindelar, K.L. Stuart, Y. Umezawa, B. Vercek, and B.R. Worth. It is a particular pleasure to acknowledge the dedicated efforts of these colleagues. I am grateful to Prof. G. B. Marini-Bettolo, Universita Cattolica, Rome, for samples of polpunonic acid, methyl polpunonate and tingenone, and to Dr. Mildred Broome, Arthur D. Little & Co., Cambridge, Mass. for the KB assays. Financial aid for the studies on Tripterygium was provided by the National Institutes of Health (Contract NO1-CM-87236) and the Natural Sciences and Engineering Research Council of Canada.

5. REFERENCES

1 H. E. Street, Tissue Culture and Plant Science 1974, Academic, New York, 1974.
2 W. Barz, E. Reinhard, and M. H. Zenk, (Eds.), Plant Tissue Culture and Its Biotechnological Application, Springer-Verlag, Berlin, 1977.
3 T. A. Thorpe (Ed.), Frontiers of Tissue Culture 1978, University of Calgary, Calgary, 1978.
4 E. J. Staba, Plant Tissue Culture as a Source of Biochemicals, CRC Press, Boca Raton, 1980.
5 K. H. Neumann (Ed.), Primary and Secondary Metabolism of Plant Cell Cultures, Springer-Verlag, Berlin, 1985.
6 F. Constabel and W. G. W. Kurz, Adv. Appl. Microbiol., 25 (1979), 209-240.
7 J. P. Kutney, Pure and Appl. Chem., 56 (1984), 1011-1024.
8 M. Misawa, Adv. Biochem. Eng./Biotechnology, 31 (1985), 59-88.
9 O. L. Gamborg and D. E. Eveleigh, Can. J. Biochem., 46 (1968), 417-418.
10 I. A. Veliky, Phytochemistry, 11 (1972), 1405-1406.
11 I. A. Veliky and K. M. Barber, Lloydia, 38(2) (1975), 125-130.
12 L. Nettleship and M. Slaytor, Phytochemistry, 13 (1974), 735-742.
13 R. H. Dobberstein and E. J. Staba, Lloydia, 32 (1969), 141-147.
14 E. J. Staba, in: M. K. Seikel and V. C. Runeckles (Eds), Recent Advances

in Phytochemistry, Vol. 2, Appleton-Century-Crofts, New York, 1969, 75-106.

15 D. P. Carew, Nature (London), 207 (1965), 89-89.

16 K. Yoshikawa, M. Suzuki, and M. Maruoka, Japan Patent (Kokai), 73-80789, 1973.

17 N. Crespi Perellino and A. Guicciardi, Farmitalia Carlo Erba Laboratories, Milan, Italy, personal communication.

18 N. Crespi Perellino, A. Guicciardi and G. Malyszko. Farmitalia Carlo Erba Laboratories, Milan, Italy, personal communication.

19 J. P. Kutney, T. Hibino, E. Jahngen, T. Okutani, A. H. Ratcliffe, A. M. Treasurywala, and S. Wunderley, Helv. Chim. Acta, 59 (1976), 2858-2882.

20 J. P. Kutney, in: E. E. van Tamelen (Ed.), Biorganic Chemistry, Substrate Behavior, Vol. 2, Academic Press, New York, 1978, p. 197-228.

21 J. P. Kutney, T. Honda, P. M. Kazmaier, N. H. Lewis, and B. R. Worth, Helv. Chim. Acta, 63 (1980), 366-374.

22 N. Langlois, F. Gueritte, Y. Langlois, and P. Potier, J. Am. Chem. Soc., 98 (1976), 7017-7024.

23 K. L. Stuart, J. P. Kutney, and B. R. Worth, Heterocycles, 9 (1978), 1015-1022.

24 K. L. Stuart, J. P. Kutney, T. Honda, and B. R. Worth, Heterocycles, 9 (1978), 1391-1395, 1419-1427.

25 J. P. Kutney, L. S. L. Choi, T. Honda, N. G. Lewis, T. Sato, K. L. Stuart, and B. R. Worth, Helv. Chim. Acta, 65 (1982), 2088-2101.

26 A. I. Scott, F. Gueritte, and S. L. Lee, J. Am. Chem. Soc., 100 (1978), 6253-6255.

27 W. G. W. Kurz, K. B. Chatson, F. Constabel, J. P. Kutney, L. S. L. Choi, P. Kolodziejczyk, S. K. Sleigh, K. L. Stuart, and B. R. Worth, Phytochemistry, 19 (1980), 2583-2587.

28 J. P. Kutney, L. S. L. Choi, P. Kolodziejczyk, S. K. Sleigh, K. L. Stuart, B. R. Worth, W. G. W. Kurz, K. B. Chatson, and F. Constabel, Phytochemistry, 19 (1980), 2589-2595.

29 J. P. Kutney, L. S. L. Choi, P. Kolodziejczyk, S. K. Sleigh, K. L. Stuart, B. R. Worth, W. G. W. Kurz, K. B. Chatson, and F. Constabel, Heterocycles, 14 (1980), 765-768.

30 W. G. W. Kurz, K. B. Chatson, F. Constabel, J. P. Kutney, L. S. L. Choi, P. Kolodziejczyk, S. K. Sleigh, K. L. Stuart, and B. R. Worth, Helv. Chim. Acta, 63 (1980), 1891-1896.

31 J. P. Kutney, L. S. L. Choi, P. Kolodziejczyk, S. K. Sleigh, K. L. Stuart, B. R. Worth, W. G. W. Kurz, K. B. Chatson, and F. Constabel, J. Natural Prod., 44 (1981), 536-540.

32 F. Constabel, S. Rambold, K. B. Chatson, W. G. W. Kurz, and J.P. Kutney, Plant Cell Reports, 1 (1981), 3-5.

33 J. P. Kutney, L. S. L. Choi, P. Kolodziejczyk, S. K. Sleigh, K.L. Stuart, B. R. Worth, W. G. W. Kurz, K. B. Chatson, and F. Constabel, Helv. Chim. Acta, 64 (1981), 1837-1842.

34 W. G. W. Kurz, K. B. Chatson, F. Constabel, J. P. Kutney, L. S. L. Choi, P. Kolodziejczyk, S. K. Sleigh, K. L.Stuart, and B. R. Worth, Planta Medica, 42 (1981), 22-31.

35 J. P. Kutney, B. Aweryn, L. S. L. Choi, P. Kolodziejczyk, W. G. W. Kurz, K. B. Chatson, and F. Constabel, Heterocycles, 16 (1981), 1169-1171.

36 F. Constabel, S. Rambold, J. P. Shyluk, D. LeTourneau, W. G. W.Kurz, and J. P. Kutney, Z. Pflanzenphysiol., 105 (1981), 53-58.

37 J. P. Kutney, B. Aweryn, L. S. L. Choi, P. Kolodziejczyk, W. G. W. Kurz, K. B. Chatson, and F. Constabel, Helv. Chim. Acta, 65 (1982), 1271-1278.

38 J. P. Kutney, B. Aweryn, K. B. Chatson, L. S. L. Choi, and W. G. W. Kurz, Plant Cell Reports, 4 (1985), 259-262.

39 J. P. Kutney, C. A. Boulet, L. S. L. Choi, W. Gustowski, M. McHugh, J. Nakano, T. Nikaido, H. Tsukamoto, G. M. Hewitt and R. Suen, Heterocycles, 27 (1988), 613-620.

40 J. P. Kutney, C. A. Boulet, L. S. L. Choi, W. Gustowski, M. McHugh, J. Nakano, T. Nikaido, H. Tsukamoto, G. M. Hewitt and R. Suen, Heterocycles, 27 (1988), 621-628.

41 J. P. Kutney, B. Botta, C. A. Boulet, C. A. Buschi, L. S. L. Choi, J. Golinski, M. Gumulka, G. M. Hewitt, G. Lee, M. McHugh, J. Nakano, T. Nikaido, J. Onodera, I. Perez, P. Salisbury, M. Singh, R. Suen and H. Tsukamoto, Heterocycles, 27 (1988), 629-637.

42 H. Yokohama, E. P. Hayman, W. J. Hsu, and S. M. Poling, Science, 197 (1977), 1076-1078.

43 S. L. Lee, K. D. Cheng, and A. I. Scott, Phytochemistry, 20 (1981), 1841-1843.

44 H. Miyasaka, M. Nasu, T. Yamamoto, and K. Yoneda, Phytochemistry, 24 (1985), 1931-1933.

45 L. Witte, J. Berlin, V. Wray, W. Schubert, W. Kohl, G. Hofle, and J. Hammer, Planta Medica, 49 (1983), 216-221.

46 N. Ishikura, K. Nabeta and H. Sugisawa, Phytochemistry, 23 (1984), 2062-2063.

47 I. Koch-Heitzmann, W. Schultze, and F. C. Czyan, Z. Naturforsch., 40c (1985), 13-14.

48 J. P. Kutney, M.H. Beale, P. J. Salisbury, R. D. Sindelar, K. L. Stuart, B. R. Worth, P. M. Townsley, W. T. Chalmers, D. J. Donnelly, K. Nilsson, and G. G. Jacoli, Heterocycles, 14 (1980), 1465-1467.

49 J. P. Kutney, G. M. Hewitt, T. Kurihara, P. J. Salisbury, R. D. Sindelar, K. L. Stuart, P. M. Townsley, W. T. Chalmers, and G. G. Jacoli, Can. J. Chem., 59 (1981), 2677-2683.

50 J. P. Kutney, L. S. L. Choi, R. Duffin, G. Hewitt, N. Kawamura, T. Kurihara, P. Salisbury, R. Sindelar, K. L. Stuart, P. M. Townsley, W. T. Chalmers, F. Webster, and G. G. Jacoli, Planta Medica, 48 (1983), 158-163.

51 S. M. Kupchan, W. A. Court, R. G. Dailey, C. J. Gilmore, and F. Bryan, J. Am. Chem. Soc., 94 (1972), 7194-7195.

52 B. N. Zhou, G. Q. Song, and C. Q. Hu, Acta Parmaceutica Sinica, 17 (1982), 146-150.

53 Research Group of Lei-Gong-Teng, Chinese Dermatology Journal, 15 (1982) 199-201.

54 J. Zheng, L. Xu, L. Ma, D. H. Wang and J. Gao, Acta Academiae Medicinae Sinicae, 5 (1983) 1-8.

55 J. Zheng, J. Liu, L. Hsu, J. Gao and B. Jiang, Acta Academiae Medicinae Sinicae, 5 (1983), 73-78.

56 W. L. Ngan, Y. Z. Auyeung, and S. Y. Cheng, Chinese Herb Medicine, 3 (1984), 123-125.

57 J. P. Kutney, R. D. Sindelar, and K. L. Stuart, J. Chromatogr., 214 (1981), 152-155.

58 T. Murashige, and F. Skoog, Physiologia Plantarium, 15 (1962) 473-497.

59 A. C. Hildebrandt and R. V. Schenk, Can. J. Botany, 50 (1972), 199-204.

60 L. S. Dujack. S. J. Pancake and P. K. Chen, in: J. D. Nelson and C. Grassi (Eds.), Current Chemotherapy and Infectious Disease. Proceedings of 11th International Conference on Antimicrobial Agents and Chemotherapy, The American Society for Microbiology, Washington, 1980, Vol. II.

DISCOVERY AND DEVELOPMENT OF NEW DRUGS FOR SYSTEMIC OPPORTUNISTIC INFECTIONS

CHARLES D. HUFFORD and ALICE M. CLARK

1. INTRODUCTION AND BACKGROUND

Systemic antifungal drug discovery has only recently received new attention primarily as a result of the association of opportunistic systemic mycoses with acquired immunodeficiency syndrome (AIDS). The polyene antibiotic, amphotericin B (AMB) was the first antifungal antibiotic to be developed for the treatement of systemic mycoses (refs. 1-5). There are, however, significant drawbacks to its therapy. The increasing incidence of disseminated fungal infections due to the AIDS epidemic has prompted renewed interest in the development of new systemic antifungal antibiotics. Acquired immunodeficiency syndrome (AIDS) is an immune disorder resulting from destruction of cell-mediated immunity by the virus known as Human Immunodeficiency Virus (HIV). Patients with AIDS are particularly susceptible to the development of a number of life-threatening systemic opportunistic infections. Two such diseases which are considered to be reasonably effective predictors of AIDS are Kaposi's sarcoma and *Pneumocystis carinii* pneumonia. In addition, certain other opportunistic infections also serve this purpose. Among these are disseminated atypical mycobacteriosis and the disseminated mycotic infections candidiasis and cryptococcosis (refs. 6-8). While opportunistic disseminated mycoses have been associated in the past with immune deficiency, the occurrence of atypical mycobacteriosis was, until recently, very rare. In the past, the management of disseminated opportunistic infections in immunosuppressed patients has involved the reduction of immunosuppressive therapy coupled with antibiotic chemotherapy. Since the course of immunosuppression in AIDS patients cannot currently be halted or reversed, the only recourse for these individuals is treatment of the infections, even though such infections are a reflection of a more complicated underlying immune disorder. Unfortunately immunosuppressed patients appear to be more resistant to conventional antibiotic therapy than normal individuals (ref. 9).

The major opportunistic mycoses associated with AIDS are candidiasis, cryptococcosis, and aspergillosis. In addition, there is an increasing incidence of other systemic mycoses not considered as opportunistic, i.e., histoplasmosis, blastomycosis, and coccidiodomycosis, among AIDS patients.

Candidiasis is normally caused by the yeast *Candida albicans,* which in normal healthy individuals can cause localized infections such as vulvovaginitis and thrush. However, in immunocompromised individuals, such as those who have AIDS or who have been on immunosuppressive chemotherapy, candidiasis can and does occur as a severe, life-threatening systemic infection. In many cases, oral or esophageal candidiasis (thrush) may serve as one of the initial indications of AIDS and may develop into disseminated candidiasis. In addition, disseminated candidiasis is being increasingly associated with other species of *Candida* such as *C. parapsilosis*, *C. guillermondii*, and *C. tropicalis* (refs. 10-12).

Cryptococcosis, which occurs in approximately 5% of AIDS patients, is usually in the form of meningitis due to *C. neoformans*; however, nonmeningeal cryptococcosis (pneumonia, disseminated infection) may also occur (refs. 6-8, 13-15). Currently only four clinically useful antifungal agents are indicated for the treatment of systemic mycotic infection and these fall into three structural categories: a) polyene antibiotics, b) flucytosine, and c) synthetic azoles (refs. 1-5).

Amphotericin B (AMB) continues to be the most effective agent available for the therapy of disseminated mycoses (refs. 1-5). There are, however, significant drawbacks to its therapy. Immediate problems associated with its administration include phlebitis, febrile reaction, headache, nausea, vomiting, malaise, diarrhea, epigastric pain, and anorexia. The most serious complication of AMB therapy is renal damage, which can be permanent in patients receiving large doses of the drug (ref. 16). This is especially significant in light of the fact that the kidney is the primary organ infected in experimental disseminated candidiasis.

The combination of AMB and flucytosine, a synthetic nucleoside analog, is also recommended for the treatment of certain disseminated mycoses. Flucytosine is rarely used alone for systemic candidiasis primarily because of the high incidence of preexisting or subsequent development of resistance by *C. albicans* strains (refs. 2-5, 9). Common side effects of flucytosine include nausea, vomiting, diarrhea, and cramps. The most serious adverse effect of flucytosine is leukopenia or thrombocytopenia. The combination of AMB and flucytosine is designed to reduce the dosage of AMB, therefore reducing its dose-related toxicity, and to eliminate the development of resistance to flucytosine. However, it has been noted that the flucytosine toxicity may increase when it is used in combination with AMB (ref. 14). This may be particularly relevant in patients with renal dysfunction since flucytosine is excreted by the kidney (refs. 2-5). In addition, it is estimated that 5% of *C. neoformans* exhibit primary resistance to flucytosine and that secondary resistance can develop during the course of therapy (ref. 6). The current recommended therapy for cryptococcal meningitis is AMB (0.3 mg/kg/day, i.v.) in combination with flucytosine (150 mg/kg/day, p.o. bid) or, ketoconazole (400 mg/day, p.o.) as an alternative for nonmeningeal cryptococcosis. However, response rates with either therapy remain unacceptably low and relapse rates are quite high (refs. 6, 8, 15). In addition, flucytosine produces bone marrow suppression in a number of AIDS patients and cannot be tolerated for the course of therapy. There is still a significant controversy regarding the most appropriate therapy of cryptococcal infections (ref. 17).

The synthetic imidazole antifungal agents were first introduced over 10 years ago as broad spectrum, topically effective antifungal agents. Ketoconazole was the first orally absorbed antifungal agent effective for the treatment of certain systemic mycoses. Although ketoconazole was initially hailed as a milestone in systemic antifungal therapy, with the treatment of more patients, several significant adverse reactions have been noted. These include gynecomastia, endocrine effects and hepatotoxicity (refs. 2, 5, 18, 19). It also appears that there is a high rate of recurrence of candidiasis following cessation of ketoconazole therapy, and that ketoconazole may not be effective in the treatment of systemic mycoses in immunocompromised patients (ref. 2). There have also been reports of azole-resistant strains of *Candida* (ref. 20), however, ketoconazole does appear to be effective for the treatment of certain limited-extent infections such as chronic mucocutaneous candidiasis (ref. 18).

In addition to the azoles, another class of synthetic antifungal agents, the allylamines, is currently receiving attention. Naftifine, the prototype of the class, is a broad spectrum antifungal particularly effective as a topical agent for dermatomycoses (ref. 21). Since the discovery of the antifungal activity of naftifine during routine screening, numerous analogs have been prepared and structure activity relationships studied (ref. 22). However, there are currently no allylamine derivatives clinically available.

Prior to the AIDS epidemic, disseminated infection with atypical mycobacteria (nontuberculosis or mycobacteria other than tuberculosis, MOTT) was very rare. However, atypical mycobacteriosis has been recently recognized as a common opportunistic infection associated with AIDS, occurring in as much as 50% of such patients (refs. 23-28). More precise statistics regarding the occurrence of this disease are difficult to obtain since it is often not recognized until post-mortem examination. The most common cause of atypical mycobacteriosis is *Mycobacterium avium-intracellulare* (MAI) which is actually a complex of two closely related strains (*M. avium* and *M. intracellulare*) (refs. 6, 29). Most clinicians have initially attempted to treat mycobacteriosis utilizing the existing agents for tuberculosis. This has met with limited success due primarily to the high level of resistance of the atypical mycobacteria to many such drugs, including multiple drug regimens involving four, five, and six agents. The current therapy for mycobacteriosis is a drug regimen including isoniazid, rifampin, ethambutol, streptomycin, capreomycin, ethionamide, cycloserine, pyridoxine, and pyrazinamide and, in resistant cases, the experimental drugs ansamycin and clofazimine. There are numerous adverse effects associated with these drugs and response rates still remain very low (refs. 6, 8, 23-29).

2. NEW DRUG DISCOVERY RATIONALE

It appears that the majority of past and current efforts to develop new, clinically effective antimicrobial agents which offer significant improvement over existing agents have relied primarily on one of five approaches, most of which rely on the use of existing agents in some manner:

a) derivatization of existing agents
b) synthesis of additional analogs of existing synthetic agents
c) use of combination therapy of existing agents with other drugs
d) improvement of delivery of existing agents to target site
e) isolation of antibiotics from natural souces

Derivatization of the antifungal antibiotic AMB to its methyl ester (AME) afforded a compound which retained antifungal activity (although less active than AMB), but appeared to lack the toxicity, particularly nephrotoxicity, of AMB (refs. 30-32). However, subsequent studies have shown total dose-related leukoencephalopathy leading to neurological disorders in patients receiving prolonged therapy with AME (ref. 33).

Ansamycin is a spiropiperidyl derivative of rifampin which was shown to exhibit greater *in vitro* activity against MAI than the parent rifampin (refs. 34, 35); nevertheless, it has not provided sustained remission of MAI infections in AIDS patients (ref. 28). Another new, more lipophilic spiropiperidyl derivative of rifamycin-S, rifabutin, has recently been reported to exhibit improved *in vitro* activity against MAI, as well as other MOTT. Controlled clinical trials have not yet been completed with the drug; however, limited data following use in humans is available. While the drug provided clinical improvement for patients with disseminated MAI and resulted in reduction of bacilli in the blood, it was not effective in eradicating the infection (ref. 33). Controversy still exists regarding the most appropriate dosage regimen and, the potential utilization of this drug as a prophylactic measure to prevent development of MAI infections in AIDS patients is likely to be further investigated (ref. 36).

Some of the most intensive efforts to develop new antifungal agents have centered on the synthesis of analogs of the existing synthetic antifungal azole and allylamine type agents. While initial studies suggest that the *in vitro* and *in vivo* activity of some new azoles such as itraconazole and fluconazole are promising (refs. 37, 38), it remains to be seen whether they share the same toxicities and cross-resistance with existing azole antifungals. Certainly, it would appear from early studies that fluconazole is effective against cryptococcal meningitis. However, the azoles in general are fungistatic agents and use in immunocompromised patients will likely require lifetime daily therapy. The notable exception is that itraconazole is apparently fungicidal against *Aspergillus*.

The development of new synthetic antimycobacterial agents has recently focused on the discovery that the known fluoroquinolone antibacterials exhibit *in vitro* activity against *Mycobacteria* species (refs. 39-43). While a number of fluoroquinolones were evaluated against a variety of species of *Mycobacteria*, the MIC values for MAI were generally the highest, suggesting the greatest intrinsic or primary resistance.

A third approach to the development of more effective agents for systemic opportunistic infections is not the development of new agents *per se,* but the utilization of the synergistic relationships between existing agents with other drugs. While as many as six or seven antimycobacterial drugs are used in combination for the therapy of MAI infections in AIDS patients, none of these has really provided good results. With regard to the antifungal agents, the approach has generally been to utilize AMB in combination with specific agents to increase the uptake of such agents into the cell, e.g., tetracycline, rifampin, or flucytosine. While *in vitro* data and animal studies suggest such combinations would be useful in the therapy of systemic mycoses, only the clinical efficacy of the AMB-flucytosine combination has been demonstrated (ref. 44). Polack (ref. 45) has recently investigated the use of combination therapy in experimental candidiasis, cryptococcosis, and aspergillosis in mice and concluded that the combination of 5-fluorocytosine and itraconazole was synergistic in candidiasis and aspergillosis, but not in cryptococcosis.

Considerable effort has been directed toward the preparation of derivatives of AMB which will enhance the delivery or passage of the drug into the target cells. In principle, this would allow the use of lower dosages and consequently reduce the dose-related toxicity. The most significant advancement in this area is the work on liposome encapsulation of AMB (refs. 46-48). Lopez-Berestein, *et al.* (ref. 46) reported that liposome-encapsulated AMB was as effective as free AMB but less toxic, *i.e.*, that liposome encapsulation improved the therapeutic index in a mouse model of disseminated candidiasis. These results were confirmed (ref. 47). And clinical studies (ref. 48) also demonstrated efficacy of the liposomal-AMB preparation. Problems related to stability, production, and sterilization of the liposomes may present the major obstacles to their use in humans in the near future (ref. 48).

3. DISCOVERY AND DEVELOPMENT OF NEW PROTOTYPE ANTIBIOTICS

While all of the approaches cited above are worthwhile in that they seek to utilize existing agents and information in the most effective manner, there still exists an urgent need for the discovery and development of totally new, **prototype** antibiotics for systemic opportunistic infections which do not share the same toxicities and resistance of the existing agents. Clearly, the therapy of opportunistic infections is complicated by several factors, including the relative ineffectiveness of available agents, the relatively severe toxicities of such agents, the development of resistance to existing agents and the clinical status of the patient suffering from an underlying immune disorder. All of these factors contribute to the obvious necessity for new effective and less toxic antifungal and antibacterial agents for the treatment of these diseases.

Although intensive efforts are currently directed toward the synthesis of additional analogs of the currently available agents and the use of combination regimens, it seems unlikely that such efforts will provide any new products or combinations that do not also suffer from similar toxicity or resistance problems. It is necessary to pursue avenues of research which will provide totally new **prototype** agents that do not fall into one of these structural classes. Natural products have, in the past, provided such prototype bioactive compounds and it seems logical that the search for new antibiotics should also pursue this route. The major advantage of this approach over chemical synthesis or modification of existing agents is the likelihood of identifying new **prototype** drugs with quite different chemical structures, and hence, less likelihood of similar toxicities and cross-resistance. Several exellent reviews are available which cover the available antifungal and antimycobacterial agents for the therapy of systemic infections (refs. 2-6, 8).

Over the last several decades the search for antibiotics has been limited to microorganisms and in particular the streptomycetes and some fungi. It seems reasonable to assume that if new antifungal and antimycobacterial antibiotics are to be found that have different structures with different or supplemental activities from the ones in current use, then a source other than the more traditional microorganisms must also be investigated. In particular, the higher plants appear to be a logical choice, chiefly because of their seemingly infinite variety of novel organic molecules, often referred to as secondary metabolites. Antifungal agents are widely distributed among the higher plants, but very few have been evaluated for their activity against human pathogenic fungi and essentially none of these have been evaluated in animal models of disseminated mycoses. In a review of the medicinal properties of many higher plants, Oliver-Bever (ref. 49) lists thirteen species of plants with reported fungicidal activity. Two other reports have appeared since that time which also cite the antimicrobial activity of crude plant products, but no attempt was made to identify the active constituents (refs. 50, 51). Therefore, it seems logical and worthy to pursue new sources of antifungal and antimycobacterial agents which may provide a greater diversity of active **prototype** compounds and, consequently, less similarity to existing agents. It is the specific goal of this chapter to review efforts to discover and develop prototype antibiotics from a variety of "untapped" sources to provide, if not an ultimately clinically useful drug, **prototype** agents for the development of clinically useful drugs for the therapy of systemic opportunistic infections.

4. GENERAL APPROACH

The approach utilized successfully in our laboratories has been to detect new "leads", primarily from higher plants and unusual microorganisms, followed by bioassay-directed isolation and identification of the active constituents, and finally, preliminary evaluation of *in vivo* efficacy and toxicity of pure active compounds in established animal models of disseminated infections.

5. METHODOLOGY

5.1 Sources of samples for *in vitro* screening

Our approach has been to concentrate our efforts on sources of compounds which in the past have either been overlooked as potential sources of new antibiotics or, at the very least, have not received significant attention. These include higher plants and unusual microorganisms. In addition, numerous novel natural products are reported in the literature each year, but for a variety of reasons, are often never evaluated for any type of biological activity. Therefore, collaborative efforts with other investigators have also provided additional new compounds. By taking this "triad" approach, an almost endless supply of new samples from a variety of sources is guaranteed, and consequently a high probability of new "leads" are identified which may ultimately afford **prototype** agents.

5.2 *In vitro* evaluation

It is the *in vitro* screening process that is the true discovery phase in prototype drug discovery and development. The *in vitro* evaluation of extracts, fractions and compounds is conducted in two phases: a qualitative analysis and a quantitative analysis. The agar well-diffusion assay is a standard, rapid, sensitive assay which simply qualitatively assesses the presence or absence of activity. New leads are identified as a result of the demonstration of positive inhibitory activity against the various test organisms. Once the activity has been confirmed for active extracts, then a judgement is made as to the priority of the "lead" relative to other extracts which are candidates for large scale extraction and bioassay-directed fractionation. For pure compounds which show good qualitative activity, quantitative activity (MIC) is determined.

5.2.1. Qualitative *in vitro* screening. All extracts, fractions and pure compounds are evaluated for antimicrobial activity against *Cryptococcus neoformans* ATCC 52657, *Mycobacterium intracellulare* ATCC 23068 (human isolate) and *Candida albicans* NIH B311 by use of the agar well-diffusion assay previously described (refs. 52, 53). The organisms should be maintained in the laboratory by four different procedures: a) as a number of lyophilized pellets, one of which is revived every 4-6 months to insure viability of the strain, b) as a suspension at -70°C, c) on stock agar slants of Sabouraud-dextrose agar (SDA) for *C. albicans* and *C. neoformans* and Lowenstein-Jensen agar for *M. intracellulare,* which are maintained at 4°C and transferred every month, and d) as stock agar slants immersed in sterile mineral oil. The viability of subcultures of each of these storage procedures should be checked periodically by subculture, morphological observation, wet-mount microscopic examination, routine determination of MIC values of standard agents (e.g., AMB, ketoconazole, rifampin), and determination of lethal dose following i.v. infection of mice.

For qualitative *in vitro* antifungal evaluation, yeast cultures are grown in Sabouraud-dextrose broth (SDB) for 24 hrs. at 37°, at which time the cells are harvested by centrifugation, washed and suspended in sterile 0.9% saline (PSS) to give a final concentration of 10^5 CFU per ml (adjusted using a hemocytometer). Using sterile cotton swabs, culture plates of SDA are streaked with the suspension (10^5 CFU/ml) of the test organism. Cylindrical plugs are removed from the agar plates by means of a sterile cork borer to produce wells with a diameter of approximately 11 mm. To the well is added 100 µl of solution or suspension of an extract, fraction, or pure compound. Crude extracts and fractions are tested at a concentration of 20 mg/ml, whereas pure compounds are tested at 1 mg/ml. When solvents other than water, ethanol, methanol, dimethylsulfoxide (DMSO), dimethylformamide (DMF), or acetone are required to dissolve extracts or compounds, solvent blanks are included. Antifungal activity is recorded as the width (in mm) of the clear zone of inhibition surrounding the agar well. Although no attempt is made to rigidly quantitate the assay, conditions are held sufficiently constant to ensure that successful fractionation methods are evidenced by an increase in the size of the zones of inhibition as purification of active constituents is achieved. In addition, a general assessment of relative activity is achieved since the antifungal agents AMB and ketoconazole are included as standards in **each** assay.

For qualitative *in vitro* antimycobacterial evaluation, *M. intracellulare* ATCC 23068 should be grown in Lowenstein-Jensen (L-J) medium for 48 hrs. at 37°C. The concentration of the inoculum is adjusted turbidometrically to 10^5 CFU/ml. The remainder of the assay is conducted as described above with rifampin used as a positive control.

5.2.2. Quantitative *in vitro* evaluation. For pure chemical compounds which show significant activity in the qualitative screen, quantitative evaluations are performed to determine the minimum inhibitory concentrations (MIC) of the compound. The method used to determine the MIC is the two-fold serial broth dilution assay.

All compounds are initially tested using a concentration of 100 µg/ml in the first tube. Two-fold serial dilutions through ten tubes will give a low-range value of 0.2 µg/ml and, if necessary, this can be decreased further either by continued dilution or by lowering the initial concentration. The compound is added to sterile media as a solution. If a solvent other than water or ethanol is required to dissolve the compound, a solvent blank is included. The inoculum for the MIC determination is prepared as previously described for qualitative screening. Each tube is inoculated with 10 μl of inoculum.

A number of methods have been utilized for the quantitative *in vitro* evaluation of antimycobacterial activity (refs. 34, 39, 54-58). These generally employ either broth or agar dilution techniques or radiometric determination of growth. It has been documented that most strains were more susceptible to test drugs in broth medium and the values obtained more accurately reflect the MIC since the incubation period is shorter (ref. 58). The determination of MIC values of new compounds against *M. intracellulare* are typically conducted in either 7H12 broth medium (refs. 54, 58), Mueller-Hinton broth (refs. 57, 59) or Lowenstein-Jensen broth (refs. 39, 55). Following duplicate two-fold serial dilution of the test compound in appropriate medium, each tube is inoculated with 10 μl of *M. intracellulare* inoculum prepared from a 48 hr. culture (37°) in L-J medium and adjusted turbidometrically to 10^5 CFU/ml. The suspension should be utilized quickly after preparation to avoid clumping of the bacteria. If necessary, the suspension should be sonicated briefly to break up clumps. Samples from alternate tubes are removed daily for plating on 7H11, L-J or Mueller-Hinton agar plates for determination of total CFU. The MIC value is taken as the lowest concentration that inhibits 99% of the bacterial population following incubation for 4-7 days. A growth control tube containing no drug should also be included. Alternately, the MIC values can be assessed based on the optical density of antibiotic-containing tubes as compared to the growth control tube.

It is well documented that there is no standard technique for the determination of MIC values for antifungal agents and, consequently, these values vary greatly depending on the techniques used, inoculum size, medium, etc. (refs. 60-63). For this reason, caution should be used when interpreting results of MIC determinations. It is important to recognize that values obtained will be useful only in determining **relative** activities of antifungal agents and primarily for assessment of priority for *in vivo* efficacy studies. For this reason, the antifungal agents AMB and ketoconazole should be included as standards in each screen to determine the relative activities of the new antifungal agents.

MIC determinations of antifungal antibiotics is accomplished in SAAMF broth (ref. 64), SDB (ref. 65), or yeast-nitrogen broth according to the two-fold serial broth dilution assay described above. Inoculum is prepared as described for qualitative evaluation and 10^4 CFU of *C. neoformans* and 10^6 CFU of *C. albicans* are used as inoculum for antibiotic-containing broth. Following incubation at 37°C for 48 hr. the MIC is taken as the lowest concentration inhibiting growth of the organism based on visual assessment. The minimum fungicidal concentration (MFC) can also be determined by subculturing from each dilution tube onto Sabouraud-dextrose agar plates.

5.2.3. Evaluation of Synergistic Combinations. The documented synergistic or antagonistic activity of known antifungal agents is a significant factor in assessing the potential utility of combination therapy. AMB and flucytosine are often used in combination for the treatment of disseminated mycoses (refs. 2-5). Potentiation of AMB activity has also been demonstrated with other agents such as tetracycline (ref. 66), minocycline (refs. 67, 68), and rifampin (refs. 69, 70), to name a few. The *in vitro* synergistic or antagonistic effects of new anticandidal agents can be determined in order to identify likely candidates for combination therapy studies in animals. Uno, *et al.* (ref. 71) have described a simple procedure for determination of synergistic or antagonistic effects, in which a paper strip is soaked in a solution (100 μg/ml) of a known antimicrobial agent. Shorter strips (up to 4) soaked in solutions of agents known or suspected to be synergistic or antagonistic are then placed perpendicular to the longer strip containing the antifungal agent on the surface of the agar plate seeded with test organism. The plates are then incubated at 37° for 24-48 hrs and synergism is indicated by an outward distortion of the pattern of inhibition from the paper strip. Antagonism is indicated by an inward distortion of the zone of inhibition.

For combinations that appear promising based on the qualitative paperstrip assay, a quantitative assessment of synergistic activity can be accomplished using a broth checkerboard assay (refs. 72-74). In this assay two-fold serial dilutions of the two drugs in yeast nitrogen broth are prepared. All possible combinations of drug concentrations within the range of dilutions used are inoculated with one calibrated loopful of a suspension (10^6 CFU/ml) of the test organism. The concentration of each antibiotic in the combination that results in inhibition of the test organism is expressed as the fractional inhibitory concentration (FIC), *i.e.*, a fraction of the MIC of the same antibiotic used alone. If the sum of the FIC values for each antibiotic is less than 0.5 (i.e., a four-fold decrease in MIC) the combination is considered to be synergistic.

5.3. Evaluation of *in vivo* efficacy and toxicity

In order to glean preliminary information regarding the chemotherapeutic ratio and possible dose ranges, acute toxicity, as LD_{50} is determined prior to initiation of efficacy studies. Compounds are initially evaluated for efficacy and toxicity by the intraperitoneal (i.p.) route. The oral (p.o.) and intravenous (i.v.) routes may be used for additional evaluation if a new drug initially shows promising therapeutic potential following i.p. administration. Depending on the solubility and route of administration, drugs are dissolved in physiologic saline, water, ethanol, or a mixture of 10% ethanol, 40% propylene glycol and nonpyrogenic sterile water, or they may be suspended in 0.25% methyl cellulose, 5% gum acacia or 0.3% agar for oral administration. Soluble drugs are filter sterilized and those which must be suspended are prepared aseptically.

5.3.1. Median Lethal Dose. The median lethal dose (LD_{50}) of new compounds is established initially via the i.v. route by the method of Litchfield and Wilcoxon. Groups of 10 ICR mice (5 mice per sex) are administered the drug and observed for overt signs of toxicity for 24 hrs. and for morbidity and lethality for 14 days. Doses are doubled or halved until the maximum dose is found which produces no lethality and a minimal dose is found which produces 100% lethality within a 14 day period. Doses are determined within that range which produce deaths within the 20 to 80% lethality range. After determining three such doses within the 0 - 100 % mortality range, an estimated LD_{50} (within 95% confidence intervals) is produced using the method of Litchfield and Wilcoxon.

5.3.2. Candidiasis. Experimental infections with *C. albicans* have typically utilized mice, guinea pigs, and rabbits. Disseminated infection is achieved by i.v. injection of sublethal doses of *C. albicans* and in all three species the primary organ infected is the kidney. The agent is rapidly cleared from the blood of experimental animals (refs. 75, 76), as it is in disseminated infections in man. The mouse model has been used extensively for studying *Candida* infections and for *in vivo* assessement of potentially active drugs. Disseminated infection of *C. albicans* in mice is achieved by i.v. injection of at least 0.5 to 1 x 10^6 CFU. The primary target organ for disseminated infection in outbred mice is the kidney. An inoculum of 5 x 10^5 CFU will cause death due to disseminated infection in outbred mice within 10-14 days. Six groups of 10 ICR outbred mice (20-22 g, equal numbers of male and female) are challenged i.v. with a suspension of 10^6 CFU of *C. albicans* in PSS, prepared from 5-7 hr. cultures in SDB. Three groups receive varying doses of the test drug, i.p., at 7 hr. post-infection. One group receives AMB (i.v.) as positive control. One group receives only vehicle (untreated control), while one group serves as uninfected toxicity control (maximum test drug dosage). Animals are housed in microisolator cages equipped with HEPA filters and sacrificed at 24 or 48 hrs. post-infection. The kidneys are aseptically removed, homogenized in PSS, diluted, and aliquots of the dilutions are plated in triplicate on SDA plates which are incubated at 37°C for 24 hrs. The number of recoverd CFU in treated vs. vehicle control groups is compared by Wilcoxon nonparametric rank sum test with $P < 0.05$ as a test of significance.

As an alternative or supplemental measure of efficacy, the survival of mice infected with a lethal dose of C. albicans and treated daily with test drug can be measured. In this assay, the preparation of inoculum and infection of test animals is the same as described above. However, rather that sacrificing the animals at 24-48 hrs. post-infection, test drug is administered daily for 12-14 days and animals are observed for mortality. Comparison of prolonged survival of drug-treated vs. vehicle-treated infected animals provides a measure of efficacy.

5.3.3. Cryptococcosis. Experimental infections with *Cryptococcus neoformans* have typically utilized mice, rabbits, guinea pigs or rats. The mouse model of cryptococcosis has also been used extensively for studying *Cryptococcus* infections (ref. 77) and for *in vivo* assessment of potentially active drugs. Six groups of 10 ICR outbred mice (20-22 g, equal numbers of male and female) are challenged i.v. with a suspension of 10^6 CFU of *C. neoformans* in PSS, prepared from 18-24 hr. cultures in SDB. Three groups receive varying doses of the test drug, i.p., at 24 hr. post-infection. As described above, one group receives AMB as positive control, one group receives only vehicle (untreated control), and one group serves as uninfected toxicity control (maximum test drug dosage). Animals are housed in microisolator cages and monitored for morbidity and mortality for 60 days. Survival in treated vs. nontreated groups is compared by the generalized Wilcoxon test of Life tables, using $P \leq 0.05$ as a test of significance. For animals which succumb during the study, the brains and spleens are aseptically removed, homogenized in PSS, and plated on SDA to verify disseminated cryptococcal infections.

For compounds which show efficacy (significant increase in survival time), the efficacy following oral administration will also be evaluated. In addition, the number of CFU recovered from the brain and spleen following i.v. treatment will be determined and compared to vehicle-treated infected control groups by the Wilcoxon nonparametric rank sum test.

5.3.4. Mycobacteriosis. An excellent review of the animal models available for the study of atypical mycobacteriosis is available (ref. 78). A mouse model is also available for evaluation of *in vivo* efficacy in mycobacteriosis. An inoculum of *M. intracellulare* will be prepared by cultivation in 7H9 culture broth for 3-5 days and adjusted turbidimetrically equivalent to McFarland No. 1. As recommended (ref. 78), 1 ml volumes of this suspension should be stored at -70°C until used. Groups of 10 C57BL/6 mice (17-20 g, equal numbers male and female) are infected (i.v.) with 10^6 to 10^7 organisms per mouse. Three control groups and three drug-treatment groups are included as described for the cryptococcosis and candidiasis models. Rifampin can be utilized as a positive control and drug treatment should begin on the day of infection and continue for at least 21 days. Comparison of survival in treated vs. non-treated groups can also be made by the generalized Wilcoxon test of Life tables, using $P \leq 0.05$ as test of significance. The lungs, liver and spleen of dead animals should be homogenized and plated on 7H11 agar to verify disseminated mycobacteriosis.

6. SELECTED EXAMPLES

The preceding discussions have served to outline the rationale and general methods used in our approach to discover and develop new antibiotic prototype compounds. A few examples will be discussed in detail which will serve to illustrate the success of our approach.

6.1. Oxoaporphine Alkaloids Active in Disseminated Candidiasis

Although microorganisms have traditionally served as the primary source for new antibiotics, it has recently been shown that higher plants also serve as sources for a number of diverse antimicrobial agents. One such example is the antifungal oxoaporphine alkaloid, liriodenine **1** (Fig. 1), a constituent of the heartwood of *Liriodendron tulipifera,* commonly known as the tulip tree. Using a bioassay-directed fractionation procedure, liriodenine **1** was isolated and identified as the major antifungal constituent of *L. tulipifera* (ref. 53).

In addition to having good activity against *C. albicans* it was also active against a wide range of gram-positive bacteria and fungi (Table 1) and the important plant fungal pathogens *Helminthosporin teres, Botrytis fabae* and *Piricularia oryzae* . A number of other oxoaporphine alkaloids of *L. tulipifera* were also evaluated for antimicrobial activity but all were less active than liriodenine (Fig. 1) (ref. 53).

TABLE 1

Minimum inhibitory concentration values (μg/ml) of Liriodenine **1** against selected microorganisms.

Organism (ATCC #)	MIC
Staphylococcus aureus (6538)	3.1
Bacillus subtilis (6633)	0.39
Herellea vaginicola (19683)	8
Staphylococcus epidermidis (155)	2
Mycobacterium smegmatis (607)	1.56
Saccharomyces cerevisiae (9763)	6.2
Trichophyton mentagrophytes (8757)	1
Trichophyton rubrum (14001)	1
Candida albicans (10231)	6.2

Another important finding was the observation that the methiodide salts of liriodenine **1** and oxoglaucine **2** were much more active *in vitro* against *C. albicans* ref. 53) than the parent compounds. These findings sparked further interest in assessment in the therapeutic potential of these alkaloids.

Based on the MIC values, the most active of the three related antifungal oxoaporphine alkaloids is liriodenine methiodide with an MIC value of 0.78 μg/ml, while liriodenine and oxoglaucine methiodide are approximately equal in activity with MIC values of 3.12 μg/ml for each. By comparison, the MIC value of amphotericin B, the current drug of choice for disseminated candidiasis, is 0.39 μg/ml for this strain of *C. albicans*.

Prior to the initiation of *in vivo* efficacy studies, a determination of acute toxicity, as median lethal dose (LD_{50}), of each compound was undertaken. The LD_{50} of liriodenine methiodide is 14.3 mg/kg following i.v. administration and between 50 and 100 mg/kg following i.p. administration (due to limitation in sample supply, complete LD_{50} values were obtained for each alkaloid following i.v. administration only, with ranges determined for i.p. administration.) The reported LD_{50} for amphotericin B following i.v. administration is between 1 and 4 mg/kg (refs. 31, 46). Liriodenine appears to be even less toxic (as LD_{50}) than liriodenine methiodide, with LD_{50} values of 120 mg/kg following i.v. administration and >250 mg/kg following i.p. administration. Oxoglaucine methiodide has an LD_{50} of 12.2 mg/kg following i.v. administration and between 75 and 100 mg/kg following i.p. administration.

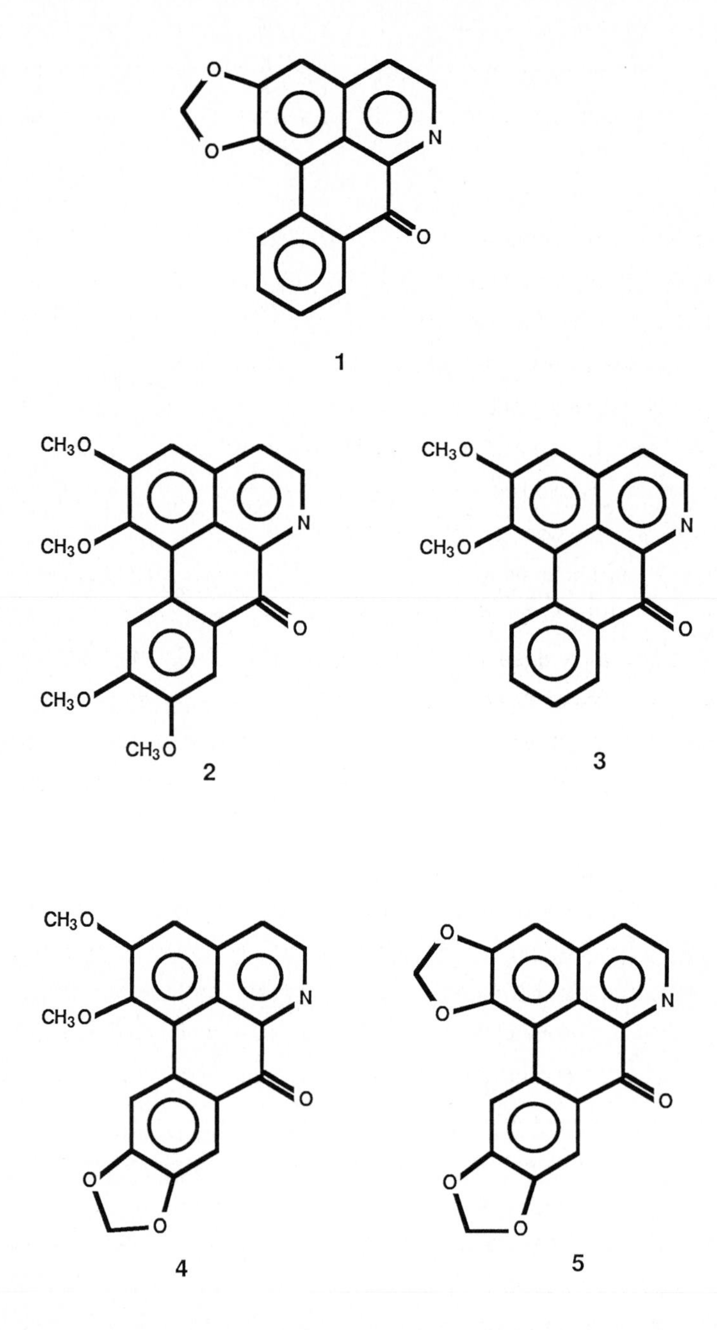

Fig. 1. Structural formulae for oxoaporphine alkaloids liriodenine **1**, oxoglaucine **2**, lysicamine **3**, oxonantenine **4**, and cassameridine **5**.

The *in vivo* efficacy of each compound was determined in mice infected with a lethal dose of *C. albicans* NIH B311 (via i.v. injection) (ref. 79). Mice treated with liriodenine methiodide and liriodenine were found to consistently exhibit significant reduction in the number of recovered CFU as compared to vehicle-treated control groups.

Following single dose i.p. administration of liriodenine methiodide seven hours post-inoculation, significant reduction ($P < 0.05$) in the number of CFU recovered from the kidneys (following sacrifice at 24 h post-infection) was observed with all dosages as compared to vehicle-treated control mice (Table 2). As indicated in Table 2, the significance level of the 0.1 mg/kg dose was not within 95% confidence intervals; however, this dose was between two doses which did show significant reduction ($P < 0.05$). Since this seemed inconsistent, two additional studies were performed in which liriodenine methiodide was administered at a dose of 0.1 mg/kg body weight. In both studies, the number of recovered CFU was significantly reduced in liriodenine methiodide-treated animals as compared to vehicle-treated control animals ($P = 0.004$, 90% reduction and $P = 0.026$, 97% reduction). In early studies, AMB was used as a positive control at a dose of 0.1 mg/kg, at which no significant reduction in the recovered CFU as compared to vehicle-treated control mice was observed. In subsequent studies, it was found that a dose of 0.5 mg AMB/kg consistently led to significant reduction (>90%) in recovered CFU in treated mice vs. vehicle-treated infected mice.

There was also significant reduction in the number of recovered CFU per mg kidney tissue following a single i.v. dose of liriodenine methiodide (Table 2). It is interesting to note that the optimum effects of liriodenine methiodide were observed at doses of 0.5 mg and 1.0 mg/kg. At higher doses of 2.0 mg and 1.5 mg/kg there appears to be less efficacy, as evidenced by $P > 0.05$ and lower percentage reduction in the number of recovered CFU. These results were verified by duplicate evaluation, and were also observed with liriodenine.

Following multiple dose i.p. administration of liriodenine methiodide, remarkably good efficacy was observed, as determined by reduction in recovered CFU (Table 3). It would therefore appear that a multiple dose regimen is more efficaceous than single dose treatment.

Liriodenine methiodide appears to be effective by either i.v. or i.p. single dose administration, particularly at doses of 0.1 and 1.0 mg/kg. However, a multiple dose regimen by the i.p. route appears to be the most effective, particularly at a dose of 0.5

mg/kg. At this dose, liriodenine methiodide is not significantly different from AMB in its ability to reduce the number of CFU recovered from kidney.

While the parent compound liriodenine **1** was less active *in vitro* than its methiodide salt, it was nevertheless considered a promising candidate for *in vivo*

TABLE 2

Counts (CFU/g) of *Candida albicans* in Kidneys Following Single Dose Treatment with Liriodenine Methiodide*

Dose, mg/kg (n)	Route of Administration	CFU/g x 10^6 (range)	P	% Reduction
Control (9)	i.p.	8.13 (0.46 - 30.6)	-	
1.0 (5)	i.p.	1.20 (.58 - 1.91)	0.041	85.2
0.1 (5)	i.p.	1.53 (0.54 - 2.5)	0.095	81.2
0.05 (5)	i.p.	1.26 (0.03 - 3.91)	0.021	84.5
0.01 (5)	i.p.	3.29 (1.55 - 5.48)	NS	59.5
AMB, 0.1 (3)	i.p.	3.28 (2.24 - 3.81)	NS	59.7
Control (5)	i.v.	28.8 (4.33 - 60.1)	-	
2.0 (6)	i.v.	22.7 (1.52 - 36.4)	NS	21.2
1.5 (5)	i.v.	21.9 (2.22 - 45.9)	NS	24.0
1.0 (5)	i.v.	4.05 (2.89 - 5.99)	0.032	85.9
0.5 (4)	i.v.	3.52 (1.71 - 5.54)	0.032	87.8
0.1 (5)	i.v.	13.7 (3.36 - 22.3)	NS	52.4
AMB, 0.5 (4)	i.v.	2.10 (0.08 - 5.56)	0.016	92.7

*Mice were injected i.v. with 10^6 CFU of *Candida albicans* B311. Liriodenine was administered 7 h post-infection.

TABLE 3

Counts (CFU/g) of *Candida albicans* in Kidneys Following Multiple Dose Intraperitoneal Treatment with Liriodenine Methiodide*

Dose, mg/kg (n)	CFU/g x 10^6 (range)	P	% Reduction
Control (4)	278 (1.3 - 121)	-	
1.0 (6)	1.52 (0.79 - 3.48)	0.026	
0.5 (6)	0.62 (0.05 - 1.0)	0.00	99.8
0.1 (5)	1.29 (0.53 - 1.99)	0.048	99.5
0.01 (6)	365 (0.5 - 929)	NS	00.0
AMB, 0.5 (5)	0.04 (0.01 - 0.07)	0.008	99.9

*Mice were injected i.v. with 10^6 CFU of *Candida albicans* B311. Liriodenine methiodide was administered 2, 24, and 42 h post-infection.

evaluation in light of its very low acute toxicity (LD_{50}) and relatively good MIC value. The *in vivo* efficacy of liriodenine was determined in the same manner as liriodenine methiodide. Following single dose i.p. administration (7 h post-infection), liriodenine also exhibited relatively good efficacy at 1.0 and 0.5 mg/kg body weight (Table 4).

Following single dose i.v. administration of liriodenine, substantial reduction in the number of recovered CFU was observed at doses of 0.5 and 0.1 mg/kg (Table 4). As was the case with liriodenine methiodide, higher doses of liriodenine appear to be less efficaceous than lower doses. Duplication of these studies verified this phenomenon, i.e., liriodenine was effective over a narrow dose range (0.1 - 1.0 mg/kg). While the cause of this phenomenon is unknown, one could speculate that at the higher doses liriodenine and liriodenine methiodide may exhibit some subacute toxicity which could be manifested as an apparent loss of antifungal efficacy. Clearly, extensive further studies would be required to establish the cause of this observation.

TABLE 4

Counts (CFU/g) of *Candida albicans* in Kidneys Following Single Dose Treatment with Liriodenine*

Dose, mg/kg (n)	Route of Administration	CFU/g x 10^7 (range)	P	% Reduction
Control (6)	i.p.	5.62 (0.04 - 13.4)	-	-
1.0 (6)	i.p.	0.77 (0.04 - 1.48)	0.032	86.3
0.5 (6)	i.p.	0.32 (0.04 - 1.00)	0.021	94.3
0.1 (7)	i.p.	1.36 (0.09 - 3.51)	NS	75.8
AMB, 0.5 (7)	i.p.	0.36 (0.003 - 1.74)	0.049	99.4
Control (5)	i.v.	14.7 (1.4 - 32.7)	-	-
5.0 (4)	i.v.	7.03 (5.9 - 8.36)	NS	52.2
1.0 (9)	i.v.	9.30 (7.53 - 11.5)	NS	36.7
0.5 (5)	i.v.	0.84 (0.11 - 1.59)	0.016	94.3
0.1 (5)	i.v.	3.56 (1.51 - 9.44)	0.075	75.8
AMB, 0.5 (4)	i.v.	0.019 (0.005 - 0.032)	0.008	99.9
Control (5)	p.o.	11.6 (0.77 - 23.7)	-	-
30.0 (6)	p.o.	2.58 (0.28 - 5.24)	0.041	77.8
20.0 (5)	p.o.	4.81 (1.82 - 8.35)	0.041	58.5
12.5 (6)	p.o.	4.17 (0.71 - 10.5)	0.063	64.1
KTZ, 5.0 (6)	p.o.	4.22 (2.23 - 5.91)	0.063	63.6

*Mice were injected i.v. with 10^6 CFU of *Candida albicans* B311. Liriodenine was administered 7 h post-infection.

The efficacy of liriodenine was also evaluated following oral administration, using ketoconazole as a positive control. While a range of doses were evaluated (50, 30, 20, 12.5, 6.25, and 3.12 mg/kg), the only doses which afforded verified reduction in recovered CFU were 30, 20, and 12.5 mg/kg (Table 4).

While oxoglaucine methiodide appeared to be as active as liriodenine *in vitro*, it did not appear to be as effective *in vivo*. Following i.p. administration (7 h post-infection), oxoglaucine methiodide did not exhibit significant reduction at any of the dosage levels evaluated (10.0, 5.0, 2.0, 1.5, 1.0, 0.5, and 0.1 mg/kg)

Based on the data observed from each of these experiments, it can be concluded that liriodenine methiodide is therapeutically effective against systemic *C. albicans* infection in mice. Maximum activity of liriodenine methiodide against *C. albicans* infection was observed at doses between 0.1 mg/kg and 1.0 mg/kg body weight regardless of the treatment regimen. However, the efficacy of liriodenine methiodide at these doses appears to be greatly enhanced if the drug is administered at 2 h, 24 h, and 48 h post-infection.

Liriodenine **1** also exhibits some level of efficacy with the added advantage of being orally effective. Additional studies to determine the efficacy of these drugs at additional dosae levels and particularly at varying dosage intervals are currently in progress.

6.2 Anticandidal Activity of the Alkaloids Eupolauridine and Onychine

Cleistopholis patens (Benth.) Engl. and Diels (Annonaceae) is a large tree found throughout West Africa. Two previous studies on the stem and root bark have resulted in the isolation of sesquiterpenes and alkaloids (refs. 80, 81). In our laboratories, alcoholic extracts of the root bark of *C. patens* showed significant anticandidal activity and bioassay-directed fractionation of the rootbark afforded an active azafluoranthrene alkaloid (ref. 82).

The dried ground root bark was extracted by percolation with n-hexane (inactive) and then with 95% EtOH. The active alcoholic extract was partitioned between $CHCl_3$ and H_2O. The active $CHCl_3$ extract was chromatographed over silicic acid using $CHCl_3$ and a stepwise gradient of increasing percentage of MeOH-$CHCl_3$ as eluent. Fractions were pooled on the basis of tlc analysis. The anticandidal activity was concentrated in a fraction (500 mg) that was eluted with 5% MeOH in $CHCl_3$. Further purification of this fraction over neutral alumina using EtOAc-n-hexane as eluting solvent resulted in the isolation of the active component as a yellow crystalline material having mp 153-155°. This substance was identified as the previously reported alkaloid, eupolauridine **6** (Fig. 2) (refs. 81,83,84), by comparison with an authentic reference sample.

Eupolauridine **6** exhibited a significant zone of inhibition in the qualitative assay against three test strains of *C. albicans.* The MIC of eupolauridine was found to be 1.56 μg/ml for each of the three strains in yeast-nitrogen broth. Based on its good MIC value, eupolauridine was considered a promising potential new antifungal drug.

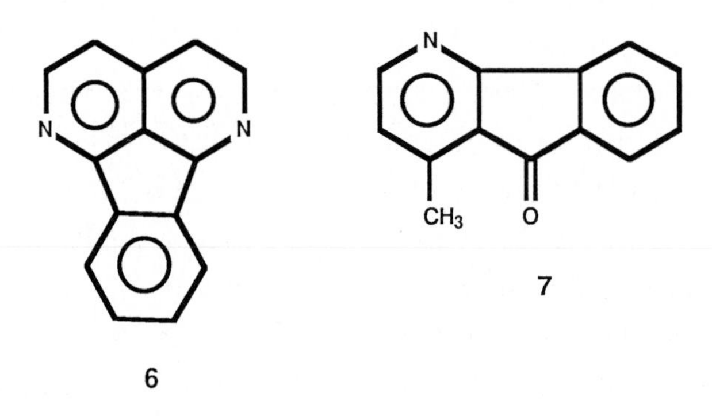

Fig. 2. Structural formulae for eupolauridine **6** and onychine **7**.

Since only small amounts were isolated and much larger quantities were necessary for *in vivo* evaluation, eupolauridine **6** was prepared by synthesis as previously reported (ref. 83). All of the synthetic intermediates were also evaluated for anticandidal activity. One of the intermediates was also shown to be active against *Candida.* This intermediate is also the known alkaloid onychine which has been reported from *C. patens* (ref. 81) and originally reported from *Onychopetalum amazonicum* (ref. 85). The original structure reported for onychine and reported in *C. patens* (ref. 81) and elsewhere (ref. 86) has been shown to be incorrect (refs. 87, 88), and the correct structure is represented by **7** (Fig. 2). While the structure reported for onychine (refs. 87, 88) appears to be based on solid chemical evidence, a COSY spectrum of dihydroonychine (refs. 81, 85) clearly showed long-range couplings between H-9 and the protons of the methyl group independently confirming structure **7** for onychine.

Onychine is comparable to eupolauridine in its anticandidal activity, with an MIC value of 3.12 μg/ml against *C. albicans* B311 in yeast-nitrogen broth. Therefore, onychine is also considered a potential candidate for further development. Studies are currently in progress to evaluate the *in vivo* efficacy and toxicity of both eupolauridine and onychine.

6.3 Antifungal Activity of *Trillium* Glycosides

Extracts of the rhizome and above-ground portion of *Trillium grandiflorun* (Michx.) Salisb. (Liliaceae) showed significant *in vitro* activity against *Candida albicans*. A bioassay-directed fractionation approach resulted in the isolation of the two active components (ref. 89).

The rhizome and above-ground portions of *T. grandiflorum* were separately extracted with *n*-hexane (inactive) followed by 95% EtOH. The active EtOH extracts of both the rhizomes and the above ground portion were partitioned between EtOAc and H_2O. The active H_2O layers were then extracted with *n*-BuOH. The active *n*-BuOH extracts (rhizomes and above ground) were chromatographed over silica gel to give two pooled fractions from which two active components, designated TG-I and TG-II, were isolated. TG-I predominates in the rhizomes while TG-II predominates in the above ground portion. The isolated yield of TG-I is 0.0094% from the above ground portion and 0.12% from the rhizomes while TG-II is 0.113% from the above ground portion and 0.004% from the rhizomes.

TG-I had ^{1}H-nmr and ir data indicative of a complex polyhydroxylated natural product. The ^{13}C-nmr spectrum was particularly informative since it showed 45 carbon signals which could be attributed to a saponin glycoside containing three sugars (δ C 100.1 d, 101.6 d and 102.7 d). The aglycone showed signals at δ C 140.8 s, 121.4 d and 109.1 s characteristic of δ 5-spirostent type sapogenins (ref. 90). Hydrolysis of TG-I produced an aglycone identical with diosgenin **8** (Fig. 3). The ^{13}C-nmr signals reported for diosgenin and its various glycosides (ref. 90) were compared with those of TG-I, thus allowing the remaining signals to be assigned to the three sugars. This data suggested that the three sugars were rhamnose (two) and glucose (one). A review of the literature revealed a report of a saponin glycoside isolated from *Trillium tschonoskii* having this constitution (ref. 91). A direct comparison of TG-I with dioscin **9** (Fig. 3) showed the two samples to be identical. The ^{13}C-nmr spectral data were also

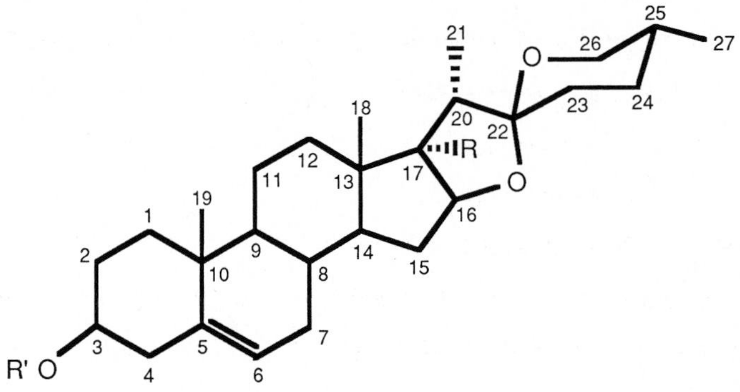

21 25 O 26 27 23 24 18 20 22 R 12 11 13 17 O 19 16 1 9 14 2 10 8 15 5 7 R' O 3 4 6

8 R = R' = H

10 R = OH , R' = H

9 R = H , R' = Rham - Gluc —
                                         |
                                       Rham

11 R = OH , R' = Rham - Rham - Gluc —
                                                     |
                                                   Rham

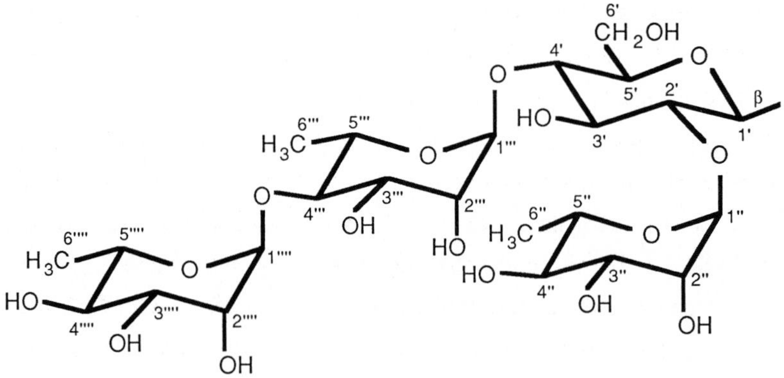

6' CH 2 OH 4' O O 5' 2' β HO 3' 1' 6''' 5''' H 3 C O 1''' O O 4''' 3''' 2''' 6'' 5'' 1'' 6'''' 5'''' OH HO H 3 C O H 3 C O HO 1'''' 4'' 3'' 2'' HO 3'''' 2'''' OH OH 4'''' OH OH

Rham — Rham — Gluc —
                         |
                       Rham

Fig. 3. Structural formulae for TG-I **9** and TG-II **11**.

identical (ref. 92).

TG-II was established as a saponin glycoside containing the aglycone pennogenin **10** (Fig. 3) and four sugars (3 rhamnose and 1 glucose) on the basis of ^{13}C-nmr spectral data (51 signals). The signals at δ C 89.7 s, 89.9 d, 109.5 s, 121.1 d and 140.5 s were characteristic for pennogenin. Pennogenin rhamnosyl chacotrioside **11** (Fig. 3) has been previously reported (ref. 91, 93) and comparison of an authentic sample with TG-II showed them to be identical. While the ^{13}C-nmr spectral data for **11** has not been reported, the ^{13}C-nmr shift assignments for pennogenin and related glycosides (ref. 90) and diosgenin **8** plus the same sugar arrangement as in **11** (ref. 94) have been reported and are totally consistent with data reported for TG-II **11**.

TG-I **9** exhibited good inhibition against *C. albicans* B311 in a qualitative agar well-diffusion assay. The MIC of TG-I (determined in yeast nitrogen broth) was found to be 1.56 µg/ml for *C. albicans* B311, 3.12 µg/ml for *C. albicans* ATCC 10231, and 6.25 µg/ml for *C. albicans* WH. TG-II **11** also exhibited good activity against *C. albicans* in the agar well-diffusion assay. The MIC values of TG-II were 6.25 µg/ml for *C. albicans* B311 and 12.5 µg/ml for *C. albicans* ATCC 10231 and WH (ref. 89).

Based on these data, both TG-I and TG-II were further evaluated for *in vivo* efficacy and toxicity. The LD_{50} of TG-I was estimated at 38 mg/kg following i.p. administration and between 10 and 20 mg/kg following i.v. administration. TG-II exhibited similar acute toxicity with a LD_{50} value of 28.0 mg/kg following i.p. administration and between 5 and 10 mg/kg following i.v. administration.

The *in vivo* efficacy of each compound was determined in the mouse model of disseminated candidiasis. No significant ($P < 0.05$) reduction in the number of recovered CFU was observed following a single i.p. dose (0.1-25 mg/kg) of either TG-I or TG-II. In the same study, a single dose of AMB (0.5 mg/kg) was found to reduce the number of recovered CFU by 97.5% ($P = 0.063$).

It is well documented that saponins similar to **9** and **11** (Fig. 3) exhibit a variety of biological activities, including antibacterial and antifungal activity (ref. 95). Therefore, TG-I **9** and TG-II **11** were also evaluated for antimicrobial activity against a number of other fungi and bacteria. Both compounds were found to exhibit some *in vitro* activity against four other genera of fungi. The MIC values of TG-I and TG-II for the yeasts *Cryptococcus neoformans* and *Saccharomyces cerevisiae* and for the filamentous fungi *Aspergillus flavus* and *Trichophyton mentagrophytes* are summarized in Table 5. Neither compound showed any significant antibacterial activity against *Staphylococcus aureus, Bacillus subtilis, Escherichia coli, Pseudomonas aeruginosa,* or *Mycobacterium smegmatis.* To our knowledge this was the first report of the antifungal activity and acute toxicity of **9** and **11** (ref. 89).

TABLE 5

Antifungal Activity of TG-I **9** and TG-II **11**

Organism	MIC (μg/ml)		
	TG-I	TG-II	Amphotericin B
Candida albicans NIH B311	1.56	6.25	0.39
Candida albicans ATCC 10231	3.12	12.5	1.56
Candida albicans WH-D	6.25	12.5	0.78
Cryptococcus neoformans ATCC 32264	25	12.5	1.56
Saccharomyces cerevisiae ATCC 9763	25	12.5	25
Aspergillus flavus ATCC 9170	100	100	50
Aspergillus flavus ATCC 26934	50	100	100
Trichophyton mentagrophytes ATCC 9972	6.25	3.12	50

6.4 3-Dimethylallylindole, An Antifungal Metabolite of *Monodora tenuifolia*

A new prenylated indole, 3-dimethylallylindole **12** (Fig. 4) has also been isolated from the stem bark of *Monodora tenuifolia,* a plant used in West African traditional medicine for a variety of ailments (ref. 96). Extraction of the stem bark with ethyl acetate gave a red oily residue which was chromatographed consecutively over silica gel 60 and neutral alumina to give a light yellow oil which crystallized with difficulty from *n*-hexane (mp 45-47°). The compound exhibited significant anticandidal activity in the agar well-diffusion assay against all three test organisms, showing moderately good zones of inhibition against *C. albicans* 10231, B311, and WH. The UV spectrum showed absorptions bands at ν_{max} 291, 282, and 225 nm which are characteristic of the indole chromophore. The ^{1}H NMR spectrum, in addition to showing aromatic protons characteristic of an indole (δ 7.25, 3H, 7.70, 1H; 7.03, 1H), showed an olefinic proton resonating as a triplet (δ 5.53, J = 7 Hz) and a six-proton singlet (δ 1.70) for a dimethylallyl functionality. The mass spectrum showed the parent ion peak at m/z 185 (85% RA) with significant fragments at m/z 155 (32% RA, M^+ - 2 CH_3) and m/z 117 (73% RA, M^+ - C_5H_8). These data suggested an indole substituted with a γ,γ-dimethylallyl group. The ^{13}C NMR spectral data further confirmed that **12** is a prenylated indole. The placement of the substituent at C-3 follows from a comparison of its ^{13}C NMR data with those reported for indole, 2-methyl indole, and 3-methyl indole. It has been observed that alkylation of indoles at position 2 or 3 causes a characteristic downfield shift of 10-13 ppm in the ^{13}C NMR resonance of the substituted carbon atom. C-3 resonates at 115.8 ppm which is shifted 13 ppm downfield relative to the corresponding carbon in indole.

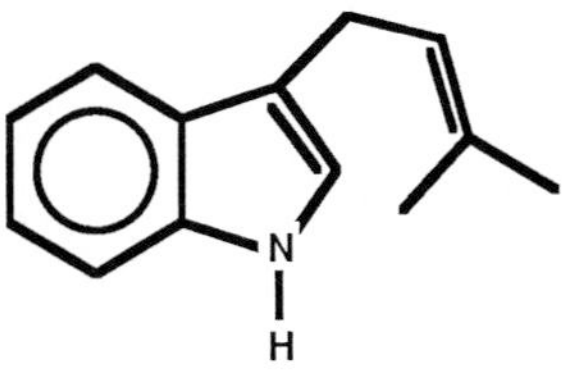

N
H
12

Fig. 4. Structural formula for 3-dimethylallylindole **12**

3-Dimethylallyl indole **12** has been reported previously as a synthetic product from studies on the insertion of isoprene units into indole systems. An authentic sample of **12** was prepared according to literature procedure and was identical in all respects to the natural product isolated from *M. tenuifolia*. The isolation of **12** from *M. tenuifolia* was the first report of the natural occurrence of this compound (ref. 96).

The MIC values of 3-dimethylallylindole **12** against the three test strains of *C. albicans* (determined in both YNB and mycophil broth) were somewhat disappointing. Table 6 summarizes the results from the MIC determinations.

6.5 Canellal, An Antifungal Metabolite of *Canella winterana*

Canellal **13** (Fig. 5) is a sesquiterpene dialdehyde isolated from the plant *Canella winterana* (ref. 97). It was been shown in our laboratories (unpublished

TABLE 6

MIC Values of 3-Dimethylallylindole Against *Candida albicans*

Organism	MIC (μg/ml)*	
	YNB	Mycophil
C. albicans 10231	12.5	25
C. albicans B311	12.5	25
C. albicans WH	12.5	25

*After 24 hrs. incubation

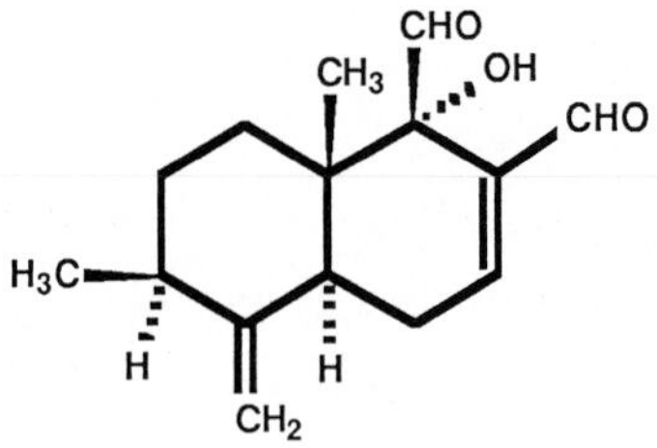

CHO
CH_3 OH
CHO
H_3C
H H
CH_2

13

Fig. 5. Structural formula for Canellal **13**.

results) to possess good *in vitro* antifungal activity against the yeasts *Candida albicans* with a MIC value of 1.56 μg/ml. Recently, additional quantities of canellal **13** were isolated for use in *in vivo* stuides. Exhaustive percolation of the trunk bark of *Canella winterana* with hexanes afforded a hexane extract which was partitioned between hexanes and acetonitrile. Chromatography of the acetonitrile fraction over silica gel using 1% acetone/chloroform as eluent afforded crystalline canellal **13.** The LD_{50} of canellal following i.p. administration is between 2.5 and 5.0 mg/kg, thus suggesting that canellal is at least as toxic as amphotericin B (based solely on LD_{50}). Studies to establish the *in vivo* efficacy of canellal are ongoing; however, preliminary results suggest that the *in vivo* activity of canellal will not be sufficient to overcome the probable toxicity of the compound.

7. SUMMARY AND CONCLUSIONS

Acquired Immunodeficiency Syndrome (AIDS) is characterized by a breakdown in the form of serious opportunistic infections. Treatment of such infections is often inadequate for a variety of reasons, including the lack of effective antimicrobial therapy. The opportunistic infections most commonly associated with AIDS are parasitic (pneumocystosis, toxoplasmosis, cryptosporidiosis), fungal (candidiasis, cryptococcosis), bacterial (mycobacteriosis), and viral (herpes simplex and cytomegalovirus).

Historically, most bacterial infections and localized fungal infections have been effectively treated with one of the numerous clinically available antibiotics. However, the need for new, more effective and less toxic antibiotics for the treatment of disseminated fungal and mycobacterial infections is obvious in light of the significant toxicities and failure rates of the currently available agents. The discovery of new antibiotics has in the past successfully relied primarily upon the isolation of such agents from natural sources. The major advantage of this approach over chemical synthesis or modification of existing agents is the likelihood of identifying new prototype drugs with quite different chemical structures, and hence, less likelihood of similar toxicities and cross-resistance. Although microorganisms have traditionally served as the primary source of new antibiotics, it has recently been shown that higher plants also serve as sources for a number of diverse antimicrobial agents.

The objective of our research is to discover new prototype antibiotics with potential utility specifically for the treatment of opportunistic disseminated mycoses and mycobacteriosis. This goal is accomplished by the inital *in vitro* evaluation of antifungal and antimycobacterial activity of extracts of higher plants and microorganisms. Plant and microbial extracts which show good activity are fractionated and purified using a bioassay-directed scheme. Pure active compounds with significant minimum inhibitory concentrations (MIC) are evaluated for *in vivo* efficacy in established animal models of disseminated mycosis and mycobacteriosis in order to determine their potential clinical utility. Several examples have been presented which outline the successful use of these methods.

REFERENCES

1. T.K. Daneshmend and D.W. Warnock, Clin. Pharmacokin., 8 (1983) 17.
2. G. Medoff, J. Brajtburg, G.S. Kobayashi, and J. Bolard, Ann. Rev. Pharmacol. Toxicol., 23 (1983) 303.
3. J.L. LeFrock and B.R. Smith, Clin. Pharmacol., 30 (1984) 162.

4. K. King, Med. J. Austr., 143 (1985) 287.
5. L.S. Young, Rev. Inf. Dis., 7,Suppl. 3 (1985) S380.
6. M.M. Furio and C.J. Wordell, Clin. Pharm., 4 (1985) 539.
7. E. Whimbey, J.W.M. Gold, B. Polsky, J. Dryjanski, C. Hawkins, A. Blevins, P. Brannon, T.E. Kiehn, A.E. Brown, and D. Armstrong, Ann. Int. Med., 104 (1986) 511.
8. D. Armstrong, J.W.M. Gold, J. Dryjanski, E. Whimbey, B. Polsky, C. Hawkins, A.E. Brown, E. Bernard, and T.E. Kiehn, Ann. Int. Med., 103 (1985) 738.
9. AMA Drug Evaluation, American Medical Association, Chicago, 1983 (5th ed) pp 1779-1788.
10. D.G. Ahearn and J.B. Lawrence, J. Clin. Microbiol., 20 (1984) 187.
11. W.G. Merz, J. Clin. Microbiol., 20 (1984) 1194.
12. L. deRepentigny and E. Reiss, Rev. Inf. Dis., 6 (1984) 301.
13. A. Zuger, E. Louie, R.S. Holzman, M.S. Simberkoff, J.J. Rahal, Ann. Int. Med., 104 (1986) 234.
14. E. Drouhet and B. Dupont, Rev. Inf. Dis., 9 Suppl. 1 (1987) S4.
15. W.E. Dismukes, G. Cloud, H.A. Gallis, T.M. Kerkering, G. Medoff, P.C. Craven, L.G. Kaplowitz, J.F. Fisher, C.R. Gregs, C.A. Bowles, S. Shadomy, A.M. Stamm, R.B. Diasio, L. Kaufman, S.-J. Soon, W.C. Blackwelder, N. Engl. J. Med., 317 (1987) 334.
16. E.K. Kastrup and J.R. Boyd (Eds), Facts and Comparisons, G.H. Schwach, Publisher, St. Louis, (1980) p. 356.
17. D.J. Drutz, Rev. Inf. Dis., 9 (1987) 417.
18. E.B. Smith and J.C. Henry, Pharmacotherapy, 4 (1984) 199.
19. P. Mosca, P. Bonazzi, G. Novelli, A.M. Jezequel, and F. Orlandi, Br. J. Exp. Path., 66 (1985) 737.
20. J.F. Ryley, R.G. Wilson, and K.J. Barrett-bee, Sabouraudia, 22 (1984) 53.
21. G. Petranyi and N.S. Ryder, Science, 224 (1984) 1239.
22. A. Stutz, A. Georgopoulos, W. Granitzer, D. Petranyi, and D. Berney, J. Med. Chem., 29 (1986) 112.
23. C.C. Hawkins, J.W.M. Gold, E. Whimbey, T.E. Kiehn, P. Brannon, R. Cammarata, A.E. Brown, and D. Armstrong, Ann. Int. Med. , 105 (1986) 184.
24. J.B. Greene, G.S. Sidhu, S. Lewin, J.F. Levine, H. Masur, M.S. Simberkoff, P. Nicholas, R.C. Good, S.B. Zolla-Pazner, A.A. Pollock, M.L. Tapper, and R.S. Holzman, Ann. Int. Med., 97 (1982) 539.
25. P. Zakowski, S. Fligiel, O.G.W. Berlin, B.L. Johnson, J. Am. Med. Assoc., 248 (1982) 2980.
26. M.-C. Poon, A. Landay, E.F. Prasthofer, and S. Stagno. 1983. Ann. Int. Med., 98 (1983) 287.
27. A.M. Macher, J.A. Kovacs, V. Gill, G.D. Roberts, J. Ames, C.H. Park, S. Straus, H.C. Lane, J.E. Parrillo, A.S. Fauci, and H. Masur, Ann. Int. Med., 99 (1983) 782.
28. G.L. Woods and J.A. Washington, Rev. Inf. Dis.., 9 (1987) 275.
29. O. Zak and M.A. Sande, (Eds), Experimental Models in Antimicrobial Chemotherapy, Academic Press, New York, 1986, Vol. 3.
30. D.P. Bonner, W. Mechlinski, and C.P. Schaffner, J. Antibiot., 25 (1972) 26l.
31. G.R. Kiem, J.W. Poutsiaka, J. Kirpan, and C.H. Keysser, Science, 179 (1973) 584.
32. G.R. Kiem, P.L. Sibley, Y.H. Yoon, J.S. Kulesza, I.H. Zaidi, M.M. Miller, J.W. Poutsiaka, Antimcrob. Ag. Chemother., 10 (1976) 687.
33. W.G. Ellis, R.A. Sobel, and S.L. Nielsen, J. Inf. Dis., 146 (1982) 125.
34. C. D. Bruna, G. Schioppacassi, D. Ungheri, D. Jabes, E. Morvillo, and A.

Sanfilippo, J. Antibiot., 36 (1983) 1502.
35. A. Sanfilippo, C. Della Bruna, L. Marsili, E. Morvillo, C.R. Pasqualucci, G. Schioppassi, D. Ungheri, J. Antibiot., 33 (1980) 1193
36. R.J. O'Brien, M.A. Lyle, and D.E. Snyder, Rev. Inf. Dis., 9 (1987) 519.
37. J. VanCutsem, F. VanGerven, R. Zaman, and P.A.J. Janssen, Chemotherapy, 29 (1983) 322.
38. P.F. Troke, R.J. Andrews, K.W. Brammes, M.S. Marriott, and K. Richarson, Antimicrob. Ag. Chemother., 28 (1985) 815.
39. C.H. Collins and A.H.C. Uttley, J. Antimicrob. Chemother., 16 (1985) 575.
40. J.D. Gay, D.R. DeYoung and G.D. Roberts, Antimicrob. Ag. Chemother., 26 (1984) 94.
41. S. Davies, P.D. Sparham, and R.C. Spencer, J. Antimicrob. Chemother., 19 (1987) 609.
42. O.G.W. Berlin, L.S. Young, and D.A. Bruckner, J. Antimicrob. Chemother., 19 (1987) 611.
43. K.A. Trimble, R.B. Clark, W.E. Sanders, J.W. Frankel, R. Cacciatore, and H. Valdez, J. Antimicrob. Chemother., 19 (1987) 617.
44. G. Medoff, Rev. Inf. Dis., 9 (1987) 403.
45. A. Polak, Chemother., 33 (1987) 381.
46. G. Lopez-Berestein, R. Mehta, R.L. Hopfer, K. Mills, L. Kasi, K. Mehta, V. Fainstein, M. Lune, E.M. Hersh, and R. Juliano, J. Inf. Dis., 147 (1983) 939.
47. C. Tremblay, M. Barza, C. Fiore, and F. Szoka, Antimicrob. Ag. Chemother., 26 (1984) 170.
48. G. Lopez-Berestein, V. Fainstein, R. Hopfer, K. Mehta, M.P. Sullivan, M. Keating, M.G. Rosenblum, R. Mehta, M. Luna, E.M. Heish, J. Reuben, R.L. Juliano, and G.P. Bodey, J. Inf. Dis., 151 (1985) 704.
49. B. Oliver-Bever, J. Ethnopharmacol., 9 (1983) 1.
50. A. Laurens, S. Mboup, M. Tignokpa, O. Sylla, and J. Masquelier, Pharmazie., 40 (1985) 482.
51. J.-C. Guerin and H.-P. Reveillere, Ann. Pharm. Franc., 43 (1985) 77.
52. A.M. Clark, F.S. El-Feraly, and W.-S. Li, J. Pharm. Sci., 70 (1981) 951.
53. C.D. Hufford, M.J. Funderburk, J.M. Morgan, and L.W. Robertson, J. Pharm. Sci., 64 (1975) 789.
54. L.B. Heifets, M.D. Iseman, J.L. Cook, P.J. Lindholm-Levy, and I. Drupa, Antimicrob. Ag. Chemother., 27 (1985) 11.
55. E. Marinis and N.J. Legakis, J. Antimicrob. Chemother., 16 (1985) 527.
56. M.H. Cynamon, Antimicrob. Ag. Chemother., 28 (1985) 440.
57. J.M. Swenson, R.J. Wallace, V.A. Silcox, and C. Thornsberry, Antimicrob. Ag. Chemother., 28 (1985) 807.
58. L.B. Heifets, M.D. Iseman, P.J. Lindholm-Levy, and W. Kanes, Antimicrob. Ag. Chemother., 28 (1985) 570.
59. V. Ausina, M.J. Condom, B. Mirelis, M. Luquin, P. Cou, and G. Prats, Antimicrob. Ag. Chemother., 29 (1986) 951.
60. A.L. Hume and T.M. Kerkering, Drug Intell. Clin. Pharm., 17 (1983) 169.
61. G. Medoff and G.S. Kobayashi, N. Engl. J. Med., 302 (1980) 145.
62. D. Drutz, Rev. Inf. Dis., 9 (1987) 392.
63. J.N. Galgiani, Antimicrob. Ag. Chemother, 31 (1987) 1867.
64. E.P. deFernandez, M.M. Patino, J.R. Graybill, and M.H. Tarbit, J. Antimicrob. Chemother., 18 (1986) 261.
65. S. Shadomy, S.C. White, H.P. Yu, W.E. Dismukes, and The NIAID Mycosis Study Group, J. Inf. Dis., 152 (1985) 1249.
66. M. Huppert. H.S. Sung, and K.R. Vukovich, Antimicrob. Ag. Chemother., 5

(1974) 473.
67. J.R. Graybill and L. Mitchell, Sabouraudia, 18 (1980) 137.
68. M.A. Lew, K.M. Beckett, and M.J. Levin, Antimicrob. Ag. Chem., 14 (1978) 465.
69. J. Arroyo, G. Medoff, and G.S. Kobayashi, Antimicrob. Ag. Chemother., 11 (1977) 21.
70. N.K. Fujita and J.E. Edwards, Antimicrob. Ag. Chemother., 19 (1981) 196.
71. J. Uno, M.L. Shigematsu, and T. Arai, Antimicrob. Ag. Chemother., 24 (1983) 552.
72. L.D. Sabath, Antimicrob. Ag. Chemother., (1967) 210.
73. M.C. Berenbaum, J. Inf. Dis., 137 (1978) 122.
74. J.P. Anhalt, L.D. Sabatj, and A.L. Barry, in: E.H. Lennette, A. Balows, W.J. Hausler, and J.P. Truant, (Eds), Special Tests, Bactericidal Activity and Activity of Antimicrobics in Combination. Manual of Clinical Microbiology, American Society for Microbiology, Washington, 1980, pp. 481-483.
75. D.L. Hurley and A.S. Fauci, J. Inf. Dis., 131 (1975) 516.
76. S. Rabinovich, B.D. Shaw, T. Bryant, and S.T. Donta, J. Infect. Dis., 130 (1974) 28.
77. J.R. Graybill, in: O. Zak and M.A. Sande (Eds), Animal Models for Treatment of Cryptococcosis. Experimental Models in Antimicrobial Chemotherapy, Academic Press, New York, 1986, vol 3, pp 131-148.
78. P.R.J. Gangadharam, in: O. Zak and M.A. Sande, (Eds), Animal Models for Nontuberculosis Mycobacterial Diseases. Experimental Models in Antimicrobial Chemotherapy, Academic Press, New York, 1986, vol 3, pp 1-94.
79. A.M. Clark, E.S. Watson, M. K. Ashfaq, and C.D. Hufford, Pharm. Res., 4 (1987) 495.
80. S.A-E. Atti, A. Ammar, C.H. Phoebe, Jr., P.L. Schiff, and D.J. Slatkin, J. Nat. Prod., 45 (1982) 476.
81. P.G. Waterman and I. Muhammed, Phytochemistry, 24 (1985) 523.
82. C.D. Hufford, S. Liu, A.M. Clark, and B.O. Oguntimein, J. Nat. Prod., 50 (1987) 961.
83. B.F. Bowden, K. Picker, E. Ritchie and W. C. Taylor, Aust. J. Chem., 28 (1975) 2681.
84. M. Leboeuf and A. Cave', Lloydia, 39 (1976) 459.
85. M.E.L. De Almeida, R. Braz, M. V. Von Bulow, 0. R. Gottlieb and J.G.S. Maia, Phytochemistry, 15 (1976) 1186.
86. M.O.F. Goulart, A.E.G. Santana, A. B. De Oliveira, G. G. De Oliveira, J.G.S. Maia, Phytochemistry, 25 (1986) 169.
87. J. Koyama, T. Sugita, Y. Suguta and H. Irie, Heterocycles, 12 (1979) 1017.
88. D. Tadic, B.K. Cassels, M. Leboeuf and A. Cave', Phytochemistry, 26 (1987) 537.
89. C.D. Hufford, S. Liu, and A.M. Clark, J. Nat. Prod., 51 (1988) 94.
90. P.K. Agrawal, D.C. Jain, R.K. Gupta and R.S. Thakur, Phytochemistry, 24 (1985) 2479.
91. T. Nohara, F. Kumamoto, K. Miyahara and T. Kawasaki, Chem. Pharm. Bull., 23 (1975) 1158.
92. S.B. Mahato, N.P. Sahu and A.N. Ganguly, Indian J. Chem., 19B (1980) 817.
93. T. Nohara, K. Miyahara and T. Kawasaki, Chem. Pharm. Bull., 23 (1975) 872.
94. M. Miyamura, K. Makano, T. Nohara, T. Toshiaki and T. Kawasaki, Chem. Pharm. Bull., 30 (1982) 712.
95. S.B. Mahato, A.N. Ganguly, and N.P. Sahu, Phytochemistry, 21 (1982) 959.
96. A.O. Adeoye, B.O. Oguntimein, A.M. Clark, and C.D. Hufford, J. Nat. Prod., 49 (1986) 534.
97. F.S. El-Feraly, A.T. McPhail, and K.D. Onan, J.C.S. Chem. Comm. (1978) 75.

SUBJECT INDEX